Supernovae: A Survey of Current Research

NATO ADVANCED STUDY INSTITUTES SERIES

*Proceedings of the Advanced Study Institute Programme, which aims
at the dissemination of advanced knowledge and
the formation of contacts among scientists from different countries*

The series is published by an international board of publishers in conjunction
with NATO Scientific Affairs Division

A	Life Sciences	Plenum Publishing Corporation
B	Physics	London and New York
C	Mathematical and Physical Sciences	D. Reidel Publishing Company Dordrecht, Boston and London
D	Behavioural and Social Sciences	
E	Engineering and Materials Sciences	Martinus Nijhoff Publishers The Hague, London and Boston
F	Computer and Systems Sciences	Springer Verlag Heidelberg
G	Ecological Sciences	

Series C – Mathematical and Physical Sciences

Volume 90 – Supernovae: A Survey of Current Research

Library of Congress Cataloging in Publication Data

NATO Advanced Study Institute (1981 : Cambridge, Cambridgeshire)
      Supernovae : a survey of current research.

      (NATO advanced study institute series. Series C, Mathematical and
physical sciences ; v. 90)
      "Published in cooperation with NATO Scientific Affairs Division."
      Includes indexes.
      1. Supernovae–Congresses.   I.   Rees, Martin J., 1942–
II.   Stoneham, Ray J.   III.   Title.   IV. Series.
QB843.S95N37   1981   523.8′446   82–9075
ISBN 978-94-009-7878-2   ISBN 978-94-009-7876-8 (eBook)
DOI 10.1007/978-94-009-7876-8

Published by D. Reidel Publishing Company
P.O. Box 17, 3300 AA Dordrecht, Holland

Sold and distributed in the U.S.A. and Canada
by Kluwer Boston Inc.,
190 Old Derby Street, Hingham, MA 02043, U.S.A.

In all other countries, sold and distributed
by Kluwer Academic Publishers Group,
P.O. Box 322, 3300 AH Dordrecht, Holland

D. Reidel Publishing Company is a member of the Kluwer Group

Supernovae:
A Survey of
Current Research

Proceedings of the NATO Advanced Study Institute
held at Cambridge, U.K., June 29 - July 10, 1981

edited by

MARTIN J. REES
Institute of Astronomy, Cambridge, U.K.

and

RAY J. STONEHAM
Institute of Astronomy, Cambridge, U.K.

D. Reidel Publishing Company

Dordrecht : Holland / Boston : U.S.A. / London : England

Published in cooperation with NATO Scientific Affairs Division

TABLE OF CONTENTS

PREFACE

The theme of the conference held at the Institute of Astronomy in the summer of 1981 was 'Supernovae'. The topic was interpreted very broadly: observations in all wavebands were discussed, along with theories for the explosion mechanism and the light curves; there were papers on supernova remnants and pulsar statistics; other sessions dealt with the use of new techniques for improving supernova searches, and with the importance of supernovae for cosmogonic and cosmological studies. This book contains texts based on all the main review lectures, together with a number of shorter papers which describe new results presented at the conference.

The Scientific Organising Committee, responsible for arranging the programme, consisted of J. Audouze, G.E. Brown, J. Danziger, F. Pacini, M.J. Rees (Chairman) and J.W. Truran. The conference was well attended, with over 100 visitors to Cambridge as well as many local participants. We are grateful to all those who helped with the practical organisation of the meeting, especially Dr Michael Ingham (Secretary of the Institute of Astronomy) and Mrs Norah Tate. We thank all the authors of the papers in this volume for the trouble they took in preparing written versions of their excellent lectures, and for the efforts they made to meet our 'final' deadline: we wish especially to thank Drs W.D. Arnett and J.M. L∘ttimer for help with the editorial work. This book could not have appeared without the efficient typing by Mrs Sally Roberts, and the help of Mr Richard Sword with the illustrations — we are indebted to them both.

M.J. Rees
R.J. Stoneham

Professor Hans Bethe celebrated his 75th birthday
(July 2nd, 1981) in Cambridge during the
Supernovae conference.

LIST OF PARTICIPANTS

J.H. Applegate - NORDITA, Denmark
W.D. Arnett - Chicago & MPI Munich
T. Axelrod - Lawrence Livermore Laboratory, California
R.H. Becker - Columbia University, New York
P. Benvenuti - ESA, Madrid
H.A. Bethe - Cornell University & NORDITA
W.P. Blair - University of Michigan
R.D. Blandford - Caltech
R. Bond - University of California, Berkeley
G. Borner - MPI, Munich
D. Branch - University of Oklahoma
W. Brinkmann - MPI, Munich
G.E. Brown - SUNY, Stony Brook & NORDITA
A. Burrows - SUNY, Stony Brook
R. Canal - University of Barcelona
B.J. Carr - IOA, Cambridge
C.R. Canizares - MIT
M.G.M. Cawson - IOA, Cambridge
P. Charles - Oxford University
R.A. Chevalier - University of Virginia
C. Chiosi - University of Padua
D.H. Clark - Rutherford Laboratories
D.D. Clayton - Rice University, Houston
S.A. Colgate - New Mexico Institute of Mining and Technology &
     Los Alamos, New Mexico
J. Cooperstein - NORDITA, Denmark
R. Corbet - MSSL, University College, London
A.A. Da Costa - Manchester University
J. Danziger - ESO, Munich
M. Dennefeld - Institut d'Astrophysique, Paris
L.K. DeNoyer - Cornell University, New York
M.A. Dopita - Mt. Stromlo Observatory, Australia
M.F. El Eid - Göttingen, West Germany
R.I. Epstein - NORDITA, Denmark
A.C. Fabian - IOA, Cambridge
R.A. Fesen - NASA, Goddard Space Flight Center, Maryland
W.A. Fowler - Caltech

C. Fransson - NORDITA, Denmark
G.M. Fuller - Caltech
C. Galas - University of Calgary, Alberta
G. Glen - McMaster University, Ontario
R. Harkness - Oxford University
D.J. Helfand - Columbia University, New York
R.C. Henry - University of Michigan
W. Hillebrandt - MPI, Munich
F. Hoyle - Cumbria
S. Ikeuchi - Hokkaido University, Japan
J. Isern - University of Barcelona
M.D. Johnston - MIT
F. Kahn - Manchester University
J.T. Kare - Lawrence Berkeley Laboratory, University of California
D. Kazanas - NASA, Goddard Space Flight Center, Maryland
E.J. Kibblewhite - IOA, Cambridge
R.P. Kirshner - University of Michigan
J. Labay - University of Barcelona
J.S. Larsen - University of Copenhagen
J.M. Lattimer - SUNY, Stony Brook
E. Liang - Lawrence Livermore Laboratory, California
R. Lieu - Imperial College, London
A.G. Lyne - Jodrell Bank, Cheshire
C.F. McKee - University of California, Berkeley
F. Matteucci - University of Padua
T.J. Mazurek - SUNY, Stony Brook
W.S. Meikle - Imperial College, London
C. Michell - Institut Henri Poincare, Paris
E. Müller - MPI, Munich
R.A. Muller - Lawrence Berkeley Laboratory, University of
    California, Berkeley
K. Nomoto - NASA, Goddard Space Flight Center, Maryland
W.W. Ober - MPI, Munich
J.P. Ostriker - Princeton University
F. Pacini - Florence
N. Panagia - Bologna
G. Pearce - University of Keele
C.R. Pennypacker - Lawrence Berkeley Laboratory, University of
    California, Berkeley
C.J. Pethick - NORDITA, Denmark
A. Priete-Martinez - Frascati
D.G. Ravenhall- University of Illinois
M.J. Rees - IOA, Cambridge
E. Schatzmann - Observatory of Nice
S.R. Schurmann - University of Michigan
F. Seward - Center for Astrophysics, Cambridge, Massachusetts.
P. Shapiro - University of Texas, Austin
J.M. Shull - University of Colorado
W. Sieber - MPI, Bonn

M. Signore - Meudon, France
R.A. Sramek - NRAO, New Mexico
R.J. Stoneham - IOA, Cambridge
R.G. Strom - Radiosterrenwacht, Dwingeloo
P. Sutherland - McMaster University, Ontario
A.E. Szymkowiak - NASA, Maryland
G.A. Tammann - Basle University, Switzerland & ESO, Munich
V. Trimble - University of California, Irvine
S. Tsuruta - University of Montana
D.L. Tubbs - Los Alamos, New Mexico
J.M. van der Hulst - University of Minnesota
R.V. Wagoner - Stanford University, California
T.A. Weaver - Lawrence Livermore Laboratory, California
W.B. Weaver - Monterey Institute, California
K.W. Weiler - NSF, Washington
J.C. Wheeler - University of Texas, Austin
R. White - Columbia University, New York
P.F. Winkler - Middlebury College, Vermont
R. Wood - Royal Greenwich Observatory, Sussex
S.E. Woosley - Lick Observatory, Santa Cruz
A. Yahil - SUNY, Stony Brook & NORDITA

INTRODUCTION

V.L. Trimble

Department of Physics, University of California, Irvine

1. THE HEROIC AGE

Supernovae, in the modern sense of stellar explosions releasing $\gtrsim 10^{49}$ ergs of electromagnetic radiation over a year or two, came gradually to light between 1920 and 1934. Lundmark (1920), a pioneer here as in many other things, seems to have been the first to realize that, if the spiral nebulae were extragalactic star systems, then some 'novae' were very bright. He noted that his 650,000 light year distance for the Andromeda Nebula implied $M_v = -15$ for S Andromeda (the 'nova' of 1885, now called SN 1885a), and suggested that some stars could flare up to luminosities thousands of times larger than those of the nebula concerned. Curtis (1921) took the next step. In the context of the historic Curtis-Shapley debate on the nature of the spiral nebulae, he stated that 'the dispersion of the novae in spirals and probably also in our own galaxy may reach at least 10^m, as is evidenced by a comparison of S And with the faint novae found recently in this spiral. A division into two magnitude classes is not impossible'. To this period belongs also the first suggestion (Lundmark 1921) of an association between the Crab Nebula and the Chinese 'guest star' of 1054.

Baade and Zwicky (1934a,b,d) drew the definitive distinction between 'common novae', with peak brightnesses $M_v \gtrsim -11$, and 'super-novae', with peak absolute magnitudes near -13, based on the old ($H_0 = 536$ km/sec/Mpc) extragalactic distance scale. The word super-novae was coined in connection with a Caltech lecture course in 1931 (Zwicky 1940), first appeared in public at the December 1933 American Physical Society meeting (Baade and Zwicky 1934c), and lost its hyphen in 1938 (Zwicky 1938a,b). The

M. J. Rees and R. J. Stoneham (eds.), Supernovae: A Survey of Current Research, xv–xxiii.
Copyright © 1982 by D. Reidel Publishing Company.

defining papers included an assortment of prescient suggestions:
(a) that the total energy released was $3 \times 10^{51} - 10^{55}$ ergs, far
in excess of the 'nuclear packing fraction', (b) that 'a
super-nova represents the transition of an ordinary star into a
neutron star consisting mainly of neutrons' and having
gravitational packing energy > 0.1 mc^2, (c) 'that super-novae
emit cosmic rays, leading to a very satisfactory agreement with
some of the major observations of cosmic rays' (including the
right flux, if one super-nova exploded in the galaxy each 1000
years, releasing 10^{53-54} ergs in cosmic rays, which flowed freely
out of the galaxy), and (d) that 'ionized gas shells are expelled
from them at great speeds' so that the cosmic rays should contain
nuclei of heavy elements.

Points (a), (b), and (d) stand today. Point (c) requires
modification only to the extent that current estimates of the
supernova rate (~ 1/30 years, Tammann 1982) and the cosmic ray
confinement time (~ 10^7 years, Garcia-Munoz et al. 1977) reduce
the production requirements to ~ 10^{49} ergs/event. The total
energy suggested in point (a) depended critically on Baade and
Zwicky's belief that the visible light was coming from a
photosphere only about 10^{13} cm across (like that of a 'common
nova'), so that the observed flux implied a black body
temperature of 10^{5-6} K and enormous amounts of flux in the
then-unobservable UV and X-ray regions.

Baade and Zwicky's conclusions were based on 12 supernovae (6
in the Virgo cluster, 6 elsewhere) accidentally discovered
between 1900 and 1930, plus S And, Z Cen (SN 1895b in NGC 5253),
and Tycho's 'new star' of 1572, with at most a few points on the
light curve of each, and no spectral data worth mentioning. More
objects and more information on each were clearly needed!

Zwicky (1965, and reminiscences in many other places) began
deliberate, organized searching for supernovae in 1934, with a
3 1/4 inch Wollensack lens camera, mounted on the roof of
Robinson Astrophysics Lab at Caltech, and in 1936 at Palomar with
the 18 inch Schmidt telescope acquired for this purpose. The 48
inch Palomar Schmidt search, lineal descendent of his project,
finally closed down in 1975, 281 supernovae later, 122 of them
having been discovered by Zwicky himself (Kowal et al. 1976 and
previous reports in that series). Charles Kowal (private
communication) found his 100th supernova a couple of years ago.
No other observer is currently in the running for a record.

Initially, Zwicky catalogued supernovae by numbering the
early ones in order of occurrence, then continuing the sequence
as he and his colleagues found additional events. A few
discoveries on old plate material required one extensive
renumbering to keep the catalogue in historical order. With the

advent of the Palomar Sky Survey, discoveries out of chronological order multiplied rapidly. Zwicky et al. (1963) eventually proposed the modern system, in which events are assigned a year corresponding to the time of their maximum light output and a letter, reflecting the order of their discovery (not occurrence) for that year. So far, only one year (1954) has yielded more than 26 supernovae. The two extras were dubbed 1954aa and 1954ab. Automated search procedures (as discussed in this volume) may soon vastly increase the annual 'score'.

2. SPECTRA AND THEIR INTERPRETATION

A few spectrograms had been recorded by chance before the Palomar search began producing results with 1937a. Humason (1936) and Baade (1936) discussed the spectra of 1926a and 1936a (both now thought to have been Type II) and believed that they were seeing Balmer lines and, probably, NIII at $\lambda 4650$, with widths of some 6000 km/sec. Payne-Gaposchkin (1936a,b) described the spectra of S And and Z Cen (the former reported verbally by visual observers, the latter recorded on a single objective prism plate: their types are unknown). She concluded that supernova spectra were really rather like those of common novae, but with much broader lines ($\sim 10,000$ km/sec). Thus, she said (Payne-Gaposchkin 1936c), supernovae ought to have photospheric temperatures like those of common novae ($\lesssim 20,000$ K), radii larger in proportion to the larger line widths (10^{14-15} cm), and total light outputs of only $\sim 10^{48}$ ergs. These numbers, allowing for the change in extragalactic distance scale, agree reasonably well with modern values.

But though Payne-Gaposchkin won the battle, she lost the war. For Zwicky (1936a,b,c) bounced back with the argument that these velocities, plus the amount of mass needed to keep something that big optically thick for a year or more, implied kinetic energies much larger than the visible light output, not to mention, he scolded, the large energies associated with the neutron star, non-visible radiation, and cosmic rays. In addition, he derived relationships among line widths, maximum brightness, total visible light output, and duration of outburst that seemed to apply to both novae and supernovae and strongly implied that the latter must be enormously more energetic events. That Nova Lacertae 1936 produced no discernible increase in the cosmic ray flux received at the earth (Barnothy and Forro 1936) seemed at the time also to point to a large energy difference, though the modern reader, thinking of diffusion time scales in the galactic magnetic field, is no longer impressed.

In 1937 came the brightest ($m_v \sim 8.4$) supernova of this century, 1937c in the dwarf irregular galaxy IC 4182. Popper (1937) took a hard look at the spectrum and bravely declared that he could not make head or tail of it. Nor could anyone else for almost 30 years (Pskovskii 1969: Mustel 1972: Branch and Patchett 1973: Gordon 1972).

But 1937c remained visible for almost two years, and, with SN 1937d in NGC 1003, formed the basis for the first systematic discussions of supernova light curves (Baade and Zwicky 1938) and spectroscopy (Minkowski 1940). Both these (and a couple of other objects for which fragmentary data existed at the time) happened to be Type I events. (Zwicky did not find a Type II until his 36th supernova.) Minkowski concluded that, although the spectra consisted of wide emission bands of unknown origin, they were at least very similar for all supernovae at a given time after maximum light. Also, the dispersion in maximum luminosity seemed to be very small (Baade 1938), leading Wilson (1939) and Zwicky (1939) to suggest the use of supernovae as distance indicators for external galaxies. Baade also noted that late-type spiral galaxies had produced 72% of the (dozen or so) supernovae known to him. The modern number (e.g. Tammann 1974) is 66%.

The simple picture collapsed with the discovery of 1940c in NGC 4725, which Minkowski (1940) described as having a spectrum 'entirely different from any nova or supernova previously observed' (at least by him). He proposed, therefore (Minkowski 1941), a provisional separation into two groups, Type I of which he knew (but did not list) nine examples, with the IC 4182 event as prototype, and Type II with five examples and the NGC 4725 event as prototype. SN 1941a in NGC 4559 belonged, he believed, to neither class. The basic distinction (based primarily on spectra) and the uncertainty about whether additional fundamentally different classes exist persist to the present (Oke and Searle 1974).

In simplest terms, what counts is the presence (Type II) or absence (Type I) of hydrogen lines in the spectrum near maximum light. As a result, no pre-telescopic supernova can confidently be assigned to either class. A possible connection between outburst type and remnant structure (Weiler and Panagia 1978) agrees with the traditional assignment of Tycho's and Kepler's new stars to Type I (Baade 1943, 1945) and requires the supernovae of 1054 and 1181 to have been Type II (cf. Chevalier 1977). This is consistent with what little is known about their light curves.

3. SUPERNOVA REMNANTS AND THEORIES

Very few supernova remnants were identified prior to the advent of radio astronomy. Apart from the Crab, Hubble (1937) suggested the Cygnus Loop. Baade (1943, 1945) found faint emission filaments at the position of Kepler's event, but nothing at the Tycho position, where faint filaments were finally located by Minkowski (1959) on the basis of a radio position (Baldwin and Edge 1957).

The Crab Nebula received detailed attention from Baade (1942) and Minkowski (1942). They agreed that the 'south preceeding star' (i.e. the pulsar, Cocke et al. 1969) was quite likely to be the exciting star, based on its colour (late B) and absence of spectral lines. They estimated a mass of 1 $M_\odot$ for it, on the assumption that the nebular excitation must be due to thermal ultraviolet radiation from the star. Minkowski noted the faintness of the Balmer lines in the nebular spectrum and correctly attributed it to hydrogen deficiency (Davidson 1979). He calculated a nebular mass of 15 $M_\odot$ and so 16 $M_\odot$ for the pre-supernova star. The modern number is 9 $M_\odot$, but with considerable mass loss before the explosion (Davidson et al. 1981).

On the theoretical front, Zwicky's (1938a, 1939) attempts to calculate neutron star binding energies, surface redshifts, and so forth, were overtaken by Oppenheimer and Volkoff's (1939) more complete formulation of the problem. The latter yielded a maximum mass of 0.7 $M_\odot$ (neglecting effects of the nuclear force) and binding energy $\sim 10^{53}$ ergs (about 10% of mc^2). Zwicky had found 100 $M_\odot$ (an independent discovery by difficult methods, of the Eddington argument for the maximum mass of a star) and 0.58 mc^2 (roughly analogous to the maximum gravitational redshift from the surface of a stable configuration, as calculated by Bondi 1964).

The various physical processes included in current models of supernovae and their remnants were first suggested by a host of workers, many still actively engaged in the field. Gamow and Schönberg (1941) attributed the triggering of the collapse of a stellar core to neutrino production; while Burbidge et al. (1957, better known as B^{2}FH) blamed photodisintegration of iron, the dominant constituent of the cores of their massive, evolved stars. Hoyle and Fowler (1960) drew attention to the possibility of triggering a supernova by a thermonuclear explosion in a degenerate stellar core, as well as to the energy contributed by nuclear burning in the outer layers of a star whose core collapses. Whipple's (1939) stellar collision model seems to have left no descendents in the supernova field but has some distant offspring in the realm of quasar models (Woltjer 1964: Gold et al. 1965).

The problem of transporting the energy of a supernova explosion from where and when it is released to where and when we see it has not yet been solved to everyone's satisfaction. Baade et al. (1956) proposed radioactive decay of Cf^{254} made in the explosion as a way of stretching out the energy release. More

recently, Colgate and McKee (1969), following a suggestion from
J. Truran, have cast Ni^{56} in the same role for Type I events (see
also Pankey, 1962). Type II light curves, on the other hand, can
be fit by a shock wave moving out through the extended, massive
envelope of the progenitor supergiant. Colgate and White (1966)
suggested that neutrinos produced in the collapsing core would
deposit momentum outside it and make such a shock. More recent
calculations (Wilson 1980) blame core bounce (occurring when the
core reaches nuclear densities and its equation of state suddenly
stiffens) for the shock.

Current theories of core collapse to neutron densities and
the subsequent bounce are reviewed in this volume by Brown and
Bethe; this is considered the basic mechanism for Type II
supernovae. As described by Wheeler and others, there is most
uncertainty about the basic mechanism for Type I supernovae. Some
of Hoyle and Fowler's pioneering ideas now find application to
Very Massive Objects (VMOs), which may have existed at earlier
epochs even if none are exploding now.

The existence of non-thermal (synchrotron) radiation in
supernova remnants was predicted by Shklovskii (1953) and
verified by Dombrovsky (1954) via the detection of large
polarization of the optical continuum radiation from the Crab
Nebula. Shklovskii (1960) also calculated the evolution of an
isolated, expanding supernova remnant; while van der Laan (1962)
called attention to the importance of interactions with the
interstellar medium, particularly for shell sources. Chevalier,
McKee and Blandford review the theory of remnants in this book;
several other papers emphasise how much has been learnt from new
observational techniques, particularly in the X-ray band.

4. HISTORICAL LITERATURE

The first published review of supernovae was by Zwicky
(1940); others (Hubble 1941: Bertaud 1941) quickly followed.
Additional reviews now largely of historical interest include
ones of supernovae-in-general by Payne-Gaposchkin (1957) and
Zwicky (1958, 1965); of stellar evolution leading up to supernova
explosions by Cameron (1960); of the explosions by Schatzman
(1965); and of supernova remnants by Minkowski (1964). Conference
proceedings and monographs prior to this volume covering a wide
range of supernova topics include Shkovskii (1968), Brancazio and
Cameron (1969), Davies and Smith (1971), Cosmovici (1974),
Schramm (1977) and Wheeler (1980).

REFERENCES

Baade, W. 1936. Pub.Astron.Soc.Pacific, 48, 226.
Baade, W. 1938. Astrophys.J., 88, 285.
Baade, W. 1942. Astrophys.J., 96, 188.
Baade, W. 1943. Astrophys.J., 97, 119.
Baade, W. 1945. Astrophys.J., 102, 309.
Baade, W., Burbidge, G.R., Hoyle, F., Burbidge, E.M., Christy,
    R.F. and Fowler, W.A. 1956. Pub.Astron.Soc.Pacific, 68,
    296.
Baade, W. and Zwicky, F. 1934a. Proc.Nat.Acad.Sci. 20, 254.
Baade, W. and Zwicky, F. 1934b. Proc.Nat.Acad.Sci. 20, 259.
Baade, W. and Zwicky, F. 1934c. Phys.Rev., 45, 138.
Baade, W. and Zwicky, F. 1934d. Phys.Rev., 46, 76.
Baade, W. and Zwicky, F. 1938. Astrophys.J., 88, 411.
Baldwin, J.E. and Edge, D.O. 1957. Observatory 77, 139.
Barnothy, J. and Forro, M. 1936. Nature 138, 344.
Bertaud, M. 1941. Astronomie 55, 73 and 103.
Bondi, H. 1964. Proc.Roy.Soc.A 281, 39.
Brancazio, P.J. and Cameron, A.G.W. 1969. Supernovae and Their
    Remnants (Gordon and Breach).
Branch, D. and Patchett, B. 1973. Mon.Not.R.astr.Soc., 161,
    71.
Burbidge, E.M., Burbidge, G.R., Fowler, W.A. and Hoyle, F. 1957.
    Rev.Mod.Phys., 29, 547 (B^{2}FH)
Cameron, A.G.W. 1960. Mem.Soc.R.des Lieges, 15th Ser, Vol.III,
    p.163.
Chevalier, R.A. 1977. In D.N. Schramm (ed.) Supernovae (D.
    Reidel:Dordrecht), p.53.
Cocke, W.J., Disney, M.J. and Taylor, D.J. 1969. Nature 221,
    525.
Colgate, S.A. and McKee, C. 1969. Astrophys.J., 157, 623.
Colgate, S.A. and R.H. White. 1966. Astrophys.J., 143, 626.
Cosmovici, C.B. (ed.) 1974. Supernovae and Their Remnants
    (Astrophys.Space Sci.Lib.No.45: D. Reidel, Dordrecht).
Curtis, H.D. 1921. Bull.Nat.Res.Council 2, part 3, p. 171.
    (The Curtis-Shapley debates)
Davidson, K. 1979. Astrophys.J., 228, 179.
Davidson, K., Gull, T.R., Maran, S.P., Stecher, T.P., Parise,
    R.A., Harvel, C.A., Fesen, R.A., Kafatos, M. and Trimble,
    V.L. 1981. Submitted to Astrophys.J.
Davies, R.D. and Smith, F.G. (eds.) The Crab Nebula (IAU
    Symp. no. 46: D. Reidel, Dordrecht).
Dombrovsky, V.A. 1954. Doklady Akad.Nauk.USSR, 94, 121.
Gamow, G. and Schönberg, M. 1941. Phys.Rev. 59, 539.
Garcia-Munoz, M., Mason, G.M. and Simpson, J.A. 1977.
    Astrophys.J., 217, 859.
Gordon, C. 1972. Astron.Astrophys., 20, 79.

Gold, T., Axford, I. and Ray, E.C. 1965. In I. Robinson, A. Schild and E. Schucking (eds.) Quasi-Stellar Sources and Gravitational Collapse (U.Chicago Press), p. 93.

Hoyle, F. and Fowler, W.A. 1960. Astrophys.J., 132, 565.

Humason, M.L. 1936. Pub.Astron.Soc.Pacific, 48, 110.

Hubble, E. 1941. Pub.Astron.Soc.Pacific, 53, 141.

Hubble, E. 1937. Mt.Wilson Annual Report, 1936-1937.

Kowal, C, Huchra, J. and Sargent, W.L.W. 1976. Pub.Astron.Soc. Pacific, 88, 521.

van den Laan, H. 1962. Mon.Not.R.astr.Soc., 124, 125 and 179.

Lundmark, K. 1920. Svenska.Vetenkapsakad.Handlingar 60, no.8.

Lundmark, K. 1921. Pub.Astron.Soc.Pacific, 33, 234.

Minkowski, R. 1939. Astrophys.J., 89, 156.

Minkowski, R. 1940. Pub.Astron.Soc.Pacific, 52, 206.

Minkowski, R. 1941. Pub.Astron.Soc.Pacific, 53, 224.

Minkowski, R. 1942. Astrophys.J., 96, 199.

Minkowski, R. 1959. Paris Symp.Radio Astronomy (Paris 1958. IAU Symp. no.3), p. 315.

Minkowski, R. 1964. Ann.Rev.Astron.Astrophys. 2, 247.

Mustel, E.R. 1972. Sov.A.J., 16, 10.

Oke, J.B. and Searle, L. 1974. Ann.Rev.Astron.Astrophys., 12, 315.

Pankey, T.Jr. 1961. Possible Thermonuclear Activities in Natural Terrestrial Materials. Ph.D. Thesis Howard University: University Microfilms, Ann Arbor.

Payne-Gaposchkin, C.H. 1936a. Astrophys.J., 83, 245.

Payne-Gaposchkin, C.H. 1936b. Astrophys.J., 83, 173.

Payne-Gaposchkin, C.H. 1936c. Proc.Nat.Acad.Sci., 22, 332.

Payne-Gaposchkin, C.H. 1957. The Galactic Novae (North Holland) p. 259-285.

Popper, D.M. 1937. Pub.Astron.Soc.Pacific, 49, 283.

Pskovskii, Yu.P. 1969. Sov.A.J., 12, 750.

Schatzman, E. 1965. In L.H. Aller and D.B. McLaughlin (eds.) Stellar Structure (U. Chicago Press, vol. VIII of the Kuiper compendium), p. 327.

Schramm, D.N. 1977. Supernovae (Astrophys.Space Sci.Lib. No. 66: D.Reidel, Dordrecht).

Shklovskii, I.S. 1953. Doklady Akad.Nauk.USSR 91, 475.

Shklovskii, I.S. 1960. Sov.A.J., 4, 243.

Shklovskii, I.S. 1968. Supernovae (Wiley Interscience).

Tammann, G. 1974. In C.B. Cosmovici (ed.) Supernovae and Their Remnants (D.Reidel: Dordrecht), p. 155.

Tammann, G. 1982. This volume.

Weiler, K.W. and Panagia, N. 1978. Astron.Astrophys., 70, 419.

Wheeler, J.C. 1980 (ed.) Proceedings Texas Workshop on Type I Supernovae (University of Texas: Austin).

Whipple, F. 1939. Proc. Nat.Acad.Sci., 25, 118.

Wilson, J.R. 1980. Ann.N.Y.Acad.Sci., 336, 358.

Wilson, O.C. 1939. Astrophys.J., 90, 634.

Woltjer, L. 1964. Nature 201, 245.

Zwicky, F.   1936a.   Proc.Nat.Acad.Sci., 22, 457.
Zwicky, F.   1936b.   Proc.Nat.Acad.Sci., 22, 557.
Zwicky, F.   1936c.   Pub.Astron.Soc.Pacific, 48, 191.
Zwicky, F.   1938a.   Phys.Rev., 54, 242.
Zwicky, F.   1938b.   Pub.Astron.soc.Pacific, 50, 215.
Zwicky, F.   1939.   Phys.Rev., 55, 726.
Zwicky, F.   1940.   Rev.Mod.Phys., 12, 71.
Zwicky, F.   1958.   Handbuch der Physik, 51, 766.
Zwicky, F.   1965.   In L.H. Aller and D.B. McLaughlin (eds.)
     Stellar Structure (U. Chicago Press).
Zwicky, D., Berger, J., Gates, H.S. and Rudnicki, K.   1963.
     Pub.Astron.Soc.Pacific, 75, 236.

An earlier version of this chapter has been submitted as part of a much longer article to Reviews of Modern Physics.

AN OPTIMIST'S GUIDE TO SUPERNOVAE

Robert P. Kirshner

Department of Astronomy
The University of Michigan
Ann Arbor, Michigan 48109
U.S.A.

There are some hints that a satisfactory agreement
between observations and models is developing in several areas
of supernova research. The light curves and spectra of Type I
supernovae look so much like the models that they may well have
their origin in the synthesis of ^{56}Ni in the explosion of a
compact star. Type II supernovae are found where massive stars
form, have light curves that correspond to 10^{51} ergs suddenly
deposited inside a red supergiant, and may in fact be the
explosions of massive stars. Some young supernova remnants,
such as Cas A, show abundance patterns that bear a striking
resemblance to those seen deep inside models of massive stars
on the verge of destruction. Observations of old supernova
remnants provide energy estimate of 10^{51} ergs -- just the amount
needed for the models of the outbursts. While the usual
cheerful anarchy still prevails, something must be right
with the general picture.

A. INTRODUCTION

Supernovae deserve to be understood. The brilliant
destruction of a star is the outward sign of a catastrophe
that marks the end of stellar evolution. It may be accompanied
by the synthesis of new elements and the formation of a
neutron star or black hole. The energy released by the
explosion produces spectacular effects in the interstellar gas
and the debris from supernovae enriches the raw material for
new generations of stars.
    We have learned a few facts about supernovae, and we have
some ideas of what might cause these phenomena. One of the

1

M. J. Rees and R. J. Stoneham (eds.), Supernovae: A Survey of Current Research, 1–12.

most satisfying features of supernova research in the past few
years is that some of the observations and theories which formerly
ran on parallel paths are beginning to intersect. In this
introductory contribution, I would like to show some of the
places where the facts and the models are developing links
that suggest we may be nearing the truth.

B.   SUPERNOVA EXPLOSIONS

Once the extragalactic distance scale was established,
the extra bright nova of 1885 seen in the center of M31, and
other extraordinary events in other galaxies,were recognized
as a distinct set of objects: the supernovae. Today, photographic searches with small Schmidt cameras turn up about 10
new stars, in other galaxies each year. These discoveries are
communicated by the IAU telegram network to interested observers.
In recent years, as this book shows, the interested group has
expanded beyond optical astronomers to include radio astronomers
at the VLA, ultraviolet workers using the IUE satellite, and
X-ray astronomers using the Einstein satellite.
The optical spectra provide the basis for classifying
supernovae into two principal types: Type I (SN I) and Type II
(SN II) (Oke and Searle 1974). The SN I have substantial
undulations in their spectra, even near maximum light when an
underlying continuum is present, and they have a characteristic
sharp dip at about 6150 Å during the first two weeks. The
identification of this feature with Si II allows the expansion
velocity to be determined as about 11 000 km/sec.
The spectra of SN I are remarkably uniform. They evolve
with time in a way that is so consistent that the age of
a SN I can be determined by comparing it to the well-observed
cases like SN 1972e in NGC 5253 (Kirshner et al. 1973a,b,1975).
The outstanding facts about SN I spectra are that no hydrogen
line are significant and that the overall energy distribution
near maximum light can be represented by a continuum of moderate
temperature (of order 15 000 K). The immense luminosity at
modest temperature means that the radius of the photosphere of
a SN I must be nearly 10^{15} cm at maximum light, and growing
at 11 000 km/sec.
The observable feature of SN I that has elicited the most
theoretical interest is the evolution of their luminosity with
time-- the light curve. The luminosity rises sharply to a peak
near 10^{43}erg/sec. During this phase of maximum light most of
the optical light emerges in the continuum. As the photosphere
expands and cools, the luminosity drops off sharply to about
10^{42} erg/sec in a month. After that, the optical luminosity
begins a long exponential decline with a half-life of about
60 days that is known to continue at least 700 days past
maximum light. As the luminosity goes into this exponential
decline, the optical continuum fades and the spectrum shifts

to a broad group of emission bands. A plausible identification
for these bands came from adding up the emission from 216
lines of [Fe II] (Kirshner and Oke 1975). This idea has been
developed by Meyerott (1978, 1980) and by Axelrod (1980) whose
synthetic spectra provide a good case that the late time
spectrum of SN I comes from Fe^+ and Fe^{++}. The absence of
hydrogen lines and these strong iron lines makes it quite
likely that the atmospheres of SN 1 are rich in iron.

This is in excellent accord with the theory for producing
the energy in SN I . The exponential decay of the luminosity
has been the inspiration for several theoretical ideas based
on radioactivity. The modern incarnation stems from the model
by Colgate and McKee (1969), which explored the possible
significance of the decay chain $^{56}Ni \rightarrow {}^{56}Co \rightarrow {}^{56}Fe$. In recent
models (Arnett 1979, Colgate,Petschek, and Kriese 1980, Chevalier
1981) this radioactive decay contributes energy both at the
peak of luminosity, which is dominated by considerations
of energy balance and thermal diffusion in the expanding star,
and for the exponential tail where the decay energy goes into
heating an electron gas that collisionally excites optical
emission lines.

These models, as reviewed by Arnett in this volume, start
with the plausible suggestion that the star which explodes to
become a SN I is a compact star-- probably a white dwarf.
This idea is broadly consistent with the fact that SN I
are found in all types of galaxies, including ellipticals
where the present rate of star formation is low. The conventional
wisdom is that SN I come from relatively low mass stars,
although Oemler and Tinsley (1979) have argued that the SN I
seen in spirals come from short-lived stars.

In any event, the detonation of a compact star nudged over
the Chandrasekhar mass should lead through nuclear burning
to the synthesis of substantial amounts of iron peak elements
and particularly the doubly magic nucleus ^{56}Ni. The subsequent
decays emit positrons and gamma rays. The positrons are
presumably annihilated on the spot, while the gamma ray energy
deposition changes as the transparency varies. The idea in
these models is that the Ni→Co decay makes up for adiabatic
losses as the star expands from white dwarf to 10^{15}cm because
of its rapid 6 day half-life. As the atmosphere turns transparent to optical photons, the diffusion picture breaks down.

At this point, enough Co has accumulated so that the
subsequent Co→Fe decay becomes important. The energy deposited
by this decay with a 77-day half-life goes into heating the
electrons of the expanding nebula. Axelrod's (1980) calculations show that the excited iron ions provide a very good,
if not iron clad, match to the observed spectra in the exponential part of the SN I light curve.

This combination of observation and modeling has so many

points of contact that it may represent a correct picture of
SN I (see Wheeler, 1980 and Meyerott and Gillespie,1980 for
more detail). Yet there are many improvements to be made.
First, the picture of the energetics is incomplete, since it
is based only on the optical data. However, the latest
ultra-violet results from IUE, presented here by Panagia,
seem to show that the UV flux is low, so the light curve in
the optical may actually represent the luminosity and tell how
much radioactive nickel needs to be synthesized. This model for
SN I   predicts that the late-time spectrum should show
significant shifts in the cobalt-to-iron ratio as the decay
proceeds. The hints are that this does happen, as described
by Axelrod (1980),yet, as the contibution by Branch in this
volume shows, alternate readings of the data are possible.

Another prediction of this model is that young galactic
remnants of SN I should have large amounts of iron-peak debris.
Based on contemporary accounts of brightness and color, Tycho's
1572 supernova and Kepler's 1604 event are thought to be Type I,
although , of course, we have no spectra. X-ray spectroscopy of
these objects does not provide any evidence that they have the
predicted iron (Becker et al.1980a, 1980b). The best observation of all would be to detect the gamma rays that escape from
the supernova during the Ni→Co→Fe chain. The Gamma Ray Observatory will be able to detect these gamma rays from a SN I in the
Virgo cluster and could provide conclusive evidence on the
nucleosynthesis of iron peak elements in Type I supernovae.

C. TYPE II SUPERNOVAE

The other spectroscopic class of supernova has a smooth continuum near maximum light and a Balmer series that develops as
the atmosphere expands. Type II supernovae are seen only in
spirals and within spirals, they seem to be concentrated near
the spiral arms (Moore 1973, Maza and van den Bergh 1976).
This is taken as evidence that SN II   result from massive stars
that evolve so rapidly that they never appear far from sites of
star formation.

For massive stars, the course of quiet stellar evolution to
the brink of core collapse includes the synthesis of elements
up to iron in advanced stages of burning inside a supergiant
envelope. Models for these stars have a large envelope of
unburned hydrogen, a zone where the helium produced in hydrogen
burning is found, a layer made mostly of oxygen produced in
helium burning, then a silicon rich zone with sulfur, argon,
and calcium, all products of oxygen burning, and finally an iron
core. The observational evidence is consistent with this kind of
model, but unlike the SN I   the observations of the outburst
do not provide detailed support for any particular version of

the nuclear turmoil in the core.

The observations confirm that SN II   have substantial hydrogen-rich envelopes (Kirshner and Kwan 1974, Branch et al. 1981), but there is no way to see far enough in to determine the nuclear burning stage of the exploded star. The smooth continuum observed at maximum may hide the most interesting events of SN II , but it does allow the photospheric temperatures to be determined rather well. As is the case for SN I, the temperatures are modest, and the radii rather large-- at maximum a temperature of about 15 000 K is indicated with a radius of order 10^{14} cm. A month later, the temperature is about 6000 K and the radius reaches a maximum value of over 10^{15} cm.

One interesting application of the ability to measure the temperature is that it provides a way to measure extragalactic distances (a problem considered more generally by Tammann in this volume). Since the temperature and the observed flux density give a measure of the angular size, and this measurement can be repeated as the photosphere expands, we can measure an angular expansion rate. The velocity of expansion can be measured by looking at blueshifted absorption lines formed near the phtosphere. Using these velocities (from 5000 to 10 000 km/sec) together with the angular expansion rate makes it possible to estimate the distance to an individual super- nova (Kirshner and Kwan 1974, Branch et al. 1981). The refinement of this method and improved efficiency of supernova detection (as detailed in this volume) may make a significant contribution to the problem of the extragalactic distance scale.

The general idea of a bomb inside a red supergiant fits well with the observations of luminosity, temperature, and velocity of SN II . As described in Chevalier's (1981) review, this type of model, with no continued energy input after the sudden deposition of 10^{51} ergs in the center, works only in stars with extended atmospheres. The models, however, do not tell very much about the events at the center of the star: just the energy. There is no simple way based on observations of the outsides of SN II   to resolve the problems of core bounce and shock propagation that are so warmly discussed in this book.

There is hope, though, that observations can be more specific about the events in SN II. If you wait long enough, the expanding atmosphere will turn transparent, and the inner regions of the star will be revealed. There is substantial flux emitted by SN II   a year or more after the explosion. Detailed studies of these objects may provide evidence on the possibility that radioactivity or a neutron star provides the energy at late times.

D. YOUNG REMNANTS AND LIFE AFTER HYDROGEN BURNING

    While the cicumstantial evidence that SN II   come from
massive stars seems good, there is no direct evidence on the
details of nuclear burning from observing the outburst. That
evidence comes from the spectra of young supernova remnants.
Supernova remnants that are younger than 1000 years are dominated
by the debris ejected by the explosion rather than the interstellar gas they encounter. By studying young remnants
(aided by the historical record discussed by Clark in this
volume), we have a chance to inspect the innards of a star
that exploded, and to test the picture of stellar evolution.
    One young remnant has provided particularly clear evidence
for nucleosythesis: Cassiopeia A. Cas A is a powerful non-thermal
radio source, a strong thermal X-ray emitter, and appears optically
as a fragmentary ring of nebulosity. Detailed work by Kamper and
van den Bergh (1976) shows that the optical filaments are of two
types: the "quasi-stationary flocculi" which are slowly expanding
and the fast moving knots which have space velocities of order
8000 km/sec. The quasi-stationary flocculi are probably fragments
of the outer envelope of the star lost in a stellar wind that
preceeded the explosion. The fast moving knots, which look
as if they have been traveling without deceleration since 1667,
are presumably debris from the explosion itself.
    Spectra of the fast knots show clear evidence of very unusual
abundances: unlike other nebulae and most supernova remnants,
they show no lines of hydrogen, helium or nitrogen (Kirshner and
Chevalier 1977, Chevalier and Kirshner 1978). Oxygen lines are
seen that come from neutral, once ionized, and doubly ionized
atoms, so it seems very unlikely that some bizarre ionization
situation can be responsible for the missing hydrogen lines.
In addition to oxygen, the fast moving knots have strong emission
lines of sulfur, argon and calcium. With a shock model, the
line strengths can be interpreted as abundances. The result is
that about half the mass in a typical fast moving knot is oxygen.
There can be no doubt that this material, which must have been
nearly all hydrogen and helium when the star was formed, has
undergone extensive nuclear processing.
    Individual knots show varying amounts of oxygen compared to
sulfur, argon, and calcium, as though the ejecta were not
thoroughly mixed. As an extreme example, Figure 1 shows the
spectrum of a knot which has only oxygen lines.
    The pattern of abundances observed in Cas A corresponds well
to the adundances expected in the inner layers of a massive
star. The computation by Weaver, Zimmerman, and Woosley (1978)
illustrated in Figure 2 shows that a massive star which has
evolved to the brink of collapse has substantial zones ·in which
oxygen is the dominant element, and also has regions where the
products of burning oxygen:silicon, sulfur,argon,and calcium

are the most abundannt species. Since silicon does not have
optical emission lines that we would expect to see, it seems
fair to characterize the abundances in Cas A as oxygen plus
variable amounts of oxygen-burning products. Although there may
be other ways to produce these abundances, it seems plausible
to identify the fast moving debris in Cas A with poorly mixed
products from deep within a massive star.

The idea that these elements were produced in a massive star
has independent confirmation based on the X-ray data presented
by Fabian et al. (1980) who show that about 15 solar masses of
hot plasma is required to produce the observed X-rays.

Cas A is not the only supernova remnant with striking
abundances in fast-moving, undiluted debris. Knots with no
hydrogen emission, but strong oxygen lines, have been identified
in the galactic remnant G 292.0+1.8 (Goss et al. 1980), in two
remnants in the Large Magellanic Cloud, N 132D (Lasker 1978,
1980) and 0540-69.3 (Mathewson et al. 1980) and in the extraordinary
remnant in the galaxy NGC 4449 (Balick and Heckman 1978, Blair and
Kirshner 1980).

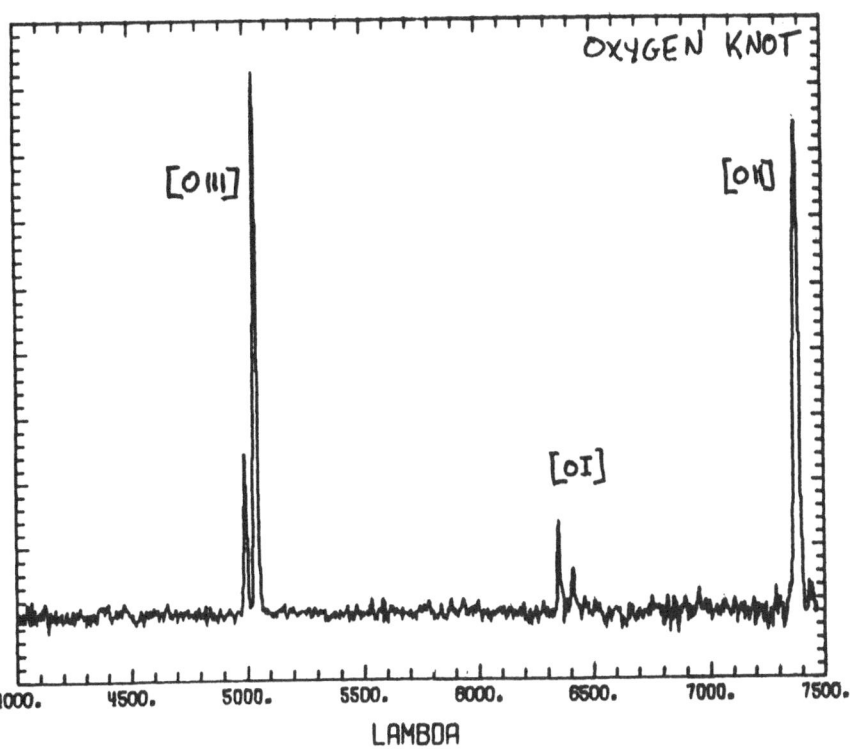

Figure 1. The spectrum of a fast moving knot in Cas A. In this
extreme case, oxygen is the only element observed.

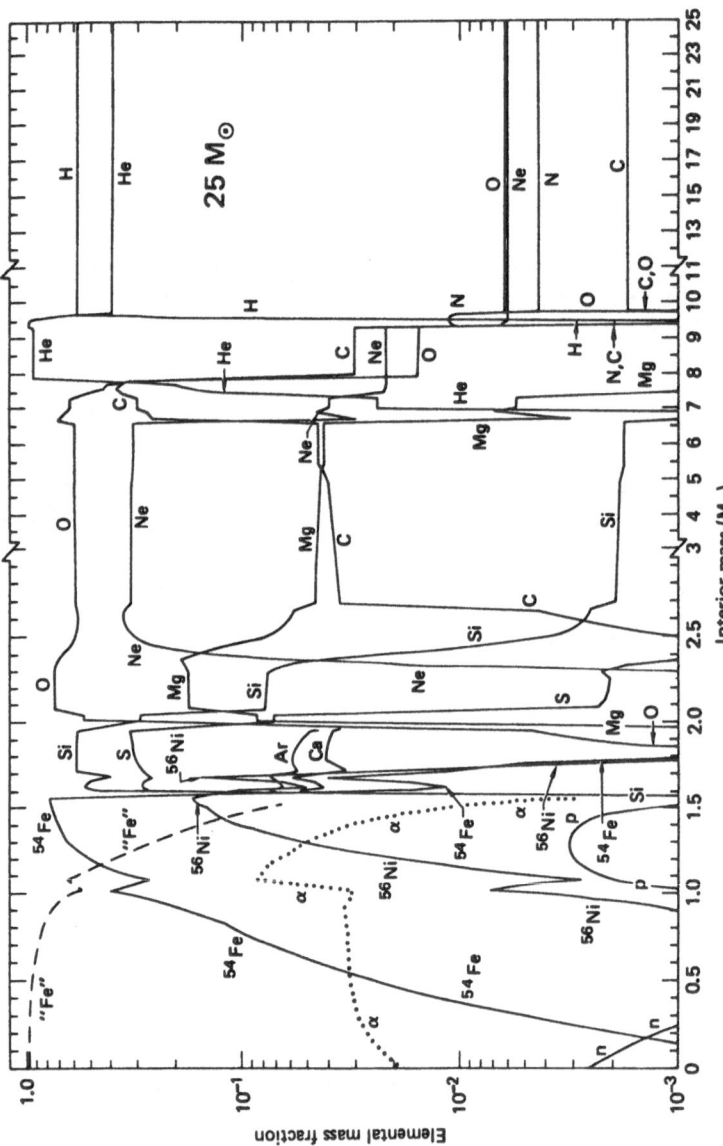

Figure 2. The chemical structure of a 25 solar mass star at the onset of collapse (Weaver, Zimmerman and Woosley 1978). Note the large zones of oxygen and the oxygen burning products sulfur, argon, and calcium.

The NGC 4449 object, about 5 Mpc away, has about 25 times
the radio luminosity of Cas A (Seaquist and Bignell 1978),
shows powerful lines of neutral, once ionized and doubly ionized
oxygen but no hydrogen lines, and has an expansion velocity of
3500 km/sec. Recent observations with the Einstein X-ray
Observatory show that the remnant is also a powerful X-ray
source (Blair, Kirshner, and Winkler 1982).

E. OLD REMNANTS

In the case of SN II explosions, the light emitted near
maximum can be successfully modeled even though the details
of the energy source remain obscure-- only the energy matters.
In the same way, old supernova remnants can be understood even
when the details of the explosion that caused it remain
uncertain. Once the remnant encounters many times its ejected
mass, the remnant is a shock phenomenon in the interstellar
medium, and retains no memory of its origin as a stellar event.
The crucial link between old remnants and the events that
cause them is the energy. It is worth noting that the spectacular
optical outburst that we see in a supernova has a total energy
output of only 10^{49} ergs, while the amount of energy used in
modeling the outburst is near 10^{51} ergs. The bulk of the energy
in a supernova explosion is in the kinetic energy of the expanding
debris. A solar mass of material moving at 10^4 km/sec has a
kinetic energy of 10^{51}ergs. The question of interest is, do
old supernova remnants provide the same estimate for the initial
energy?
From straightforward models , the initial energy is related
to the observable properties of an old supernova remnant by:

$$E = 2 \times 10^{46} \, n \, v^2 \, R^3 \text{ erg. (McKee and Cowie 1975)}$$

The observational problem is to measure the radius R (here in
parsecs) and n v^2 (the particle density in cm^{-3}, and the
expansion velocity in units of 100 km/sec). The ambient density
and the shock velocity are not easily observed, principally
because the optical filaments are the dense clouds encountered
by the shock. Even so, nv^2, which is the pressure in the shock
wave, should not be very different from the pressure in the dense
cloud that we see. For remnants in our galaxy, it is hard to
estimate R accurately. Some studies of proper motion and radial
velocity can be used to give distances. These measurements need
to be repeated: we are still using Hubble's(1937) measurement of
the proper motion to get the distance to the Cygnus Loop! Lower
limits to the distances can come from 21 cm absorption lines and
a kinematic model for the Galaxy, but the whole problem of
distances to galactic SNR is difficult.

Another approach is to find and study supernova remnants in nearby galaxies. The Magellanic Clouds (Dopita 1979), M33 (Dopita, D'Odorico, and Benvenuti 1980a,b) and M31 (Dennefeld and Kunth 1981) have been explored: further work by Helfand and Dopita that brilliantly exploits new X-ray observations is covered in this volume.

For remnants in another galaxy, the diameters give relative sizes and the distance to the galaxy sets the scale. The pressure can be derived from optical observations by measuring the density sensitive ratio 6717/6731 of [S II]. Since the emission from S^+ ions always takes place near 10 000 K, this gives the thermal pressure. Figure 3, from the work by Blair, Kirshner, and Chevalier(1981) shows the result, including data on our own galaxy from the work of Daltabuit, D'Odorico, and Sabbadin (1976).

The general result is satisfying-- the big remnants have initial energies near 10^{51} ergs, which is the same number used to model the details of the outburst that took place 10^4 years earlier. The curious relation between the energy derived this way and radius is unexpected: some of the possible explanations are given by Blair's contribution to this book.

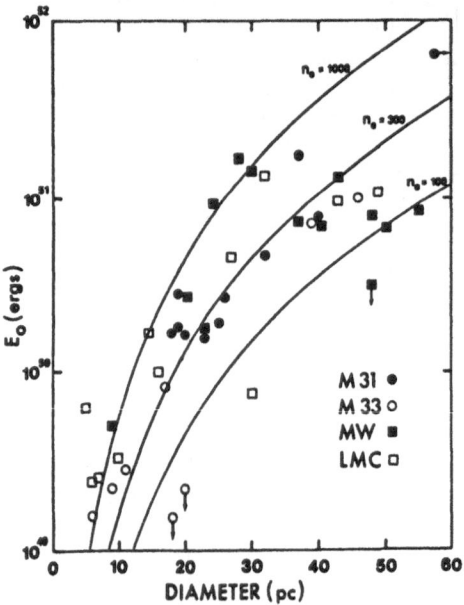

Figure 3. Energies for extragalactic supernova remnants, after Blair, Kirshner, and Chevalier (1981).

F. CONCLUDING REMARKS

I have tried to show that theoretical ideas about supernovae have enough intersections with the observations that
we are beginning to develop a unified picture that probably
has some elements of truth. By omitting areas of the subject
where confusion dominates, I have neglected to mention a large
fraction of the work on supernovae. To gain a fuller appreciation
of the myriad loose ends that remain to be raveled, the reader
is invited to turn the pages of this book.

ACKNOWLEDGEMENTS

I am grateful to the organizers of this meeting for doing
such a spendid job. This work was supported in part by an
Alfred P. Sloan Research Fellowship, NSF grants AST 76-17600
and AST 80-5050 and NASA grant NAG-8341.

REFERENCES

Arnett, W.D.: 1979, Ap.J.Lett. 230, L37.
Axelrod, T.S.: 1980, Ph.D. Thesis, UC Santa Cruz.
Balick, B. and Heckman, T.: 1978, Ap.J.Lett. 226,L7.
Becker, R.H., Boldt, E.A., Holt,S.S., Serlemitsos, P. and White, N.E.:
    1980a, Ap.J.Lett. 235,L5.
    1980b, Ap.J.Lett. 237,L77.
Blair, W.P. and Kirshner, R.P.: 1980, Ap.J. 236,135.
Blair, W.P., Kirshner, R.P., and Chevalier, R.A.: 1981, Ap.J.
    247,879.
Blair, W.P., Kirshner, R.P. and Winkler, P.F.: 1982, Ap.J. in press
Branch, D., Falk, S.W., McCall, M.L., Rybski, P., Uomoto,A.K.,
    and Wills, B.J.: 1981, Ap.J. 244, 780.
Chevalier, R.A.: 1981,Fund. Cos. Phys., 7,1.
Chevalier, R.A.: 1981, Ap.J. 246,267.
Chevalier, R.A. and Kirshner, R.P.: 1978, Ap.J. 219,931.
Chevalier, R.A. and Kirshner, R.P.: 1979, Ap.J. 233,154.
Colgate, S.A. and McKee, C.: 1969, Ap.J. 157,623.
Colgate, S.A., Petschek, A.G., and Kriese, J.T.: 1980, Ap.J. Lett.
    237,L81.
Daltabuit, E., D'Odorico, S., and Sabbadin, F.: 1976, Astr. Ap.
    52,93.
Dennefeld, M. and Kunth,D.: 1981, A.J. 86,989.
D'Odorico, S., Dopita, M.A., and Benvenuti, P.: 1980, Astr. Ap.
    Suppl. 40,67.
Dopita, M.A.: 1979, Ap.J. Suppl. 40,455.
Dopita, M.A., D'Odorico, S., and Benvenuti, P.: 1980, Ap.J. 236,628.
Fabian, A.C. et al.: 1980, M.N.R.A.S. 193,175.
Goss, W.M. et al.: 1980, M.N.R.A.S. 193,901.

Hubble, E.P.:   1937, Carnegie Yearbook 36, 189.
Kamper, K. and van den Bergh, S.: 1976, Ap.J. Suppl. 32,351.
Kirshner, R.P. and Chevalier, R.A.:   1977, Ap.J. 218,142.
Kirshner, R.P. and Kwan, J.:   1975, Ap.J. 197,415.
Kirshner, R.P. and Kwan, J.:   1974, Ap.J. 183,27.
Kirshner, R.P. et al.: 1973a, Ap.J. Lett. 180,L97.
Kirshner, R.P. et al.:   1973b, Ap.J. 185,303.
Kirshner, R.P. and Oke, J.B.:   1975, Ap.J. 200,574.
Lasker, B.: 1978, Ap.J. 223,109.
Lasker, B.: 1980, Ap.J. 237,765.
Mathewson, D.S. et al.:   1980, Ap.J.Lett. 242,L73.
Maza, J. and van den Bergh, S.:   1976, Ap.J. 204,519.
McKee, C.F. and Cowie, L.L.:   1975, Ap.J. 195,715.
Meyerott, R.E.:   1978, Ap.J. 221,975.
Meyerott, R.E.:   1980, Ap.J. 239,257.
Meyerott, R.E. and Gillespie, G.H. (eds.):   1980,"Supernovae
    Spectra", American Institute of Physics, New York.
Moore, E.:   1973, Pub. Astron. Soc. Pacific 85,564.
Oemler, A. and Tinsley, B.M.: 1979, A.J. 84,985.
Oke, J.B. and Searle, L.:   1974, Ann.Rev.Astron.Astrophys. 12,315.
Seaquist, E.R. and Bignell, R.C.:   1978, Ap.J.Lett. 226,L5.
Weaver, T.A., Zimmerman, G.B. and Woosley, S.E.:   1978, Ap.J. 225,
    1021.
Wheeler, J.C. (ed.) :   1980, "Texas Workshop on Type I Supernovae",
    University of Texas Press: Austin.

THE FATE OF MASSIVE STARS:

COLLAPSE, BOUNCE AND SHOCK FORMATION

G.E. Brown

NORDITA, Blegdamsvej 17
DK-2100 Copenhagen Ø, Denmark
                and
Dept. of Physics, State University of New York
Stony Brook, NY 11794

1. INTRODUCTION

This lecture, together with those of Hans Bethe, and Amos
Yahil and James Lattimer which follow, deal with Supernova theory.
Some past work on this subject will be summarized, but chiefly
we shall report on the work of the BABBLY group:
        Bethe,
        Applegate,
        Baym,
        Brown,
        Lattimer and
        Yahil.
I shall deal with the collapse of massive stars, bounce and shock
formation. Much of the discussion will be conducted by "engineering methods", summarizing general features of extensive computer
calculations, with some background theory to connect these features. Amos Yahil and James Lattimer will give these general features a theoretical foundation, in discussing self-similar collapse
and hydrostatic models of the core and the shock. Hans Bethe will
conclude, with a discussion of shock propagation, neutrino diffusion and formation of compact remnants. The work of the BABBLY
group has appeared in references 10 (BBAL), 13 (BAB) and 20(B^3).

Supernovae of type II are generally believed to occur at the
end of the evolution of massive stars (1-10) in the range $\gtrsim$ 8 to
perhaps 100 solar masses ($M_\odot$). When the nuclear fuel in the core
of the star is exhausted, the center of the star begins to undergo
gravitational collapse, and falls inward until the core rebounds,
or "bounces"; the inner core of the star is believed to remain as

13

M. J. Rees and R. J. Stoneham (eds.), Supernovae: A Survey of Current Research, 13–33.
Copyright © 1982 by D. Reidel Publishing Company.

a neutron star while the outer parts of the star are driven off
in the supernova explosion.

The fundamental problem is to understand how the gravitational energy released on infall is transferred to the outer layers
of the star, expelling them. For a long time it was believed that
neutrinos, formed in electron capture and emitted from the core
of the star, would transmit their outward momentum to the outer
layers and thereby cause the explosion (2). This mechanism became unlikely when the weak neutral currents were discovered theoretically and experimentally, because these currents tend to trap
neutrinos in the core (11,12). Since then interest has focussed
on the idea that the outward propulsion of the mantle and envelope
of the star is due to a hydrodynamic shock which is generated by
the gravitational collapse of the central core and the subsequent
rebound (5-10,13).

The purpose of this lecture is to analyze the basic physics
of collapse of more massive stars and the subsequent formation of
the shock. In this analysis we shall draw heavily upon computer
studies of collapse, particularly by J. Wilson (5), his "Munich"
calculation, and by Mazurek, Cooperstein and Kahana (6) as guides
to determining the crucial underlying physics. While we do not
agree with certain parts of these calculations, they have been
extremely valuable in showing in a very detailed way a model of
the collapse, and shock formation and propagation.

In our theory we shall assume spherical symmetry of the star;
in other words rotation is assumed to have negligible effect.
Whether this is realistic for all or most of the stars is not
known. However, spherical symmetry is evidently the simplest assumption to make, and it appears capable of leading to a satisfactory explanation of the outgoing shock. Calculations with spherical symmetry also serve as a starting point for subsequent per-
turbative calculations of the influence of rotation.

2. PRE-SUPERNOVA CONDITIONS

In the core of the star just prior to collapse, ^{28}Si burns
and is transformed into ^{56}Ni which then, by electron capture,
changes into a variety of neutron-rich nuclei with masses in the
range 50 to 60. The composition of this presupernova core has
been investigated in detail by Weaver, Zimmerman and Woosley (4)
and by Arnett (3). The average number of electrons per nucleon,
Y_e, is found to be

$$Y_e^{(i)} = 0.41 \text{ to } 0.43 \qquad (2.1)$$

(the superscript i stands for initial). At the center of the star,

Arnett finds a central density and temperature

$$\rho_c = 3.7 \times 10^9 \text{ g/cm}^3, \quad kT_c = 0.69 \text{ MeV}. \tag{2.2}$$

Entropy will play a crucial role in the collapse, and we now evaluate the initial entropy.

The initial entropy per nucleus for translational motion is

$$S/kN = 2.5 + \ln \{ (V/N)(MkT/2\pi\hbar^2)^{3/2} \} = 16.7 \tag{2.3}$$

where N is the number of nuclei in volume V, and M the average mass, which we take to be that of ^{56}Fe. To this we add the entropy of the excited states of the nucleus and the entropy due to the α particles and neutrons present due to thermal dissociation of nuclei. At the temperature and density (2.2) the α's and n's contribute 3.6 and the excited nuclear states contribute 4.8. Thus the initial entropy per nucleon, obtained by dividing by 56, is 0.45. For the electrons, the Fermi energy is initially $\mu_e = 6.0$ MeV = 8.6kT. Thus,

$$(S/k)_{\text{per electron}} = \pi^2 \frac{kT}{E_f} = 1.15, \tag{2.4}$$

and this gives a total entropy per nucleon of $0.45 + 1.15(.42) = 0.93$. This initial entropy is low, and we shall now show that the increase in entropy during collapse is small, about 0.3 units, so that the collapse approximately follows an adiabat with S/k = 1. This low entropy simplifies the collapse scenario considerably.

During collapse, entropy increase proceeds only through the weak interactions. Although the dynamical time for the important part of the collapse is only about 1 ms, this is a long time compared with equilibration times of the strong and electromagnetic interactions. Indeed, we shall find that even the weak interactions, which are a factor of 10^{12} weaker than the strong interactions, equilibrate by a density of 5×10^{11} g/cm^3 to the extent that entropy production beyond this density in the collapse is small.

It follows from arguments which we shall present below that the entropy increase due to electron capture is given by

$$T(dS)_{\text{per nucleon}} = - dY_e (\mu_e - \hat{\mu} - \frac{3}{5}\Delta), \tag{2.5}$$

where $\hat{\mu} \equiv \mu_n - \mu_p$ is the difference in neutron and proton chemical potentials and Δ is the maximum energy of emitted neutrinos, before neutrino trapping. Eq. (2.5) can be easily understood. Of the energy μ_e from capturing an electron, $\hat{\mu}$ is used up in paying symmetry energy to convert a proton into a neutron in a neutron-rich nucleus. The neutrino, assumed to escape freely in the region of densities before trapping, carries off, on the average,

3/5 of the maximum remaining energy available. As long as neutrinos are not trapped

$$\Delta = \mu_e - \hat{\mu} - \Delta_n \qquad (2.6)$$

where Δ_n is the excitation of the daughter nucleus, roughly 3
MeV. We show below that the rate of electron capture to be employed is considerably less, and $\Delta \simeq 10$ MeV is somewhat higher,
than in BBAL (10). Thus the bracket in (2.5) = $(2 \Delta/5 + \Delta_n) \simeq$
7 MeV. From an initial Y_e of 0.42 the diminished electron capture rate takes Y_e to 0.36 as we show later. Thus, since the
average temperature is $T \simeq 1.4$MeV, 0.3 units of entropy are generated by electron capture. This increase will take the entropy
per nucleon up to about 1.23. To a reasonable approximation,
the central zone then follows the unit adiabat during collapse.

3. ELECTRON CAPTURE RATE

We now discuss the rate of electron capture and "neutronization" of the nuclear matter. Capture takes place chiefly on nuclei, not on free protons, because of the paucity of the latter.
In BBAL a Fermi gas model was used for the nuclei; it was pointed
out later (14) that in the region of A relevant for the collapse,
the electron capture is forbidden, and one must resort to first-
forbidden Gamow-Teller transitions. Calculation of these (14,15)
indicated that the BBAL rate was a factor of about 15 too high,
the more correct rate, obtained from the weak interaction cross
section and phase space considerations, being

$$dY_e/dt = - 0.14 \, X \, \rho_{10} \, Y_e^2 \, \sec^{-1} \qquad (3.1)$$

with X a phase space factor

$$X = \Delta^4/(4 \, m_e^2 \, (k_f^2/3m_n) \, \mu_e) , \qquad (3.2)$$

where k_f and m_n are the neutron Fermi momentum and mass. Eq.
(3.2) can be understood in the following way:
    (i) a Δ^2 comes from the weak interaction cross section; it
        represents the phase space of the neutrinos.
    (ii) one Δ comes from the excitation energy of the proton
        hole, another from that of the electron hole.

In order to obtain $dY_e/d\rho$, we next determine $d\rho/dt$ from the
equation for homologous collapse of a core of mass M and radius R:

$$d^2R/dt^2 = - GMF/R^2 . \qquad (3.3)$$

Here the fraction of lepton support pressure which has been removed at the stage of collapse corresponding to a given Y_e is

$$F = (4/3)(Y_e^{(i)} - Y_e)/Y_e^{(i)}. \tag{3.4}$$

An average value of this, $\overline{F} \sim 0.11$, applies for $Y_e = 0.36$. Thus, the collapse of the homologous core proceeds much more slowly than free fall, being strongly slowed down by the pressure from the many trapped leptons. Solution of (3.3) for F=1 and use of $\rho = 3M/4\pi R^3$ leads to the free fall equation

$$d(\ln\rho)/dt = 224 \, \rho_{10}^{\frac{1}{2}} \, \sec^{-1}. \tag{3.5}$$

Taking an average F converts this to

$$d(\ln\rho)/dt = 224 \, (\overline{F} \, \rho_{10})^{\frac{1}{2}} \, \sec^{-1} = 74 \, \rho_{10}^{\frac{1}{2}} \, \sec^{-1} \tag{3.6}$$

with $\overline{F} = 0.11$. Thus, the density-doubling time in homologous collapse is about three times that of free fall. Dividing (3.1) by (3.6) we obtain

$$dY_e/d(\ln\rho) = - (0.14/74) \, X \, \rho_{10}^{\frac{1}{2}} \, Y_e^2. \tag{3.7}$$

This is a factor of 11.2 less than in BBAL, resulting, as noted earlier, from the blocking of the transitions used there (14,15).

The factor X depends sensitively on Δ, eq. (2.6), i.e., on the difference $\mu_e - \hat{\mu}$. This difference, aside from the small Δ_n, tells one how far the reactions are out of beta equilibrium; Δ is often termed the "out-of-whackness". It is therefore desirable to convert (3.7) into an equation depending on the symmetry energy $\hat{\mu}$ by writing Y_e in terms of $\hat{\mu}$. The nuclear volume symmetry energy

$$E_s = \beta \, (N - Z)^2/A^2 , \tag{3.8}$$

with $30 < \beta < 36$ MeV, can be differentiated with respect to Z/A:

$$\hat{\mu} = \partial E_s/\partial(.5 - Z/A) = 8\beta(0.5 - Y_e) \tag{3.9}$$

where Y_e has been identified with Z/A, which is correct as long as the number of nucleons is small. We will see that ignoring surface and Coulomb energies here is a valid approximation. Thus

$$d\hat{\mu}/dY_e = - 8\beta \simeq 250 \text{ MeV}, \tag{3.10}$$

the value used in BBAL. Since ρ can be expressed in terms of Y_e and μ_e (eq. 3.14) and since it can be shown that $d(\ln Y_e)/d\ln\rho$ is less than 10% of $d(\ln\mu_e^3)/d\ln\rho$, we have

$$(\mu_e/3)d\hat{\mu}/d\mu_e \simeq d\hat{\mu}/d\ln\rho = -250 \text{ MeV } dY_e/d\ln\rho , \tag{3.11}$$

using (3.10) in the last equality. Coupled with (3.7), we obtain

$$d\tilde{\mu}/d\mu_e = 0.05 \ \rho_{10}^{\frac{1}{2}} \ Y_e^2 \ \Delta^4/\mu_e^2 \ . \tag{3.12}$$

This important equation gives the rate at which $\tilde{\mu}$ follows μ_e.

As the density, and hence μ_e, increases, the right-hand side of (3.12) would become very large if Δ were a constant fraction of μ_e. The differential equation would then indicate that $\tilde{\mu}$ increases much faster than μ_e, but this is impossible because of (2.6). Hence Δ/μ_e must decrease with increasing μ_e (density), which means that $\tilde{\mu}$ follows μ_e more and more closely. Therefore $d\tilde{\mu}/d\mu_e \to 1$ for large μ_e. Setting the left side of (3.12) to 1,

$$\Delta = 2.11 \ \rho_{10}^{1/8} \ (\mu_e/Y_e)^{\frac{1}{2}} \ \text{MeV}. \tag{3.13}$$

Inserting the value of μ_e, assuming complete degeneracy,

$$\mu_e = 11.1 \ (\rho_{10} \ Y_e)^{1/3} \ \text{MeV}, \tag{3.14}$$

we obtain

$$\Delta = 7.03 \ Y_e^{-1/3} \ \rho_{10}^{1/24} \ \text{MeV}. \tag{3.15}$$

Using our standard value $Y_e = 0.36$ we find $\Delta = 9.88 \ \rho_{10}^{1/24}$ MeV. The dependence on the density is extremely weak; even at 5×10^{11} g/cm^3, $\Delta = 11.6$ MeV. The total lag $\mu_e-\tilde{\mu}$ is $\Delta+\Delta_n$ (eq. 2.6), and hence 14.6 MeV. For the density 5×10^{11}g/cm^3, which we shall next show is the density at which neutrinos are trapped, from the above Δ and $Y_e=0.36$, we find, using (2.6) and (3.14) that $\tilde{\mu} = 14.5$ MeV.

Knowing $\tilde{\mu}$, one can work backwards and obtain a Y_e consistent with it. However, eq. (3.9), although adequate for obtaining $d\tilde{\mu}/dY_e$, does not include the nuclear Coulomb or surface energies. It was found in BBAL that the formula

$$\tilde{\mu}= 207 \ (0.45 - Y_e)(1.32 - Y_e) \ \text{MeV} \tag{3.16}$$

included these well. We find from (3.16) that $Y_e^{(f)} = 0.37$ (f = final). This should be corrected slightly downwards to take into account effects of neutron drip (compare Tables 9 and 10 of BBAL), giving us a $Y_e^{(f)}$ of 0.36, or, possibly, 0.35.

The increase in Y_e from the BBAL value of $Y_e^{(f)} = 0.31$ is chiefly due to the 15 times smaller electron capture rate (14,15) found once blocking is taken into account. From (3.12) one sees that the value of Δ at trapping changes from the BBAL value by $(15)^{1/4} \simeq 2$ with inclusion of this rate. The above argument then gives the smaller $Y_e^{(i)} - Y_e^{(f)}$, or larger $Y_e^{(f)}$. This shows that the electron capture rate just at trapping is chiefly responsible in determing $Y_e^{(f)}$.

4. NEUTRINO TRAPPING

The neutrino mean free path is (10), assuming $\sin^2\theta_W = 0.25$,

$$\lambda_\nu \simeq 10\text{km}(100/\rho_{10})(\overline{N}^2 X_h/6A + X_n + 5X_p/6)^{-1}(10\text{MeV}/\varepsilon_\nu)^2 \quad (4.1)$$

where X_h, X_n and X_p are the fractions of mass in heavy nuclei, neutrons and protons, respectively, and $\overline{N}$ is the average number of neutrons in each heavy nucleus. If we take $Y_e = 0.36$, from Table 1 of BBAL we find $A = 105$, thus $\overline{N} = 67$. With a neutrino energy $\varepsilon_\nu = 10\text{MeV}$, $\rho_{10} = 50$ implies that $\lambda_\nu = 2.9\text{km}$. In a typical dynamic time $t \sim 1\text{ms}$ these neutrinos diffuse over a distance

$$\langle r^2 \rangle^{\frac12} = (\lambda_\nu ct/3)^{\frac12} = 17 \text{ km.} \quad (4.2)$$

This corresponds to a neutrino drift velocity $v_d = 1.7\text{x}10^9\text{cm/sec}$. From (3.3) we see that the edge of the homologous core will have an infall velocity

$$u = - (2\,\overline{F}\,G\,M/r)^{\frac12} . \quad (4.3)$$

It is found (13) that the density at the edge of the homologous core and beyond is given by (see also Yahil and Lattimer herein)

$$\rho_{10} = 3/r_7^{\,3} \quad (4.4)$$

where $r_7 = r/10^7\text{cm}$. Thus a density of $5\text{x}10^{11}\text{g/cm}^3$ corresponds to $r_7 = 0.39$ and (4.3) gives an infall velocity $u = -2.45\text{x}10^9\text{cm/sec}$, where we have used $\overline{F} = 0.11$ and $M = 0.8\ M_\odot$.

The neutrino drift velocity v_d gives the average velocity of the neutrinos relative to the nuclear material which does the scattering. The net velocity of the neutrinos relative to the star center is $u + v_d$ and this will be negative for the above parameters. Therefore, neutrino trapping will occur at a density less than $5\text{x}10^{11}\text{g/cm}^3$. Once neutrino trapping sets in, the total number of leptons $Y_\ell = Y_e + Y_\nu$ in a given material element remains constant. In the trapping region we soon reach beta equilibrium:

$$\mu_e - \mu_\nu \ = \hat\mu = \mu_n - \mu_p, \quad (4.5)$$

i.e., electron capture is fast enough to preserve statistical equilibrium.

5. SCHEMATIC COLLAPSE SCENARIO

Given the low entropy, nucleons have no choice but to remain mostly in nuclei, since free neutrons would carry a large entropy per particle, ~ 8 at the point of trapping. The pressure contri-

bution from nuclei is small, so the total baryon pressure P_b is
only a few percent of the pressure P_e from the electrons (see Table 4 of BBAL). The ratio P_b/P_e becomes less with increasing ρ.

Our equation of state is thus exceedingly simple. For den-
sities to nuclear matter density $(\rho_o = 2.5 \times 10^{14} \text{g/cm}^3)$, the pressure
is determined almost completely by the relativistic degenerate
leptons and is consequently that of a relativistic gas of fermions.
The adiabatic index Γ is slightly less than 4/3, the decrease
from 4/3 coming chiefly from electron capture and because, just
below nuclear matter density, P_b is negative (due to Coulomb effects). Collapse continues then at least up to nuclear matter
density. As the nuclei are squeezed together, they begin to
"feel each other", and merge into uniform nuclear matter. Initially nuclear matter is formed at normal nuclear density, for
which the contribution to the pressure from the nucleons is zero.
Because of the stiffness of nuclear matter the density cannot be
increased very much further before bounce takes place. Throughout
much of the region of nuclear matter compression, the increase in
nuclear energy is determined by the compression modulus of nuclear
matter. The effective Γ in these super-nuclear regions is ~ 2.5.

We thus have a very simple and relatively unambiguous collapse scenario in which Γ is slightly less than 4/3 over the entire region of densities up to nuclear matter density, and then is
substantially larger, ~ 2.5 up to bounce. Bounce occurs because
of the exceedingly stiff equation of state above nuclear matter
density. The bounce is analogous to that of a very stiff spring,
as we shall discuss later. During the bounce, a shock wave will
be formed slightly outside the homologous core, which retains its
low entropy. The shocked matter beyond the core will, however,
have a high entropy, $\sim 5-10$, and the nuclei will be split up into
nucleons. The equation of state of the shocked matter is again
simple, essentially that of a $\Gamma = 5/3$ gas of nucleons.

6. DYNAMICS OF THE COLLAPSE

The maximum mass of the star that degenerate electron pressure can support against gravity is the Chandrasekhar mass, given
for zero temperature matter by

$$M_{ch} = 5.76 \langle Y_e^2 \rangle M_\odot = 11.5 \langle Y_e^2 \rangle \ 10^{33}g \qquad (6.1)$$

where $\langle Y_e^2 \rangle$ is a suitable average over the object. The core of a
presupernova star, being supported by electron pressure, is thus
limited in mass; as time goes on, the mass in which the ^{28}Si reaction has gone to completion increases, and on a timescale of a
few days reaches M_{ch}, at which point gravitational collapse sets
in. With Y_e given by (2.1), the T=0 Chandrasekhar mass is 1 $M_\odot$.

However, the non-negligible temperature (2.2) increases the pressure above that of a cold electron gas, and thus increases the
critical mass by a factor $1 + (\pi T/\mu_e)^2$, equivalent to an additive
correction $0.584 s_e^2 M_\odot$, where s_e is the electron entropy per nucleon. Weaver et al. (4), by considering in addition convective
mixing of Fe and Si which brings hotter material into the core,
find a pre-collapse core mass of $1.5 M_\odot$.

Gravitational collapse starts through photodisintegration of
nuclei, before electron capture becomes important in relieving the
pressure, the chief breakup going into α particles and neutrons.
For any given density ρ and temperature T this breakup is easily
calculated from the Saha equation. With the initial central conditions of Weaver, Zimmerman and Woosley (4), and Arnett (5), the
breakup is only a few percent, but as the collapse proceeds and
T increases, the amount of breakup grows rapidly. The Q value
for the reaction $^{56}Fe \rightarrow 13\alpha + 4n$ is 124.4 MeV, so the breakup uses
up energy. Since the thermal pressure and energy reside mainly in
the electrons, the electron temperature and pressure do not rise
as rapidly as they would in the absence of breakup; that is, the
breakup causes an effective decrease in the electron pressure.
It is straightforward to show (10) that for a fractional breakup
δ of ^{56}Fe, this decrease in pressure is equivalent to that produced in a cold electron gas by an effective change in Y_e,

$$\Delta Y_e = \Delta\epsilon/n\mu_e = - 0.20 \; \delta/(\rho_{10} Y_e)^{1/3}, \qquad (6.2)$$

where $\Delta\epsilon$ is the energy of breakup, n is the baryon density and μ_e
is the electron chemical potential. For typical conditions early
in the collapse ($\rho_{10}=1$, T=1MeV) a breakup $\delta \sim 0.3$ gives $\Delta Y_e \sim$
-0.08 for $Y_e = 0.42$. As ρ increases in the collapse, the amount
of breakup decreases, and the effective ΔY_e becomes smaller.
Since $p \propto Y_e^{4/3}$, an effective drop of Y_e by 20% translates into a
26% drop in pressure. The effective decrease in electron pressure occurs rather suddenly as nuclear breakup sets in, reducing
the Chandrasekhar mass by

$$\Delta M_{ch}/M_{ch} = 2 \; \Delta Y_e/Y_e; \qquad (6.3)$$

the core mass soon exceeds the maximum mass that can be stable,
and collapse ensues. Electron capture begins in just the range
of densities where nuclear breakup decreases, and leads to a real
decrease in Y_e comparable with the above effective change in Y_e;
the accompanying pressure reduction is quite smooth.

In the collapse the inner portions of the star fall in slowly, compared to sound propagation speed, and remain in good communication with each other. Farther out the infall proceeds supersonically. As we shall discuss in detail below, the collapse
of the inner core becomes homologous, i.e., the density distribu-

tion remains effectively similar to itself in the collapse, while
the outer material is in quasi-free fall, i.e., falling "freely"
under a reduced effective gravity.

Figure 1, from Arnett (ref. 3, model 4660), shows the velocity of the infalling material, u, as a function of its distance
r from the center, at a time about 1 ms before bounce. Out to
about 40 km, the velocity is nearly proportional to r, with $-u/r$
$\simeq$ 300 to 400 sec^{-1}; thus the inner part of the star shrinks homologously. The velocity reaches a maximum at r $\sim$ 75 km and then
falls off essentially as $r^{-\frac{1}{2}}$, but remains supersonic until r $\sim$
1000 km. Similarly, in Wilson's (5) calculation, Figure 2, one
sees a homologous core with $-u/r \sim$ 600 to 750 sec^{-1} (dashed lines)

The formation of a homologous core is expected when the adiabatic index Γ is close to $4/3$, since at that value the equation
of state of the matter has no absolute length scale. A core with
$\Gamma = 4/3$ can have a state of neutral equilibrium with pressure
balancing gravity independently of the core length scale. As
Goldreich and Weber (17) have shown, such cores can undergo homologous collapse. Let us introduce the scale parameter $\alpha(t)$ by
writing the position of a given mass point as $r(t) = \alpha(t)r_0$, where
r_0 is the initial position of the mass point; in homologous collapse $u/r = \dot{\alpha}/\alpha$. We already employed a rough description of homologous core collapse in the discussion following eq. (3.3).

In Arnett's calculations Y_e at the edge of the homologous
core at the time of Fig. 1 is 0.32, and smaller in the interior
of the core. Eq. (6.1) with $Y_e = 0.32$ gives $M_{hc} = 0.59M_\odot$, while
Arnett finds a core mass of $0.56M_\odot$. Finite T effects with $s_e =$
0.45 increase M_{ch} by $0.12M_\odot$, but this effect is compensated by
the decrease of Y_e in the interior of the core. In Wilson's calculation $Y_e = 0.37$ at trapping, for which (6.1) predicts $M_{ch} =$
$0.79M_\odot$, compared with his result of $0.77M_\odot$. Again this result
must be corrected upward by $\sim 0.12M_\odot$ by finite T, but downward by
a similar amount since Y_e in the core center is only 0.29. With
Wilson's smaller amount of initial electron capture, the transition from the original equilibrium core to the homologously collapsing core should be smoother, and the Goldreich-Weber theory
more applicable.

The core responds to the initial underpressuring by starting
to move inward in quasi-free-fall; then through a rearrangement
of the internal pressure and density distribution it approaches a
homologous configuration. The Goldreich-Weber cores have a maximum infall rate, corresponding to a maximum initial underpressure.
Under actual collapse conditions, however, the initial underpressure is substantially larger than that corresponding to their
λ_{max}. Consequently only the inner part of the core, which moves
subsonically, is able to establish a homologous configuration,

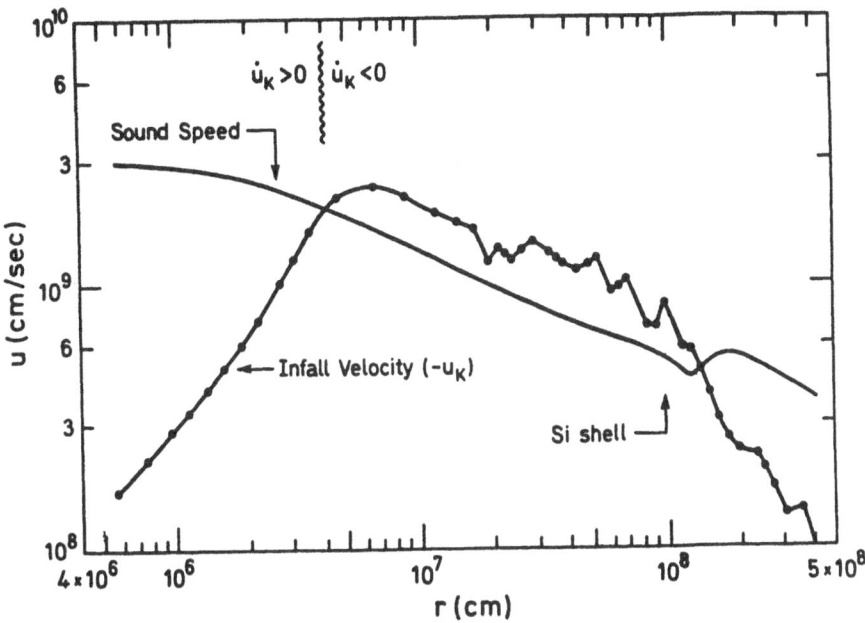

Figure 1. Material velocity u and sound speed as a function of r, about 1 ms before bounce (from Arnett, ref. 3).

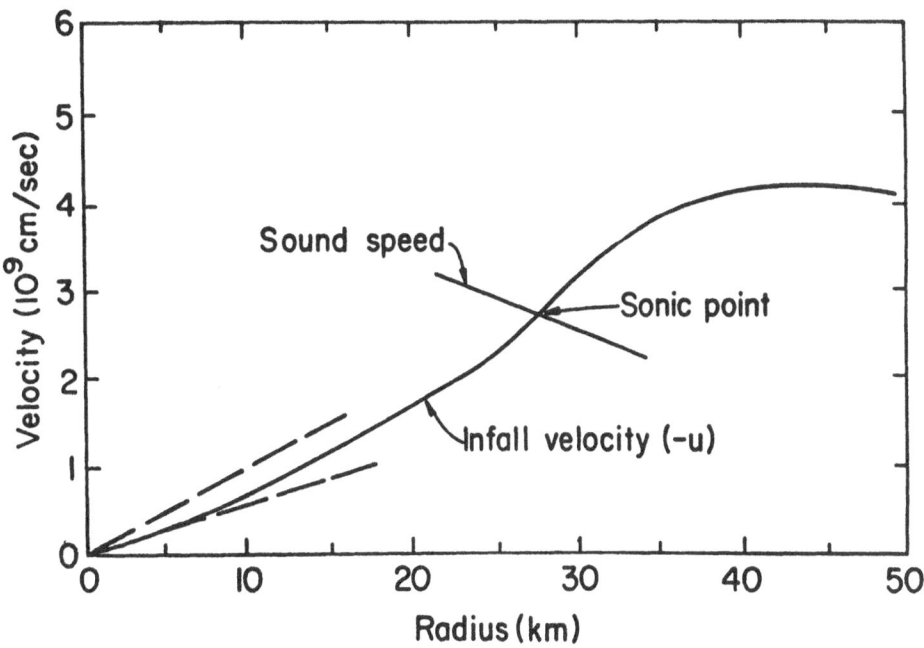

Figure 2. Material velocity and sound speed as a function of r at time "0" (see text; from Wilson, ref. 5).

while the exterior matter continues to move in quasi-free-fall.
As we see in Figs. 1 and 2, the infall speed, $|u|$, is less than
the sound speed a within the homologous core, but the homologous
motion ends at the <u>sonic</u> point, r_{sp}, at which

$$u + a = 0 . \qquad (6.4)$$

In Arnett, $r_{sp} \sim 40$ km, and in Wilson, $r_{sp} \sim 27.7$ km.

The significance of the sonic point is that since in a moving medium a sound signal travels in space at velocity $u + a$, a
signal generated interior to the sonic point cannot move outward
in space through the sonic point; material outside r_{sp} cannot receive signals from inside r_{sp}. Thus while the material interior
to r_{sp} is able, through rearrangement of p and ρ, to establish homology, the material outside r_{sp} does not sense the homologous
motion, and continues in quasi-free fall. As we shall see later,
the shock also forms at r_{sp}, i.e., at the homologous core edge.

Outside the homologous core the infall velocity as found in
the computer calculations reaches a maximum and then falls as $r^{-\frac{1}{2}}$;
in Wilson's calculation the velocity in units of 10^9 cm/sec $u_9^2 \sim$
$10/r_7$ for times near bounce. By contrast the free-fall velocity

$$u_{ff}^2 = 2GM(r)/r = 26.7 \times 10^{18} M(r)/(r_7 M_\odot) \text{ cm}^2/\text{sec}^2 \quad (6.5)$$

so that for a central mass of 1 $M_\odot$, $u^2/u_{ff}^2 \sim 0.37$.

Yahil (see Yahil and Lattimer, this volume) has shown that
the region outside of the homologous core can also be described
by a self-similar solution. This solution represents a "wind"
blowing outwards from the system of coordinates moving with the
Goldreich-Weber homologously collapsing core. The division between the two self-similarly collapsing regions, homologous core
and wind, is given roughly by the <u>sonic point</u> (6.4); more precisely, by the point of minimum velocity in the collapse (see Figs. 1
and 2). He has also derived the coefficients in eq. (4.4) and the
eq. for u_9^2 preceding eq. (6.5).

7. EQUATION OF STATE

We shall now discuss the equation of state in more detail,
expanding upon our schematic treatment of §5. The most satisfactory equation of state of hot, dense matter has been derived by
Lamb, Lattimer, Pethick and Ravenhall (18; LLPR), using a Skyrme
interaction between nucleons. They minimized the free energy of
a fixed amount and volume of matter at temperature T and lepton
number Y_ℓ. Figure 3 shows their results for the adiabats for entropy per nucleon s = 1 to 5, in the ρ - T plane (dots refer to

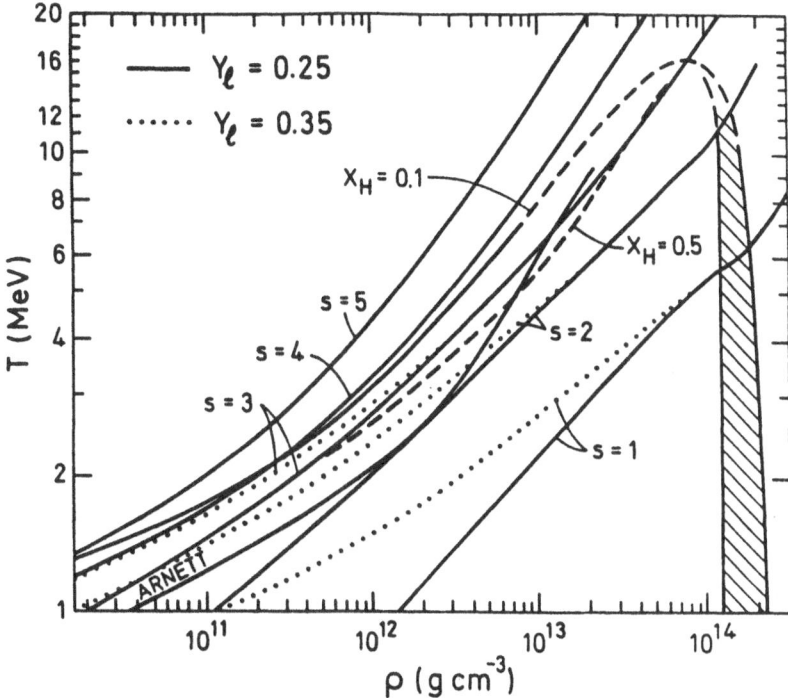

Figure 3. Phase diagram for hot, dense matter, according to Lamb et al. (18). See text for details.

$Y_\ell = 0.35$, lines to $Y_\ell = 0.25$). Also shown is the (heavy dashed) line at which half the matter is in heavy nuclei, $X_H = 0.5$; above this line, most of the matter is in the form of free nucleons. In the shaded area, most of the space is occupied by nuclear matter, with small "bubbles" of dilute neutron matter interspersed. To the right of this area is uniform nuclear matter. The curve labelled Arnett shows the trajectory of his collapsing core (3).

As discussed in §2, the entropy is expected to remain between 1 and 1.5, and $Y_\ell \sim 0.36$. We see that the adiabat for this case stays well below the line $X_H = 0.5$ for all densities up to 10^{14} g/ cm^3; in other words, the matter consists mostly of heavy nuclei, as a consequence of the fact that complex nuclei have an enormous number of highly excited energy levels which can accommodate a large entropy.

In the stages of collapse before ρ_0 is obtained, the contribution to the pressure from the nuclei becomes negative, as already mentioned in §5. There is a large nucleus-nucleus Coulomb (lattice) correlation at high densities, which lowers the Coulomb energy considerably. The surface energy is related to the Coulomb energy by a virial theorem (19).

$$\omega_{surf}A^{2/3} = 2 \ \omega_{Coul}A^{5/3} \tag{7.1}$$

so that the surface energy will be decreased and the nuclei grow in size. At $\rho \sim \rho_0/2$, the large nuclei fill so much space that it becomes favorable to go over to the bubble phase. With ρ increasing towards ρ_0 the bubbles are squeezed out. This lowers the energy per nucleon, since the surface and Coulomb energies are positive. Thus, the nuclear pressures are negative right up to ρ_0; the total pressure is less than that for a $\Gamma = 4/3$ gas, so the core collapse is accelerated throughout this region. There is thus no tendency for the collapse to be halted before ρ_0.

On the basis of the LLPR equation of state, BBAL showed that the infall of matter cannot be stopped until the central density of the star rises well above ρ_0. In Wilson's calculation, the maximum density reached is $3.2 \times 10^{14} g/cm^3$, but we believe that a higher value $\sim 1.5 \ \rho_0$ is more realistic. The subsequent development does not depend critically on the exact equation of state above ρ_0, although the higher the density at bounce the higher is the energy put into the shock. (Of course, were the equation of state too soft above ρ_0, the matter would not necessarily bounce, but could continue to fall in to form a black hole.)

After matter is hit by the shock, its entropy rises greatly, with $s \gtrsim 5$ being a typical post-shock value (5). Fortunately, the shock affects essentially only matter with $\rho \lesssim 10^{13} g/cm^3$, where the equation of state is quite well known. Fig. 3 shows that below this density nuclei are completely dissociated into nucleons, and T is of the order of 10 MeV.

8. START OF THE SHOCK

When the central density is below ρ_0, the collapse of the inner core is homologous. However, once the central density reaches ρ_0 the matter in the center suddenly becomes substantially stiffer; the increased pressure decelerates the innermost material and destroys the homology. Pressure signals generated in the interior due to the stiffened equation of state propagate outward through the matter, slowing it until it reaches r_{sp} (6,4); however, as discussed in §6, pressure waves cannot move in space beyond r_{sp}, as long as the density and velocity are continuous there. Since r_{sp} moves slowly compared to a, the pressure inside r_{sp} continues to build up; eventually (as is seen in the computer studies) a discontinuity in p, i.e., a shock, is formed with high p inside and low p outside. The pressure discontinuity is accompanied by one in velocity. Shock formation, as this discussion indicates, occurs near r_{sp}, i.e., near the edge of the homologous core, rather than near the center of the star. The shock forms in mass coordinate roughly at the position of maximum magnitude in infall vel-

ocity at the time of the last good homology. This result follows
from general considerations of self-similar collapse (Yahil and
Lattimer, this volume).

The sonic point in Wilson is $r_{sp} = r_0 = 27.7$ km at time "0"
$= 262.83$ ms, the last time at which the inner core motion is homologous (the "last good homology"); the mass within r_0 at this time
is $M_1 = 0.77M_\odot$, which is essentially the homologous core mass.
Figure 4 gives Wilson's results for the velocity distribution in
space at four times. It is seen that at time 0, the velocity is
continuous, while at time I, 263.2 ms, evidence of a discontinuity
in velocity is noticeable at 27.7 km; by time II, 263.69 ms, or
0.5 ms later, a well-formed shock is apparent, at the same radius.
Only after another 0.95 ms (time III, 264.64 ms) has the shock
moved out to $\sim$ 76 km.

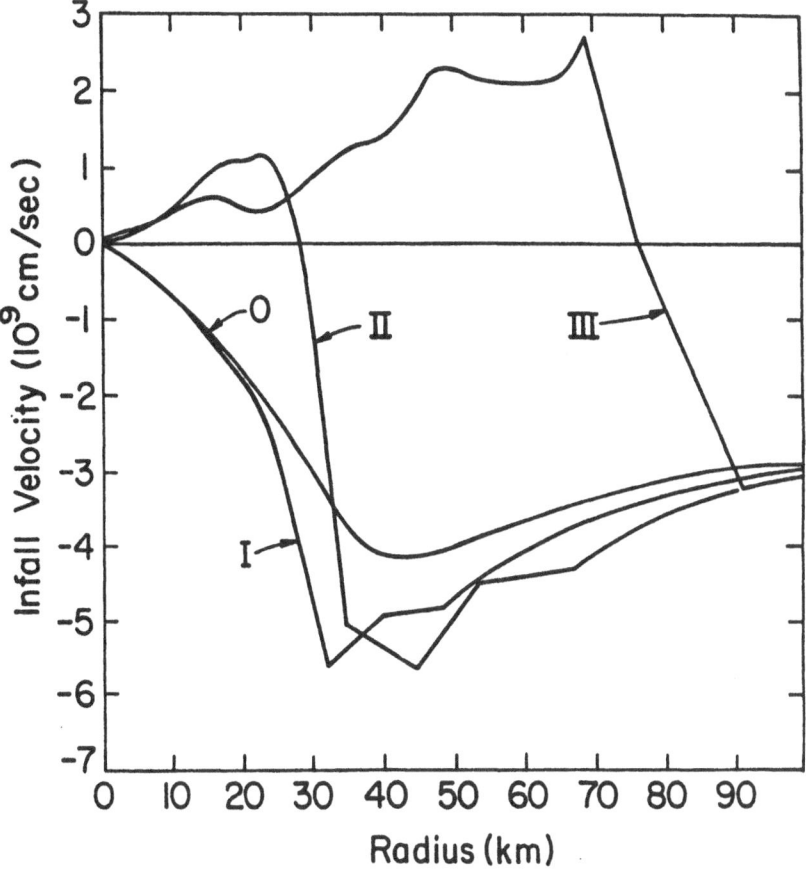

Figure 4. Velocity in 10^9 cm/sec, versus radius, in
km, at four times (see text), from Wilson (5).

While no signal can propagate in space through the sonic point, material flows in through it. The first pressure signal from the decelerating core reaches r_{sp} in a sound travel time, ~ 0.7 ms, after t_0. (The fact that a discontinuity begins to form in Wilson's calculation by t_I, only 0.37 ms after the last good homology, can be attributed to growing deviations of his equation of state from $\Gamma = 4/3$ within the homologous core, not only at the center, but in particular to the excessive breakup of nuclei into   particles.) During this time interval, the material that at t_0 lies between r_0 and ~ 44 km, about 0.12 $M_\odot$, travels in through r_0. This relatively small mass, whose infall velocity is near maximum at t_0, brings with it a large kinetic energy, which as we shall see below is an important component of the initial shock energy. By time II of Fig. 4, an additional 0.17 $M_\odot$ has flowed in, bringing with it further kinetic energy. The original homologous core of 0.77 $M_\odot$ has contracted at t_{II} to a minimum radius of about $r_2 = 17$ km $= 0.62$ r_0. The time of contraction from r_0 to r_2 is about 0.08 ms, so the average speed of the edge of the homologous core is 1.3×10^9 cm/sec, consistent with its initial infall speed of 2.6×10^9 cm/sec. (Since $r_2 > r_0/2$, the average speed of the outgoing pressure wave is obviously greater than the infall speed, in fact about 5/3 of it; this result is plausible, but, unfortunately, the time values of Wilson's available computer printouts are not sufficiently closely spaced to check this latter speed.) Between time 0 and II the sonic point moves slowly outward to 31.4 km. By time II the increased interior p has reversed the core motion; the material moves, as can be seen in Fig. 4, slowly outward, while the outer material still falls rapidly towards the center. Rather than oscillate back and forth, the kinetic energy of the central core is, as discussed in Appendices B and C of B^3 (20), quickly transferred to the shock.

From the standpoint of the shock, dilute material from the outside moves through the shock, thereby increasing its density by a factor 4 to 7, and becoming strongly heated. As discussed above, only material outside the homologous core is heated in this way; the homologous core itself remains unheated and retains its original low entropy s $\lesssim 1.5$. T and ρ behind the shock are determined by the Hugoniot relations. Denoting quantities referring to the unshocked material by the subscript 1, and to those of the shocked material by 2, we have

$$(u_2 - u_1)^2 = (p_2 - p_1)(1/\rho_1 - 1/\rho_2) . \qquad (8.1)$$

If ρ_1 and $u_2 - u_1$ are given, and ρ_2/ρ_1 can be estimated (even if only roughly) this relation can be used to calculate p_2, especially since in a strong shock $p_1 \ll p_2$. The change of internal energy per unit mass is

$$E_2 - E_1 = \tfrac{1}{2}(p_2 + p_1)(1/\rho_1 - 1/\rho_2) \simeq \tfrac{1}{2}(u_2 - u_1)^2 \qquad (8.2)$$

where in the second line we have used $p_1 \ll p_2$. Thus the internal energy added (very nearly) equals the kinetic energy of the shocked material in the frame in which the unshocked matter is at rest. Equation (8.2) can be used to calculate T_2, as we now do.

The velocity of the shock itself, relative to the unshocked material, is

$$U = \rho_2 (u_2 - u_1)/(\rho_2 - \rho_1), \qquad (8.3)$$

so that its absolute velocity is

$$U' \equiv U + u_1 = (u_2\rho_2 - u_1\rho_1)/(\rho_2 - \rho_1). \qquad (8.4)$$

Thus the shock begins to move outward when $u_2 = u_1\rho_1/\rho_2$. For a strong shock in a medium in which most of the pressure is the thermal pressure of non-relativistic particles (nucleons), $\rho_2/\rho_1 \simeq 4$; therefore $u_2 \simeq u_1/4$. Thus the shock begins to move outward when the inside material is still moving inward, at a time shortly after time II of Fig. 4. (A stationary or inwardly moving shock is often called an accretion shock. We do not believe that the distinction between an accretion and an outwardly-moving shock is very important.)

The outward motion of this shock is rapid; in fact, between times II and III, an interval of 0.95 ms, the shock moves from 27.7 km to 76 km, at an average speed $U_{Av} = 5 \times 10^9$ cm/sec. The average outer (infall) velocity is $u_{1\ Av} = -5.6 \times 10^9$ cm/sec, giving $U = 10.6 \times 10^9$ cm/sec,

$$u_2 - u_1 = 3\ U/4 = 8.0 \times 10^9 \text{cm/sec, and} \qquad (8.5)$$

$$E_2 - E_1 = \tfrac{1}{2} (u_2 - u_1)^2 = 32 \times 10^{18} \text{erg/g} = 33 \text{ MeV/b.} \quad (8.6)$$

About one-third of this energy goes into light particles (electrons, neutrinos, γ-rays), and about 8 MeV of the rest is needed to dissociate nuclei into nucleons. If we consider the remaining 15 MeV/b as thermal energy of the nucleons, the corresponding $T_2 = 10$ MeV, just about the value computed by Wilson at time II.

The high temperature in the nucleons behind the shock is essential for the shock propagation, as Lichtenstadt, Sack and Bludman (8) have pointed out. Since $p/\rho E = \gamma-1$ is 1/3 for the relativistic components of the gas, but 2/3 for the thermal energy of the non-relativistic gas of nucleons, the more a given energy in the gas resides in the non-relativistic particles, the more pressure to drive the shock that energy delivers.

9.   INITIAL ENERGY OF THE SHOCK

    Ways to calculate, from the computer output, the energy ini-
tially put into the shock are discussed in B^3 (2Ω) and are generally consistent. Yahil and Lattimer will discuss these matters
from the standpoints of self-similar collapse and hydrostatic
models in a following lecture. Our discussion here will be from
an "engineering" point of view; these methods appear to work well
in analyzing computer results to date.

    In Wilson's calculations the mass of the homologous core is
initially 0.77 M$_\odot$ at the moment of the last good homology t_0. Between t_0 and the start of shock formation at t_I, an extra 0.12 M$_\odot$
flows through r_0. Thus the mass of unshocked material is $\sim$ 0.89
M$_\odot$. We refer to this as the unshocked core. In computer calculations it can be identified by the fact that the entropy in this
region does not (or, should not) change during and after bounce.

    The initial energy of the unshocked core at the beginning of
collapse is small, typically a fraction of 1 foe (in the following
we use the acronym foe (for fifty-one-ergs) as an abbreviation for
10^{51} ergs). This energy is so small, that we can consider it for
our present purposes to be zero. During the infall, the zones external to the core do pdV work on it, amounting, in Wilson, to 3.5 foe
up to t_0 (see fig. 2). Neutrino emission before trapping carries
off $\sim$ 0.3 foe, leaving a core energy of about 3.2 foe at this
stage. Most of this energy, 2.9 foe, is in kinetic energy of the
core. Additional pdV work of 1.45 foe is done on the core in the
0.86 ms from the time of the last good homology up to t_{II}, when
the shock has formed and the core has minimum radius (see fig. 3);
the core energy is $\sim$ 4.7 foe at this stage.

    The core then rebounds, doing pdV work on the material out-
side, supplying energy to the shock which is pushed out from r_{sp}
by the rebounding core, and travels rapidly outwards, reaching
76 km by t_{III}. The pressure in the expansion phase, at given r
inside the shock, is considerably greater than in the compression
phase because the pressure wave from the center has come to the
surface; the core therefore does not have to move very far out to
do the same pdV work as was done during the total collapse time.

    As a consequence of the work done by the rebounding core on
the outside material, the core loses most of its kinetic energy;
rather than continue to oscillate, the core quickly becomes hydrostatic. In Appendix B of B^3 it is shown how this rapid damping, seen in the computer calculations, can be simply understood
in terms of an oscillating spherical piston in an external gas,
losing energy by emission of pressure waves into the gas. In this
model, the time t_0 corresponds to the oscillator moving through
its eventual equilibrium position, with maximum kinetic energy.

We can estimate the work done by the core in expanding by
computing pdV directly, which in Wilson's calculations is $\sim$ 3.4
foe up to t_{III} (although we have only two output points, with
quite different pressures, so this number is rough).

Our picture is, thus, a simple one. The unshocked core of
initially zero energy has work done on it by the outer zones
during the collapse. The sum of gravitational plus internal energies of this core remains zero, to an accuracy of $\lesssim$ 1 foe, so
that the work goes chiefly into building up the kinetic energy of
the core, up to the time of the last good homology. The core gains
somewhat further in energy, by additional pdV work done on it by
the outer zones, as it is brought to rest, building up its internal energy. At minimum radius, the core resembles a tightly compressed spring, ready to push outwards. During the subsequent
outward push, the core does work on the outer zones. Because of
the presence of the shock during this time, the pressures are
much greater, and the core moves only a few km in order to do
roughly the same amount of work on the external zones, as was
done on it during the entire infall, starting from $r \sim 500$ km.
The shock acts like a black body absorber for pressure waves (Appendix C of B[3]); there is little back reflection. As they are absorbed, the core motion is quickly damped, in less than a dynamical time $\sim$ 1 ms, and it comes to rest. Once hydrostatic equilibrium is reached, the energy bookkeeping can conveniently be kept
by using the virial theorem, as Yahil, Lattimer and Bethe will
develop in the following lectures. The core comes to rest at essentially the radius corresponding to r_0, which it had at t_0.

In Appendix E of B[3] and Table 13 of B[3] one can see that in
the Mazurek-Cooperstein-Kahana shock (6) the kinetic energy of
the unshocked core at t_0 is 6.08 foe, and the pdV work done by
the core as it moves outward from its minimum radius to its hydrostatic position is 6.91 foe. The energy bookkeeping becomes
quite accurate when one adds to the 6.08 foe another 1.81 foe
representing the pdV work done in the core, last good homology to
minimum radius, and notes that the internal plus gravitational
energy is 1.1 foe higher at hydrostatic position than it was at
t_0. Note that the pdV work done by the core upon rebound puts
6.91 foe into the initial shock energy, double the 3.4 foe in
Wilson's calculation. As shown in B[3], the energy put into the
shock should go roughly as 1/R, with R the radius of shock formation, so the above difference in energies is understandable.

The MCK calculation (6) should give a better number for the
initial shock energy, since their central zone correctly goes
well above ρ_0. Wilson's collapse stopped mainly through nuclear
breakup, taking place well before ρ_0, due to errors in handling
the dissociation energy.

No computer calculations to date have the negative nuclear pressure in the last stages before nuclear matter density, as predicted by the LLPR equation of state. This should accelerate the core further in the last stages, and give an increased initial shock energy. Furthermore, the artificial viscosity in the computer calculations (incorrectly) reaches slightly into the unshocked core, heating it a bit and dissociating nuclei just inside the outer edge. The latter represents a loss of energy from the shock; it is ~ 1 foe in Wilson. My personal guess as to the correct initial shock energy is ~ 10 foe , and I believe that computer calculations will, with improvement, progress towards this number. This is ~ 3 times the energy in Wilson's shock!

10. PROPAGATION OF THE SHOCK TO THE NEUTRINO SPHERE

As the shock progresses outwards, the net kinetic energy of material moving into the shock is added to it because $|u_1| > u_2$. On the other hand, ~ 8 Mev/b is used up in dissociating the nuclei which move through the shock. In Wilson's computations, 0.206 $M_\odot$ is shocked before the neutrino sphere, where neutrinos can begin to escape, is reached. The kinetic energy brought into the shock is 5.61 foe , the dissociation energy, is 3.16 foe ; about 0.8 foe must be furnished to provide the kinetic energy of the matter behind the shock. Thus, there is a net gain in shock energy of 1.65 foe, as the shock progresses outwards from formation to the neutrino sphere. As the shock passes the neutrino sphere, it loses energy through neutrino emission in addition to the dissociation energy that must be paid. Just these stages, particularly the period in which the shock just passes the neutrino sphere, are critical for computer calculations, and shocks in most recent computer calculations have died in this period.

I promised Hans Bethe to bring the shock in good health up to the neutrino sphere and he will take over at this point, to discuss neutrino diffusion and dissipation, and the further progress of the shock outwards. Needless to say, I have benefited greatly from working with the rest of the BABBLY group, and the report here should be considered as my view of our joint work on these stages of supernovae.

REFERENCES

1. Burbidge, E.M., Burbidge, G.R., Fowler, W.A. and Hoyle, F.: 1957, Rev. Mod. Phys. 29, 547.
2. Colgate, S.A. and White, R.H.: 1966, Astrophys. J. 143, 626.
3. Arnett, W.D.: 1977, Astrophys. J. 218, 815; 1980, Proc. N.Y. Acad. Sci. 336, 366.
4. Weaver, T.A., Zimmerman, G.B. and Woosley, S.E.:1978, Astro-

phys. J. 225, 1021.

5. Wilson, J.: 1980, in "Ninth Texas Symposium on Relativistic Astrophysics", Proc. N. Y. Acad. Sci. 336, 358.

6. Mazurek, T., Cooperstein, J. and Kahana, S.: 1981, in Dumand-80, ed. V.J. Stenger (Honolulu: University of Hawaii Dumand Center), p. 142.

7. Van Riper, K.A. and Lattimer, J.M.: 1981, Astrophys. J. 249, 270.

8. Lichtenstadt, I., Sack, N. and Bludman, S.A.: 1980, Phys. Rev. Letters 44, 832.

9. Lichtenstadt, I., Sack, N. and Bludman, S.A.: 1980, Astrophys. J. 237, 903.

10. Bethe, H.A., Brown, G.E., Applegate, J. and Lattimer, J.M.: 1979, Nucl. Phys. A324, 487.

11. Mazurek, T.J.: 1975, Astrophys. Space Sce. 35, 117; 1976, Astrophys. J. Letters 207, L87.

12. Sato, K.: 1975, Prog. Theor. Phys. 54, 1352.

13. Bethe, H.A., Applegate, J.H. and Brown, G.E.: 1980, Astrophys. J. 241, 343.

14. Fuller, G.M., Fowler, W.A. and Newman, M. J.: 1980, Astrophys. J. Suppl. 42, 447; 1981a, preprint; 1982b, preprint; Fuller, G.M.: 1981, preprint.

15. Zaringhalam, A., in preparation.

16. Treiner, J.: 1981, Thesis, Université de Paris Sud, Orsay.

17. Goldreich, P. and Weber, S.: 1980, Astrophys. J. 238, 991.

18. Lamb, D.Q., Lattimer, J. M., Pethick, C.J. and Ravenhall, D.G.: 1978, Phys. Rev. Letters 41, 1623.

19. Baym, G., Bethe, H.A. and Pethick, C.J.:1971, Nucl. Phys. A 175, 225.

20. Brown, G.E., Bethe, H.A. and Baym, G.:1981, Nucl. Phys. A, to be published.

This research was supported in part by USDOE grant DE-AC02-76ER13001 at the State University of New York.

SUPERNOVA SHOCKS AND NEUTRINO DIFFUSION

H. A. Bethe

NORDITA, Blegdamsvej 17,
DK-2100 Copenhagen Ø, Denmark, and
Laboratory of Nuclear Studies, Cornell University,
Ithaca, N.Y. 14853

The general features of an outgoing supernova shock are discussed. The effects of neutrino production and emission from the
shock, nuclear dissociation, and electron capture behind the shock
are computed. An analytic model of neutrino diffusion is made.
The virial theorem is used to calculate the shock velocity and the
radius beyond which matter may reach escape velocities.

1. THE NEUTRINO SPHERE

The dominating physical phenomena of the explosive phase of
a supernova are the outgoing shock and the neutrino diffusion including production and absorption. We are considering only spherically symmetric explosions because they are the simplest case
possible, and only supernovae of type II. The shock, as shown
by computer calculations and discussed in physical terms in G.
Brown's talk yesterday, is formed at the surface of the "homologous core", at a distance of perhaps 10-20 km from the center of
the star. The inner core is never touched by the shock and remains therefore at the original, low entropy of 1 to 2 units (k_B
per nucleon). In most of the core, therefore, the nuclear matter
is in the form of complex nuclei, typically of $A > 100$, which
makes neutrino diffusion very slow. It has been estimated (1)
that it takes neutrinos about 100 ms to diffuse out of the core,
very long compared with dynamic times. The time for all neutrinos
to leave the core is a factor of several longer, because as the
original neutrinos leave, they are replenished by neutrinos from
electron capture (see eq. 12). Typically the number of electrons
per baryon Y_e is several times the number of neutrinos Y_ν at the
time the high density core is formed and the shock is generated.

35

M. J. Rees and R. J. Stoneham (eds.), Supernovae: A Survey of Current Research, 35–52.
Copyright © 1982 by D. Reidel Publishing Company.

The material in the <u>shock</u>, by contrast, has high entropy, perhaps about 7 units. This means that nuclei are completely dissociated into nucleons, at least as long as the density is high. At lower density, 10^{10} g/cm^3 or less, nucleons recombine into particles (see §7). The material <u>outside</u> the shock is still in its original state of low entropy, with essentially all nucleons in heavy nuclei. At least at distances r > 100 km, the heavy nuclei are in the neighborhood of Fe^{56}, with Z/A ≈ 0.46.

The neutrino scattering mean free path is given by the theory of Lamb and Pethick (as modified in ref. 1):

$$\lambda_\nu = \lambda_1/(X_N + (N^2/6A)X_H)$$ (1)

where

$$\lambda_1 = 10 \ (10/\varepsilon_\nu)^2/\rho_{12} \quad km$$ (2)

with ρ_{12} the material density in units of 10^{12} g/cm^3 and ε_ν the mean neutrino energy in MeV. The denominator in (1) is the coherence factor, N the average neutron number and A the average mass number of nuclei in the unshocked matter, X_H the fraction (by weight) of heavy nuclei and X_N that of nucleons. If the nuclei are indeed Fe^{56}, and if $X_H = 0.8$ (which is typical) and $X_N = 0$, then the denominator of (1) is 2.2. At higher densities higher values of this coherence factor occur, about 5 to 10. In the shock, the mean free path of μ- and τ-neutrinos is given by λ_1/X_N while that of electron neutrinos is shorter, and will be discussed later.

The neutrino energy increases with the density, so that the denominator of (2) is a strong function of the density. For most purposes, $\rho\lambda$ is more significant than λ itself, being the mean free path in g/cm^2, or the inverse of the opacity. Because of the denominator in (1), $\rho\lambda$ is smaller outside the shock than inside. This difference is enhanced if we consider the random walk distance

$$L = (c\tau\lambda/3)^{\frac{1}{2}}$$ (3)

where τ is the dynamic time, i.e. the time in which there is a large change   of, say, the shock radius. Therefore, the matter <u>outside</u> the shock will, in general, determine whether neutrinos can escape or remain confined in the shock. We define the neutrino sphere r_ν by the condition

$$\int_{r_\nu}^{\infty} dr/\lambda_\nu(r) = 1$$ (4)

i.e., there is one neutrino mean free path between r_ν and infinity. For some problems it is better (2) to use 8/3 on the right hand side of (4). As long as the shock radius, r_s, is less than

r_ν, neutrinos remain confined in the shock. Thus an important moment in the history of the supernova is when the shock reaches the neutrino sphere: Then the "gate" at the shock front suddenly opens and neutrinos from inside the shock can stream out. With the neutrino energies and density distributions usually occurring, the neutrino sphere is at about 75 ± 10 km and the density of the (unshocked) material outside r_ν is about $5 \times 10^{10} g/cm^3$. This is about $1/10$ of the density at which neutrinos are trapped, cf. §3.

When the shock is well inside r_ν, neutrinos will diffuse out of the shock, but because of the short λ_ν in the unshocked material, their drift velocity in that material is small. Since the shock moves quite rapidly ($U = 5$ to $7 \times 10^9 cm/sec$), the neutrino drift velocity is less than U, so that neutrinos which have come out of the shock will be engulfed again by the shock; we have merely a neutrino precursor outside the shock. Only when $r_s \gtrsim r_\nu$ can neutrinos move freely away from the shock.

The neutrinos in the inner core diffuse very slowly. The material density is high, the neutrino energy is large because μ_ν is large, and the denominator in (1) is also large because we have nuclei of rather high N. In Bethe et al. (1) it was estimated that the escape time of the core neutrinos is about 100 ms; in this time, the shock has travelled ~ 1000 km, and the change of composition of the core will have no influence on it.

2. SHOCK ENERGY

The energy in the shock, or any other part of the star, can be defined in two different ways:
1. We can take the gravitational energy of each material element, $-GM(r)/r$, and add to it the internal energy of the nucleons, electrons, etc. This is the most logical and absolute definition which we shall use in §9. It has the disadvantage that we cannot tell from its sign whether the shock will successfully emerge from the star.
2. We can take just the internal and kinetic energy in the matter, and try to compare their sum at two, not too distant times, choosing times when the gravitational energy of the given piece of matter is not very different. This definition is often very useful (3).

We have calculated the energy in the shock, according to the second definition, at the time when the shock reaches the neutrino sphere, which we call the "gate time". This energy has been calculated in at least two computer runs, one by Wilson (4) of the Lawrence Livermore National Laboratory (his "Munich" run), the other by Mazurek, Cooperstein and Kahana (5) of Stony Brook and Brookhaven (to be referred to as MCK). At the neutrino gate time

they give the same shock energy, which is important because they
use quite different assumptions.

Wilson uses his own equation of state which gives somewhat
too little compression of the central core of the star, and consequently too little initial energy in the shock. On the other
hand, according to his own statement (which we have confirmed),
he assumed too small energy losses due to dissociation of nuclei.
Wilson's calculation, however, is very valuable because he treated
neutrino diffusion with great accuracy and detail, and we shall
have much occasion to refer to it.

MCK use an equation of state similar to that of Lamb, Lattimer, Pethick and Ravenhall (6, referred to as LLPR) which we believe to be the best one existing; they get an initial shock energy of 7×10^{51} ergs. They also account accurately for the energy
expended in dissociating nuclei when the shock hits the material.
Their calculation is inaccurate as regards neutrino diffusion, but
this should not make much difference until "neutrino gate time".
The actual determination of the shock energy is discussed in the
paper by G.E. Brown in this volume.

3. NEUTRINO DIFFUSION ON INFALL

Neutrinos are generally produced near the center of the star
and diffuse outwards. The diffusion may be described by a "drift
velocity". The drift velocity is relative to the surrounding medium because the nuclei in the medium scatter the neutrinos. The
drift velocity is

$$v_d = j_\nu / n_\nu = - c \, \lambda_\nu \, d\ln(n_\nu)/3dr \qquad (5)$$

where j_ν is the neutrino flux (number per $cm^2 sec$) and n_ν the neutrino density (number per cm^3). For many purposes, it is better
to make n_ν the energy density and j_ν the energy flux; then (5)
still holds, except that the energy average in λ_ν needs altering.

At high material density ρ and high neutrino energy, λ_ν becomes very small, and so will v_d. In the infall, the material has
a large negative velocity u. The neutrino drift relative to the
center of the star, $u + v_d$, is then negative in regions of high ρ,
i.e., these neutrinos drift to smaller r where the density is
still higher, and they are therefore trapped. The reverse is true
at low ρ. The limit of trapping is then determined by the condition $u + v_d = 0$. In Bethe et al. (1) this is shown to correspond
to $\rho \approx 5 \times 10^{11} g/cm^3$. The physics of trapping during the infall is
thus seen to be very simple.

A similar argument was applied at the end of §1 to the neutrino diffusion out of the shock. As long as v_d in the material

outside the shock is smaller than the shock velocity u_s relative
to the velocity u_1 (< 0) of that material, $v_d < u_s - u_1$, the neutrinos remain attached to the shock as a precursor. The neutrino
diffusion behind the shock is considerably more complicated, as
we shall show in §5 and 6.

4. NEUTRINO PRODUCTION BY PLASMA AND ELECTRON CAPTURE

    At high temperature, neutrino pairs can be produced rapidly
by collisions of positrons and electrons (7-9), in the process

$$e^- + e^+ \rightarrow \nu + \bar{\nu} \tag{6}$$

without the help of any nucleon. The rate is about

$$Q_e = 1.0 \times 10^{25} \; T^9 \; f \; \text{erg/cm}^3\text{-sec} \tag{7}$$

where T is the temperature in MeV; the rate is nearly independent
(2) of the electron chemical potential μ_e. The factor f is (9)

$$f = 1 + 0.20 \; n \tag{8}$$

where n is the number of essentially massless neutrinos other
than e-neutrinos; we assume n=2 for μ- and τ-neutrinos.

    The rate (eq. 7) may be compared with the energy in each
species of neutrino or antineutrino in thermal equilibrium,

$$W_\nu^{\;\circ} = 0.60 \times 10^{26} \; T^4 \; \text{erg/cm}^3. \tag{9}$$

Therefore the rate of filling the available ν_e quantum states is

$$\zeta_e = Q_e/2W_\nu^{\;\circ} = 0.08 \; T^5 \; \text{sec}^{-1}. \tag{10}$$

For ν_μ and ν_τ, the rate is 5 times slower. The "dynamic time" is
of the order of 1 ms, therefore thermodynamic equilibrium will be
reached if and only if $T > T_{eq}$, where

$$(T_{eq})_{\nu_e} = 6.6 \; \text{MeV} \;, \quad (T_{eq})_{\nu_\mu} = 9.1 \; \text{MeV}. \tag{11}$$

    Electron neutrinos can also be produced by electron capture,

$$p + e^- \rightarrow n + \nu_e \tag{12}$$

for which the rate is

$$\zeta_e = n_{\nu \; eq}^{-1} \; dn_\nu/dt = c \; \sigma_{ep} \; n_p n_e/n_{\nu \; eq} \tag{12a}$$

where n_p and n_e are the proton and electron densities, and σ_{ep} is the cross section of (12). But detailed balancing requires

$$\sigma_{ep} \, n_p \, n_e = \sigma_{\nu n} \, n_n \, n_{\nu \, eq} \tag{12b}$$

where n_n is the density of neutrons, and $\sigma_{\nu n}$ is the cross section for the inverse reaction of (12). Therefore the capture rate is

$$\zeta_c = c \, \sigma_{\nu n} \, n_n \, . \tag{13}$$

Including the effect of blocking of the neutrino capture by occupied electron states, the cross section of this process is about

$$\sigma_{\nu n} = 5 \times 10^{-44} \, <\varepsilon_e^2> \, cm^2 \tag{13a}$$

where ε_e is the energy of the resulting electron in MeV. For a Fermi distribution with $\mu_e = 0$ (not a bad approximation),

$$<\varepsilon_e^2> \, = 6.9 \, T^2. \tag{14}$$

Assuming that 2/3 of the nucleons are neutrons (§6), we get

$$\zeta_c = 42 \, \rho_{10} \, T^2. \tag{15}$$

This is equal to the plasma production rate ζ_e when

$$k = \rho_{10}/(T/10 \text{ MeV})^3 = 1.9 \, . \tag{16}$$

Typical values of k in the material behind the shock are k = 20 to 100; therefore ζ_c/ζ_e = 10 to 50; the production of ν_e by electron capture (eq. 12) dominates over the plasma production (eq. 6).

Bethe et al. (2) have calculated the effect of neutrino outflow on the energy of the shock. They make the simplifying assumption that T, and hence the neutrino energy density, is uniform behind the shock. This would be correct if the neutrino diffusion behind the shock were very rapid. In the next section we shall show that this is not the case, but the general features of ref. 2 are likely to be correct.

Most of the neutrinos accumulated in the shock are emitted soon after the "gate" opens, essentially as black body radiation. A more accurate discussion of the rate of emission will be given in the next section. When the neutrinos are emitted, the pressure behind the shock decreases, perhaps by about 30%, but T does not change. The energy in μ- and τ-neutrinos together is about equal to that in e-neutrinos. Some emission takes place before gate opening, and some after the end of the black body phase.

In Table 1, we give some features of the neutrino emission, according to ref. 2. We believe now, after studying the details

of the neutrino diffusion (§5 and 6), and the Wilson (4) computer
output, that the neutrino losses in Table 1 are somewhat too high,
because of our assumption of uniform T behind the shock. But ac-
cepting the results of Table 1, the final shock energy will be
between 2.5 and 5×10^{51} ergs.

Initial shock energy (10^{51} ergs)	3.0	4.7	7.3	11.3
Gate opening T (MeV)	5.6	6.4	7.1	7.8
Gate opening radius (km)	63	72	80	87
Neutrino loss (10^{51}ergs)	0.5	1.4	3.3	6.2
Final energy (10^{51}ergs)	2.5	3.3	4.0	5.1

Table 1. Energy loss by neutrinos

5. NEUTRINO DIFFUSION BEHIND THE SHOCK

We wish to solve the diffusion equation for the neutrinos
(e, μ and τ) behind the shock. The material in this region has
a very non-uniform ρ so that the usual solution to the diffusion
equation   does not apply. However, an analytical solution of that
equation exists if the density of the medium varies as $1/r^2$. Fortunately, the actual density distribution is not very far from
this variation, being approximately proportional to $1/r^3$ far behind, and more nearly constant close to the shock.

We thus assume the material density to be

$$\rho = B/r^2 \qquad (17)$$

between r_c, the radius of the unshocked inner core, and r_s, the
shock radius. B is given by the mass between these radii,

$$(M(r_s)-M(r_c))/4\pi = \int_{r_c}^{r_s} \rho(r) \, r^2 dr = B(r_s - r_c) . \qquad (18)$$

B is of the order 10^{25}g/cm, and is not very sensitive to the location of the shock. The diffusion equation is (using n for n_ν)

$$\partial n/\partial t = (c/3r^2) \, \partial(r^2 \, \lambda_\nu \, \partial n/\partial r)/\partial r . \qquad (19)$$

It is convenient to write n as a sum of normal modes, where

$$\partial n_k/\partial t = - \alpha_k \, n_k \qquad (19a)$$

for each normal mode. The higher modes decay rapidly with time,
so that in practice we are only concerned with the lowest mode,
k=1, no matter what the initial distribution n(r,t=0). We denote
the decay constant (eigenvalue) of this lowest mode simply by α.

We insert in (19) the neutrino mean free path,

$$\lambda_\nu = 1/\kappa\rho = r^2/\kappa B \qquad (20)$$

where κ is the opacity. Using eq. (1) for λ_ν, we find

$$\kappa = 1.0 \times 10^{-18} \, (\varepsilon_\nu/10 \text{ MeV})^2 \, D \text{ cm}^2/\text{g} \qquad (20a)$$

where D=1 for ν_μ or ν_τ in a medium consisting of nucleons, while $D_e \simeq 3$ for ν_e. The advantage of κ is that it does not depend explicitly on ρ. In the MCK computation of late 1980, when the shock reaches the neutrino sphere,

$$r_s = 77 \text{ km}, \ r_c = 19 \text{ km}, \ M(r_s)-M(r_c) = 5.7 \times 10^{32} \text{g}; \qquad (21)$$

therefore B = 0.78×10^{25} g/cm. The neutrino energy ε_ν = 2.6 T_ν, where T_ν is taken to be the "equilibrium temperature" (11), so for μ and τ-neutrinos,

$$\varepsilon_\nu/10\text{MeV} = 2.38, \ \kappa_\mu B = 4.4 \times 10^7 \text{cm}. \qquad (21a)$$

For electron neutrinos, we find

$$\varepsilon_\nu/10 \text{ MeV} = 2.00, \ \kappa_e B = 9.3 \times 10^7 \text{ cm}, \qquad (21b)$$

using D=3.

Inserting (20) and (19a) in (19),

$$r^{-2} \, d(r^4 \, dn/dr)/dr = - 3 \, \alpha \, \kappa B \, n/c \ . \qquad (22)$$

This can be solved by substituting n = r^k, giving

$$k(k + 3) = - 3 \, \alpha \, \kappa B/c, \qquad (22a)$$

whose solution is

$$k = - 3/2 \pm ((9/4) - 3 \, \alpha \, \kappa B/c)^{\frac{1}{2}} . \qquad (23)$$

Usually, the expression under the square root is negative, so

$$k = - 3/2 \pm i \, m \qquad (23a)$$

and

$$n = r^{-3/2} \cos(m \ln(r/r_0)) \qquad (24)$$

in which the two constants m and r_0 must be determined so as to satisfy the boundary conditions. Then from m and (23) we can determine the decay constant α of the mode,

$$\alpha = c \ (m^2 + 2.25)/3\kappa B \ . \tag{24a}$$

From the values of κB, eqs. (21a,b), it is clear that α is of order 10^3 sec^{-1}, i.e., the diffusion of neutrinos out of the shock takes on the order of 1 ms.

The boundary conditions relate to the ratio of the ν current,

$$j = - (c \ \lambda_\nu/3) \ \partial n/\partial r \tag{25}$$

to the neutrino density, n. Inserting (20), we have

$$3j \ \kappa B/(cnr_s) = -d(\ln n)/d(\ln r) = 3/2 + m \ \tan(m \ \ln r/r_0). \tag{26}$$

At the core boundary the current must vanish, because essentially no neutrinos come out of the unshocked core. Thus

$$3/2 + m \ \tan(m \ \ln r_c/r_0) = 0 \ . \tag{26a}$$

This shows that $r_0 > r_c$. At the shock front, the condition depends on whether neutrinos can escape out to infinity: Before "neutrino gate time", the neutrino current merely provides neutrinos for the material which is newly shocked. If u_s-u_2 is the velocity of the shock relative to the material behind, then $j/n = u_s-u_2$, and

$$3/2 + m \ \tan(m \ \ln r_s/r_0) = 3(u_s-u_2) \ \kappa B/cr_s \ . \tag{27}$$

Together with (26a), this determines m, since r_s and r_c are known. After "gate time", the neutrinos stream out as black body radiation, and $j/n = c/4$. Thus (27) still holds, but with the change

$$u_s - u_2 \rightarrow c/4 \ . \tag{27a}$$

As the shock approaches the neutrino sphere, the neutrino precursor becomes increasingly long, and the boundary condition lies between (27) and (27a).

With the numbers given in (21) and (21a,b), $\kappa B/r_s$ is between 6 and 12. Typical values of u_s-u_2 are $1-2 \times 10^9$ cm/sec. So if condition (27) is valid, the right hand side of (27) is of order 1, while for black body emission (condition 27a) it is 5 to 10. We have solved eq. (27), using eqs. (21) and (21a,b) for input parameters. For a rather wide variation of u_s-u_2, the result for α does not change very much. In all cases, the time, $1/\alpha$, for neutrino diffusion is between 1 and 3 ms (see Table 2), of the same order as the dynamical time. Of course, the electron neutrinos diffuse more slowly than the μ- and τ-neutrinos because they undergo the capture process (eq. 28a) in addition to elastic scattering. The neutrino diffusion times found by Wilson (4) in his "Munich" calculations are of the same order as those in Table 2.

m	r_0/r_c	ν_μ u_s-u_2 (10^9cm/s)	α (ms^{-1})	ν_e u_s-u_2 (10^9cm/s)	α (ms^{-1})
0.0	∞	1.78	0.51	0.85	0.24
1.0	2.64	3.4	0.74	1.63	0.35
1.3	1.94	5.9	0.90	2.8	0.42
1.36	1.85	7.5	0.93	3.6	0.43
1.5	1.69	(12.7)	(1.02)	6.0	0.48
1.53	1.66	(15.9)	(1.06)	7.5	0.50

Table 2. Neutrino diffusion in the shock

6. ELECTRON CAPTURE IN THE SHOCK

For about a year, it has been widely believed that in the material which has just been engulfed by the shock, electrons are captured in large numbers, thereby "neutronizing" the material. This capture was to occur after the shock has crossed the neutrino sphere, and the neutrinos were assumed to stream out immediately because the optical depth at the shock front was less than 1. This strong neutrino emission implied a large loss of energy, and that loss occurred right at the shock front, thereby weakening the shock rapidly and greatly. These computations then led to an early death of the shock.

This line of argument is incorrect. If there is a strong emission of neutrinos from the shock, J_ν , there must also be a substantial density of neutrinos just behind the shock, since the maximum possible emission is black body, thus

$$\rho Y_\nu c/4 \geq J_\nu . \qquad (28)$$

The finite Y_ν then leads to neutrino capture,

$$n + \nu_e \to p + e^- , \qquad (28a)$$

the inverse of (12), and this limits the amount of electron capture.

But in fact the limitation of electron capture is even stronger. We know from the discussion of the preceding section that once the shock is beyond the neutrino sphere there is already a current of neutrinos from the inside of the shocked region which saturates the black body limit, eq. (27a). These neutrinos were already present (trapped) in the material, and their loss was already taken into acount, e.g., in ref. 2. The neutrinos which are newly formed by capture of electrons in the newly shocked material then serve only one purpose: there must be enough of them so that the reaction (28a) balances reaction (12).

Thus we have the following relations in the newly shocked material. First,

$$Y_{ei} = Y_{ef} + Y_{\nu} , \qquad (29)$$

where Y_{ei} is the initial electron concentration in the material outside the shock, usually about 0.42, Y_{ef} is the final Y_e and Y_{ν} is the neutrino concentration behind the shock. Second, there must be balance between reactions (12) and (28a),

$$Y_{ef} \, Y_p \, \sigma_{ep} = Y_{\nu} \, Y_n \, \sigma_{\nu n} . \qquad (30)$$

But $\sigma_{\nu n} = 2 \, \sigma_{ep}$ because neutrinos have definite chirality while electrons can have either chirality. Further, $Y_p \simeq Y_{ef}$ because the concentration of e^+ is small compared with that of electrons. Of course, $Y_n = 1 - Y_p$ because essentially the entire material is in the form of nucleons. Therefore, using (29) and (30),

$$Y_p^2 = 2 \, (1 - Y_p)(Y_{ei} - Y_p) . \qquad (31)$$

If $Y_{ei} = 0.42$, this simple quadratic equation for Y_p has solution

$$Y_p = 0.335, \quad Y_{\nu} = 0.085 . \qquad (31a)$$

Therefore only a small fraction of the electrons are captured when they enter the shock. The value of $Y_{ef} = Y_p$ in (31a) agrees well with the computations of Wilson (4). The value has also been used by us in §4, eq. (14).

For anti-neutrinos, (30) is replaced by

$$Y_e + Y_n = 2 \, Y_p Y_{\bar{\nu}} . \qquad (32)$$

Since $Y_n \simeq 2 \, Y_p$, eq. (31a), we have $Y_{\bar{\nu}} \simeq Y_{e^+}$. The concentration of e^+ is very small compared to that of e^-; a good approximation is

$$Y_{e^+}/Y_e = \exp(-1.8 \, \eta_e) \qquad (32b)$$

where $\eta_e = \mu_e/T$ is about 2 most of the time. Using this,

$$Y_{e^+} = 0.009 \simeq Y_{\bar{\nu}} \simeq 0.1 \, Y_{\nu} . \qquad (33)$$

This again is in good accord with the ratio of $\bar{\nu}_e$ and ν_e fluxes in the computation of Wilson (4).

Electron capture behind the shock stops when either (a) the density becomes so low that the capture process (12) cannot occur sufficiently in the time available, or (b) nucleons combine into α-particles (8) which favor Y_e near 0.5. Both events occur at densities (outside the shock) of order $10^9 g/cm^3$, i.e. r ≃ 300km.

These calculations show that it is important, in numerical calculations, to record Y_ν and not merely assume that neutrinos, generated or scattered at a certain place r, will have a chance $e^{-\tau}$ of escaping to infinity if τ is the optical depth. The inverse reaction (28a) is likely to be very important.

Another possible cause of error in numerical computations arises from the artificial viscosity. This causes the shock to have a precursor in which the temperature is lower than in the shock, though higher than in the unshocked material outside. In this precursor, nuclei may be partially dissociated into nucleons, and the electron capture process (12) may take place. Because the temperature is relatively low, the neutrinos will have long mean free paths, and will flow out easily. Moreover, higher energy neutrinos coming from inside may have collisions in this precursor region, which reduce their energy and make them in turn flow out easily. All these effects, which drain energy from the shock, are spurious.

7. MOMENTUM AND ENERGY IN NEUTRINOS

As neutrinos flow out, they have a pressure distribution which decreases toward the outside. The pressure gradient will accelerate the material, especially where the density is small, and will give it a more outward velocity. The equation of motion,

$$\dot{u} = - \rho^{-1} \, \partial p_\nu / \partial r = - (3\rho)^{-1} \, \partial w_\nu / \partial r \quad , \tag{34}$$

where w_ν is the energy density of neutrinos, gives the change in velocity. Now as long as the neutrinos form a "precursor", they move with the shock velocity relative to the material outside the shock; therefore

$$w_\nu = w_\nu (r - (u_s - u_1)t) \quad . \tag{34a}$$

Therefore, for a given material element, the change of velocity from the moment it first "sees" neutrinos, to the time it is reached by the shock wave, is

$$\Delta u = \int \dot{u} dt = (3\rho_s)^{-1} \, w_\nu(r_s)/(u_s - u_1) \quad . \tag{35}$$

The energy density of neutrinos at the shock is

$$w_\nu(r_s) = 3 \, W_\nu/(4\pi \, r_s^3 \, R) \quad x10^{51} ergs/cm^3 \tag{35a}$$

where R is the ratio of the average neutrino energy density inside the shock to that at the shock front. Calculations of the type shown in §5 give $R \simeq 3$. W_ν is the total neutrino energy in units of 10^{51}ergs. The material density just outside the shock is (2)

$$\rho_s = C \, r_s^{-3} \times 10^{31} g/cm^3 \; ; \tag{35b}$$

$C \simeq 3$ according to computer results (also see Yahil and Lattimer, this volume). Inserting (35a,b) into (35),

$$(u_s - u_1)\Delta u = (100/4\pi CR)W_\nu \; 10^{18} cm^2/s^2 = .9 \; W_\nu \; 10^{18} cm^2/s^2. \tag{36}$$

The total energy in μ- and τ-neutrinos may be of order $W_\nu = 1.5$, and $u_s - u_1 = 5$ to 7×10^9cm/s. Then

$$\Delta u = 0.2 \text{ to } 0.3 \times 10^9 cm/s \; . \tag{37}$$

So the effect is rather small, but not negligible. It was pointed out to me by Yahil, and was included in Wilson's(4) model.

There is an additional effect of acceleration inside the shock, but it is smaller because the material density is larger. The effect of electron neutrinos is smaller because R is larger for them. All these effects consist of giving the material an outward impulse which helps in the shock propagation. There is also a (very small) effect on the energy because the neutrinos emerging from the shock get scattered by the infalling material which moves in the opposite direction and therefore increases the neutrino energy. Many of these scattered neutrinos return to the inside of the shock and can give part of their energy to that matter.

8. NUCLEAR DISSOCIATION

When the shock is first formed, the nuclei in it are completely dissociated into nucleons, and most of the energy is in the form of thermal energy of these non-relativistic nucleons. Lichtenstadt, Sack and Bludman (10) have pointed out the importance of this, because the pressure is as much as 2/3 of this thermal energy, while for relativistic particles it is only 1/3; pressure is essential for starting the shock.

However, complete dissociation costs 8 MeV per nucleon. For half a solar mass = 10^{33}g, this amounts to 8×10^{51}ergs, which is more than the shock energy. Fortunately, there is a self-regulatory mechanism: As the shock loses energy and cools down, full dissociation into nucleons stops, and instead nuclei dissociate only into α-particles. This is shown in some detail in Table 3 (computed by J. Cooperstein) which refers to an entropy of s=7, Y_e=0.464 (i.e., matter composed of ^{56}Fe which can dissociate into 13 α-

particles and 4 neutrons).

$\log_{10}\rho$	T(MeV)	X_α	$P(10^{28}$ergs/cm$^3)$
11.0	2.86	0.17	47.
10.6	2.22	0.29	13.5
10.2	1.81	0.41	4.0
9.8	1.52	0.51	1.22
9.4	1.30	0.61	0.38
9.0	1.12	0.70	0.122

Table 3. Fraction of α-particles for s=7, Y_e=0.464

Recombination into α-particles begins at T > 3 MeV and ρ > 10^{11}g/cm^3, and is reasonably complete at T = 1 MeV and ρ < 10^9g/ cm^3. Note that, on this adiabat, decrease of ρ leads to increasing condensation into α-particles, as in a cloud chamber. When nuclei dissociate only into α-particles, the energy required is only 2 MeV per nucleon, so the shock energy is sufficient to cause this much dissociation in a large mass. Some of the material in the inner part of the shock which originally is fully dissociated into nucleons, may later partially recombine into α-particles and make the released energy available to the shock, but the main effect is in the outer part of the shock.

The pressure in the adiabat s=7 behaves roughly as

$$p \sim \rho^{1.27} \quad . \tag{38}$$

Of course, at ρ > 10^{11}g/cm^3, the exponent is much greater. The most important quantity for the shock is

$$\beta = \rho\varepsilon/p \tag{39}$$

where ε is the energy per unit mass. For s=7, β ranges from 2.8 at 10^{12}g/cm^3 to 5.3 at 10^9g/cm^3. The density ratio at the shock,

$$\rho_2/\rho_1 = 2\beta + 1 \tag{40}$$

is 11.6 if $\rho_2 = 10^9$g/cm^3 (the subscripts 1 and 2 refer to material outside and inside the shock, respectively). Much of the energy is then in radiation and electrons, and the energy and pressure of these components has been included.

A very useful approximate formula is thus obtained,

$$\varepsilon = 3\times10^{18}\text{erg/g} + 3p/\rho \quad , \tag{41}$$

which is valid for 10^9g/cm^3 < ρ < $10^{11.4}$g/cm^3. Outside these limits, ε is smaller than in (41). The shock velocity is given by

$$\tfrac{1}{2}(u_2 - u_1)^2 \approx \varepsilon_2 \qquad\qquad (42)$$

if we neglect $\varepsilon_1 \ll \varepsilon_2$ and $p_1 \ll p_2$. For our case s=7, the difference in velocities $u_2 - u_1 = 8.5 \times 10^9$cm/s and 3.6×10^9cm/s at $\rho_2 = 10^{12}$g/cm^3 and 10^9g/cm^3, respectively.

9. THE VIRIAL THEOREM

Multiplying the equation of motion by r, and integrating by parts, one obtains the virial theorem

$$- \int_0^R d^3r\ G\rho M(r)/r + 3 \int_0^R d^3r\ p(r) =$$
$$4\pi R^3\ p(R) + \int_0^R d^3r\ \rho r\ \ddot{r}\ . \qquad\qquad (43)$$

This is especially useful if the last term is negligible, i.e., if the material inside R is hydrostatic. According to the computer results, this occurs some time after the shock has gone through the mass point $M(R)$. If high accuracy is desired, the last term can be computed.

The first term is the gravitational energy. It is most useful to express the second term by the internal energy. This is easy if we can write ε in the form

$$\varepsilon = \varepsilon_1 + (\gamma' - 1)^{-1}\ p/\rho. \qquad\qquad (44)$$

This is in fact just the form of the approximate formula (41), with $\gamma' = 4/3$. Then, neglecting the last term in (43), we get

$$E_{grav} + E_{int} + \int^{M(R)} (3\gamma' - 4)\ \varepsilon\ dm -$$
$$3(\gamma' - 1)\ \varepsilon_1\ M(R) = 4\pi\ R^3\ p(R) \qquad\qquad (45)$$

where $dm = \rho d^3r$, and $M(R)$ is the mass included by R. The first two terms are the total energy within R. The theorem holds, of course, also for the difference between R_1 and R_2; this is useful because the simple formula (44) holds only in limited ranges of p/ρ. Then we find, writing $E_{tot} = E_{grav} + E_{int}$,

$$4\pi R_2^3 p(R_2) - 4\pi R_1^3 p(R_1) = E_{tot}(R_2) - E_{tot}(R_1) +$$
$$(3\gamma'-4)(E_{int}(R_2)-E_{int}(R_1))-3(\gamma'-1)\varepsilon_1(M(R_2)-M(R_1)) \qquad (46)$$

The total energy of the star, out to a large radius R_2, is the initial energy, usually only about $-\tfrac{1}{2}\times10^{51}$ergs, minus the energy lost in neutrinos, usually about 2 to 3×10^{51}ergs (4). The total energy in the smaller radius R_1 can often be obtained from a relatively simple hydrostatic calculation, as can $R_1^3 p(R_1)$

(see, for example, Yahil and Lattimer, this volume). In the particular case when $\gamma' = 4/3$, as in (41), the first term in the second line of (46) is zero, and the second term is $-\varepsilon_1(M(R_2)-M(R_1))$. Then the pressure at R_2 can be deduced from known quantities at R_1.

The product $R^3p(R)$ is particularly useful because the shock velocity is very nearly given by

$$(u_2 - u_1)^2 = p_2/\rho_1 \tag{47}$$

(for a more accurate formula, see eq. (50)). Further, the outside density is approximately given by (35b), with $\rho_s = \rho_1$, $r_s = R$; therefore

$$(u_2 - u_1)^2 = p_2 R^3/Cx10^{31} \quad \text{(cgs units).} \tag{47a}$$

Thus the shock velocity is directly given by (46). Moreover, the last term in (46) is likely to not be very large in the regions of low density, so $u_2 - u_1$ should decrease only slowly.

10. ESCAPE

I believe that the shock will survive the energy losses by neutrino emission and by dissociation. If it does, then at some distance r_s the velocity behind the shock, u_2, will become greater than the escape velocity, which is given by

$$v_{esc}^2 = 2.65x10^{18} \, (M/M_\odot) \, r_8^{-1} \tag{48}$$

with r_8 the distance in units of 10^8cm = 1000 km. The u_2 could be obtained from (47a) if pR^3 were known. We prefer here to obtain it from the energy in the shock which has a more intuitive meaning, and which we believe to be between 2 and $4x10^{51}$ ergs. We assume this energy E_s to be uniformly distributed in space, then the energy density is

$$\rho_2\varepsilon_2 = 3 \, E_s/(4\pi \, r_s^3) \, . \tag{49}$$

The pressure, at large r_s, is mostly in relativistic particles, electrons or γ-rays, and is $p_2 = \rho_2\varepsilon_2/3$. The change of velocity across the shock is given by

$$(50)$$

$$(u_2-u_1)^2 = p_2(\rho_1^{-1}-\rho_2^{-1}) = 6p_2/(7\rho_1) = 2.7x10^{18}E_{51} \, cm^2/s^2$$

if C in (35b) is 3, and $E_{51} = E_s/10^{51}$ergs. The infall velocity u_1, which is of course negative, can be obtained from computer outputs (Brown, this volume). We have used the results of Wilson (4) for u_1 as well as for the relation between M and r_8. On this basis, we have computed Table 4.

Shock energy $(10^{51}$ ergs)	Remnant mass $(M_\odot)$	Escape radius (km)
2.0	1.60	1330
2.5	1.56	1080
3.0	1.53	930
4.0	1.49	730

Table 4. Gravitational escape

According to the table, the separation between escaping material and remnant occurs at 1000 km ± 30%. The mass of the remnant is rather accurately determined, 1.55 $M_\odot$ ± 10%. When the remnant loses its thermal energy, it shrinks into a neutron star. In this process, it acquires substantial negative gravitational energy, so that its gravitational mass is about 10% smaller, 1.4 $M_\odot$ ± 10%.

In summary, the major conclusions reached in this lecture are:
i) Shock energies are typically in the range $2.5-5\times10^{51}$ ergs,
ii) Neutrino diffusion times behind the shock are 1-2 ms for ν_μ and ν_τ, and from 2 to 3 ms for ν_e, whether or not the shock has reached the neutrino sphere.
iii) There is little ($\Delta Y_e \sim 0.09$) electron capture immediately behind the shock after it has passed the neutrino sphere.
iv) Momentum deposition by ν_μ and ν_τ leads to a change in matter velocity ahead of the shock of $2-3\times10^8$ cm/sec.
v) Nuclear dissociation weakens shocks, but full dissociation into nucleons ceases, and partial dissociation into α's ensues before a shock cools down significantly.
vi) The virial theorem is used to determine the remnant mass from a supernova. A remnant gravitational mass of ~ 1.4 $M_\odot$ is derived.

REFERENCES

1. Bethe, H.A., Brown, G.E., Applegate, J. and Lattimer, J.M.:1979 Nucl. Phys. A324, 487.
2. Bethe, H.A., Applegate, J.H. and Brown, G.E.: 1980, Astrophys. J. 241, 343.
3. Brown, G.E., Bethe, H.A. and Baym, G.:1981, Nucl. Phys. A, in press
4. Wilson, J.:1980, Proc. N.Y. Acad. Sci. 336, 358.
5. Mazurek, T., Cooperstein, J. and Kaharia, S.:1981, in Dumand-80, ed. V.J. Stenger (Honolulu:Univ. of Hawaii Dumand Center), p. 142.
6. Lamb, D.Q., Lattimer, J.M., Pethick, C.J. and Ravenhall, D.G.: 1978, Phys. Rev. Lett. 41, 1623.
7. Beaudet, G., Petrosian V. and Salpeter, E.E.:1967, Astrophys. J. 150, 979.

8. Dicus, D.A.:1972, Phys. Rev. D6, 941.
9. Soyeur, M. and Brown, G.E.:1979, Nucl. Phys. A324, 464.
10. Lichtenstadt, I., Sack, N. and Bludman, S.A.:1980, Phys.
    Rev. Lett. 44, 832; 1980, Astrophys. J. 237, 903

SUPERNOVAE FOR PEDESTRIANS

Amos Yahil and James M. Lattimer

Astronomy Program, State University of New York,
Stony Brook, NY 11794, and
NORDITA, DK-2100 Copenhagen Ø, Denmark

1. INTRODUCTION

It is now generally agreed that stars which are massive
enough to burn carbon nonexplosively all evolve to a similar configuration with a degenerate iron core, and an overlying "onion
skin" mantle of lighter elements. A dynamical instability in the
iron core, due to partial photodissociation plus electron capture,
then leads to its implosion. W. Hillebrandt and G. Brown (both
this volume) have summarized previous research. Most calculations
of this process have employed numerical hydrodynamics. But detailed calculations of the rate of electron capture and neutrino
losses, before neutrino trapping occurs at a density $10^{12} g/cm^3$,
show that the collapse is nearly adiabatic, with only a small
change in entropy (1). This suggests that the description of the
collapse can be simplified by approximating the equation of state
by a polytropic law

$$p = K \rho^{\gamma} . \qquad (1)$$

Here p is the total pressure, ρ is the density, γ is the adiabatic
index, and K (whose value is a function of entropy only) is a constant, both in space and time. In §2 we provide self-similar solutions to the collapse of a stellar core whose equation of state
is given by eq. (1). These solutions show how the iron core breaks
up into an inner core, which collapses homologously, and an outer
core whose supersonic infall velocity is at about half the escape
velocity and at a constant Mach number $\sim$ 2 - 3.

The collapse of the inner core is brought to a sudden halt
when the central density reaches nuclear density. A pressure wave

53

M. J. Rees and R. J. Stoneham (eds.), Supernovae: A Survey of Current Research, 53–70.
Copyright © 1982 by D. Reidel Publishing Company.

then moves out transferring energy with it. At about the edge of
the inner core (defined here to be the point of maximum infall
velocity) a shock develops, and propagates into the outer core.
In §3 we estimate the amount of energy available to this shock
by calculating the negative (binding) energy left in the unshocked
inner core. This is possible because the inner core quickly reaches hydrostatic equilibrium, and the energy of the hydrostatic configuration is simple to calculate. We show that this binding energy is about 5×10^{51} ergs, independent of the details of the nuclear
part of the equation of state.

The success of a shock in propagating through the outer core
depends on the ram pressure which the infalling outer core exerts
on the shocked material. In §4 we provide a "hydrostatic" model
for shock propagation by neglecting the acceleration term in the
Euler equation of motion. This is a conservative estimate for the
shock propagation, since the shocked material is in reality decelerated. Our calculation shows a clear division between successful
and unsuccessful shocks, which depends on the strength of the ram
pressure due to the outer core.

All the calculations offered in this paper are idealizations
and approximations, which are not substitutes for detailed hydrodynamical calculations. But, by offering a pedestrian view of the
supernova process, they may provide qualitative concepts by which
the numerical work can be interpreted.

2. SELF-SIMILAR SOLUTIONS FOR STELLAR COLLAPSE

In our self-similar model for stellar collapse we approximate
the equation of state by the polytropic law eq. (1). The essence
of the self-similar model is the existence of only two dimensional
parameters in the problem, namely K and G. It is assumed, and this
is borne out by numerical calculations, that any additional parameter, such as the initial central density, affects only transients,
and memory of it is lost, at least in that part of the flow in which
the density greatly exceeds the initial central density. If that
is the case, then only one dimensionless combination of radius r
and time t can be formed:

$$X = K^{-\frac{1}{2}} G^{(\gamma-1)/2} r (-t)^{\gamma-2} \tag{2}$$

where the origin of time is chosen to be the catastrophic moment
when the central density becomes infinite. (In reality rebound
occurs at nuclear density, but this is not taken into account by
the polytropic equation of state.) All hydrodynamical variables
must therefore be functions of X only, except for overall scale
factors, which can be determined by simple dimensional arguments.
Thus, the solutions are self-similar, i.e., they have the same

spatial structure at all times.

A self-similar stellar collapse theory has been presented by Goldreich and Weber (2), henceforth referred to as GW, for the case $\gamma=4/3$. They restrict their attention to the inner core which collapses homologously, and they show that this inner core cannot have a mass exceeding $m_{ic}= 4.76(K/G)^{3/2}$. This mass is just 4.5% greater than the Chandrasekhar mass M_{ch} for the same K. But the electron fraction drops from $Y_e=0.46$ for Fe to $Y_e=0.35$ at neutrino trapping (3), and since $K \propto Y_e^{4/3}$, the mass of the inner core at trapping is only about half of the total mass of the original core.

The main limitation of the GW model is its neglect of the outer core, which is unable to collapse homologously. First, the edge of the inner core, which we define to be the point of maximum infall velocity, and the outer core beyond, are not in free-fall as in GW. In the numerical calculations the infall velocity of the outer core is only about half the free-fall velocity. A typical relation (4) is

$$v^2 = 10^{26}/r \ cm^3/s^2; \quad \rho = 3x10^{31}/r^3 \ g. \tag{3}$$

This retardation of the collapse is made possible by the finite density profile which results in a nonnegligible pressure gradient. The ensuing Mach number of the flow is fairly constant ~ 2. All these features are absent in GW, in which the very existence of the outer core is ignored. Furthermore, the pressure of the outer core enhances the rate of collapse of the inner core. A suitable dimensionless collapse parameter is

$$\Omega_o = \lim_{r \to 0} \ 8 \ \pi \ G \ \rho \ r^2/3 \ v^2 \tag{4}$$

which is analogous to the cosmological density parameter, and is the inverse of the parameter λ of GW. In the GW theory $\Omega_o>153$, but numerical calculations show it to be 2-3 times smaller.

The inadequacies of the GW model have led us to search for a generalization to their theory, which can provide a simultaneous description of both the inner (homologous), and outer (nonhomologous) parts of the core. Such a generalization is possible for adiabatic indices in the range $6/5 \lesssim \gamma \lesssim 4/3$. The GW model is then a limiting case, $\gamma = 4/3$, of a family of solutions.

The effective adiabatic index is, in fact, smaller than $4/3$. A change of Y_e from 0.42 at $10^9 g/cm^3$ to 0.35 at $10^{12} g/cm^3$ is equivalent to a polytropic equation of state with $\gamma =1.30$ and constant K. Furthermore, after neutrino trapping, and until nuclear density, the adiabatic index is also smaller than $4/3$ due to nuclear lattice energy (5).

As emphasized above, the hydrodynamical variables in a self-similar solution are, except for scale factors, functions of the similarity variable X only. The required scale factors, which are powers of K, G and -t, can all be deduced by dimensional arguments:

$$\rho = G^{-1}(-t)^{-2}D(X) \quad , \tag{5}$$

$$v = K^{\frac{1}{2}}G^{(1-\gamma)/2}(-t)^{1-\gamma}V(X) \ , \tag{6}$$

$$m(r) = K^{3/2}G^{(1-3\gamma)/2}(-t)^{4-3\gamma}M(X) \ , \tag{7}$$

$$\mathcal{E}(r) = K^{5/2}G^{(3-5\gamma)/2}(-t)^{6-5\gamma}E(X) \ . \tag{8}$$

Equations (5-8) relate the usual variables ρ, v, the enclosed mass m(r), and the enclosed energy $\mathcal{E}(r)$ to the dimensionless hydrodynamical variables D, V, M, and E, where M and E are defined by

$$M(X) = 4\pi \int_0^X x^2 D(x)dx \ , \tag{9}$$

$$E(X) = 4\pi \int_0^X x^2 D(x)\{\tfrac{1}{2}V(x)^2 + D(x)^{\gamma-1}/(\gamma-1) - M(x)/x\}dx \ . \tag{10}$$

The asymptotic behavior of the dimensionless variables as $X \to \infty$ can also be deduced by dimensional arguments. At catastrophe time, t=0 in our notation, the central density becomes infinite, but the flow is quite regular at all other radii. We cannot describe the spatial distribution at t=0 in terms of the similarity variable X, because the entire X-axis shrinks to r=0, eq. (2). We can, however, consider the r dependence of a hydrodynamical variable, say ρ, in the limit $t \to 0$. In order to avoid a singularity, or an equally inadmissable zero, we see that the overall factor $(-t)^{-2}$ in eq. (5) must be cancelled by the time dependence implicit in D(X). This entails the asymptotic behavior

$$D(X) \propto X^{-2/(2-\gamma)} \ . \tag{11}$$

Similarly,

$$M(X) \propto X^{(4-3\gamma)/(2-\gamma)} \tag{12}$$

$$V(X) \propto X^{-(\gamma-1)/(2-\gamma)} \tag{13}$$

$$E(X) \propto X^{(6-5\gamma)/(2-\gamma)} \ . \tag{14}$$

These asymptotic relations are valid in the outer core, and together with eqs. (5-8) define the (time independent) r dependence of the hydrodynamical variables there. Eq. (3) immediately is obtained for $\gamma \simeq 4/3$, except for the constants of proportionality, which remain to be determined.

Another limit of interest is that of the total energy. From

eq. (14) we see that it diverges for $\gamma < 6/5$. This underscores the criticism levelled by Shu (6), using other arguments, against attempts (7,8) to find a self-similar solution for isothermal collapse, $\gamma = 1$. His criticism, however, does not apply in the range $6/5 \leq \gamma \leq 4/3$ which interests us. On the contrary, we see that for those adiabatic indices the self-similar solutions have a zero total energy. (In the marginal case $\gamma = 6/5$ the energy turns out to be nonzero but finite.) Indeed, the requirements of zero total energy, together with regularity at the origin, uniquely determine the self-similar solution for each γ. We therefore have a full solution in the inner core, joining smoothly to the outer core. In particular, we can calculate the constants of proportionality in the asymptotic relations (11)-(14) as well as the dimensionless collapse parameter Ω_o, eq. (4).

The method of calculation of the self-similar solution is standard (9), and details are given elsewhere (10). Using the defining eqs. (5) - (8), the hydrodynamical partial differential equations (in our case the energy equation is replaced by the equation of state) are converted into the similarity equations, which are ordinary differential equations in the similarity variable X. The similarity equations possess a singularity analogous to the one encountered in steady-state accretion (11) and winds (12). The critical solution is the one which is regular at that point, and has the correct behavior as $X \to 0$ (regularity) and $X \to \infty$ (see above). This critical solution automatically satisfies the requirement of zero total energy.

The full X dependence of the various dimensionless variables is plotted in figures 1-5 for the illustrious case $\gamma = 1.30$. The dependence of the solutions on γ is elaborated in tables 1 and 2. Table 1 lists various velocity parameters as a function of γ. The subscripts o, m, and a stand for the origin, the point of maximum infall velocity, and asymptotic value as $X \to \infty$ respectively. The dimensionless speed of sound, defined in complete analogy with V, eq. (6), is denoted by A; V_{ff} is the dimensionless free-fall velocity; and ξ is the Mach number. Table 2 lists various density parameters. The collapse parameter Ω_o is related to the central density D_o by $\Omega_o = 6 \pi D_o$. M_{ch} is the dimensionless mass of a static polytrope having the same central density as our collapsing one (13). The enclosed gravitational energy is defined as

$$W(X) = 4\pi \int_0^X x\, D(x)\, M(x)\, dx \quad , \tag{15}$$

consistent with the definition of E(X), eq. (10).

The general features of all the models are rather similar. The maximum velocity point, which we define as the edge of the inner core, occurs around X=3. The mass of the inner core decreases with time, eq.(7), but only very slowly if $\gamma \simeq 4/3$. The maximum in-

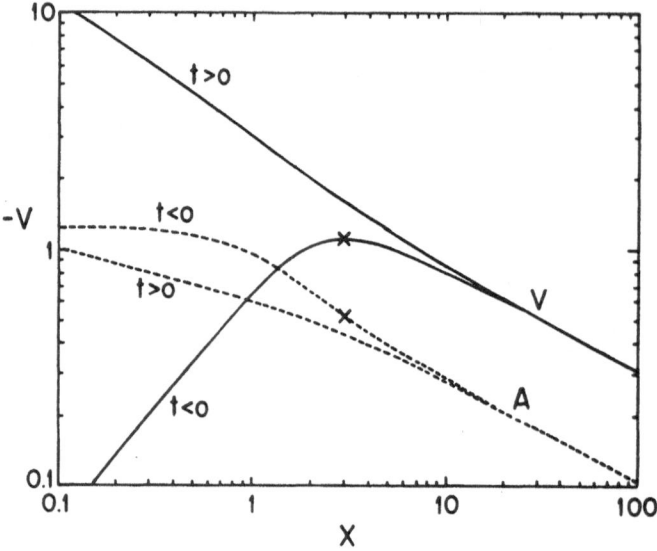

Figure 1. Dimensionless infall velocity -V and sound speed A for γ=1.30. Both precatastrophe (t<0) and post catastrophe (t>0) solutions are shown. A cross at the point of maximum infall velocity marks the edge of the inner core. The same notation holds in figures 2-5.

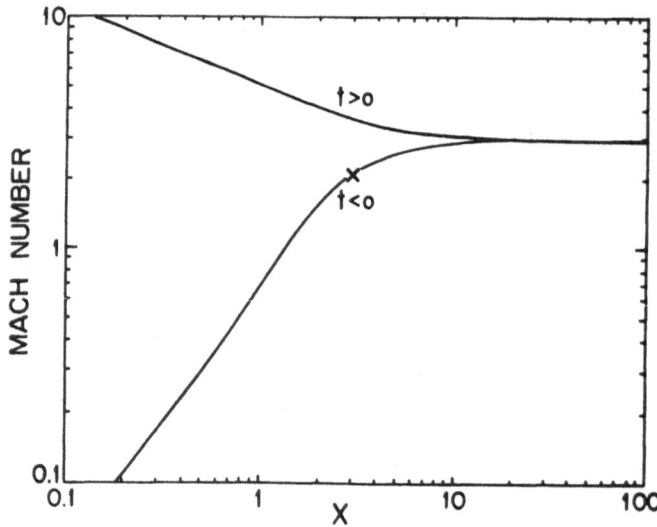

Figure 2. Mach number as a function of X for γ=1.30. Note the constant asymptotic value in the outer core.

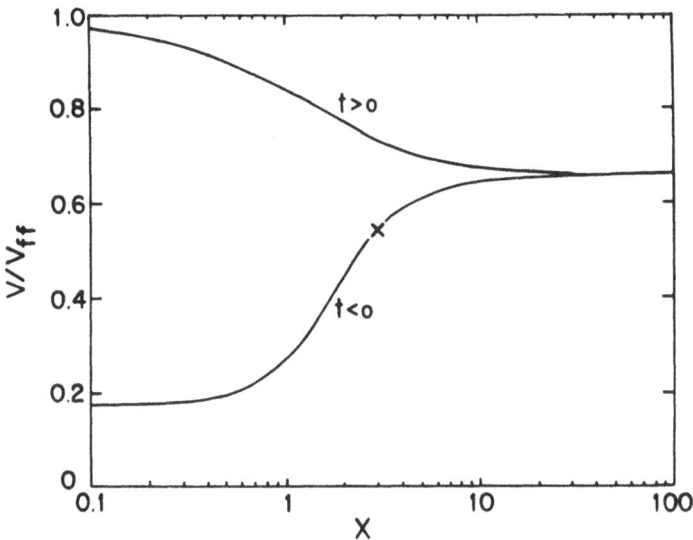

Figure 3. The ratio of the infall velocity to the
 free-fall velocity as a function of X for γ=1.30.
Note the constant asymptotic value in the outer core,
and the approach to free-fall in the postcatastrophe
phase.

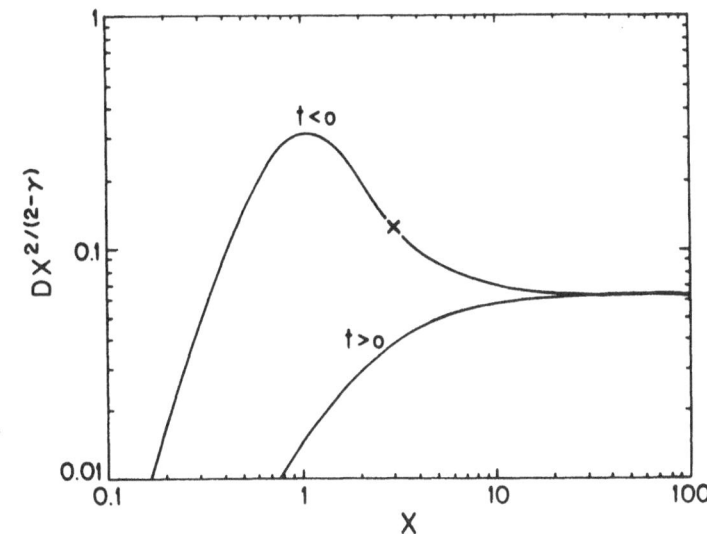

Figure 4. The density factor $DX^{2/(2-\gamma)}$ as a function of
X for γ=1.30. Note the constant asymptotic value in the
outer core, and the drop in the density in the postcatastrophe stage.

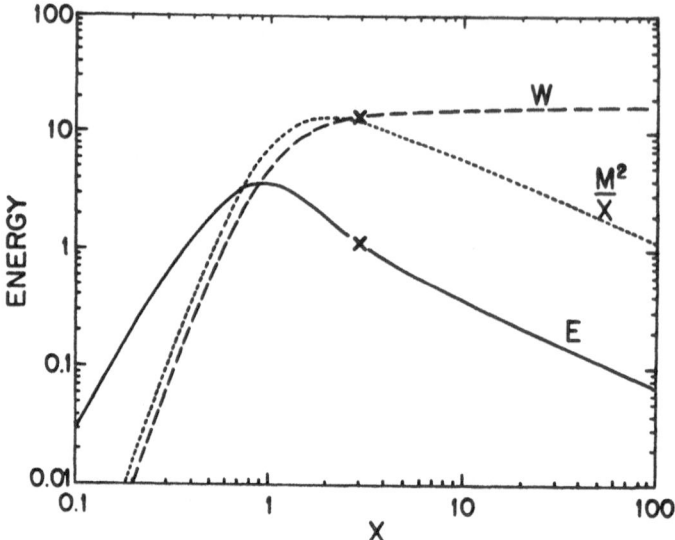

Figure 5. Dimensionless enclosed energies as functions
of X for γ=1.30. The postcatastrophe solution is not
shown, because the central point mass renders the enclosed energies meaningless.

γ	X_m	$-V_m$	A		V/V_{ff}			ξ	
		m	o	m	o	m	a	m	a
1.20	4.90	1.22	1.04	0.65	0.30	0.56	0.62	1.86	2.28
1.21	4.58	1.19	1.05	0.64	0.29	0.55	0.62	1.84	2.29
1.22	4.34	1.16	1.06	0.63	0.28	0.55	0.62	1.84	2.30
1.23	4.09	1.14	1.07	0.62	0.27	0.54	0.62	1.83	2.33
1.24	3.87	1.12	1.09	0.61	0.25	0.54	0.62	1.83	2.36
1.25	3.66	1.11	1.11	0.60	0.24	0.53	0.62	1.84	2.40
1.26	3.51	1.10	1.13	0.59	0.23	0.53	0.62	1.86	2.46
1.27	3.36	1.09	1.15	0.58	0.21	0.53	0.63	1.89	2.53
1.28	3.21	1.09	1.18	0.56	0.20	0.53	0.64	1.93	2.63
1.29	3.06	1.09	1.20	0.55	0.19	0.54	0.65	1.99	2.76
1.30	2.95	1.10	1.24	0.52	0.17	0.54	0.66	2.09	2.94
1.31	2.88	1.12	1.28	0.49	0.16	0.56	0.68	2.26	3.21
1.32	2.77	1.16	1.34	0.46	0.14	0.59	0.72	2.55	3.69
1.33	2.77	1.28	1.45	0.37	0.12	0.66	0.80	3.45	5.11
4/3	2.77	1.85	1.64	0	0.08	1	-	∞	-

Table 1. Velocity parameters for Precatastrophe Collapse

fall velocity is V~1, which is about half the free fall
velocity at that point, and the ratio between the two does not
increase substantially in the outer core, where it tends to an
asymptotic value ~ 0.5-0.7. (The infall does approach free-fall

γ	D_o	Ω_o	M_m	M_{ch}	DX^3/M_m	$DX^{2/(2-\gamma)}_a$	W_m	M^2/X_m	E_m
1.20	0.57	11	11.54	8.03	.058	.301	.246	25.0	27.2	13.10
1.21	0.63	12	10.56	7.39	.055	.287	.229	22.6	24.3	10.55
1.22	0.69	13	9.77	6.82	.053	.273	.211	20.7	22.0	8.53
1.23	0.75	14	9.05	6.40	.050	.258	.193	19.0	20.0	6.89
1.24	0.83	16	8.43	6.04	.047	.243	.176	17.6	18.4	5.57
1.25	0.92	17	7.87	5.72	.045	.227	.158	16.5	16.9	4.48
1.26	1.03	19	7.42	5.48	.041	.209	.140	15.6	15.7	3.57
1.27	1.16	22	7.09	5.26	.037	.191	.121	14.9	14.6	2.80
1.28	1.31	25	6.63	5.06	.033	.171	.102	14.3	13.7	2.16
1.29	1.50	28	6.28	4.90	.029	.149	.083	13.8	12.9	1.62
1.30	1.75	33	5.98	4.78	.024	.125	.064	13.6	12.1	1.15
1.31	2.10	40	5.71	4.67	.019	.095	.045	13.5	11.3	0.73
1.32	2.66	50	5.42	4.60	.013	.062	.025	13.6	10.6	0.38
1.33	4.00	75	5.07	4.56	.004	.020	.006	14.4	9.5	0.08
4/3	8.11	153	4.76	4.53	0	0	-	17.0	8.2	0

Table 2. Mass and Energy Parameters for Precatastrophe Collapse

as $\gamma \to 4/3$, together with the disappearance of the outer core.) The Mach number is already ~ 2 at the edge of the inner core, and increases by about 50% in the outer core, to approach a constant.

The models differ more substantially in the density of the outer core. This density is zero in the GW limit of $\gamma=4/3$, and becomes progressively larger as γ decreases. As a result, the collapse is much faster (smaller Ω_o) for lower values of γ. In addition, the mass difference between the static star and the collapsing inner core increases for lower γ. Another related parameter is the total energy enclosed by the inner core, which varies from 0 in the GW model to 50% of W_m for $\gamma=6/5$.

As emphasized above, the collapse does not end at the catastrophe time t=0. At that point the asymptotic power-law relations hold everywhere, and form initial conditions for a self-similar solution valid for $t \to 0$. This postcatastrophe solution can be calculated analogously to the precatastrophe solution, by replacing -t by +t in eqs. (5)-(8). But there is now no singular point to force a critical solution, and the initial conditions determine it instead. An important difference between our model and other such calculations (6,14), however, is that the initial conditions are not arbitrary, but are obtained from the asymptotic form of the precatastrophe solution.

The variables plotted in figures 1-4 for $\gamma=1.30$ are also plotted for the postcatastrophe solution. We see that the velocity approaches the escape velocity as $X \to 0$. This limit corresponds, for a given radius r, to a sufficiently long time after catastrophe.

In the precatastrophe stage the outer core cannot go into free-fall because the inner core is collapsing too slowly, and retards the collapse of the outer core. This obstacle is removed at t=0, after which the outer core builds up momentum, and approaches free-fall. The free-fall collapse of the outer core is made possible by a reduction in the pressure gradient relative to gravity. This is due to a substantial reduction in density, as can be seen in fig. 4. The asymptotic behavior of density for X<<1 is $D(X) \propto X^{-3/2}$.

If the outgoing shock which results from rebound at nuclear density is unsuccessful, then the outer core can indeed reach free-fall before impinging on what is by now a stalled accretion shock. In the case of a successful shock, however, the asymptotic free-fall regime is unlikely to be reached before the infalling outer core is hit by the outgoing shock. This is because the shock velocity is high enough, so that no mass element in the outer core has enough time to reach the asymptotic free-fall regime X<<1. In §4 we calculate the shock propagation assuming that the outer core has not had a chance significantly to change between t=0 and its encounter with the outgoing shock.

3. HYDROSTATIC CORES

The collapse phase of the supernova, we have seen, can be described by a self-similar solution, the mass of the collapsing inner core being determined by the adiabatic index Γ in the matter. The collapse is halted, however, near nuclear density $\rho_0 = 2.7 \times 10^{14}$ g/cm^3 when Γ rises suddenly above 4/3, due, basically, to nucleon degeneracy. We will parametrize the equation of state in this regime by 1) ρ_t, the transition density at which this stiffening occurs; 2) Γ_b, the value of the adiabatic index due to the baryon pressure P_b above ρ_t; and 3) the compressibility K_{nuc}. Near ρ_0, $K_{nuc} = 220$ MeV to about 15%, but Γ_b is much more highly uncertain. The stiffening of the equation of state, and the rebound of the core, is rapid enough to generate a shock wave at the outer boundary of the inner core (Brown, this volume). Dynamical calculations reveal that the rebounding core achieves hydrostatic equilibrium quickly compared to dynamical timescales, because the pressure on the inner core surface is so large (3). If this is true, we can calculate the binding energy B of the hydrostatic inner core, and apply the principle of the conservation of energy to determine, approximately, how much energy is made available to the shock. The collapsing star has very nearly zero energy, but B ~few x 10^{51} ergs. In as much as little energy has been transferred through the shock front either by neutrinos or pdV work (see below) the energy of the young shock must have the same value. It is therefore of interest to build hydrostatic models of the inner core to determine this energy.

A consequence of the fact that $\Gamma \simeq 4/3$ for $\rho < \rho_t$ is that the total energy of the collapsing inner core is nearly zero. The virial theorem shows that a $\Gamma = 4/3$ gas in hydrostatic equilibrium with no external pressure has zero energy. Another consequence of a $\Gamma = 4/3$ equation of state is that the mass of the configuration is entirely determined by K of eq. (1): For a degenerate electron gas, this mass, the Chandrasekhar mass M_{ch}, is 5.76 Y_e^2 $M_\odot$ (13), where Y_e is the number of electrons per baryon in the matter. We will generalize the definition of M_{ch} to take into account thermal and other effects - it will be the mass for which the total energy is zero. The self-similar solutions discussed in §2 show that the mass of the collapsing inner core M_{ic} is somewhat larger than M_{ch},

$$M_{ic} = (1 + f)\, M_{ch} \quad . \tag{16}$$

The value of f is determined by the average value of $4/3 - \Gamma$ in the matter; for a realistic equation of state (5) f $\simeq$ 0.10.

After bounce, the central part of the hydrostatic inner core has $\rho > \rho_t$ and hence $\Gamma > 4/3$. The total core mass $M_{ic} > M_{ch}$ and the virial theorem shows that this leads to a certain amount of binding of the core. If M_{ic} were equal to M_{ch}, none of the core would have $\rho > \rho_t$, and the total energy would be zero.

We have integrated the equation of hydrostatic equilibrium for   both a realistic equation of state (5) and for others with much different values of K_{nuc} and ρ_t. The results for B as a function of M are displayed in figure 6. In this exercise neutrinos were neglected. The standard case a assumes $Y_e = 0.35$, entropy per baryon s/k=0, and realistic values for $K_{nuc} = 210$ MeV and $\rho_t = 2 \times 10^{14}$ g/cm^3. Other cases differ from this in one of these four parameters: b has $Y_e = 0.25$, c has s/k=2, d has $K_{nuc} = 2100$ MeV and e has $\rho_t = 2 \times 10^{15}$ g/cm^3. Two interesting results are observable: B varies linearly with M near B=0 (and M near M_{ch}) in all cases, and the slope of this graph depends only on Y_e and ρ_t - it is independent of s, Γ_b and K_{nuc} at least to lowest order. Because ρ_t is known much more accurately than Γ_b or K_{nuc}, we conclude that, given Y_e in the collapsing star,the value of dB/dM near M_{ch} is well known.

The immediate consequence of this result is that B is now determinable from the self-similar solutions:

$$B = dB/dM\big|_{M_{ch}} (M_{ic} - M_{ch}) = dB/dM\big|_{M_{ch}} M_{ch}\, f \quad . \tag{17}$$

Taking $Y_e = 0.35$, fig. 6 shows that $dB/dM|_{M_{ch}} \simeq 60 \times 10^{51}$ ergs/$M_\odot$ and §2 indicates f$\simeq$0.1, so that B$\simeq 5 \times 10^{51}$ergs, in excellent agreement with the results of dynamical calculations. We reiterate that this result does not depend sensitively on the properties of the nuclear part of the equation of state.

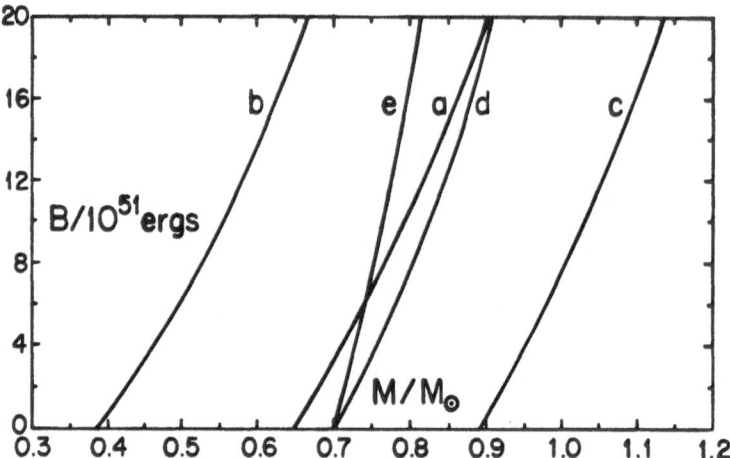

Figure 6. Binding energy B vs. mass M for several hydrostatic cases. See text for details.

It is straightforward to show that the relation (17) depends,
to lowest order, only on ρ_t and on properties of the equation of
state below ρ_t, and not on the equation of state above ρ_t. For
our purposes, a useful approximation to the equation of state is

$$\epsilon = \epsilon_\ell + \epsilon_b \; ; \quad p = p_\ell + p_b \qquad\qquad (18)$$

where ϵ is the energy density, ℓ refers to leptons and b to baryo-
ons. The lepton gas is relativistic so $p_\ell = \epsilon_\ell/3$. The baryons
contribute to the equation of state only when $\rho > \rho_t$; measuring
energies relative to Fe, we take

$$\epsilon_b = K_b x^2/2 \; ; \quad p_b = K_b x \qquad\qquad \rho > \rho_t$$
$$\epsilon_b = p_b = 0 \qquad\qquad\qquad\qquad\quad\; \rho < \rho_t \qquad (19)$$

in which $x = \rho/\rho_t - 1$. Equation (19) represents the lowest order
expansion around ρ_t of a realistic equation of state with compressibility $K_{nuc} = 9K_b/N_0\rho_t$, where N_0 is Avogadro's number. In as
much as the baryons are very incompressible, $x < 1$ even in the
center of the star. Thus, our model for a hot neutron star is a
nearly incompressible fluid "hard sphere" core surrounded by an
n=3 polytropic envelope.

The total energy of the star is the sum of potential and
internal energies

$$E = E_{grav} + E_{\ell} + E_b . \tag{20}$$

Application of the equation of hydrostatic equilibrium gives

$$E_{grav} = - 3 \int pdV = - E_{\ell} - 3 \int p_b dV , \tag{21}$$

or

$$E = E_b - 3 \int p_b dV = \int (\varepsilon_b - 3p_b)dV \approx -3 \int p_b dV \tag{22}$$

since x<<1; the binding energy of the star comes only from the hard sphere core. The variation of density, or x, with radius is obtained from the equation of hydrostatic equilibrium,

$$dp/dr = - GM(r)\rho(r)/r^2 \approx -4\pi G\rho_t^2 r/3 , \tag{23}$$

keeping only lowest order terms in x. We may write from the equation of state, to the same accuracy,

$$dp/dx = K_{\ell} + K_b \equiv K; \tag{24}$$

$$K_{\ell} = \rho_t dp_{\ell}/d\rho \big|_{\rho_t} = \mu_e Y_e \rho_t N_o/3 , \tag{25}$$

where μ_e is the electron chemical potential, if we can ignore neutrinos. Substituting (24) into (23) and integrating, we find

$$x(r) = 2\pi G p_t^2 (r_{hs}^2 - r^2)/3K \tag{26}$$

where r_{hs} is the radius of the hard sphere core in which $\rho > \rho_t$. The total energy is thus

$$E = -3\int p_b dV = -16\pi^2 K_b \rho_t^2 Gr_{core}^5 /15K . \tag{27}$$

Thus the total energy varies with the hard sphere core mass M_{hs} to the 5/3 power.

The next step is to determine how M_{hs} changes as mass is added to the star - connecting this relation with eq. (27) will determine the relation between the total energy and the mass excess $\Delta M = M_{ic} - M_{ch}$. Dimensional analysis argues that the magnitude of the mass difference ΔM_1 in the mass interior to r_{hs} between our simple neutron star (compressibility K) and a pure n=3 polytrope (compressibility K_{ℓ}), in which $\rho(r_{hs}) = \rho_t$ in both cases, is directly proportional to the difference ΔM in the total mass of the entire inner core between these two cases. Then, we can write $\Delta M_1 = \alpha\Delta M$. To lowest order in x we can calculate ΔM_1:

$$\Delta M_1 = \int_0^{r_{hs}} \rho_t (x_p(r) - x(r))dV \tag{28}$$

where x_p is the density deviation in the n=3 case. From eq. (26)

$$x_p(r) = x(r) \, K/K_\ell \ . \tag{29}$$

Therefore,

$$\Delta M_1 = 16\pi^2 G \, \rho_t^{\,3} \, r_{core}^{\,5} \, K_b/45KK_\ell \tag{30}$$

which also varies as $M_{hs}^{5/3}$. Thus

$$E = -3 \, \alpha \, K_\ell \, \Delta M/\rho_t = -B \tag{31}$$

which depends only on ρ_t and the leptonic equation of state, and
is linear in ΔM. The value of the constant α can only be found
from numerical integration, since the structure of an n=3 polytrope cannot be written in a simple analytic form. Numerical calculation shows that it is ~ 0.5, and does not depend on ρ_t or the
lepton equation of state. The slope

$$dB/dM|_{M_{ch}} = 3 \, \alpha \, K_\ell \, /\rho_t = \alpha \, \mu_e \, Y_e \, N_o \tag{32}$$

if we ignore neutrinos, and hence is proportional to $\rho_t^{1/3} Y_e^{4/3}$, as
observed. Finally, if $4/3-\Gamma$, and hence f, does not depend too
much on Y_e, eqs. (25) and (31) say that

$$B = \alpha \, N_o \, \mu_e \, Y_e f M_{ch} \simeq 60f(\rho_t/2\text{x}10^{14}\text{g/cm}^3)^{1/3}(Y_e/.35)^{10/3}$$
$$\text{x}10^{51} \text{ ergs} \tag{33}$$

which illustrates that the amount of electron capture during the
collapse, which determines Y_e, is likely to be the most important
consideration in determining the shock energy.

4. HYDROSTATIC SHOCKS

    Once the shock has been generated and is outwardly propaga-
ting, the following schematic picture emerges. The unshocked
core, with an entropy per baryon $s/k \approx 1$ and an outer boundary at
$R_{ic} \simeq 15$km, $M_{ic} \simeq 0.8 \, M_\odot$ (f $\simeq 0.1$), and $\rho \simeq 10^{13}$g/cm^3, is surrounded by a shocked region through which the shock has propagated.
The entropy in this region is about $6 < s_2/k < 12$. The location
of the shock is at R_s. Beyond R_s is the outer core which is still
infalling.

    Except for neutrino radiation and pdV work done during the
infall, no energy has passed through any radius beyond R_s. The
initial energy of the star, before collapse, integrated out to R_s
is small, probably less than $0.5\text{x}10^{51}$ergs. Therefore, the total
integrated energy from r=0 to R_s is small, just being the sum of
$-E_\nu$, where E_ν is the total of the neutrino losses through this
radius, and the total pdV work done at this radius up until the

shock reaches R_s. The neutrino losses are composed of two parts
- a burst when the shock reaches the neutrino photosphere (see
Bethe and Mazurek et al., both this volume) of size $\sim 1-2\times 10^{51}$
ergs, and the long term ($\frac{1}{2}$s) loss of neutrinos from the unshocked
core ($\sim 0.1-0.2\times 10^{51}$ergs/ms after bounce). The total neutrino
loss E_ν for R_s in the range 20 - 200 km can be expected to be
$2-4\times 10^{51}$ergs. To calculate the total pdV work done, we use the
fact that p varies as r^{-4} (see §2), so that $\int pdV \propto r^{-1}$. Assuming
that the equation of state is that for degenerate electrons with
$Y_e=0.35$, at a radius of 50km we find this work to be about 0.75x
10^{51}ergs. It is, of course, smaller at larger radii. Note that
this energy is small, but of opposite sign to E_ν.

Looking globally at the total energy up to the shock front,
integrating the Euler equation of motion, we find

$$4\pi R_s^3 p = E_{grav}(R_s)+E_{int}(R_s)+\int(3p-\varepsilon)dV-\int r\ddot{r}dM , \qquad (34)$$

where $E_{grav}(R_s)$ and $E_{int}(R_s)$ are the integrated gravitational and
internal energies from the center to R_s. Note that this includes
both the unshocked inner core, as well as the shocked outer core.

A conservative model of shock propagation can be realized if
we neglect the non-hydrostatic term, the last term in eq.(34).
This hydrostatic model will underestimate the pressure at the head
of the shock, since $\ddot{r}$ is negative, and, hence, the strength of the
shock. In any event, dynamical calculations show that the last
term is small compared to the other terms except in a narrow region behind the shock front itself. Moreover, it becomes increasingly unimportant as the shock progresses to larger radii.

In lieu of a self-consistent model for the temperature profile behind the shock, we simply assume that s_2 is a constant from
R_{ic} to R_s. In addition, in this qualitative model, we take $Y_e=.35$
to be constant everywhere in the star. We need, then, two outer
boundary conditions at R_s to determine s_2: the total energy and
the pressure p_2 must be specified. For the energy, as we have
discussed, values in the range 2 to 4×10^{51}ergs are taken. p_2 will
be assumed to be some multiple z of the ram pressure P_r of the
infalling material: We thus take p_2 to be

$$p_2 = zP_r = z\,\rho_1\,u_1^2 \qquad (35)$$

where ρ_1 and u_1 are given in eq.(3). Eq.(35) is tantamount to
requiring that the Hugoniot condition

$$(p_2-p_1)\,(1/\rho_1 - 1/\rho_2) = (u_2 - u_1)^2 \qquad (36)$$

be satisfied across the shock front. Neglecting $p_1\ll p_2$ and taking $u_2 = -g\,u_1$, we find

$$p_2 = \rho_1 u_1^2 (1 + g)^2 / (1 - \rho_1/\rho_2). \qquad (37)$$

As long as the density behind the shock $\rho_2 < 10^{12} g/cm^3$, $\rho_2/\rho_1 \gtrsim 6$, increasing with increasing R_s (15), and, as long as $u_2 > 0$, $g \gtrsim 0$ (if the shock is stationary, $g < 0$).

Choosing a constant entropy profile in the shocked region and ignoring the kinetic energy behind the shock implies that the second Hugoniot relation

$$E_2 - E_1 = \tfrac{1}{2}(p_1 + p_2)(1/\rho_1 - 1/\rho_2) \simeq \tfrac{1}{2}u_1^2(1+g)^2 \qquad (38)$$

in general cannot be exactly satisfied; however, the actual solutions are reasonable consistent with eq. (38).

The final constraint on these shock models is the mass of the unshocked core. The larger this mass, the more energy is available to the shock and, presumably, the easier it is for the shock to successfully propagate. From §2, a realistic equation of state will give $f \simeq 0.1$ and, for $Y_e \simeq 0.35$ and $s/k \simeq 1.$, the unshocked core mass will be $0.8\ M_\odot$, with $B \simeq 5 \times 10^{51}$ ergs.

Figure 7 shows the results of numerically integrating the equation of hydrostatic equilibrium for a realistic equation of state. Plotted as a function of R_s is the entropy s_2 required

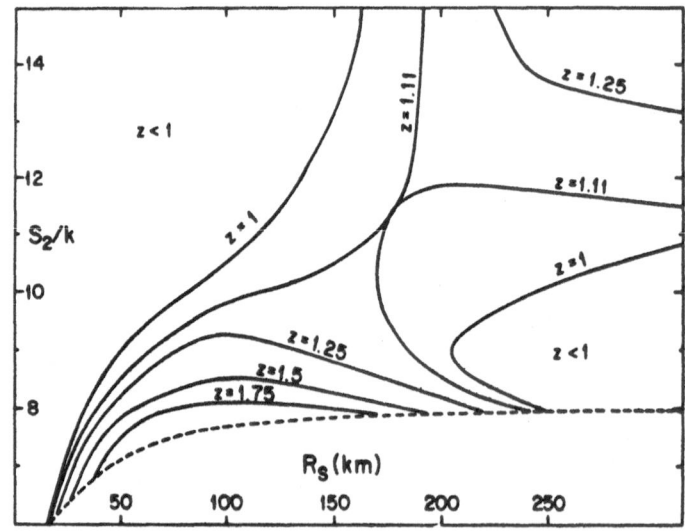

Figure 7. Hydrostatic shock models for $M_{ic} = .8\ M_\odot$, $Y_e = .35$ and $E(R_s) = -3 \times 10^{51}$ ergs. See text for details.

behind the shock to satisfy the pressure and energy boundary conditions at R_s. The total energy at the shock front $E(R_s)$ was set to -3×10^{51} ergs, and results for various values of z are shown as contours. The dashed line shows the value of s_2 below which no solution to the energy boundary condition is possible.

The qualitative result that increasing the pressure behind the shock, for a given energy and R_s, requires less entropy s_2 can be understood on the basis that lower entropies will require higher densities, and enclosed masses, to achieve the same total energy at the same R_s. A higher density implies a greater pressure behind the shock, hence a higher value for z. This argument does not necessarily hold for radii in excess of 150 km, when the situation is more complicated. The rapid increase in z for large entropies for radii greater than 200 km indicates that successful shocks rapidly accelerate in this region.

The existence of a branch cut for a z value of 1.11 divides the diagram into two parts that we consider qualitatively as being one for "successful" shocks, the other for "unsuccessful" shocks. Shock propagation is to be viewed as a progression of points from small R_s to greater R_s, with z changing constantly in response to changing ρ_2/ρ_1 ratios and shock speeds (g values). Shocks seeing high ram pressures, but having low (s/k $\lesssim$ 10) entropies, appear to die, being unable to propagate beyond 150-200 km. Lower ram pressures and higher entropies imply successful shock propagation. Because z cannot be self-consistently determined as a function of R_s, it is not possible for us to determine if a shock will be successful. However, the qualitative result that entropies of order 10/k are needed behind the shock to propagate it to large radii (> 200 km) is interesting.

Changing the energy boundary condition does not qualitatively change the topology of fig. 7, but the critical z value is much larger for smaller (negative) energy. For example, for $E(R_s)$ $=-2 \times 10^{51}$ ergs, the branch cut occurs for z = 1.55. Changing the mass point at which the entropy jumps to s_2 is completely equivalent to changing the total energy. More unshocked matter in the inner core results in more binding, and, hence, more positive **energy in the shocked outer core.** The general effect of increasing the energy in the shocked outer core is to increase, for a given R_s and s_2, the value of z, or, for a given R_s and z, the value of s_2.

5. CONCLUSIONS

The approximations used in this paper have yielded a number of qualitative conclusions concerning the supernova process:
1. The original Fe core breaks up into an inner core which col-

lapses homologously and an outer core which is infalling supersonically at about half the escape velocity and a Mach number $\sim$
2 - 3.
2. The density of the outer core varies as $\rho \propto r^{-2/(2-\gamma)}$, where
γ is an effective adiabatic index. The constant of proportionality depends crucially on the difference $4/3-\gamma$, in the sense that
the closer γ is to $4/3$, the more tenuous the outer core becomes.
3. The mass of the inner core is about 10% higher than the Chandrasekhar mass for the same Y_e.
4. The energy transferred by the inner core to the shocked outer
core at rebound is independent of the details of the nuclear equation of state, and is given by the binding energy of the hydro-
static inner core: $B \simeq 5\text{x}10^{51}(\rho_t/2\text{x}10^{14}\text{g/cm}^3)^{1/3}(Y_e/.35)^{10/3}\text{ergs}$.
5. The propagation of the shock depends on a fine balance between
the total energy behind the shock, and the ram pressure of the
infalling outer core ahead of the shock. There is a critical ram
pressure for each energy, which can be estimated by a "hydrostatic"
model for shock propagation. Successful shock propagation appears
to require entropies behind the shock $\gtrsim 10k$.

This research was supported in part by USDOE grants DE-AC02-
76ER13001 and DE-AC02-80ER10719 at the State University of New York.

REFERENCES

1. Bethe, H.A., Brown, G.E., Applegate, J. and Lattimer, J.M.:
     1979, Nucl. Phys. A324, 487.
2. Goldreich, P. and Weber, S.V.: 1980, Astrophys. J. 238, 991.
3. Brown, G.E., Bethe, H.A. and Baym, G.: 1981, Nucl. Phys. A,
     in press.
4. Bethe, H.A., Applegate, J.H. and Brown, G.E.: 1980, Astrophys.
     J. 241, 343.
5. Lamb, D.Q., Lattimer, J.M., Pethick, C.J. and Ravenhall, D.G.:
     1978, Phys. Rev. Lett. 41, 1623; Lattimer, J.M.:1981, Ann.
     Rev. Nucl. Part. Sci. 31, 337.
6. Shu, F.H.: 1977, Astrophys. J. 214, 488.
7. Larson, R.B.: 1969, MNRAS 145, 271.
8. Penston, M.V.: 1969, MNRAS 144, 425.
9. Sedov, L.: 1959, Similarity and Dimensional Methods (London:
     Cleaver-Hume Press).
10. Yahil, A.: 1982, preprint.
11. Bondi, H.: 1952, MNRAS 112, 195.
12. Parker, E.N.: 1963, Interplanetary Dynamical Processes (New
     York: Wiley).
13. Chandrasekhar, S.: 1939, An Introduction to the Study of Stel-
     lar Structure (Chicago: University of Chicago Press).
14. Cheng, A.F.:1978, Astrophys. J. 221, 320.
15. Lattimer, J.M. and Mazurek, T.J.: 1981, Astrophys. J. 246, 995.

SHOCK STAGNATION AND NEUTRINO LOSSES IN STELLAR COLLAPSE

T.J. Mazurek and J. Cooperstein

Department of Physics
State University of New York
Stony Brook, New York 11794

and

S. Kahana

Brookhaven National Laboratories
Upton, New York 11973

1. INTRODUCTION

Stellar collapse results in the formation of a hydrostatic core at about nuclear matter densities and a shock that propagates into the infalling mantle. Hydrodynamic calculations of different research groups are now converging on the result that the shock falters before ejecting the mantle and stagnates into a stationary accretion shock (cf. W. Hillebrandt's review, this volume). A prompt supernova explosion by shock ejection is thereby prevented. The primary factor behind the failure of the shock appears to be the energy absorbed by nuclear dissociation. Additional large energy losses in neutrino emission firm the conclusion that the shock cannot eject matter.

In this communication we present the results of our detailed hydrodynamic study of stellar collapse. We focus particularly on two of our calculations, one including neutrino losses and the other neglecting them. Shock behaviors in both calculations are similar. The energy loss in neutrinos is detrimental, but it is not in itself responsible for stopping the shock. However, the energy loss in neutrinos is appreciable in the hydrodynamic calculations, and needs to be examined critically. To analyze this loss semianalytically, we adopt the results of H.A. Bethe's presentation (this volume) and show that its magnitude depends criti-

M. J. Rees and R. J. Stoneham (eds.), Supernovae: A Survey of Current Research, 71–77.

cally on the speed of the shock in the transparent regions. We
show that the semianalytic estimates of energy losses in neutrinos
are entirely consistent with those of the hydrodynamic calculations
when upper limits on the shock speed given by the latter are included.

2. HYDRODYNAMIC RESULTS

First, we briefly summarize the input physics in our hydrodynamic study. The neutrino transport is treated via a diffusion
approximation (1) with suitable flux-limiting procedures to ensure
free-streaming in transparent regions. The neutrinos are assumed
to have a Fermi-Dirac distribution with the temperatures of the
matter in regions where they interact appreciably. Transport by
only the electron-type neutrinos is included. The opacity includes
contributions due to coherent scattering by nuclei, elastic scattering by nucleons, and absorption on neutrons. The equation of
state (2) includes photons, electrons, neutrinos, nucleons, alpha
particles, and one heavy nucleus. The properties of the heavy
nucleus are fit so as to reproduce results that are based on a
network of heavy nuclei. The density at which nuclei disappear
is determined approximately by demanding pressure and entropy continuity across the phase transition. Electron capture is assumed
to take place on free protons.

The hydrodynamic calculations follow the collapse of only the
central core composed of iron-peak elements. The results we present start with the core of 1.6 $M_\odot$ that forms (3) in a star with
total mass of 15 $M_\odot$. The collapse starts when the central density
is around 10^9 g-cm^{-3}. The neutrinos become trapped at a density
of $\sim 10^{11}$ g-cm^{-3}, and the collapse halts when the central density
is $\sim 4 \times 10^{14}$ g-cm^{-3}. The shock forms at the surface of the cool
inner core of ~ 0.8 $M_\odot$ at a density of $\sim 10^{13}$ g-cm^{-3}.

The subsequent evolution of the shock was followed for two
cases. In one case the neutrino transport scheme discussed above
was employed. Large neutrino losses of around 2×10^{51} ergs occurred
in the first 2 ms as the shock propagated through the transparent
regions. In the second case the neutrino transport was turned off.
Electron capture was allowed to proceed toward beta equilibrium,
but the total lepton number per baryon X_ℓ at each mass point was
fixed to its value at core bounce. In both cases the shock eventually stalled and became a stationary accretion shock.

Figure 1 shows the radial behavior with time at several Lagrangian mass points. The behavior with neutrino transport is
shown by the solid lines. Dashed lines show the results of the
calculations without neutrino transport wherever the trajectories
deviate from those with neutrino transport. In regions when the

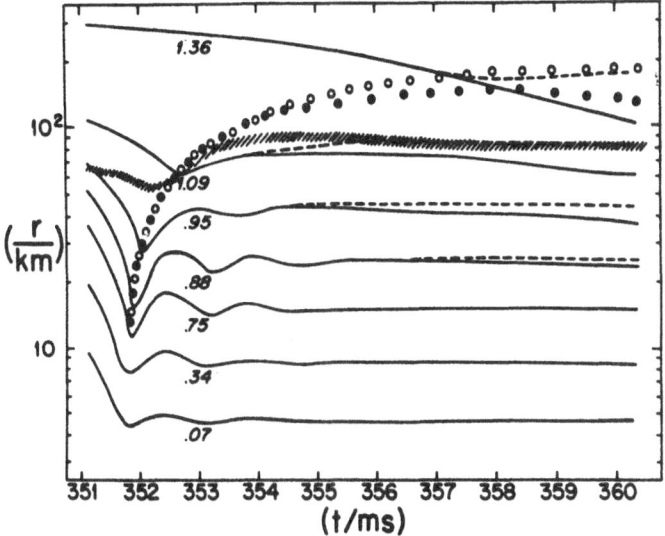

Figure 1. Radial behavior of the collapses with time at
Lagrangian mass points labeled by $M(r)/M_\odot$.

dashed and solid lines merge or where no dashed lines are shown
the evolution without neutrino transport was identical to that
which included it. It is seen from the figure that the behavior
of the innermost homologous core is unaffected by the neutrino
losses. The shocked mantle, however, shows progressively greater
deviations between the two calculations with increasing radius.
The transparent regions outside the neutrino sphere at optical
depth unity (shown by the hatched band) shows the greatest difference in post-shock behavior. The shocked mantle at large optical
depths is affected only slightly. With neutrino losses, the outer
zones reach a maximum radius during expansion and then recollapse
to form a denser shocked mantle on the homologous core. Without
the neutrino losses there is little apparent tendency to recollapse.
Instead the zones expand to a new equilibrium after the shock with
no oscillations.

    The propagation of the shock for the two cases is given by
the circles in the figure. Filled circles denote the resulting
shock with neutrino transport, opened ones denote that without
energy loss by neutrinos. It is seen that during the first 2 ms,
the shocks propagate identically in the two calculations. In particular, the shock breaks out through the neutrinosphere at 353 ms
and propagates unhindered by the neutrino losses for around 1 ms
through a distance of around 30 km. This shows that the damping of
the shock by neutrino losses does not occur precipitously when the
shock enters the neutrino sphere. Instead it occurs gradually as
the pressure defect increases behind the shock and rarefaction waves
accumulate at the shock front. At 354 ms before the effects of neu-

trino losses become noticeable, the shock has already slowed to 2×10^9 cm/s (see Table 1 below). Beyond this point, even without neutrino losses the shock rapidly slows down, becoming essentially stationary 8 ms after core bounce. With neutrino losses the shock reaches a maximum radius of about 160 km at around 6 ms after bounce and then falls downward as the pressure support is removed by neutrino energy losses. Overall, there is no dramatic difference between the two shock behaviors, and the radii at which the two shocks stall differ by less than 50 km. Hence neutrino energy losses do not play the dominant role in damping the shock in our hydrodynamic calculations. However, significant dissociation into nucleons occurs as long as the shock has an appreciable outward velocity. Hence the dominant process that weakens the shock is the energy loss of dissociation.

3. SHOCK CHARACTERISTICS AND THE LEPTON LOSSES DUE TO NEUTRINOS

Our calculations show a rapid drop in X_ℓ in regions at and around the neutrino.sphere at optical depth unity. There are large lepton losses in the transparent region between the neutrinosphere and the shock. As discussed above, this lepton loss is not the fatal foe since the shock also dies in the adiabatic calculation; however, the shock damping due to neutrino losses is significant. Thus the drop in X_ℓ to below .2 after a few ms shows that our calculations require a critical examination. We examine the drop in post-shock X_ℓ using a semi-analytic approach that incorporates the features of neutrino transport discussed by Bethe (this volume).

The mean lepton fraction in the region between the neutrinosphere and the shock is determined by the input of neutrinos from the neutrinosphere, the influx of fresh leptons across the shock, and the loss of leptons due to black-body radiation at the shock front. Assume that the emission from the neutrinosphere is that of a black-body, with the flux given by the particle density times $c/4$. According to Bethe the neutrino particle density at the neutrinosphere decays exponentially with a timescale of about 2 ms when the neutrino energies are around 20 MeV. In our calculations the neutrino energies are lower ($\lesssim 15$ MeV), which reduces the decay time to about 1 ms. The neutrino fraction X_ν in the neutrinosphere at time t after shock passage is thus

$$X_\nu^e (t) = X_\nu^e(0) \, e^{-t/\tau} , \tag{1}$$

where the superscript e denotes the neutrinosphere and $\tau \simeq 1$ ms. The input of neutrinos from the neutrinosphere is then

$$\dot{N}^e = 4 \, \pi \, r_e^2 n_B^e X_\nu^e(t) \, c/4 = \pi \, c \, r_e^2 n_B^e X_\nu^e(0) \, e^{-t/\tau} , \tag{2}$$

where n_B is the baryon density, r_e denotes the radius of the neu-

trinosphere and all other notation is standard. The influx of
fresh leptons through the shock front is

$$\dot{N}^s_+ = 4 \pi r_s^2 n_B^o X_\nu^o (D - u_o) ,$$ (3)

where the sub- and superscript o denotes preshock conditions, D is
the shock speed and u is the matter velocity. The lepton loss at
the shock front, using the sub- and superscript s to denote post-
shock conditions, is

$$\dot{N}^s_- = 4\pi r_s^2 n_B^s X_\nu^s c/4 = \pi c r_s^2 X_\nu^s n_B^o (D-u_o)/(D-u_s) .$$ (4)

To proceed, make the further simplifying assumptions that the
preshock baryon density varies as r^{-3}, and the X_ℓ^o, $(1-u_o/D)$,
$(1-u_s/D)$ and D are constants. To the extent that D decreases at
larger radii, our estimate of the mean lepton fraction will be too
large. We also assume, following Bethe, that the post shock neutrino fraction (X_ν^s) is constant. One can then use the relation
$dr_s = D \, dt$ to obtain the total neutrino losses and gains between
the neutrinosphere and the shock, and the total mass in this space:

$$\Delta N^e = \pi c r_e^2 X_\nu^e(0) n_B^e \tau [1 - \exp{-(r_s-r_e)/D\tau}] ,$$ (5)

$$\Delta N^s_+ = 4 \pi r_e^3 n_B^o X_\ell^o (1 - u_o/D) \ln(r_s/r_e) ,$$ (6)

$$\Delta N^s_- = \pi r_e^3 n_B^o X_\nu^o (c/D) \ln(r_s/r_e) (D-u_o)/(D-u_s) ,$$ (7)

$$\Delta M = 4 \pi r_e^3 m_n n_B^o (1 - u_o/D) \ln(r_s/r_e) .$$ (8)

ΔM is the mass between the shock and the neutrinosphere and m_n is
the nucleon mass. With eqs. (5 - 8), one finds for the mean $\overline{X}_\ell$
in this region assuming that n_B^s/n_B^o is constant:

$$\overline{X}_\ell = X_\ell^o - X_\nu^s \frac{n_B^s}{n_B^o} \frac{(c/4D)}{(1-u_o/D)} \{1 - \frac{X_\nu^e}{X_\nu^s} \frac{[1-\exp{-(r_s-r_e)/D\tau}]}{(r_e/D\tau)\ln(r_s/r_e)} \}$$ (9)

Equation (9) shows that $\overline{X}_\ell$ depends critically on the parameter
$r_e/D\tau$. For small $r_e/D\tau$ the input from the neutrinosphere can easily compensate for the losses at the shock front until the shock
radius becomes extremely large. For large $r_e/D\tau$, the losses dominate the input from the neutrinosphere and the mean lepton fraction drops rapidly. In the transparent region $\overline{X}_\ell$ thus depends
sensitively on shock parameters that can be determined only in
detailed hydrodynamic calculations.

To apply equation (9) to our hydrodynamic results we consider
the shock parameters as given in our adiabatic calculation. Due
to the use of the pseudoviscosity, the shock is spread out over a
few zones. To determine the characteristics of the shock from the

numerical output we use the following procedure. The shock position is taken to be the point at which the velocity is zero within the shock transition region. The preshock velocity is taken
to be the maximum infall velocity before the shock, while the
postshock velocity is the maximum expansion velocity after the
shock. The preshock conditions are determined by their values
before the shock and extrapolation to the shock position assuming
$n_B \propto r^{-3}$ and $p \propto r^{-4}$. Finally, the density contrast across the
shock is roughly constant (4) with $n_B{}^s/n_B{}^o \simeq 6$. The usual shock
relations then are used to determine D and p_s/p_o. The results are
shown in Table 1 (in cgs units).

$10^3 t$	$10^{-7} r_s$	$10^{-10} \rho_o$	$10^{-18} \dfrac{p_o}{\rho_o}$	$10^{-9} u_o$	$10^{-9} u_s$	$10^{-9} D$	$\dfrac{p_s}{p_o}$	$\overline{X}_\ell$
0	0.80	4.31	1.82	-3.3	2.3	3.4	22	.42
0.86	1.08	1.73	1.41	-2.9	1.5	2.4	17	.24
1.18	1.15	1.46	1.37	-2.8	1.2	2.0	15	.20
1.73	1.23	1.13	1.29	-2.5	1.0	1.7	13	.13
3.20	1.44	0.64	1.16	-2.4	0.0	0.5	7	-
7.58	1.83	0.23	0.91	-2.0	-0.3	0.0	6	-

Table 1. Shock Characteristics in Adiabatic Model

Table 1 shows that the average shock velocity during the
first 3 ms after the shock emerges from the neutrinosphere is $D \lesssim$
3×10^9 cm/s. The radius of the neutrinosphere is ~ 80 km. Thus
with $\tau \simeq 1$ ms, as discussed above, one has $r_e/D\tau \gtrsim 3$. Table 1 also
shows that $1-u_o/D \simeq 2$ during this time. Finally assuming $X_\nu{}^s =$
$X_\nu{}^e(0)$ and setting $D = 3 \times 10^9$ cm/s, one can use equation (9) to
write the mean lepton fraction as

$$\overline{X}_\ell \sim X_\ell{}^o - 7.5 X_\nu{}^s \{ 1 - [1-\exp-3(y-1)]/3 \ln y \} \qquad (10)$$

where $y = 1 + Dt/r_e$. Using $X_\ell{}^o = .42$ and $X_\nu{}^s = .09$ (cf. Bethe,
this volume) one obtains the values of $\overline{X}_\ell$ shown in the last column of Table 1. These agree to within 10% of the results given
by hydrodynamic calculations with neutrino transport. Thus the
large lepton losses that result in the hydrodynamic calculations
can be readily understood within the context of Bethe's discussion. Indeed, the large lepton loss is unavoidable for shock
speeds $\lesssim 3 \times 10^9$ cm/s in the transparent region. The adiabatic
calculation, which gives maximum shock speeds, results in such low
velocities. Hence a large lepton loss must occur behind the shock.

4. SUMMARY

To summarize, our results show that neutrino energy losses

have significant effects on the propagation of shocks from stellar collapse. However, they are not critical in the stagnation of the shock. Our hydrodynamical calculations without neutrino losses also lead to a stagnant accretion shock. Thus is appears that nuclear dissociation critically determines the outcome. The neutrino losses are large, however, and they modify the shock behavior significantly. With neutrino losses the shock stops advancing at a radius that is about 30% smaller than that of the adiabatic calculation.

REFERENCES

1. Mazurek, T.J.: 1975, Astrophys. Space Sci. 35, 117.
2. Mazurek, T.J., Lattimer, J.M. and Brown, G.E.: 1979, Astrophys. J. 229, 713.
3. Weaver, T., Zimmerman, G.B. and Woosley, S.E.: 1978, Astrophys. J. 225, 1021.
4. Lattimer, J.M. and Mazurek, T.J.: 1981, Astrophys. J. 246, 995.

This research was supported in part by USDOE grants DE-ACO2-76ER13001 and DE-ACO2-80ER10719 at the State University of New York.

THEORETICAL MODELS FOR SUPERNOVAE

S.E. Woosley[1,2] and Thomas A. Weaver[2]

ABSTRACT

The results of recent numerical simulations of supernova explosions are presented and a variety of topics discussed. Particular emphasis is given to i) the nucleosynthesis expected from intermediate mass (10 $M_\odot$ < M < 100 $M_\odot$) Type II supernovae and detonating white dwarf models for Type I supernovae, ii) a realistic estimate of the γ-line fluxes expected from this nucleosynthesis, iii) the continued evolution, in one and two dimensions, of intermediate mass stars wherein iron core collapse does <u>not</u> lead to a strong, mass-ejecting shock wave, and iv) the evolution and explosion of very massive stars (M > 100 $M_\odot$) of both Population I and III. In one dimension, nuclear burning following a 'failed' core bounce does not appear likely to lead to a supernova explosion although, in two dimensions, a <u>combination</u> of rotation and nuclear burning may do so. Near solar proportions of elements from neon to calcium and very brilliant optical displays may be created by 'hypernovae', the explosions of stars in the mass range 100 $M_\odot$ to 300 $M_\odot$. Above ~ 300 $M_\odot$ a black hole is created by stellar collapse following carbon ignition. Still more massive stars may be copious producers of ^{4}He and ^{14}N prior to their collapse on the pair instability.

1. Lick Observatory, Board of Studies in Astronomy and Astrophysics, University of California, Santa Cruz.
2. Lawrence Livermore National Laboratory, University of California.

M. J. Rees and R. J. Stoneham (eds.), Supernovae: A Survey of Current Research, 79–122.

1. NUCLEOSYNTHESIS IN SUPERNOVAE

The study of heavy element production in supernovae has a long and distinguished history, extending back to at least 1946 when Hoyle first suggested the synthesis of a solar set of iron isotopes by the 'e-process'. Interestingly, the proposed ejection mechanism for this nucleosynthesis was rotation, a concept to which supernova modellers are now returning (see Section 3). Since that time many people, most of whom are in attendance at this meeting, have contributed to our understanding of nuclear processes in exploding stars to the point where at least the qualitative aspects of the origin of the intermediate mass elements (carbon to nickel) are now understood. Observers too, in recent years, have found compelling evidence to support the idea that supernovae do produce heavy elements, in fact as well as in theoretical model. We refer here especially to the recent work in X-ray spectroscopy (discussed elsewhere in this volume) and the work of Kirshner and Oke (1975) and Axelrod (1980) showing evidence for freshly synthesized ^{56}Fe, and even radioactive ^{56}Co, in the spectrum of supernova 1972e.

For purposes of this workshop we shall not dwell on the historical development of nucleosynthesis theory nor on the observations, but shall instead concentrate on our own recent studies and, even so, will only summarize what has been developed in greater detail elsewhere (Weaver, Zimmerman and Woosley 1978: Weaver and Woosley 1980: Weaver, Axelrod and Woosley 1980: Woosley and Weaver 1980, 1981a,b: Woosley, Weaver and Taam 1980). It now appears likely that the nucleosynthetic products of Type I and Type II supernovae will differ markedly, a distinction that will undoubtedly have important implications for galactic chemical evolution. In general, we subscribe to the belief that Type I supernovae occur in compact, lower mass objects, such as detonating white dwarfs, and are more likely to produce heavier elements, especially iron, than are the Type II supernovae which occur in more massive stars and are held responsible for making lighter elements such as oxygen, neon, magnesium, (perhaps) carbon, etc. This hypothesis accounts nicely for the early enhancement of [O/Fe] in our Galaxy (cf. Clegg, Lambert and Tomkin 1981) providing that, at early times, there existed an enhancement of massive star formation and death.

If all elements heavier than helium can be lumped together in a single group that astronomers like to call 'metals', then the metallicity ejected by a star of mass $M > 13$ $M_\odot$ is well fit by the expression $Z = (0.5 - 6.3 \ M_\odot/M)$. Multiplying this function by the mass of the exploding star gives the heavy element production in a star of mass M. If one combines this expression with an estimate of the initial mass function (e.g. Miller and Scalo 1979), which may have been time varying in the

early galaxy, a value of 20 - 30 $M_\odot$ is obtained for the
'representative' star such that equal amounts of heavy elements
were produced in stars lighter and heavier than this mass. Of
course this is a gross oversimplification of the true situation,
since stars of differing mass will produce variable proportions
of different heavy elements. Still, it justifies the calculation,
at first pass, of nucleosynthesis in a star of $\sim$ 25 $M_\odot$.

The isotopic nucleosynthesis from a parametrized 25 $M_\odot$ Type
II supernova (Weaver and Woosley 1980: Woosley and Weaver 1981b)
is shown in Figure 1. The overproduction factor is the ratio of
the mass fraction of a given species in the (homogenized) ejecta
to its mass fraction in the sun (Cameron 1973). Thus a solar
abundance of oxygen could have been created in the Galaxy if 1
gram out of every 14 experienced conditions similar to those
characterizing the evolution of a 25 $M_\odot$ star. Dashed lines give
a range of a factor of 2 for consistent production of other
isotopes in solar proportions. As the figure shows, many abundant
isotopes lighter than sulphur can be 'properly' created in this
fashion and very likely originate in intermediate mass supernovae
like this one. The production of iron group species in this
explosion is highly sensitive to uncertain parametrization,
especially the choice of the 'mass cut', and the value indicated
in Figure 1 may be an overestimate of the actual Type II
supernova contribution. Elements lighter than silicon are made
almost entirely by pre-explosive nuclear burning and are merely
pushed off of the star when it explodes. Elements heavier than
silicon are produced by explosive neon, oxygen, and silicon
burning. Explosive carbon burning does not occur to any
appreciable extent.

There appears to be a relative deficiency of nucleosynthesis
for the elements between sulphur and the iron group. Although
these elements are produced with mutual proportions closely
resembling those in the sun, too little mass experiences the
requisite temperature range for explosive oxygen burning (3 to 4
x 10^9 K) and the absolute yield of its products is small compared
to the results of other processes. The small amount of mass
heated to this range of temperature by shock wave passage is, in
turn, a consequence of the steep density gradient just outside
the silicon shell in the pre-explosive 25 $M_\odot$ star (Weaver,
Zimmerman and Woosley 1978: Woosley and Weaver 1981b). We have
found that the pre-explosive structures of 35, 50 and 100 $M_\odot$
stars do not exhibit such a sharp decline in density around the
core and, although the calculations of core bounce and explosion
remain to be done in such stars, it seems likely that they may
compensate for the deficient production in the 31 $<$ A $<$ 56 mass
range shown in Figure 1. Alternatively one may wish to consider
producing most of these species in an early generation of very

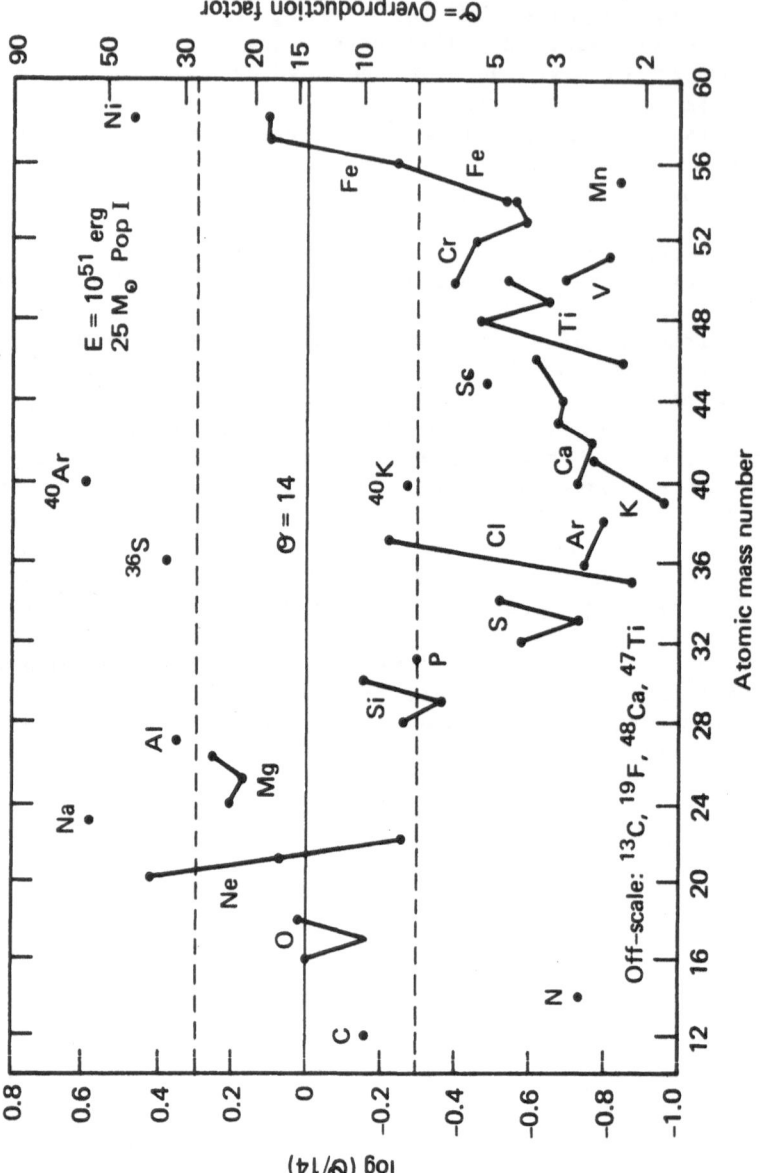

Figure 1.   Final abundances from a parameterized 25 $M_\odot$, Pop.I supernova explosion having 10^{51} erg of kinetic energy, are shown compared to solar values (Cameron 1973).   Dashed lines indicate a range of a factor of 2 about a consistent average overproduction factor of 15.   Nuclei below sulphur appear to be produced in reasonable proportions to account for their present Galactic abundances, but the relative deficiency of element production in the range S to Mn is indicative of substantial nucleosynthesis in stars more massive than 25 $M_\odot$.

massive stars (M > 100 $M_\odot$, see Section 4) or by an altogether different explosion mechanism (Section 3).

For lighter elements whose nucleosynthesis is not critically dependent upon explosion properties, the final abundance may be approximated by its value at the time of silicon ignition. We have evolved a variety of stellar models to the silicon ignition point. The elemental production of carbon, oxygen, and neon is summarized in Table 1 and the oxygen isotropic ratios are given in Table 2. Close agreement with solar values is encouraging and suggests that all these species have been produced in such massive stellar explosions. The slight overproduction of neon may be attributed to an uncertain solar abundance, uncertain nuclear cross-sections, especially for ^{20}Ne $(\alpha,\gamma)^{24}$Mg, or both.

Table 1

C, Ne, and O production in massive stars

Mass Pop	Solar Abundance	15$M_\odot$ I	20$M_\odot$ I	25$M_\odot$ I	35$M_\odot$ I	25$M_\odot$ II
^{12}C	0.41	0.38	0.42	0.27	0.39	0.45
^{16}O	1	1	1	1	1	1
^{20}Ne	0.18	1.0	0.86	0.58	0.51	0.52
OV(^{16}O)		3.9	9.0	14	21	12

Table 2

Oxygen isotope production in massive stars

Mass Pop	Solar	15$M_\odot$ I	25$M_\odot$ I	35$M_\odot$ I	25$M_\odot$ II
^{16}O	8.3(-3)	0.032	0.10	0.17	0.10
$10^4(^{17}$O$/^{16}$O)	4.0	9.2	2.5	1.1	0.019
$10^3(^{18}$O$/^{16}$O)	2.3	3.1	2.1	0.61	0.028

As Arnett (1971, 1973) pointed out some time ago, the neutron
excess, $\eta = (N-Z)/(N+Z)$, is a critical parameter for stellar
nucleosynthesis. During helium burning, this parameter is
strictly determined by the initial metallicity of the star, which
is transformed into nuclei with a net neutron excess by the
reaction chain $^{14}N(\alpha,\gamma)^{18}F(e^+,\nu)^{18}O$. During carbon burning and
subsequent stages, memory of the initial metallicity is
diminished as a complex set of weak interactions increase η. The
distribution of neutron excess with mass for the inner 12 $M_\odot$ of
two 25 $M_\odot$ completely evolved stars is shown in Figure 2. One
star, labelled Model I, had an initially solar abundance set, the
other, labelled Model II, had an initial metallicity 1% as large.
Memory of the distinct metallicities is clearly retained in the
helium shell where the values of η differ by a factor of 100. In
the carbon convective shell this difference has annealed and
amounts to only a factor of 8. Still deeper within the two stars,
the compositions and neutron excesses are essentially identical.
This distribution of neutron excess is reflected in the final
nucelosynthesis from the two stellar explosions (Table 3).
Isotopes produced by hydrogen and helium burning show large
variations, those from carbon and neon burning less so, and the
results of oxygen burning are almost indistinguishable. Recent
observations by Tomkin and Lambert (1980) of magnesium isotopic
ratios in Gmb 1830, a metal deficient sub-dwarf star with an iron
abundance 1/20 that of the sun, are in good accord with Table 3
($^{25,26}Mg$ are down by a factor of 4.5 compared to ^{24}Mg). Also in
good agreement are the elemental abundance determinations by
Peterson (1981) for Na, Al, and Mg in a number of metal poor
field stars (see Woosley and Weaver 1981b).

Important nucleosynthesis also transpires in massive stars
for the very heavy elements (A > 60). The s-process in our 25 $M_\odot$
star is similar to that studied by Lamb et al. (1977) for helium
burning in another model 25 $M_\odot$ star. Additional neutron capture
during carbon and neon burning increases the net neutron
influence by an additional factor of only ~ 10%. While this
limited sort of s-process cannot provide the proper distribution
of neutron fluxes to produce the entire solar abundance array of
s-nuclei, it does move abundance peaks around and may be
responsible for the production of several isotopes just above the
iron group.

The p- (or γ-) process (Woosley and Howard 1978) takes
place in mass shells that experience peak temperatures in the
range 2.1 to 3.2 x 10^9 K, i.e.,those regions that undergo
explosive neon burning. In the 25 $M_\odot$ star this temperature range
occurred in ~ 1 $M_\odot$ of material and, in that region,
photodisintegration reactions on s-process seed should give
overproduction factors ~ 260 (Woosley and Howard 1978). This
implies an overproduction in the entire mass of ejected material

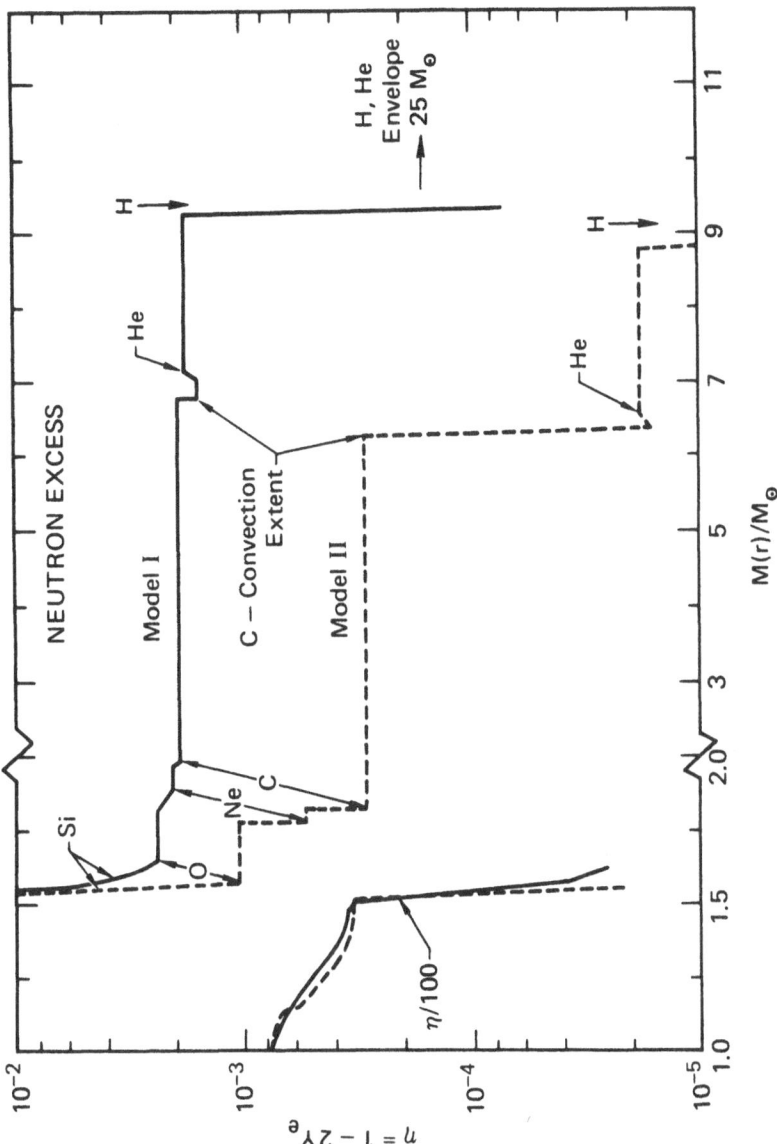

Figure 2. The neutron excess parameter, $\eta = (N_i-Z_i)X_i/A_i$, as a function of Lagrangian mass coordinate is shown for two 25 $M_\odot$ stars, one of which commenced its life with a solar (Pop. I) set of trace elements, the other wth a set 1% as great (Pop. II). Cumulative effects of hydrostatic neutronization have led to quite similar values of η in the inner 1.5 $M_\odot$ of the two stars although substantial differences still exist in the mantles, especially in the region exterior to the carbon burning convective shell. This figure is prepared just as the maximum collapse velocity in both iron cores reaches 1000 km s^{-1}.

Table 3

Variation of nucleosynthesis[a] with stellar population

Species	Z = 1.7(-2)	Z = 1.7(-4)	Species	Z = 1.7(-2)	Z = 1.7(-4)
^{1}H	0.51	0.52	^{23}Na	57.4	17.9
^{4}He	1.21	1.30	^{24}Mg	22.7	15.5
^{12}C	9.1	12.7	^{25}Mg	21.6	1.7
^{13}C	0.9	0.008	^{26}Mg	25.0	2.0
^{14}N	2.6	0.033	^{27}Al	32.4	3.2
^{15}N	0.4	0.005	^{28}Si	13.3	9.3
^{16}O	14.1	11.3	^{29}Si	5.8	0.42
^{17}O	9.8	0.058	^{30}Si	6.5	0.38
^{18}O	14.9	0.15	^{31}P	8.5	1.1
^{19}F	0.8	0.008	^{32}S	7.0	3.2
^{20}Ne	37.4	33.4	^{33}S	5.7	1.7
^{21}Ne	16.5	4.0	^{34}S	10.1	6.3
^{22}Ne	7.7	0.20	^{36}S	32.6	0.20

[a] Productions are for 25 $M_\odot$ stars with results given as an overproduction factor relative to solar values.

of $260/(25 - 1.4) = 11$, in excellent accord with the production
of other, more abundant species (Figure 1). Thus massive Type II
supernovae appear to properly produce the p-nuclei.

The site and nature of the r-process continues to be an
unsolved problem. This is not to say that we are lacking for
sites and processes that produce certain select neutron-rich
nuclei. Passage of the shock wave through neon, carbon, and
helium zones in a 25 $M_\odot$ star produces a limited r-process
(Truran, Cowan, and Cameron 1978: Howard et al. 1971), but, in
the case of explosive helium burning, the amount of mass
experiencing the strong neutron irradiation is too small to
produce an absolute yield of r-nuclei comparable to the abundance
of s-nuclei and other heavy elements that exist in the rest of
the star. Also the neutron flux seems to be less than is required
to produce a solar r-process distribution (Blake et al. 1981),
a difficulty that is even more extreme for explosive carbon and
neon burning. An alternative site for the r-process is deeper
down, near the 'mass cut' that separates outgoing ejecta from the
newly formed neutron star. There, material exists that is already
very neutron-rich. If this material were to be completely
photodisintegrated and then cooled again on a time scale short
compared to that required to reassemble the heavy nuclei, a high
neutron to seed ratio could be produced (Hoyle and Fowler 1960).
Unfortunately this seems to require more rapid cooling than
occurs in our models. We characteristically find photodisintegration products reassembling at $\sim 5 \times 10^9$ K, too high a
temperature for the r-process to proceed. Perhaps ejection by a
jet (LeBlanc and Wilson 1970) could alter these results but that
remains to be demonstrated. It may even be that the r-process, if
indeed there is a single 'r-process', does not occur
predominantly in supernovae! Recent work by Cowan, Cameron and
Truran (1981) suggests that the r-process may occur during an
off-centre helium core flash in a low mass star (following the
mixing of hydrogen into the helium core by 2-dimensional
instability). These interesting speculations point out just how
uncertain the true nature of the r-process really is.

Important clues to nucleosynthesis and constraints on
supernova mechanisms can often be found in a single nucleus. The
case of ^{48}Ca is illustrative. With an abundance 58 times that of
its neighbour ^{46}Ca, it is very difficult to produce, with any
reasonable set of neutron capture cross sections, a solar
abundance of ^{48}Ca without overproducing ^{46}Ca (this argument could
be tightened if the capture cross sections on these two isotopes
were actually measured). It is possible to produce large
quantities of ^{48}Ca, however, in nuclear statistical equilibrium
with a neutron excess of ~ 0.16 (Weaver, Zimmerman and Woosley
1978). Thus it may be that the existence of such a relatively
large amount of ^{48}Ca in the Galaxy requires at least the

occasional ejection of some very neutron rich material. The
ejection of high η material of this type may also be necessary to
the origin of ^{50}Ti, ^{54}Cr, and several other rare iron-group
species (Hainebach et al. 1974).

Important heavy element synthesis will also occur in Type I
supernovae. Currently the most successful models, (i.e., those
that agree best with observation), are based upon detonating
white dwarfs. Nucleosynthesis from one such typical explosion is
shown in Figure 3 (Woosley, Weaver and Taam 1980). Almost the
entire mass of this 1.12 $M_\odot$ white dwarf has been converted into
iron group nuclei, principally ^{56}Ni, by a passage of a detonation
wave. The detailed isotopic composition remains to be computed,
but should differ little from that obtained for the carbon
detonation model (Arnett, Truran and Woosley 1971), i.e., a solar
abundance set for the most abundant isotropes of Cr, Mn, Fe, Co,
and Ni. In the outer part of this same white dwarf, in a layer
whose pre-explosive composition was nearly pure helium explosive
burning produces ^{44}Ti (Figure 3), whose later decays to ^{44}Sc and
^{44}Ca provide both a late time energy source for the supernova
remnant and a possible γ-ray line signal (see Section 2).
Production of 1 $M_\odot$ of iron every 50 or so years by Type I
explosions would provide a reasonable, although not necessarily
unique, explanation for the current abundances of iron and ^{44}Ca
in the Galaxy, although Ostriker (1981) has pointed out that
difficulties arise if this amount of iron production in the
interstellar medium has continued in recent epochs.

2. γ-LINE ASTRONOMY

As has been realized for some time, the nucleosynthesis of
elemental species in novae and supernovae as radioactive
progenitors provides both a late time energy source for powering
the light curve and a γ-line signal that should be visible to a
proper space or balloon-borne detector. Thus far no unambiguous
signal from extra-solar radioactivity has been detected. A
strong signal from positron annihilation in the vicinity of the
Galactic centre was discovered by balloon experiments and
subsequent studies with HEAO-C found the source to be variable on
a period of months. This time variability and the ability of
produce positrons by methods other than radioactive decay make
the connection to supernova nucleosynthesis somewhat tenuous, but
the observed flux might be explained by the positrons produced
during the decay of ^{56}Co made in supernovae (Clayton 1973:
Woosley, Axelrod and Weaver 1981).

Theoretical production of γ-line candidates by a 25 $M_\odot$ Type
II supernova are summarized in Table 4. About 5 times as much
^{56}Co might also be produced in a Type I supernova along with a
comparable abundance of ^{44}Ti (Woosley, Weaver and Taam 1980).

Given the recent <u>optical</u> evidence for the presence of solar mass quantities of radioactive cobalt in Type I supernova 1972e (Axelrod 1980), it is virtually certain that at least some Type I supernovae are strong γ-line emitters.

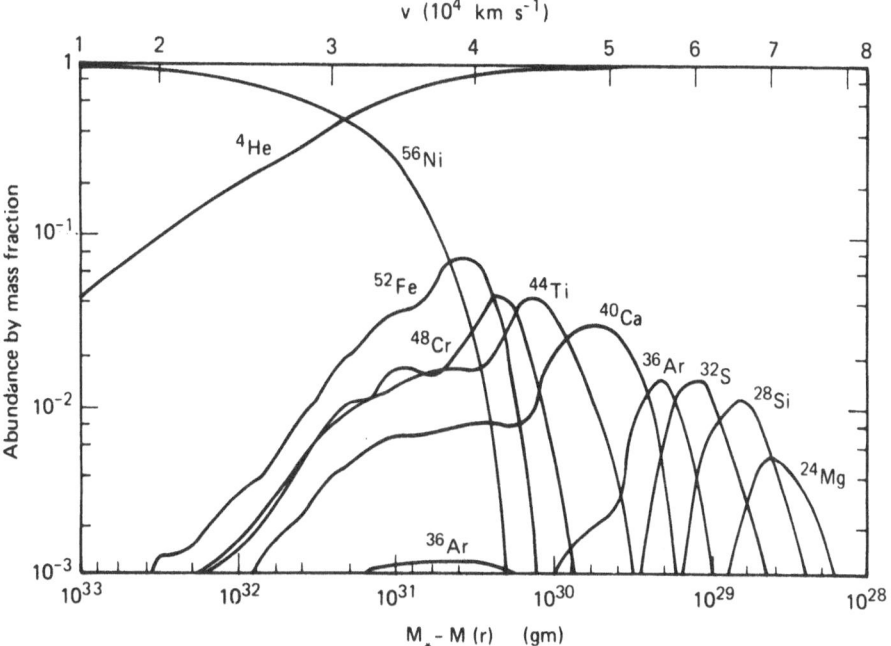

Figure 3. Nucleosynthesis in the outer layers of a detonating white dwarf model for a Type I supernova. The abundance of various species are given by mass fraction and the bottom mass scale is the mass exterior to the given Lagrangian mass shell. The upper scale is ejection velocity in units of 10,000 km s^{-1}. Most of the (totally disrupted) white dwarf now consists of rapidly expanding, radioactive ^{56}Ni with a thin layer of ^{4}He on top. Other heavy elements shown are from explosive helium burning. The ^{44}Ti shown may be the source in nature of its stable daughter, ^{44}Ca.

Table 4

γ-Radioactivities from Type II supernovae

Species	$\tau_{1/2}$ (yr)	Mass Ejected[a] ($M_\odot$)	Synthesis Process	Years visible at 10 kpc ($\phi > 2 \times 10^{-5} cm^{-2} s^{-1}$)
^{56}Co	2.16(-1)	< 2(-1)	Exp. Si Burn	< 5
^{57}Co	7.42(-1)	< 1(-2)	Exp. Si Burn	< 10
^{22}Na	2.60(0)	3(-5)	C-Burn	15
^{44}Ti	4.7(1)	< 1(-4)	NSE	< 120
^{60}Fe	3.0(5)	< 2(-5)	Exp. He Burn	Continuous[b]
^{26}Al	7.2(5)	2(-5)	Exp. Ne Burn	Continuous[b]

[a]All productions except ^{60}Fe are for a 25 $M_\odot$ supernova.
^{60}Fe production is in a 15 $M_\odot$ supernova.

[b]Strength depends on present rate of nucleosynthesis in the Galaxy.

    Because of the considerable uncertainty associated with the nucleosynthesis of the most important of the potential γ-line candidates in Type II supernovae (Table 4), it is useful to consider each isotope separately. The ^{56}Co and ^{57}Co production in Type II supernova is not well known since it depends on a choice for the interior mass cut (Woosley and Weaver 1981b). Radioactive tails to the light curves of such Type II explosions as 19691 suggest that at least some Type II supernovae produce some radioactive cobalt, but others may produce none (see also Section 3). The entry given in Table 4 would provide a full Galactic abundance of iron from Type II explosions (Figure 1). Another species that may be produced in both Type I and Type II supernovae is ^{44}Ti. In each case the yield is highly uncertain. In Type II supernovae ^{44}Ti is produced near the mass cut by very high temperature explosive silicon burning. In Type I supernovae production depends on the existence of a low density helium layer capping a detonating core. While this seems reasonable, there may be alternative models for Type I explosions that do not involve high temperature explosive helium burning.

The other 3 species in Table 4 should be synthesized only in Type II supernovae, supplemented in the cases of ^{22}Na and ^{26}Al by a contribution from ordinary novae. The ^{22}Na made by supernovae comes primarily from hydrostatic (i.e., pre-explosive) carbon burning. It is likely to be produced in all massive star explosions, but owing to its short lifetime and comparatively low yield, ^{22}Na would only be visible from a supernova in our own Galaxy, and, even then, for a relatively short time. The production of ^{60}Fe occurs as a result of the supernova shock wave passing through the helium shell and the operation of a limited r-process (Clayton 1981: Blake et al. 1981). Synthesis is sensitive to shock wave energy and stellar model parameters, but production in a model 15 $M_\odot$ supernova is really quite large (Blake et al. 1981). The amount given in Table 4 is approximately equal to that inferred by Clayton (1981) based on nucleosynthetic arguments and would, as he suggests, constitute a prime candidate for Galactic γ-line astronomy in the steady state.

Like ^{60}Fe, the species ^{26}Al should also accumulate in the interstellar medium from many supernovae. Woosley and Weaver (1980) show that the expected cumulative signal from Type II supernovae would be $\sim 10^{-4}$ photons cm^{-2} s^{-1} times the present rate of Galactic heavy element production divided by the average nucleosynthesis rate over the Galactic lifetime. That is, unless the rate of element synthesis in our Galaxy has declined by more than a factor of 10 since nucleosynthesis began, the flux of ^{26}Al γ-rays, integrated over the Galactic disk, should exceed 10^{-5} photons cm^{-2}. Novae are also expected to contribute a comparable amount of ^{26}Al (Woosley and Weaver 1980: Wallace and Woosley 1981) and red giant mass loss may also be an important source (Norgaard 1980).

In addition to absolute yields, the time dependent γ-ray transparency of the expanding supernovae is critical in determining the observability of short-lived radioactivities. Woosley, Axelrod and Weaver (1981) have carried out such an analysis for both Type I and II supernovae. Owing to their extended red giant structure and slower expansion velocities, Type II supernovae do not become transparent to their own γ-emission for about one year. The slower velocities of the heavy elements synthesized in Type II explosions (v $\sim$ 1000 kkm s^{-1}) also implies a narrower γ-line, $\Delta E_\gamma < 10$ keV. Because they presumably lack an extended hydrogen envelope and because the heavy elements are ejected with much higher velocity ($\sim$ 10,000 km s^{-1}) Type I supernovae become transparent to γ-rays at a much earlier time. Type I supernovae should present a strong signal of broad ($\Delta E_\gamma \sim$ 100 keV) γ-lines from the decay of ^{56}Co after only several weeks of expansion. Indeed, this signal should commence shortly after maximum luminosity (Weaver, Axelrod and Woosley 1980).

Numerically, the flux from 1 $M_\odot$ of freshly synthesized ^{56}Co, normalized to a distance of 20 Mpc, would be given by

$$\phi_\gamma(^{56}\text{Co}) = 4.6 \times 10^{-5}[e^{-t/114} - e^{-t/8.8}]$$

$$*(1-f_{dep})\ M_{56}\ (20\ \text{Mpc/d})^2$$

where t is the elapsed time in days since the explosion, M_{56} is the mass of radioactive ^{56}Ni initially synthesized, and f_{dep} is the time dependent γ-ray deposition efficiency (see Figure 4). Since they become transparent before ^{56}Co has substantially decayed and because they probably make more ^{56}Co to start with, Type I supernovae are much more attractive candidates for γ-line astronomy than Type II supernovae.

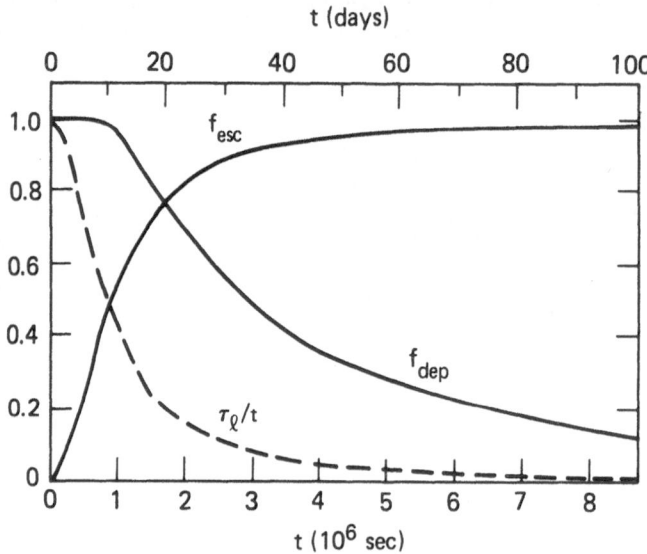

Figure 4. Deposition and escape factors and lag time for a Type I supernova based on white dwarf detonation and radioactive energy input. f_{esc} is the fraction of deposited energy from radioactive decay that avoids adiabatic decompression to escape as optical or near optical emission; f_{dep} is the fraction of nuclear decay energy deposited locally (a lower bound is 0.04); and τ_e is the mean time betwen energy deposition and escape (or decompression). The optical luminosity is given by

$$L(t) = M_{56}\ \dot{S}(t-\tau_e)\, f_{dep}(t-\tau_e)\, f_{esc}(t)$$

where M_{56} is the mass of radioactive ^{56}Ni produced and $\dot{S}$ is the energy production rate from ^{56}Co and ^{56}Ni decay. The γ-ray flux will be proportional to the quantity $(1-f_{dep}(t))$. See Weaver, Axelrod and Woosley (1981).

The major space mission of this decade dedicated to γ-ray astronomy is the Gamma Ray Observatory (GRO). Now scheduled for launch in 1987, this complement contains as one of its 4 experiments the Oriented Scintillation Spectrometer Experiment (OSSE) which should be an excellent tool for discovering and analyzing lines from radioactive decay. For a typical line energy of 1 MeV this experiment is capable of detecting γ-lines down to a flux of $\sim 2 \times 10^{-5}$ photons cm^{-2} s^{-1} with an observation time of roughly 1 week. Furthermore, sensitivity at least this good is maintained throughout the energy interval 100 keV to 10 MeV.

Barring the fortunate occurrence of a Galactic supernova during the 2 year lifetime of GRO, the most attractive targets of opportunity will be steady signals from ^{60}Fe, ^{26}Al, and (perhaps) ^{44}Ti within our own Galaxy, and ^{56}Co decay lines from Type I supernovae in other galaxies. Signals from ^{26}Al and ^{60}Fe, which come from Type II supernovae, should be associated with regions of active massive star formation and the Galactic disk. The ^{44}Ti signal would originate from the remnants of recent Type I and (perhaps) Type II supernovae. The strength of these emissions is highly uncertain, as we have discussed.

The sensitivity of OSSE is such that ^{56}Co emissions from as far away as the Virgo cluster (20 Mpc) should be visible from a Type I supernova producing 1 $M_\odot$ of iron. This possibility is appealling, not only because it is quite likely that at least one Type I will happen in Virgo during the course of 2 years, but also because the study of these lines would reveal interesting and unique information about the nature of supernova and young remnants (Woosley, Axelrod and Weaver 1981). Care must be taken, however, because the supernova rate in Virgo is not all that large. A rate of ~ 1 per year seems reasonable based on an estimated luminosity of the Virgo cluster of 2×10^{12} $L_\odot$ and a total supernova rate per year per 10^{10} $L_\odot$ of 0.008 (Tammann 1974). This is consistent with the actual discovery rate in Virgo over a 13 year search period (Tammann 1974). Perhaps this number can be doubled by including other galaxies within 20 Mpc, and maybe even multiplied by an additional factor of 2 to 5 if a relatively large fraction of nearby supernovae have gone undiscovered in the past (Tammann 1976, 1981). The point is, however, that extra-galactic γ-line astronomy can not be left to the serendipitous discovery of supernovae by OSSE. With a roughly 10 degree field of view and 10^6 s observation time, those few prime candidates may either go undetected or else decay away ($\tau_{1/2} = 78$ d) before they are observed. We must have a ground based supernova search program operational by the time GRO goes up in 1987 (and hopefully long before that time to more properly plan observational strategy).

3. 'FAILED' CORE BOUNCES IN MASSIVE STARS

Thus far our discussions of Type II supernovae have been based upon the implicit assumption that the collapse of the iron core and its subsequent rebound at nuclear density is capable of generating a strong, outgoing shock wave that will eject all material outside a mass shell of about $1.4 - 1.5\ M_\odot$. As other papers at this meeting have pointed out, this assumption is still questionable for some, if not all, masses of supernovae. (See, for example, papers by Hillebrandt and by Arnett.) In cases where core bounce does not lead directly to mass ejection one is left to contemplate the continued evolution of a red supergiant whose core has collapsed to a neutron star (rapidly in the process of becoming a black hole!). In the summer of 1978 we began a series of calculations to study this phenomenon. Discussions of our results were presented at the Aspen supernova workshop in 1979 and the Santa Barbara workshop in 1980. A 2-dimensional treatment of the subject is in progress by Bodenheimer and Woosley (1981) and was reviewed at the Texas Relativistic Astrophysics meeting (Woosley and Weaver 1981a).

Our study centres on a 25 $M_\odot$ star (Weaver, Zimmerman and Woosley 1978) for which Wilson and Bowers (1978) calculated a core bounce that did <u>not</u> lead to an explosion. The trajectory of the mass shell containing 1.35 $M_\odot$ is shown in Figure 5. Here bounce occurs at 0.2495 s with t = 0 defined by the published snapshots of Weaver, Zimmerman and Woosley (1978). The boundary of the 1.35 $M_\odot$ core initially moves outwards from a bounce radius of 1.3×10^7 cm with a velocity of $\sim 9 \times 10^8$ cm s^{-1}, reaches a maximum radius of 2.1×10^7 cm at 0.262 s, then falls back to become part of the newly formed neutron star. This motion sets up an outgoing shock in the overlying material which propagates as far as 1.50 $M_\odot$ before being overwhelmed (0.277 s) by both the inward momentum of the collapsing stellar mantle and the photo-disintegration of ^{28}Si into free neutrons and protons. The velocity of inward bound material just above the dying shock is $\sim 10,000$ km s^{-1} and photodisintegration of silicon removes about 8×10^{18} erg g^{-1}. At 0.277 s, what had been an outward moving shock with positive velocity reverses sign and becomes an accretion shock. From this time on no positive velocities are observed anywhere in the calculation.

In order to see if thermonuclear burning might have an important effect in the subsequent evolution and, in particular, to see if the energy from that burning might lead to a reversal of the collapse and still create an explosion, the calculation was continued for an additional 6 s following core bounce. To facilitate the computation, the inner 1.65 $M_\odot$, which at a time 0.410 s existed in a state of near hydrostatic equilibrium, was removed from the problem and replaced by a rigid inner boundary

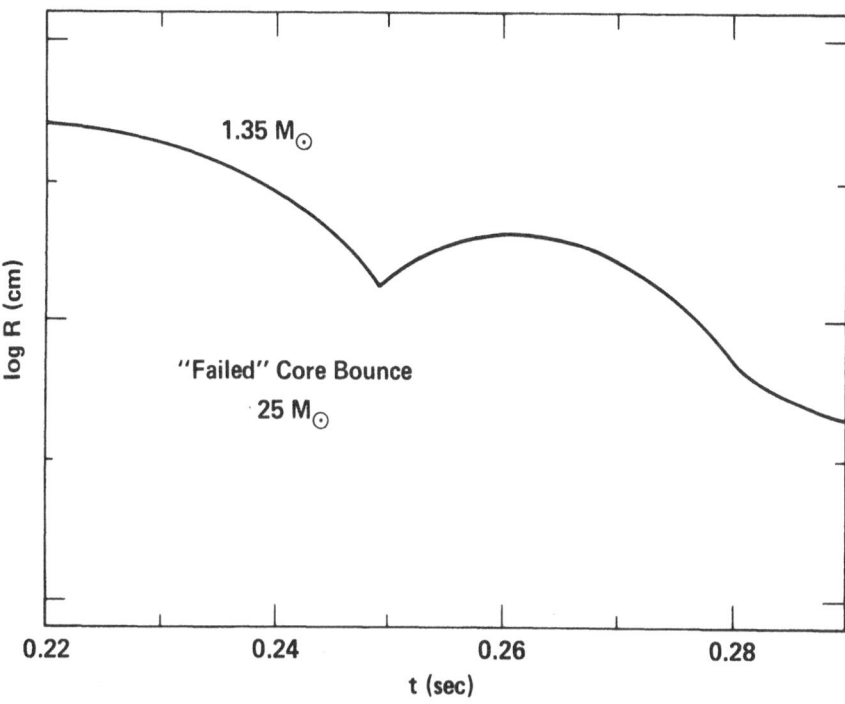

Figure 5. Radial history of that Lagrangian mass point that encloses 1.35 $M_\odot$ in a supernova core collapse calculation that did **not** produce a strong mass ejecting shock wave. Taken from a study by Wilson and Bowers (1978) of a 25 $M_\odot$ presupernova star evolved by Weaver, Zimmerman and Woosley (1978).

having the same radius as that mass shell at that time, i.e., 1.007×10^7 cm. The gravitational potential of this core continued to be carried in the calculation and the removal of the core in this fashion should have little effect on the results, especially for regions outside the accretion shock. During the next 6 s, the accretion shock moved outwards in Lagrangian mass coordinate until 3.6 $M_\odot$ was contained in the neutron star core. The radial location of the shock also moved out slowly from ~ 1.5×10^8 cm at 1 s, to ~ 2×10^8 cm at 2 s, and ~ 4×10^8 cm at 6 s. Throughout this entire time the maximum accretion velocity remained very nearly constant at 10,000 km s^{-1}. Thermodynamic conditions, composition, pressure-to-gravity ratio, and Lagrangian velocity profiles are given at several times during the evolution in Figures 6, 7, 8, 9, and 10. The actual core radius is overestimated and temperature and density of the core region underestimated in these calculations owing to the neglect of electron capture. This should have the effect of maximizing the opportunity for an explosion.

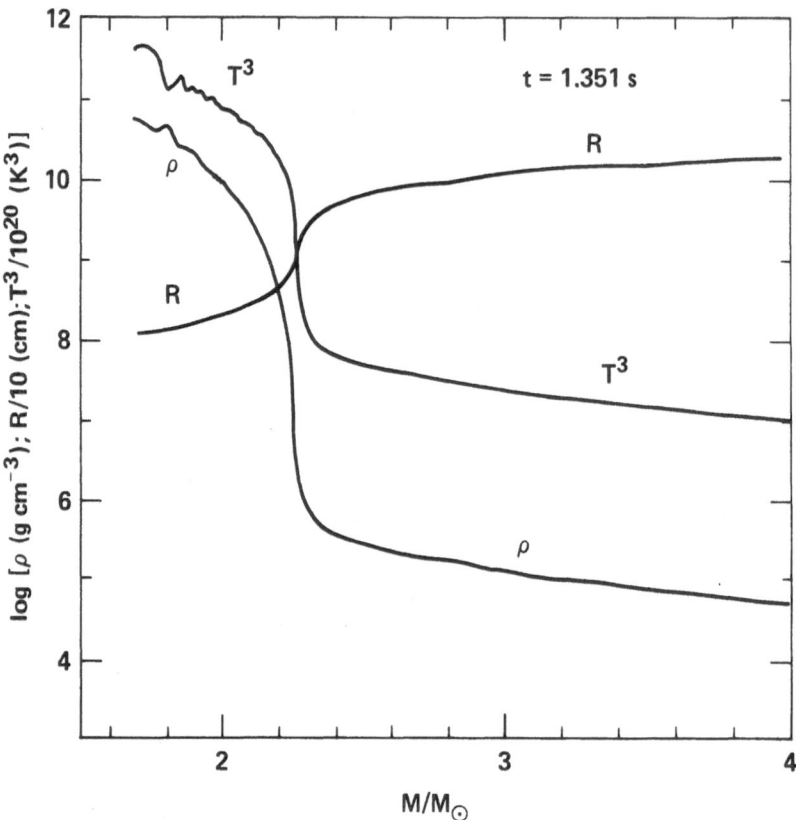

Figure 6. Temperature, density, and radius in the interior of a
25 M$_\odot$ 'failed' supernova at a time 1.351 s following core
collapse, or 1.101 s following core 'bounce'. The cube of the
temperature has been plotted on a logarithmic scale along with
the density so that the curves are parallel when the adiabatic
index is constant and close to 4/3. The inner 1.65 M$_\odot$ has been
removed from the problem and replaced by a hard boundary
condition at 1.007 x 10^7 cm.

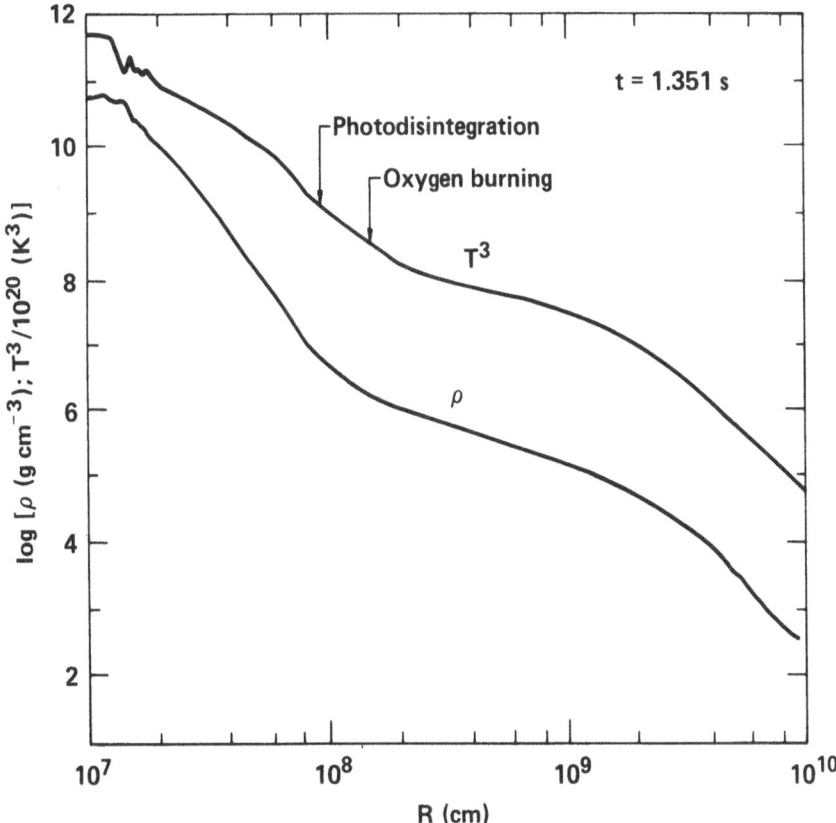

Figure 7. Same as Fig.6, but plotted on a logarithmic radial scale. The location of the oxygen burning shell and region where iron from oxygen and silicon burning undergoes photo-disintegration are indicated. The non–adiabatic cooling from photodisintegration is apparent.

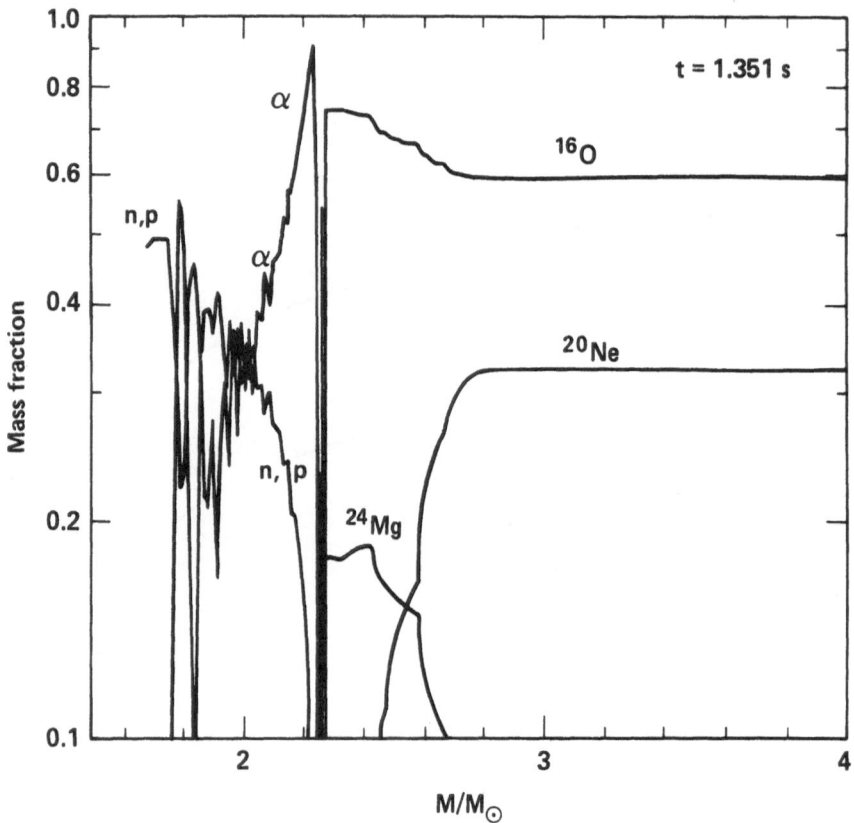

Figure 8. Compositions of the accreting and accreted material at the time corresponding to Fig. 6 and 7 (see also Fig. 10). The thin nature of the oxygen burning shell is readily apparent. Large oscillations in ·the abundances of free nucleons are amplifications of small temperature fluctuations that result from the artificial viscous damping of small core oscillations induced by the accreting matter.

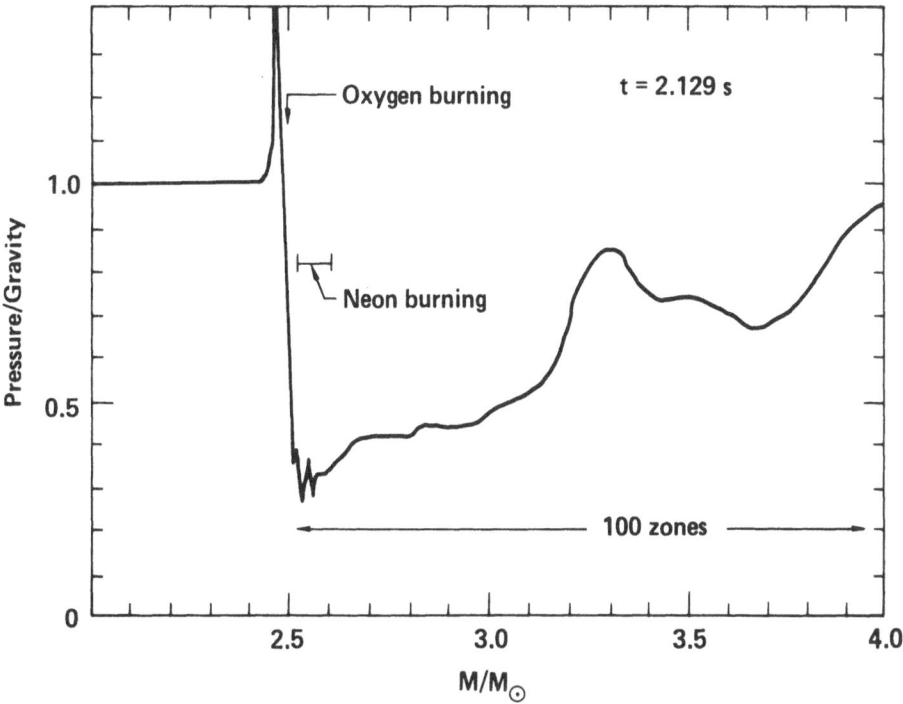

Figure 9. The ratio of pressure force to gravitational force as a function of Lagrangian mass coordinate at a time 2.129 s following core collapse. The units of pressure and gravity have been normalized in such a way as to give a ratio of unity if the given mass shell is in a state of hydrostatic equilibrium. The sharp spike at 2.47 $M_\odot$ is the accretion shock and the fluctuations just above are numerical (owing to a very low value of artificial viscosity employed in the calculation). The region above 4.0 $M_\odot$ (not shown) is still near hydrostatic equilibrium and has not yet responded to core collapse. The 'bump' round 3.3 $M_\odot$ results from relatively coarse zoning in the pre-explosive star prior to the fine rezoning indicated in the figure.

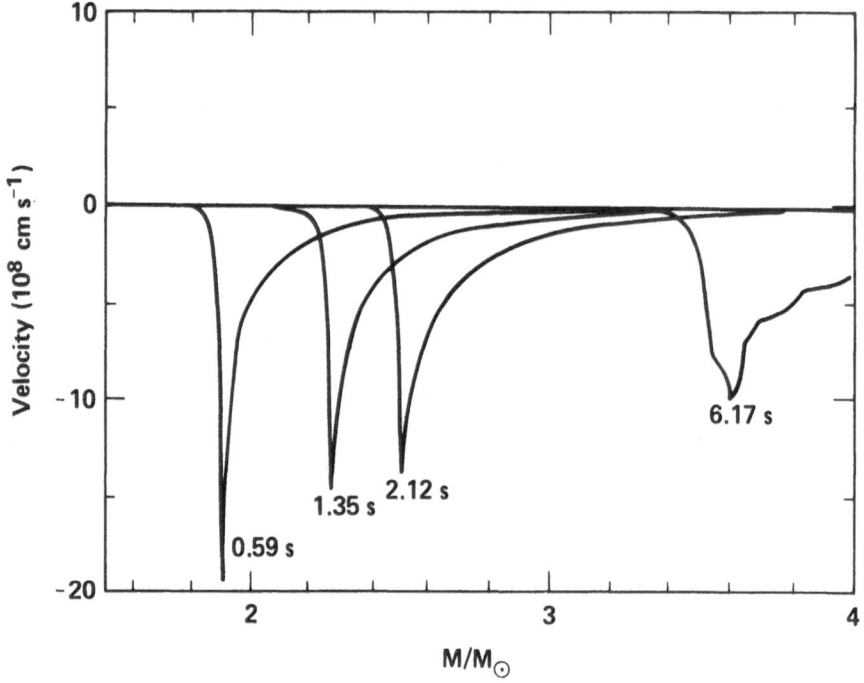

Figure 10.   Velocity profiles in units of 1000 km s^{-1} as a
function of Lagrangian mass at several times after core bounce.
The slight decrease with time of the maximum infall velocity is
a result of the increasing radial size of the hot shocked core.
Had electron capture been properly included (it was in fact
ignored in this calculation) the core would not have grown so
large, thus this figure overestimates the effect. The irregular
shape at the latest time indicated (6.17 s) is an artifact of
coarse zoning.

At no point during this evolution did nuclear burning lead to a velocity reversal (contrary to recent speculations by Applegate and Yahil 1981). It is important to note that the combustible mantle, which certainly contains sufficient nuclear energy to disrupt the star if it could be ignited instantaneously, does not collapse homologously (see Figure 9 for the ratio of actual pressure to that required for hydrostatic support at 2.129 s). While knowledge of core collapse is communicated to the mantle at sonic speed (roughly 5000 km s^{-1}) the pressure deficit is much greater, the density higher, and the dynamic response time therefore much shorter, for material closer to the core. The oxygen actually burns in a very thin shell ($<$ 0.02 $M_\odot$, Figure 8) quite close to the accretion shock. Layers farther out are supported both by the inertia of layers beneath them and by the 'spherical rocket' effect of material being accelerated down into the collapsed core. An isolated layer of oxygen falling into the remnant subsonically could, in principle, reverse its infall by nuclear burning, but two effects act to inhibit this occurrence. First is the enormous pressure of the overlying star against which the burning shell must work in order to reverse its velocity. Second is photodisintegration. While oxgyen burning provides about 4 x 10^{17} erg g^{-1} the burning of silicon to ^{54}Fe and 2 protons (the favoured products in this relatively low density environment) is endoergic by 2.3 x 10^{16} erg g^{-1} and the photodisintegration of iron, first into free alphas, then into nucleons, removes 8.5 x 10^{18} erg g^{-1}. In layers immediately beneath the thin oxygen burning shell these endoergic processes rob the gas of thermal energy that might have provided support for the star (Figure 8).

In order to explore more fully the role of photodisintegration during the post-core-bounce evolution, the above calculation was repeated employing an identical parametrization except that all nuclear energy generation, both negative and positive, was suppressed once oxygen in a zone had been fully depleted. This prescription had the desired effect of suppressing thermal energy losses to photodisintegration (obviously not a realistic procedure but useful for isolating a specific effect). No explosion resulted in this case either (or at least had not resulted at a time 2.946 s when the oxygen burning shell had reached 2.7 $M_\odot$), but the evolution was clearly qualitatively different. The distinction is illustrated in Figures 11 and 12, which show velocity and thermodynamic quantities at 1.351 s, the same time as in Figures 6, 7 and 8. The accretion velocity in the model with photodisintegration is much slower at this point, by about a factor of 3, and the inner regions of the star are not nearly so centrally condensed. Because of the shallower temperature gradient, the region where oxygen is burning is also substantially larger ($\sim$ 0.1 $M_\odot$) and there are layers, again $\sim$ 0.1 $M_\odot$, of unburned silicon and ^{56}Ni

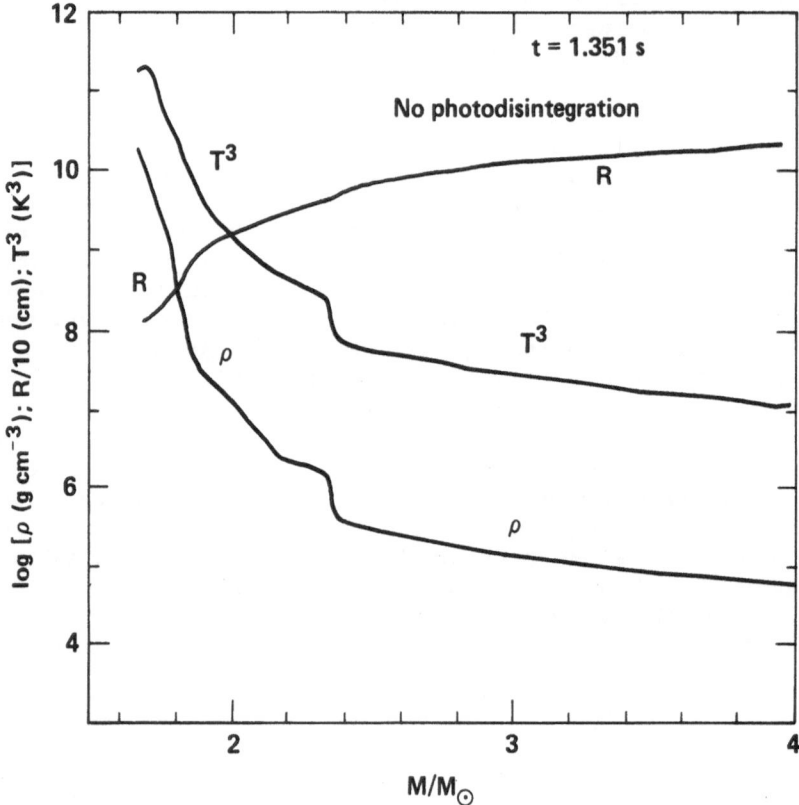

Figure 11. Same as Fig. 6 but photodisintegration has been artificially suppressed in the shock and compressively heated matter. Comparison with Fig. 6 shows that at this time a more highly extended and less centrally condensed core is formed when photodisintegration is neglected.

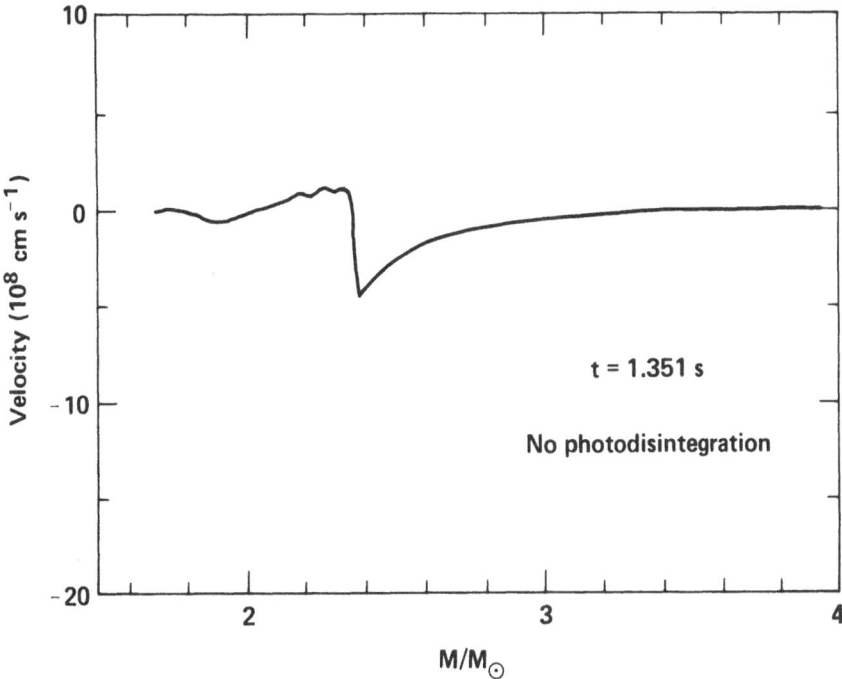

Figure 12.   Velocity profile corresponding to Fig. 11. Motions at this point are just becoming sub-sonic. Compare with Fig. 10.

that has not been photodisintegrated. These are all direct results of suppressing photodisintegration. Energy that would have gone into disrupting nuclei is now able to provide pressure to hold the star up in a more extended, cooler configuration. Continued evolution of this model saw the dying out of the accretion shock and a return to near hydrostatic equilibrium throughout the star. At a time of 2.95 s, no velocities greater than a few hundred km s^{-1} existed anywhere in the star. Not too surprisingly, suppressing the photodisintegration instability and ignoring electron capture makes it possible to construct a stable (at least on dynamic time scales!) stellar model in which a neutron star lies at the centre of a highly evolved, supergiant star. Evolution on a thermal time scale would still be unstable due to high neutrino losses near the core.

The fact that thermonuclear burning does not appear likely to produce supernovae in 1-dimensional models with failed core bounces does not categorically remove such models from consideration since <u>rotation</u> is likely to be an important effect. Oxygen fuel falling almost freely towards the collapsed core will experience an increasing centrifugal barrier to its inward

progress. Since rotational breaking occurs throughout a large
fraction of mass concurrently and not at a single Lagrangian mass
point like a shock, it is possible to stagnate large regions of
unburned fuel concurrently. If inertial overshoot occurs, then
the centrifugal barrier leads to a radial bounce, and if that
bounce is amplified by nuclear burning, an explosion may result.
Such a mechanism is inherent in Figure 11 of Fowler and Hoyle
(1964) and has been the object of recent study by Bodenheimer and
Woosley (1981, BW). The essence of the BW results is that
rotation plus the explosive burning of oxygen can lead, at least
for some choices of parametrization, to energetic mass ejection
in the equatorial plane. Details of the study, which essentially
involve a 2-dimensional recalculation of the same 25 $M_\odot$ failed
core bounce we have been describing, were presented at the Texas
Relativistic Astrophysics meeting (Woosley and Weaver 1981a), will
be the subject of a forthcoming publication (BW), and need not be
duplicated here. However, the results of one such calculation are
displayed in Figure 13. A steady state solution is found,
displayed here at a time 15.2 s following core bounce/failure, in
which matter collapsing along a radial vector roughly 60 degrees
above the equatorial plane experiences a rotational bounce,
accelerated by explosive oxygen burning, and then moves outwards
with high velocity in the equatorial plane. At this last time
step (artificially restricted by computational requirements),
more than 10^{50} erg of outward directed kinetic energy is
contained in about 0.5 $M_\odot$ of oxygen and oxygen burning ashes
moving with a velocity of about 5000 km s^{-1}, well above the local
escape velocity. Continued evolution is likely to increase this
kinetic energy as more oxygen fuel circulates and burns. A
portion of this energy will be shared with the hydrogen envelope
with the creation of a nearly spherical shock wave and typical
Type II light curve being possible results. The asymmetric
momentum of the explosion may have additional observational
consequences, however, in that the supernova remnant would retain
equatorial, but not spherical symmetry. Evidence for annular
structure in several remnants including SNR 132D (Lasker 1980),
0540-69.3 (Mathewson et al. 1981), and CAS A (Markert et al.
1981) has recently been reported. If indeed remnants like CAS A
are to be interpreted with such a model, one might anticipate
(depending on the unknown interaction of the equatorially ejected
oxygen blobs and the hydrogen envelope) observable differences in
the spatial asymmetry of oxygen knots and hydrogen 'flocculi'. It
would be interesting to know if such effects exist.

Since the Texas meeting several sensitivity tests have been
carried out by Bodenheimer and Woosley. A calculation with 1/2
the initial angular momentum (C/G = 0.06) of the one depicted
also gave an explosion. The calculation that produced Figure 13
was also repeated with identical parametrization, but with all
nuclear energy generation turned off. No mass ejection

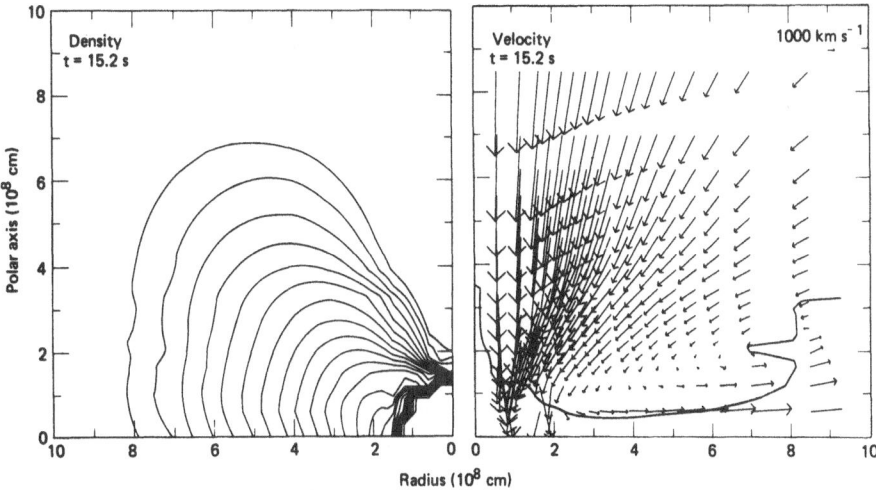

Figure 13. Density profiles and velocity fields from a two-dimensional study (Bodenheimer and Woosley 1981) of a rotating 25 $M_\odot$ presupernova star whose core has collapsed without generating a strong outgoing shock (see also Fig. 5). Logarithmic density profiles are given 10 per decade. The outermost contour shown has a density of 10^5 g cm^{-3} and the innermost, 2×10^7 g cm^{-3}. A standard velocity arrow of length 1000 km s^{-1} is also indicated. At this time the polar regions have been essentially evacuated. The entire stellar mantle exists as a thick, lobed accretion disk. A persistent velocity field has been set up involving collapse along a roughly 45° angle, nuclear burning coupled with a rotational bounce near the core, and high velocity equatorial mass ejection. The solid line in the right-hand figure is the contour of one-half oxygen depletion. Underneath and to the right of this line, the composition is that resulting from explosive oxygen burning. Above the line oxygen has not burned. The interaction of the equatorially ejected matter with the red giant envelope and the subsequent light curve remain to be calculated.

developed, only a stagnant accretion disk formed. Another calculation, which included nuclear burning but in which the inner boundary pressure was never increased (see Woosley and Weaver 1981a), also failed to produce dynamic mass ejection. However, after an elapsed time of 41 s (11000 models of 1600 zones each!), at which time the inner core contained 6 $M_\odot$ of the original 8 M_O carried in the problem, the remaining matter had developed a ratio of rotational energy to gravitational binding energy of 0.27 and would thus be unstable, on a dynamic time scale, to tri-axial deformation (Ostriker and Bodenheimer 1973). If a bar-like structure develops or if fragmentation ensues, angular momentum may be transported out of the central regions of the star and may lead to mass ejection. If so, we would have come full circle to the original (Hoyle 1946) model for supernovae!

Clearly further work is greatly needed on the whole subject of 'failed' core bounces. The results of Arnett and of Hillebrandt (discussed elsewhere in this same volume) have shown that if the core bounce mechanism is ever to function adequately in one dimension, it may do so only in iron cores having relatively small total mass (say < 1.3 $M_\odot$). This is because the energy available to the shock is set, to first order, by the mass of the 'homologous' core (0.7 $M_\odot$), a quantity that is not sensitive to the total iron core mass. On the other hand, the work that the shock must do against gravity, inward momentum, and photodisintegration in order to get out of the core does increase rapidly with iron core mass. Hence larger cores are less likely to explode.

But stars of larger mass have larger cores. For example, a 25 $M_\odot$ star has a core mass of 1.61 $M_\odot$ (Weaver, Zimmerman and Woosley 1978) while a 10 $M_\odot$ star calculated using the same physics had a core mass of < 1.4 $M_\odot$ (Woosley, Weaver and Taam 1980). Thus it may be that the explosion of only the lightest stars in the intermediate mass range (10 $M_\odot$ to 100 $M_\odot$) may be attributable to core bounce of the simplest kind while the final evolution of heavier stars is sensitive to 2-dimensional effects as we have described.

It is, of course, a problem of fundamental cosmic importance to determine the critical mass that separates black hole remnants from neutron star remnants. If it is not too high the supernova statistics would not be greatly altered. The lighter stars are, after all, the more abundant ones. On the other hand, our views concerning the nucleosynthesis of heavy elements would be radically altered if stars of mass greater than 25 $M_\odot$, say, do not explode by the core bounce mechanism. If this is the case, then maybe 2-dimensional effects such as we have just discussed will dominate, leading perhaps to new nucleosynthetic processes involving, for example, the high temperature combustion of hydrogen and helium mixtures as the portion of the red giant envelope collapses and is equatorially expelled. Otherwise, the nucleosynthesis of intermediate mass elements may require the still more massive stars that we will now discuss.

4. PAIR INSTABILITY SUPERNOVAE (HYPERNOVAE)

While the exact mechanism whereby stars lighter than ~ 100 $M_\odot$ become supernovae has always been controversial, the pair instability provides a straightforward explanation for explosions in more massive stars. The general nature of the instability is well understood and has been discussed elsewhere (Barkat, Rakavy and Sack 1967). The principal difficulty is, of course, that few, if any, stars this massive are believed to be forming nowadays. There are reasons to believe, however that more massive stars existed in the early evolution of our own and other galxies (Silk 1977), thus such stars are of interest if only for their nucleosynthesis. Also, as we shall see, the outbursts of these stars may be so energetic as to be visible to a proper (satellite-borne) detector at very great distances, perhaps even to the edge of the universe!

An important first question is the mass range for which the pair instability is likely to result in a supernova. We have evolved models of several very massive stars in order to answer this question. A 100 $M_\odot$ Pop I star studied several years ago (neglecting mass loss) evolved to silicon ignition without encountering this instability, a 150 $M_\odot$ Pop I star examined more recently collapsed on the pair instability as it attempted to ignite carbon burning ($T_9 = 1.0$ at $\rho = 3 \times 10^4$ g cm^{-3}). We conclude that the minimum mass is somewhere between the two although mass loss on the main sequence and during helium burning could substantially increase this limit.

Since the explosion energy results solely from nuclear burning, which for these almost completely convective stars scales as M, while the gravitational binding energy scales as M^2, there also exists an _upper bound_ to the mass of pair instability supernovae (we exclude here the _very_ massive stars, $M > 10^5$ $M_\odot$ that collapse on a _general relativistic_ instability). This limit has also been discussed earlier by Fraley (1968) and Wheeler (1977). In order to circumvent a long calculation of envelope structure we considered, in this case, carbon-oxygen cores with ^{16}O in a 2:1 ratio to ^{12}C. Cores of 60 $M_\odot$, 80 $M_\odot$, 100 $M_\odot$ and 200 $M_\odot$ were studied. The 60, 80, and 100 $M_\odot$ cores all exploded. Nuclear burning was unable to reverse the collapse of the 200 $M_\odot$ core, and its final evolutionary state was presumably a black hole. A 150 $M_\odot$ core is currently under study in an attempt to determine more precisely this mass limit. Because of the

uncertainty introduced by neglecting mass loss, relating this
core size to mass on the main sequence is difficult, but we can
attempt to normalize to our other models. The 150 $M_\odot$ Pop I star
had a carbon-oxygen core of 93 $M_\odot$, the cores of 200 $M_\odot$ and 500 $M_\odot$
Pop III stars (to be discussed) were 103 $M_\odot$ and 370 $M_\odot$
respectively. We conclude that stars (evolving without
substantial mass loss) that have main sequence masses
substantially in excess of about 300 $M_\odot$ will become black holes.

Two pair instability supernovae have been studied in greater
detail. One, the 150 $M_\odot$ Pop I star mentioned above is the largest
mass star likely to be forming today in our Galaxy, the other, a
200 $M_\odot$ Pop III (zero metallicity) star, might have existed in the
early universe during the time (or perhaps before the time) of
galaxy formation. A 500 $M_\odot$ Pop III star, whose final
evolutionary state is anticipated to be a black hole, was also
examined. Mass loss was neglected in these calculations or lack
of a realistic prescription for its inclusion. Papaloizou (1973)
and Talbot (1971) have shown that pulsational instability need
not lead to substantial mass loss in these stars, thus
radiatively driven winds are likely to dominate. Pulsational
instabilities are automatically suppressed by the implicit nature
of our hydrodynamics code which strongly damps oscillatory
behaviour on a time scale much shorter than the characteristic
time step the code is taking. At one point we forced the time
step down to 1000 s and were able to see the 150 $M_\odot$ star
undergoing very small scale oscillations on the main sequence
with period $\sim 4.2 \times 10^4$ s. Numerical dissipation, however,
precluded a further study of this phenomenon. Radiative mass loss
on the main sequence was also artifically suppressed by a surface
boundary pressure of 300 dyne cm^{-2} and by coarse mass zoning near
the surface. Future calculations including a realistic
prescription for mass loss would be interesting. At a time of 50%
hydrogen depletion (1.75 my) the luminosity of the 150 $M_\odot$ star
was 1.29×10^{40} erg s^{-1}: its radius, 1.8×10^{12} cm, and its
effective temperature, $\sim 50,000$ K. By the time of hydrogen
depletion ($X < 0.5\%$) the radius of the star had increased to
4.1×10^{14} cm, the luminosity to 1.65×10^{40} erg s^{-1}, close to
the Eddington value, 2.2×10^{40} erg s^{-1} for a 150 $M_\odot$ star, and
the temperature had declined to 3400 K. Given major uncertainties
with regard to mass loss, these latter values and all subsequent
photospheric properties are highly suspect, but there is a
compellingly simple explanation for the dramatic increase in
radius (Penrod 1981). The electron scattering opacity of totally
ionized helium is less than that of a solar mixture of hydrogen
and helium. Therefore a massive helium core has a higher
Eddington limit than an equivalent sphere of solar mix. In the
limit that the helium core is a large fraction of the stellar
mass and radiating near its own Eddington value the luminosity in
the overlying hydrogen shell will be super-Eddington. The excess

radiation pressure may provoke an expansion of the envelope and rapid mass loss. Our 150 and 200 $M_{\odot}$ stars do not seem to have quite reached this limiting case but are close.

The age of the 150 $M_{\odot}$ star at hydrogen depletion was 2.857 my, a value that does not vary greatly with mass for such high mass stars where luminosity and nuclear energy reservoirs both scale as M. Continued evolution saw the growth of an extended, low density ($\rho \sim 10^{-10}$ g cm^{-3}) envelope that contained an increasingly large fraction of the entire hydrogen shell. Calculations were difficult and time consuming during this stage as Lagrangian shells of small mass moved down a 10 order-of-magnitude density gradient. During this same time the surface convection region also reached down into the outer regions of the helium core mixing large quantities of helium up to the surface. At the time of helium core ignition (2.880 my, central $T = 2.21 \times 10^8$ K, central $\rho = 171$ g cm^{-3}) the core contained 105 $M_{\odot}$ and the envelope, now all at low density, contained $\sim 50\%$ helium by mass. Helium burning took an additional 380,000 years and, as mentioned earlier, carbon ignition took place under unstable circumstances.

Collapsing on the pair instability, the core of the 150 $M_{\odot}$ star reached a maximum central temperature of 3.77×10^9 K and a density of 2.02×10^6 g cm^{-3}. The nuclear energy released by explosive carbon, neon, and oxygen burning reversed this implosion giving an explosion having total kinetic energy 2.20×10^{52} erg. The resulting light curve and nucleosynthesis are shown in Figures 14 and 15 and in Table 5. Although nucleosynthesis for elements having odd Z such as Na, Al, P, Cl, and K, have not been calculated in this model, their production should also be close to solar since a Pop I set of initial seed nuclei was included and nuclear reactions during helium burning should give a 'neutron excess' appropriate to their synthesis (Woosley and Weaver 1981b).

A total of $\sim 10^{51}$ erg of energy comes out in the form of electromagnetic radiation. This brilliant display, ~ 30 times brighter than a typical Type II supernova (Weaver and Woosley 1980), results both from the greater inherent energy of this massive stellar explosion and from a greater efficiency or conversion of that energy into light. During the first 8×10^6 s the luminosity comes from internal energy diffusing out through the low-density envelope. When the radius reaches 6×10^{15} cm a wave of atomic recombination begins, commencing at the surface of the star and moving inwards in both radius and mass. Radiation released by this recombination is responsible for the emissions of the next 6×10^6 s. Owing to the large volume and mass of the envelope this continues to be a very luminous phase. The decline in luminosity at about 1.4×10^7 s occurs as the transparency wave reaches the denser core material. This brilliant display

will be greatly diminished if the star does not retain its
hydrogen envelope. The dashed line in Figure 14 shows the light
output from an identical calculation in which the hydrogen
envelope was artificially removed.

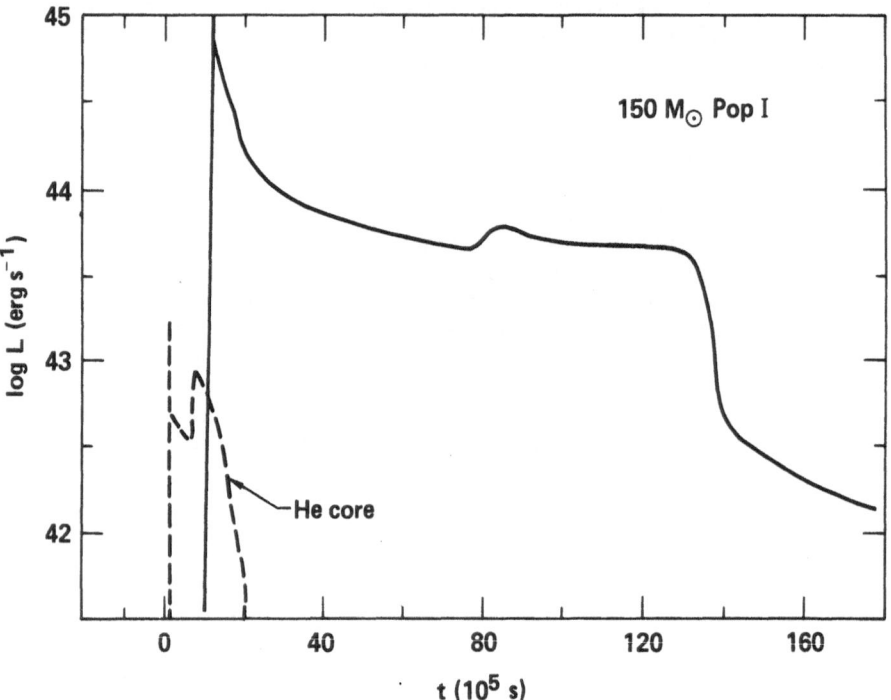

Figure 14. Bolometric light curve for a 150 $M_\odot$ hypernova. The
solid curve is obtained if the star retains (a large fraction of)
its low density hydrogen envelope. The dashed curve results if it
has lost its envelope. The turnover at 140 days occurs as the
transparency wave reaches the helium core – hydrogen envelope
interface. The bump at 85 days is artificial (due to coarse
surface zoning) but indicates the changeover from a diffusion
dominated light curve to a transparency wave. The solid line is
about 30 times brighter than a typical Type II supernova.

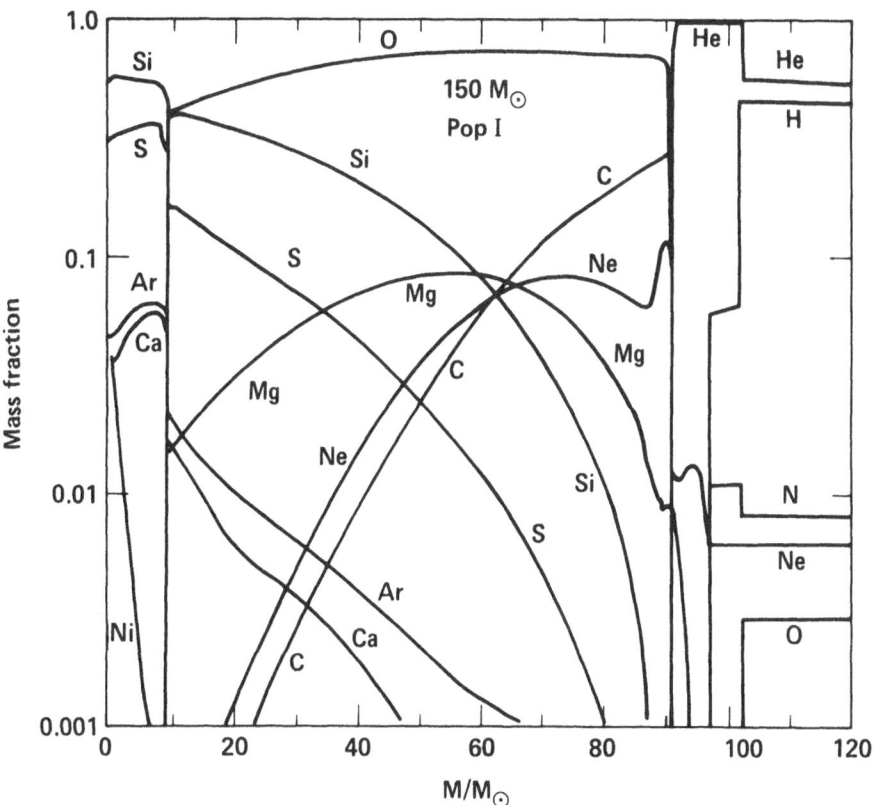

Figure 15. Final nucleosynthesis from a 150 $M_\odot$ Pop I hypernova. Products of explosive oxygen burning plus a trace of ^{56}Ni from explosive silicon burning are visible in the inner regions. The gradual slopes of the lines result both from the gradient in peak explosion temperature experienced by various mass shells and time-dependent convection, which was employed throughout the collapse and explosion phases. Material external to 120 $M_\odot$ was not carried in the core nucleosynthesis calculation, but is expected to have a composition similar to that indicated at 120 $M_\odot$.

Table 5

Nucleosynthesis in pair instability supernovae[a]

Species	^{4}He	^{12}C	^{14}N	^{16}O[b]	^{20}Ne
150 $M_\odot$ Pop I	1.6(−2)	0.24	5.2(−2)	1.0	0.28
200 $M_\odot$ Pop III	2.8(−2)	0.48	2.8(−3)	1.0	0.36

Species	^{24}Mg	^{28}Si	^{32}S	^{36}Ar	^{40}Ca
150 $M_\odot$ Pop I	0.96	4.1	2.6	1.3	1.5
200 $M_\odot$ Pop III	1.5	4.0	2.8	1.4	1.9

[a]Entries are ratios to solar abundances normalized to ^{16}O.

[b]The ejected mass fraction of ^{16}O was 0.43 in both cases.

The evolution of the 200 $M_\odot$ Pop III star was qualitatively similar to that of the 150 $M_\odot$ Pop I star. An interesting variation, however, is the manner in which hydrogen burning in the 200 $M_\odot$ star is mediated by CNO catalyst created in the star itself prior to hydrogen ignition. The star first contracts to a temperature of 1.4×10^8 K, burns a trace of helium to create about 7×10^{-9} by mass CNO, and then commences hydrogen burning by the ordinary (not β-limited) CNO-cycle (see also Ezer and Cameron 1971). Near the time of hydrogen exhaustion, the CNO mass fraction had grown to ~ 10^{-7}, created mostly during the latter stages of hydrogen exhaustion. At a time of 1/2 hydrogen depletion, the star had central temperature 1.22×10^8 K, density $\rho = 22$ g cm^{-3}, luminosity 2.0×10^{40} erg s^{-1}, radius 8.8×10^{11} cm, and effective temperature 7.8×10^4 K. Again the surface properties are highly uncertain owing to the neglect of mass loss. Hydrogen was depleted at an age of 2.74 my and this star too developed a highly extended structure (R = 5.1×10^{14} cm at helium ignition) with a helium core mass of 140 $M_\odot$ and the remainder of the star in a low density envelope of less than about 10^{-10} g cm^{-3}. As helium burning progressed, the surface convective shell ate into this core reducing its mass to ~ 105 $M_\odot$ by the end of helium burning. This convective dredge-up also increased the helium abundance in the envelope to ~ 60% (the exact value again depending on uncertain mass loss parameters). At the carbon ignition stage, the star collapsed on the pair

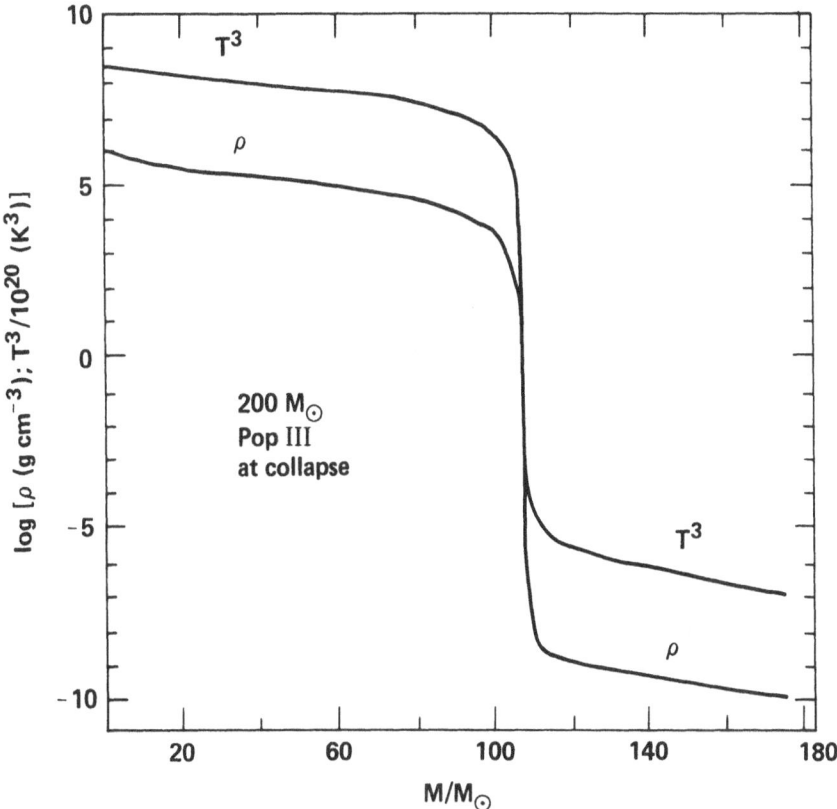

Figure 16.   Pre-collapse temperature and density of a 200 M⊙
Pop III (pure hydrogen-helium) star. Note the centrally condensed
core and low density hydrogen-helium envelope. An adiabatic
exponent close to 4/3 is apparent throughout the star leading to
parallel curves for ρ and T^3.

instability reaching peak temperature and density before velocity reversal of 3.88×10^9 K and 2.15×10^6 g cm^{-3} respectively. The thermodynamic conditions and collapse velocity profile at this time are shown in Figures 16 and 17. We note that for peak temperatures only slightly larger than this (i.e., $T > 4.0 \times 10^9$ K) the onset of endoergic nuclear reactions would have led to a continuing collapse. Thus $200M_\odot$ is not far below the most massive possible pair instability supernova. The resulting explosion here produced total kinetic energy (at infinity) of 2.62×10^{52} erg and a light curve, effective temperature history, nucleosynthesis, and final velocity profile as given in Figures 18, 19, 20, and 21 and in Table 5. Once again this exceptionally brilliant display ($\sim 10^{51}$ erg) is critically dependent on the star retaining at least a fraction of its low density hydrogen–helium envelope up to the time of its explosion. The nucleosynthetic yield of odd Z elements, although not yet calculated for nuclei heavier than ^{14}N, should be very low owing to the lack of any heavy seed nuclei in the initial abundance set and the fact that the star collapsed without undergoing a stable stage of hydrostatic carbon burning to create the necessary neutron excess required for odd–Z synthesis (Woosley and Weaver 1981b). It is also interesting that the nucleosynthetic yield calculated here for even Z elements in the carbon to calcium range is in better accord with solar ratios than obtained by Arnett (1978) for a similar 100 $M_\odot$ helium core explosion. The difference presumably reflects a more realistic treatment of nuclear physics and time-dependent connection in the present calculation.

A 500 $M_\odot$ Pop III star is also currently under study and has been evolved through core hydrogen and (most of) core helium burning. At a time of 1/2 hydrogen depletion (age 1.1 my), the central temperature and density were 1.262×10^8 K and 13.8 g cm^{-3}, the central CNO mass fraction created by helium burning, 9.3×10^{-9}, luminosity, 6.02×10^{40} erg s^{-1}, radius, 1.6×10^{12} cm, and effective emission temperature, 76,000 K. Once again, pulsations are damped by the implicit nature of the hydrodynamics code and radiative mass loss suppressed by an artificial surface boundary pressure (500 dyne cm^{-2}) and later on, during core helium burning, by an artificial surface boundary temperature (25,000 K), to inhibit recombination and pulsation. The artifical nature of the surface boundary conditions makes all but the central properties of this star highly suspect.

This particular 500 $M_\odot$ star is of special interest because of its ability to produce large amounts of primary nitrogen (^{14}N) in a Pop III object. The nitrogen is produced during the core helium burning phase as an extensively convective hydrogen burning shell dredges up the outer regions of a helium convective core where about 30%, by mass, carbon has been produced (although

at no time is there ever a complete convective link-up between
hydrogen shell and helium core). The principal thermodynamic
distinction of this star appears to be the lack of a steep
density gradient separating the core from the 'red-giant'
envelope. The entropy gradient between core and envelope is also
rather shallow - about a factor of 2. At the onset of the
convective dredge up phase, the helium core mass is about 300 $M_\odot$
and almost completely convective. In the steady state at the
hydrogen-helium/carbon interface, hydrogen burning produces a
locally super-Eddington luminosity due to both the high
temperature and the large ^{14}N abundance produced by photon
capture on dredged-up carbon. Typically the H-shell temperature
grows from ~ 45 to ~ 70 million degrees as the ^{14}N mass fraction
in the shell increases from 1% to 20%. The dredge-up of carbon is
a result of convective overshoot at the base of the H-shell. In
our calculation, convective shells are bounded on both
extremities by semi-convective zones that (over a period no
shorter than 10 times the radiative diffusion time) lead to
compositional mixing. It seems likely that such a phenomenon
exists in nature, although considerable argument may exist

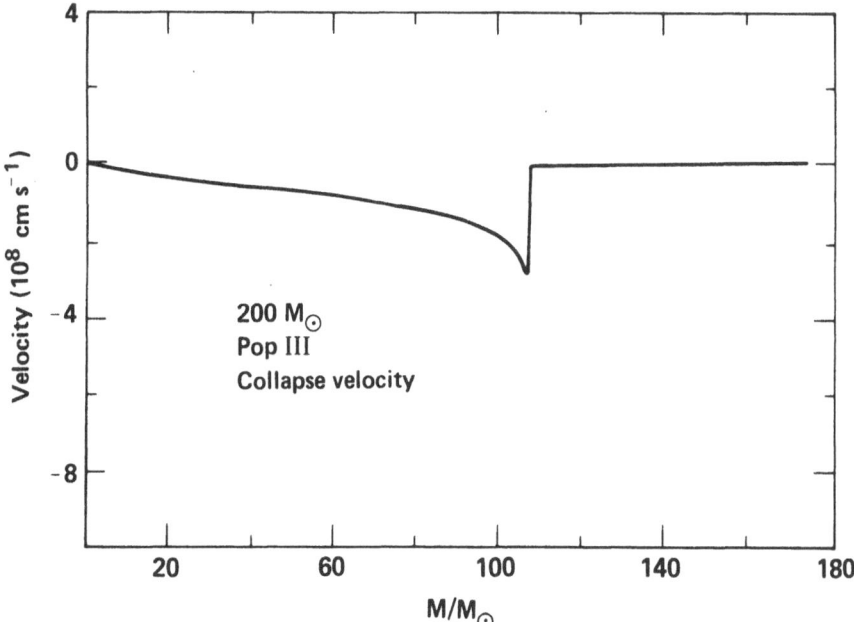

Figure 17. Velocity profile at a time when oxygen first begins
to burn explosively in a collapsing 200 $M_\odot$ Pop III star. At this
time the central temperature is 3.3 x 10^9 K and density
1.3 x 10^6 g cm^{-3}. The low density hydrogen envelope does not
participate in this collapse.

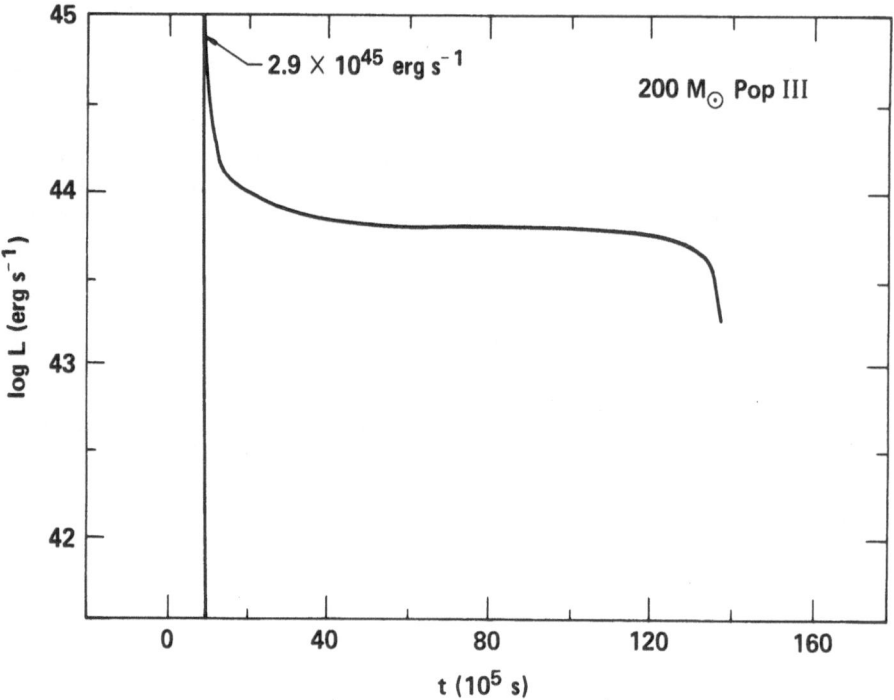

Figure 18. Bolometric curve for a 200 $M_\odot$ Pop III hypernova that retains its low density hydrogen–helium envelope. A spike, due to shock break out, a diffusion tail, a plateau phase as a transparency (recombination) wave eats into the envelope, and a sharp decline as that wave reaches the core envelope interface (see Fig. 16) are all apparent features. See Fig. 14 for likely modifications if the hydrogen envelope is lost.

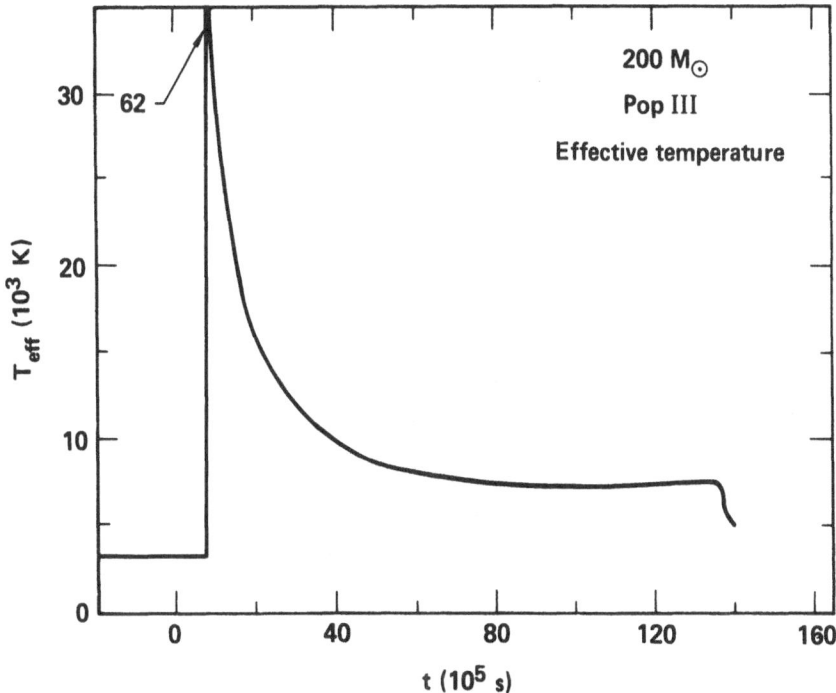

Figure 19. Effective emission temperature corresponding to Fig. 18. The same evolutionary stages are apparent.

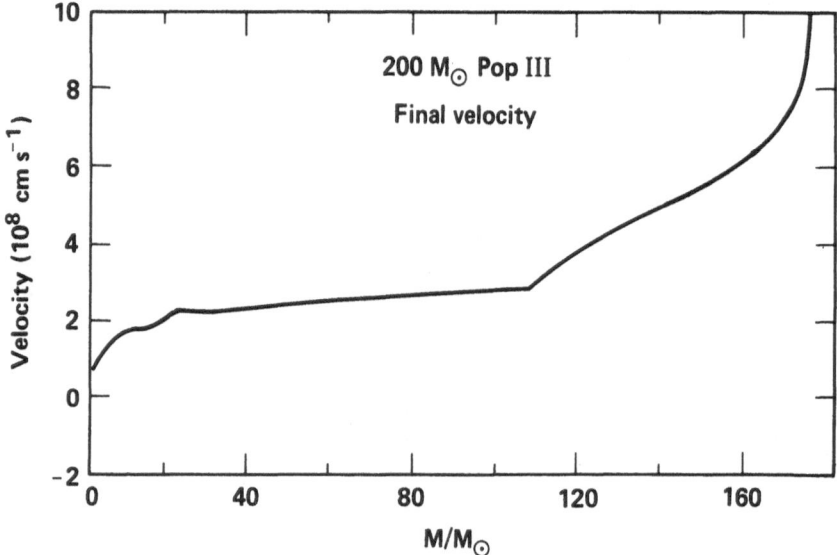

Figure 20. Final velocity profile for a '200 M$_\odot$' hypernova. The outer 25 M$_\odot$ was removed prior to the completion of the calculation due to numerical difficulties associated with the low density outer layers and the artificial handling of surface boundary conditions. The sharp increase in velocity at 175 M$_\odot$ therefore corresponds to the new 'surface' of the star and shock wave steepening is apparent. The final kinetic energy in this explosion was 2.62 x 10^{52} erg.

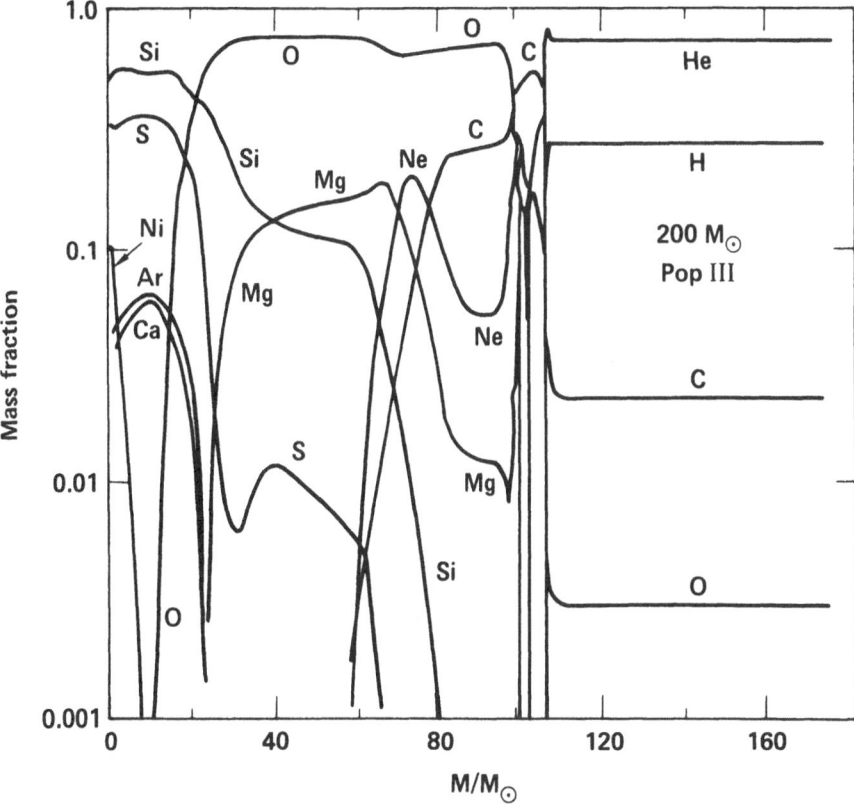

Figure 21. Nucleosynthesis from a 200 $M_\odot$ Pop III hypernova. Products of explosive oxygen and silicon burning can be seen in the inner core. This calculation differed from that shown in Fig. 15 in that numerical difficulties precluded the inclusion of convection throughout the collapse and explosion phases. Hence the curves are not as smooth. The bulk nucleosynthetic yields and explosion energetics should not be overly sensitive to this deletion. Also convection in a medium moving at a fraction of sonic velocity is a highly questionable proposition.

regarding its extent and time scale. It is noteworthy however, that this same parametrization has never before led to primary nitrogen production in any lower mass star (and stars from 8 $M_\odot$ to 200 $M_\odot$ have been studied).

Over a period of 55,000 years the entire helium-carbon core of the 500 $M_\odot$ Pop III star is dredged-up and the hydrogen burning reignites in its centre. At the time of hydrogen reignition the star's age is 2.15 my and the composition at its centre (the star is by now almost completely convective) is, by mass fraction, hydrogen, 10%, helium, 62%, carbon 2%, oxygen 3%, and nitrogen, 3%. Some 290,000 years later, hydrogen is (re-)depleted in the inner 400 $M_\odot$ and, about 300,000 years after that, the $^{14}N(\alpha,\gamma)^{18}F$ reaction ignites, eventually converting all nitrogen in the inner 370 $M_\odot$ into ^{22}Ne (making a very neutron-rich core) and finally helium burns to carbon and oxygen. A residual abundance of $\sim$ 20% ^{14}N and 70% ^4He remains in the outer envelope where it could be lost to a radiative wind (providing that envelope did not long ago disperse). The carbon-oxygen core itself will certainly become a black hole.

Throughout this study of very massive stars, which should be regarded as very preliminary, grave concern may rightfully be expessed concerning the neglect of mass loss, the artificial parametrization of surface boundary pressure and temperature, and the treatment of convective overshoot. The very interesting results of these studies warrant future detailed investigations with approximate, but realistic, corrections for these deficiences.

This research has been supported in part by the National Science Foundation (AST-81-08509) and the Department of Energy (W-7405-ENG-48).

REFERENCES

Applegate, J. and Yahil, A. 1981. Preprint. See also this volume.
Arnett, W.D. 1971. Astrophys.J., 166, 153.
Arnett, W.D. 1973. Ann.Rev.Astron.Astrophys., 11, 73.
Arnett, W.D. 1978. In Physics and Astrophysics of Neutron Stars and Black Holes, ed. R. Giaconni and R. Ruffini (North Holland Pub: Amsterdam), 356.
Arnett, W.D., Truran, J.W. and Woosley, S.E. 1971. Astrophys. J., 165, 87.
Axelrod, T.A. 1980. PhD. Thesis, Univ.Calif. Santa Cruz and Proc.Texas Workshop on Type I Supernovae, ed. J.C. Wheeler, Univ. Texas Press, 80.

Barkat, Z., Rakavy, G. and Sack, N. 1967. Phys.Rev.Letters, 18, 379.

Blake, J.B., Woosley, S.E., Weaver, T.A. and Schramm, D.N. 1981. Astrophys.J., in press.

Bodenheimer, P. and Woosley, S.E. 1981. Bull.Am.Astron.Soc., 12, 833 and in preparation for Astrophys.J.

Cameron, A.G.W. 1973. Space Sci.Rev., 15, 121.

Clayton, D.D. 1973. Nat.Phys.Sci., 224, 137.

Clayton, D.D. 1981. in Essays on Nuclear Astrophysics, ed. C.A. Barnes, D.D. Clayton and D.N. Schramm (Cambridge University Press).

Clegg, R.E.S., Lambert, D.L. and Tomkin, J. 1981. Preprint. Submitted to Astrophys.J.

Cowan, J.J., Cameron, A.G.W. and Truran, J.W. 1981. Astrophys. J., in press.

Ezer, D. and Cameron, A.G.W. 1971. Astrophys.Space Sci., 14, 399.

Fowler, W.A. and Hoyle, F. 1964. Astrophys.J.Suppl., No. 91., 9, 201.

Fraley, G.S. 1968. Astrophys. Space Sci., 2, 96.

Hainebach, K.L., Clayton, D.D., Arnett, W.D. and Woosley, S.E. 1974. Astrophys.J., 193, 157.

Howard, W.M., Arnett, W.D., Clayton, D.D. and Woosley, S.E. 1972. Astrophys.J., 175, 201.

Hoyle, F. 1946. Mon.Not.R.astr.Soc., 106, 343.

Hoyle, F. and Fowler, W.A. 1960. Astrophys.J., 132, 565.

Kirshner, R.P. and Oke, J.B. 1975. Astrophys.J., 200, 574.

Lamb, S.A., Howard, W.M., Truran, J.W. and Iben, I. 1977. Astrophys.J., 217, 213.

Lasker, B.M. 1980. Astrophys.J., 237, 765.

LeBlanc, J.M. and Wilson, J.R. 1970. Astrophys.J., 161, 541.

Markert, T.H., Canizares, C.R., Clark, G.W. and Winkler, P.F. 1981. Bull.Am.Astron.Soc., 12, 799.

Mathewson, D.S., Dopita, I.R., Tuohy, I.R. and Ford, V.L. 1981. Astrophys.J., 242, L73.

Miller, G.E. and Scalo, J.M. 1979. Astrophys.J.Suppl., 41, 513.

Norgaard, H. 1980. Astrophys.J., 236, 895.

Ostriker, J.P. 1981. Private communication.

Ostriker, J.P. and Bodenheimer, P. 1973. Astrophys.J., 180, 171.

Papaloizou, J.C.B. 1973. Mon.Not.R.astr.Soc., 162, 169.

Penrod, D. 1981. Private communication.

Peterson, R. 1981. Astrophys.J., 244, 989.

Silk, J. 1977. Astrophys.J., 211, 638.

Talbot, R.J. 1971. Astrophys.J., 165, 121.

Tammann, G. 1974. In Supernovae and Their Remnants, ed. C. B. Cosmovici, (D. Reidel: Dordrecht), 155.

Tammann, G. 1976. Proc. DUMAND Summer Workshop, ed. A. Roberts, Office of Publications, Fermi National Laboratory, 137.

Tammann, G. 1981. Private communication. See also this volume.

Tomkin, J. and Lambert, D.L. 1980. Astrophys.J., $\underline{235}$, 925.

Truran, J.W., Cowan, J.J. and Cameron, A.G.W. 1978. Astrophys.
   J., $\underline{222}$, L63.

Wallace, R.K. and Woosley, S.E. 1981. Astrophys.J.Suppl., $\underline{45}$,
   389.

Weaver, T.A., Zimmerman, G. and Woosley, S.E. 1978. Astrophys.
   J., $\underline{225}$, 1021.

Weaver, T.A. and Woosley, S.E. 1980. Ann.N.Y.Acad.Sci., $\underline{336}$,
   335.

Weaver, T.A., Axelrod, T.S. and Woosley, S.E. 1980. Proc.Texas
   Workshop on Type I Supernovae, ed. J.C. Wheeler, Univ.
   Texas Press, 113.

Wheeler, J.C. 1977. Astrophys.Space Sci., $\underline{50}$, 125.

Wilson, J.R. and Bowers, R. 1978. Private communication.

Woosley, S.E. and Weaver, T.A. 1980. Astrophys.J., $\underline{238}$, 1017.

Woosley, S.E. and Weaver, T.A. 1981a. Proc.10th Texas Symp.on
   Relativistic Astrophysics, to be published by the N.Y.
   Academy of Science.

Woosley, S.E. and Weaver, T.A. 1981b. in Essays on Nuclear
   Astrophysics, ed. C.A. Barnes, D.D. Clayton and D.N.
   Schramm (Cambridge University Press).

Woosley, S.E., Weaver, T.A. and Taam, R.E. 1980. In Proc.
   Texas Workshop on Type I Supernovae, ed. J.C. Wheeler,
   Univ. Texas Press, 96.

Woosley, S.E. Axelrod, T.S. and Weaver, T.A. 1981. Comm.Nucl.
   and Part.Phys., in press.

Woosley, S.E. and Howard, W.M. 1978. Astrophys.J.Suppl., $\underline{36}$,
   285.

COMPUTER SIMULATIONS OF STELLAR COLLAPSE AND SUPERNOVAE
EXPLOSIONS: NON-ROTATING AND ROTATING MODELS

Wolfgang Hillebrandt

Max-Planck-Institut für Physik und Astrophysik,
Institut für Astrophysik, D-8046 Garching b. Munchen,
FRG

1. INTRODUCTION AND SUMMARY

Although over the last fifteen years many authors have
investigated the question of whether or not the gravitational
collapse of the cores of highly evolved massive stars (M $\gtrsim$ 10 M$_\odot$)
can give rise to the observed (type II) supernova events, the
problem has not been solved yet, despite the fact that the
general ideas are quite simple and convincing.

It is now generally accepted that the iron-nickel cores of
these stars become dynamically unstable and collapse. But because
of their low initial entropy (~ 1 k$_B$/nucleon) they will collapse
until the central density somewhat exceeds nuclear density and a
'proto-neutron star' of about 1 M$_\odot$ forms in the centre of a
massive red giant envelope (Bethe et al. 1979: see also Brown
1982). This young neutron star has gained about 2 x 10^{53} erg of
gravitational energy, which is compensated by kinetic energy,
energy stored in leptons, excitation energy of nuclei, etc. If
only a few percent of the energy available is transferred to the
loosely bound stellar envelope, a supernova explosion will
result, leaving behind a neutron star or a black hole. But the
fact that a mechanism with a few percent efficiency only is
searched for causes many difficulties. In particular one has to
be aware of the possibility that small errors in both the
numerical treatment of the problem and the input physics may even
lead to qualitatively wrong results. In addition it may also be
that the general idea only works with some modifications, with
for example rotation included, and that only those massive stars
that have the right amount of angular momentum explode as
supernovae.

M. J. Rees and R. J. Stoneham (eds.), Supernovae: A Survey of Current Research, 123–155.
Copyright © 1982 by D. Reidel Publishing Company.

The first part of this review (section 2) will deal with the simplest models, i.e. non-rotating models. The question will be discussed whether or not a shock wave generated from the rebounding core by itself causes mass ejection. In particular two classes of models will be analyzed. The first class is usually called adiabatic because the entropy is strictly conserved and weak interaction reactions are ignored. Consequently the electron concentration stays high, $y_e \approx 0.42-0.5$, depending on the initial model, which will have an enormous effect and lead to a very energetic supernova explosions, as will be demonstrated in section 2.2.

Of course, suppressing electron captures, production of thermal neutrinos and neutrino transport is an oversimplification of the real problem and in the light of what was said earlier may even lead to qualitatively wrong conclusions. Therefore complete calculations are needed, which unfortunately so far are rather rare. Most numerical studies still use some approximations, which, however, seem to be not too crude, since the results of all recent investigations agree fairly well. This fact is not very surprising, although somewhat unexpected, as will be discussed in section 2.3. Generally speaking it is found that, starting from very similar initial models, electron captures reduce the mass of the homologously collapsing part of the core significantly. As a result, the shock from the rebounding proto-neutron-star forms deep inside at a mass of only about 0.7 $M_\odot$. Most of the shock energy is then consumed in dissociating heavy nuclei on the way out, the rest being radiated away by neutrinos once the shock has passed the neutrino photosphere. So in none of the recent computations does a supernova explosion result. There may be a few possibilities to overcome these serious difficulties, which we will also briefly discuss.

Section 3 is devoted to those computations in which the assumption of spherical symmetry is omitted. These models require a two-dimensional hydrocode at least and are therefore much more laborious and expensive. They allow, however, one to investigate physical processes, which from their very nature cannot be treated in one dimension and which may become important if the simple core-bounce models fail. If not for other reasons than from nucleosynthesis arguments one would hope that at least some of the very massive stars do explode (see e.g. Woosley and Weaver 1982).

The most straightforward and realistic extension is of course to incorporate rotation since we know that all stars do rotate and it is very likely that they do not lose much of their angular momentum during the evolution. Therefore we will first discuss rotating collapse models in section 3.1. Because recent equations of state predict an adiabatic index significantly smaller than 4/3 at high densities and low entropies, rotation will not play the dominant role during collapse unless the angular velocity is already close to its critical value in the initial model. Nevertheless it will be shown that even initially moderately rotating stellar cores will add important modifications to the simple core-bounce picture, which may finally lead to supernova explosions.

In addition it is obvious that once we consider rotating stars other processes should also be taken into account. Here in particular those coming from the presence of magnetic fields, lepton number and entropy gradients and unburned nuclear fuel may turn out to be important (section 3.2). All those latter models are rather speculative and preliminary at present, but open a variety of possibilities for future investigations. Moreover, it is not obvious from the beginning that two-dimensional effects do indeed help and produce supernova explosions in cases where the core-bounce models failed. On the contrary, it cannot be excluded that two-dimensional effects are disadvantageous in cases where the core-bounce models were successful.

2. NON-ROTATING SUPERNOVA MODELS

2.1 General considerations

Before we can start and simulate the collapse of the cores of massive stars on a computer we have to decide which physical and astrophysical input data we are going to use.

First of all a stellar model has to be chosen. Weaver et al. (1978) have evolved stellar models of 15 and 25 $M_\odot$ from the main sequence to the onset of core collapse. These models can be characterized as follows. After core-silicon depletion a silicon burning shell develops. When the core becomes dynamically unstable this shell is at 1.55 $M_\odot$ in the 15 $M_\odot$ star and at 1.6 $M_\odot$ in the 25 $M_\odot$ star. The central entropy of both models is about 1 k_B/nucleon and the electron concentration in the centre is approximately 0.44. These models are now commonly used in hydrodynamical studies. Arnett (1977a), on the other hand, has evolved He-cores in the mass range from 4 $M_\odot$ to 32 $M_\odot$, corresponding to total main sequence masses from about 15 $M_\odot$ to 80 $M_\odot$. Although on the average Arnett's model-cores are not so different from the ones computed by Weaver et al., there are some differences which may turn out to be important. First he finds a systematic trend to smaller Fe-Ni-core masses with decreasing total mass, giving a 1.45 $M_\odot$ core for the 4 $M_\odot$ He-star. Second the entropy in the centre of this particular model is significantly lower (~ 0.7 k_B/nucleon) than in the corresponding star of Weaver et al. Whether or not these, on a first view,

minor differences in the initial models will change the dynamics
of collapse will be discussed in some detail later. It is
worthwhile to mention that even smaller core masses are not out
of the range of possibilities. In particular stars with total
main sequence msses of less than about 12 $M_\odot$ (but larger than
8 $M_\odot$) will form Fe-Ni-cores of less than 1.4 $M_\odot$ (Nomoto et al.
1979: Sugimoto and Nomoto 1980).

Next an equation of state (EOS) has to be chosen. Here two
different approaches are feasible at present for the large
varieties of densities, temperatures and electron concentrations
needed. The first approach is based on the assumption that even
at high densities ($\rho \gtrsim 10^{13}$ g cm^{-3}) nuclei can be treated
essentially as a non-interacting gas of Boltzmann particles and
that their properties can be obtained by extrapolation from those
of known nuclei. The composition in nuclear statistical
equilibrium (NSE) can then be computed from a set of modified
Saha equations (which corresponds to a minimization of the free
energy for a given sample of nuclei), the modifications being due
to excluded volume and Coulomb lattice effects, and to neutron
degeneracy and nucleon-nucleon interactions at high densities.
This approach has been discussed in detail by Mazurek et al.
(1979) and El Eid and Hillebrandt (1980). It has the advantage
that at low densities ($\rho \gtrsim 10^{11}$ g cm^{-3}) and moderate temperatures
(T $\lesssim 10^{10}$ K), the nuclear composition, which is important for
computations of the electron capture rates, can be determined
very accurately, since shell and pairing effects in the nuclear
binding energy and partition function are properly taken into
account. Moreover at higher temperatures (entropies) one expects
a large variety of nuclear species coexisting in NSE and this
effect is properly included.

A major shortcoming of the Boltzmann gas approach is, of
course, that interactions between free nucleons and nuclei cannot
be treated and that modifications in the properties of nuclei due
to the presence of an external free neutron gas cannot be
included in a consistent and straightforward way. These effects
are certainly important at high densities ($\rho \gtrsim 10^{13}$ g cm^{-3}) and
can be handled consistently in the finite temperature liquid drop
model of Lamb et al. (1978). (See also the review of Lattimer
1981.) In this approach the total free energy per nucleon of a
nucleus surrounded by a vapour of free nucleons is calculated and
minimized with respect to the number of neutrons and protons both
in the nucleus and the vapour in a given unit cell. In addition a
contribution from ^{4}He to the free energy is usually included in
order to mimic the possible coexistence of heavy and light
nuclei, the major assumption still being that the composition in
NSE can be approximated by one representative nucleus and that
shell and pairing effects can be ignored.

Although it may seem that both ways of computing the EOS are incompatible, they lead to good agreement if numerical results are compared. Only for entropies around 3 k_B/nucleon the agreement is not as good, which is not surprising, since this is the region of photodisintegrations. On the other hand, the transition to homogeneous nuclear matter (near nuclear matter density) is more abrupt in the Boltzmann gas approach, indicating a shortcoming of this method. But again it turns out that the results of hydrodynamical studies are not very sensitive to these differences.

The good agreement between the different EOS computations does not course prove that they are correct. More sophisticated calculations like finite temperature Hartree-Fock (Bonche and Vautherin 1980) or Thomas-Fermi (Buchler and Epstein 1980), however, suffer from the fact that they become technically very difficult for high temperatures and cannot be extended to entropies beyond 1 or 2 k_B/nucleon. The low entropy calculations, on the other hand, show good agreement with the more simple models, so that in general the EOS up to nuclear matter density seems to be sufficiently well known.

The same cannot be said about the EOS at even higher densities. But fortunately in recent collapse calculations the central density never exceeds about 5×10^{14}g cm^{-3}, and for these conditions the assumption of a temperature independent effective nucleon-nucleon interaction is not so bad. Moreover, we do not have to worry about exotic states of nuclear matter such as pion condensates or quark matter.

Next, weak interaction rates have to be discussed. Here electron captures on heavy nuclei are still somewhat uncertain although recently some of the uncertainties have been removed (Fuller et al. 1981). On the other hand at the densities where most of the electron captures occur ($10^{10} \lesssim \rho$(g cm^{-3}) $\lesssim 10^{12}$) recent equations of state predict a free proton mass-fraction of around 10^{-3} to 10^{-4} at entropies of about 1 k_B/nucleon (El Eid and Hillebrandt 1980: Van Riper and Lattimer 1981) and the most abundant nucleus in full NSE calculations for $y_e \sim$ 0.4-0.42 turns out to be ^{48}Ca (El Eid and Hillebrandt 1980). For such a situation the electron capture rates are dominated by electron capture on free protons (Fuller 1981), for which the rates can be computed exactly and excellent approximation formulae are available (Bludman and Van Riper 1977: Takahashi et al. 1978). Even if one neglects shell effects in determining the composition in NSE, as it is done in the liquid drop formalism of Lamb et al. (1978) and Bethe et al. (1979), electron captures on free protons dominate at densities above 10^{11} g cm^{-3} (Van Riper and Lattimer 1981). Electron capture rates, therefore, are no longer a major source of uncertainties in collapse calculations.

Concerning the question of the importance of neutrino transport in hydrodynamical studies of core-collapse supernova models, the answer is less clear. The most sophisticated scheme for neutrino transport that has so far been used in collapse calculations is a flux-limited multi-energy group-diffusion approximation (Arnett 1977, 1982: Wilson 1978). While this scheme is certainly sufficiently accurate during collapse and probably even simple neutrino leakage schemes (Van Riper and Lattimer 1981) do not modify the results significantly, once the shock reaches the neutrino photosphere simple diffusion approximations may fail. This is mainly due to the fact that the shock front is normally smeared over several mass-zones by the pseudo-viscosity in the numerical codes, thus giving rise to an artificial shock precursor. Therefore the standard diffusion schemes will overestimate neutrino losses from the shock (Arnett 1982). Whether or not neutrino losses from the shock are the main damping mechanisms, however, is still an open question and subject to some dispute (see e.g. Bethe 1982: Hillebrandt and Müller 1981: Mazurek et al. 1980) and certainly calls for further investigation.

So in summarizing it seems that we know most of the input physics we need for numerical studies of core-collapse supernova models rather well. Nevertheless one should keep in mind that whether or not a supernova explosion results from core collapse may indeed depend upon minor details of the initial model, equation of state, etc., and that we still may not have reached the required level of confidence. These uncertainties will be addressed in some more detail in the following sections.

2.2 Adiabatic Models

Adiabatic collapse models, e.g. models in which electron captures and neutrino losses are neglected, have been investigated by several authors (see e.g. Van Riper 1978: Van Riper and Arnett 1978: Lichtenstadt et al. 1980). The main aim of those studies was to show what properties the equation of state must have in order to obtain supernova explosions from either a rebounding core or a reflected shock wave at the forming neutron star. In most of these models an initial configuration of an $n = 3$ polytrope of $1.4 \, M_\odot$ was assumed, and a simple parameterized EOS was used. It was then found that for a certain range of parameters (adiabatic index γ during collapse slightly below 4/3 and sufficient thermal stiffness, which means that most of the kinetic energy dissipated in the shock is transformed into thermal pressure) explosion could indeed be obtained.

Recently Arnett (1981) and Hillebrandt (1981) have investigated the question whether or not more realistic initial stellar models and realistic equations of state lead to the same results. They both found that for different stellar models (the 25 $M_\odot$ star of Weaver et al. (1978) was used by Hillebrandt, and 4 $M_\odot$ and 32 $M_\odot$ He-stars by Arnett) and slightly different equations of state, very energetic supernova explosions were obtained (see also Fig. 1). One of the main features of all those computations was, however, that the shock formed rather far outside at, at least, 1 to 1.2 $M_\odot$, which was caused by the fact that electron captures were ignored. Electron captures, on the other hand, decrease the electron concentration in the inner parts of the collapsing core significantly. Consequently the homologously collapsing part of the core will decrease in mass and therefore the shock should form deeper inside. The propagating shock then has to photodisintegrate much heavier nuclei on its way out, an effect which costs about 3×10^{51} erg for 0.2 $M_\odot$ of heavy nuclei.

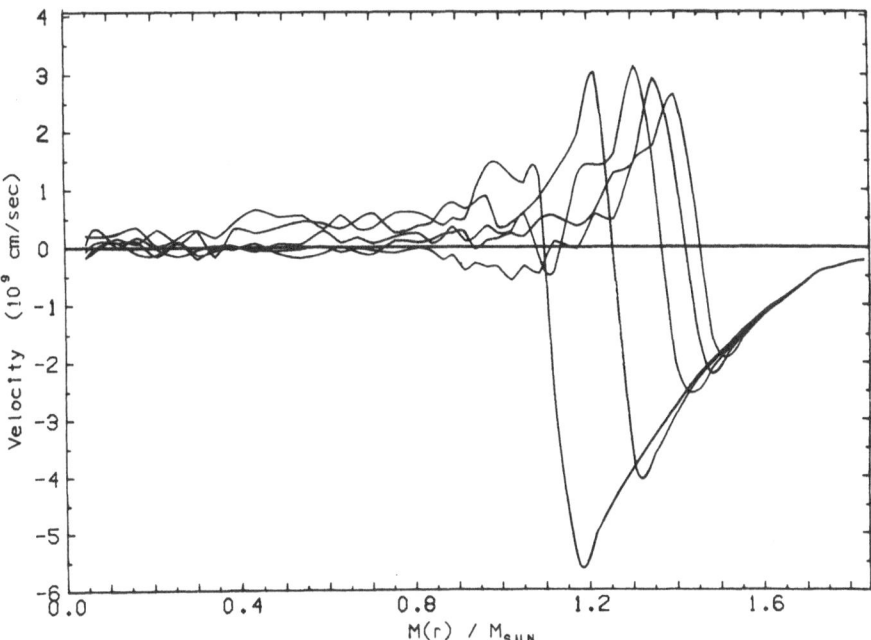

Figure 1.   Velocity versus mass for an adiabatic collapse at different times after core bounce. The computation was based on the 25 $M_\odot$ star of Weaver et al. (1978). Because electron captures were suppressed the shock forms at about 1.1 $M_\odot$. When the computation was stopped about 8 ms after bounce the outward velocity had almost reached escape velocity.

In order to analyze this effect in some more detail we have
performed a computation, starting again from the 25 $M_\odot$ star of
Weaver et al. (1978), in which we allowed for electron captures
but kept essentially all neutrinos that were produced. We found
that the mass of the unshocked core was reduced to 0.85 $M_\odot$, the
shock was considerably damped on its way out, and by the time it
reached a density of about 2 x 10^{10} g cm^{-3} (a radius of 150 km)
it changed into a standing accretion shock, mainly due to both
dissociation of heavy nuclei and creation of electron-positron
pairs (which costs about 1 MeV per pair).

This rather disappointing result, which is consistent with
computations reported by Mazurek (1981), indicates that core
collapse calculations based on the stellar models of Weaver et
al. (1978) are right on the border line between explosions and
absence of explosions, but that explosions generated from a
reflected shock are unlikely to occur if there is no additional
energy deposition behind the shock front from another mechanism.
Brown et al. (1981) have suggested that neutrino diffusion from
the core through the shocked and dissociated part of the mantle
to the shock front will supply this energy. There are, however,
some difficulties with this picture. First neutrinos cannot leave
the unshocked core on timescales required to drive the shock ($\sim$
milliseconds) because of the rather large neutrino opacities
there. For the dissociated matter behind the shock, neutrino
diffusion timescales may be sufficiently short, but in typical
computations we found that the total energy stored in neutrinos
in this matter behind the shock was about 1-2 x 10^{51} erg only.
This is about 5 to 10% of the internal energy of those mass
zones. Applying the ordinary diffusion approximation to the
results from our computations we found that at most 30% of that
neutrino energy can be brought to the shock front on timescales
of a few milliseconds. Therefore the overall energy deposition by
neutrinos behind the shock will be less than about 3-4 x 10^{50}
erg/ms over about 3 or 4 ms until the shock reaches the neutrino
photosphere, which is probably not sufficient to change the
dynamics considerably.

2.3 Non-adiabatic models

a) The collapse stage

Several groups have recently performed collapse computations
which differ in the initial models, the equations of state, weak
interaction rates and neutrino transport models. Some
calculations have been done which also include general
relativistic effects. The input physics used in all these models
is summarized in Table 1. From the rather large spread in the
input physics one would also expect significant differences in
the final results, but this turned out to be not the case,

Table 1: Characteristics of collapse computations

Author(s)	Wilson (1980)	Mazurek, Cooperstein and Kahana (1980)	Hillebrandt and Müller (1981)	Van Riper and Lattimer; Var. Riper (1981)	Arnett (1982)	Hillebrandt (1981)
Initial model	15 $M_\odot$ Weaver et al. (1978)	15 $M_\odot$ Weaver et al. (1978)	25 $M_\odot$ Weaver et al. (1978)	15 $M_\odot$ Weaver et al. (1978)	4 $M_\odot$, 32 $M_\odot$ He-cores Arnett (1977a)	4 $M_\odot$ He-cores Arnett (1977a)
Equation of state	one-nucleus approach Wilson (1978)	Nuclear network, cold neutron matter EOS	Nuclear network, finite temperature nuclear matter EOS, El Eid, Hillebrandt (1980)	one-nucleus approach, corrected Lamb et al. (1978) EOS	one-nucleus approach, adjusted to nuclear networks, cold neutron matter EOS	as Hillebrandt and Müller (1981)
Weak interaction	electron capture on protons	electron capture on protons	electron capture on protons	electron capture on protons, nuclei with and without shell blocking (Fuller et al. 1980)	electron capture on protons, nuclei with and without shell blocking (Fuller et al. 1980)	as Hillebrandt and Müller (1981)
Neutrino tranport	flux limited (equilibrium) diffusion	flux limited (equilibrium) diffusion	parameterized neutrino scheme	parameterized neutrino leakage scheme	Multigroup flux-limited neutrino diffusion	as Hillebrandt and Müller (1981)
Hydro-dynamics	Newtonian	Newtonian	Newtonian	Newtonian and relativistic	relativistic	relativistic

proving some of the statements made in section 2.1 (see also
Tables 2 and 3). Some typical results are also displayed in
figures 2 to 5.

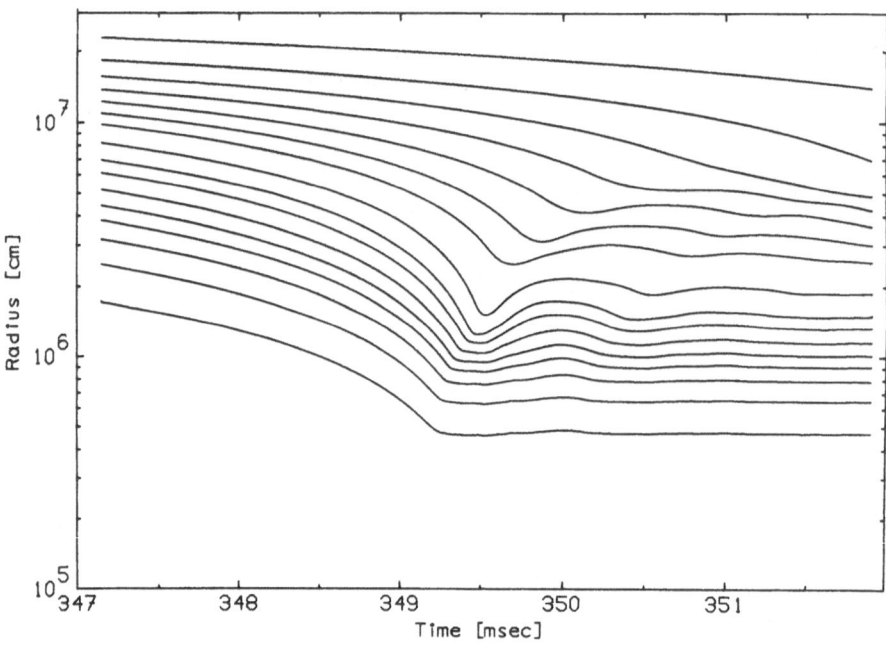

Figure 2.   Radius versus time for selected mass-zones of a non-
adiabatic collapse computation (from Hillebrandt and Muller
1981). Again the 25 $M_\odot$ star of Weaver et al. (1978) served as an
initial model. In this computation neutrino trapping was assumed
to occur at 3 x 10^{11} g cm^{-3}, whereas the neutrino photosphere was
assumed to be at 10^{10} g cm^{-3}. The time is measured from the onset
of collapse. The hydrostatic core after bounce contains about
0.8 $M_\odot$.

Table 2: Results of core collapse computations

Author(s)	Wilson (1980)	Mazurek et al. (1980)	Hillebrandt, Müller (1981)	Van Riper (Lattimer) (1981)	Arnett (1982)	Hillebrandt (1981)
ρ_c at bounce (in 10^{14} gcm^{-3})	4	4	2.3	3 - 4	4	5
S_c at bounce (in k_B/nucleon)	1.0	1.2	1.1	1.1±0.2	1.3	1.0
mass of the homologous core (in $M_\odot$)	>0.8	0.8	0.7	0.75±0.15	0.75	0.75
maximum infall velocity at bounce (in 10^9 cm s^{-1})	6	-	6	7 - 9	4	4
lepton concentration in the central zone at bounce	0.37	0.38	0.38	0.34-0.36	0.38	0.39
electron concentration in the central zone at bounce	0.28	-	0.22	0.28-0.29	0.28	0.31

Notes: 1.) - : no numbers available.

2.) The different numbers in the Van Riper-Lattimer computations were obtained for different assumptions about electron capture rates, etc.

3.) The rather low electron concentration in the computations of Hillebrandt and Müller was caused by electron capture rates which did not include neutron degeneracy.

Table 3: Shock properties

Author(s)	Wilson (1980)	Mazurek et al. (1980)	Hillebrandt Müller (1981)	Van Riper (1981)	Arnett (1982)	Hillebrandt (1981)
Explosion	yes	no	no	no	almost	yes?
Maximum shock radius (km)	∞	130	80-100	80	500	>300
and mass ($M_\odot$)	–	1.3	1.0-1.2	1.0	1.4	> 1.3
main damping of the shock	–	dissociation, ν-losses	dissociation, pairs	dissociation, ν-losses	dissociation, ν-losses, pairs	dissociation, pairs
Maximum entropy	10	8	7-8	8	8	8

Note: The only clearly successful model (Wilson, 1980) has the highest entropy in the shock, although the initial shock energy was significantly lower than in the other computations (see Brown et al., 1981). This difference was probably caused by rather high neutrino losses from the inner unshocked core.

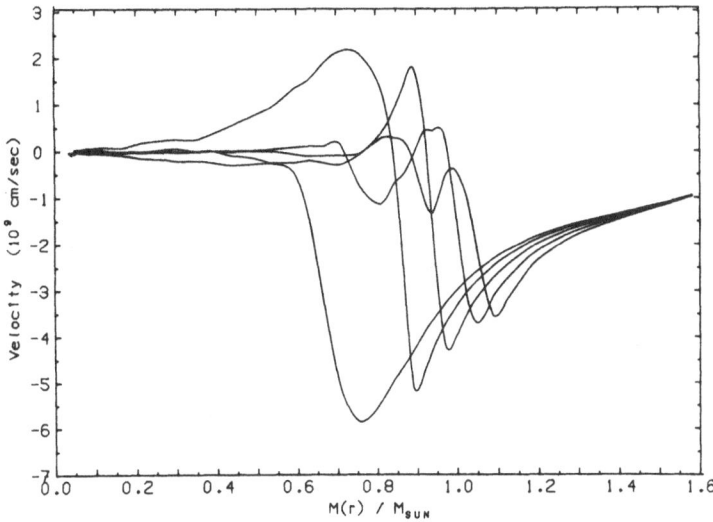

Figure 3. Velocity versus mass for different times. The model is the same as in Fig. 2. The maximum outward velocity in this computation was about 2.5×10^9 cm s^{-1}, but the shock turned into an accretion shock at about 1.1 $M_\odot$.

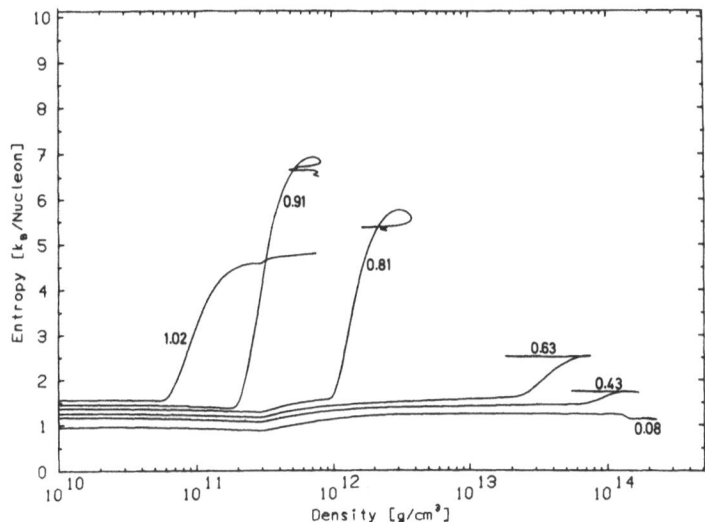

Figure 4. Entropy versus density for selected mass-zones labelled with the mass in units of $M_\odot$. The model is the same as in Fig. 2. The maximum entropy was reached at about 0.9 $M_\odot$. The little loops result from both numerical errors in the energy equation and anti-neutrino losses.

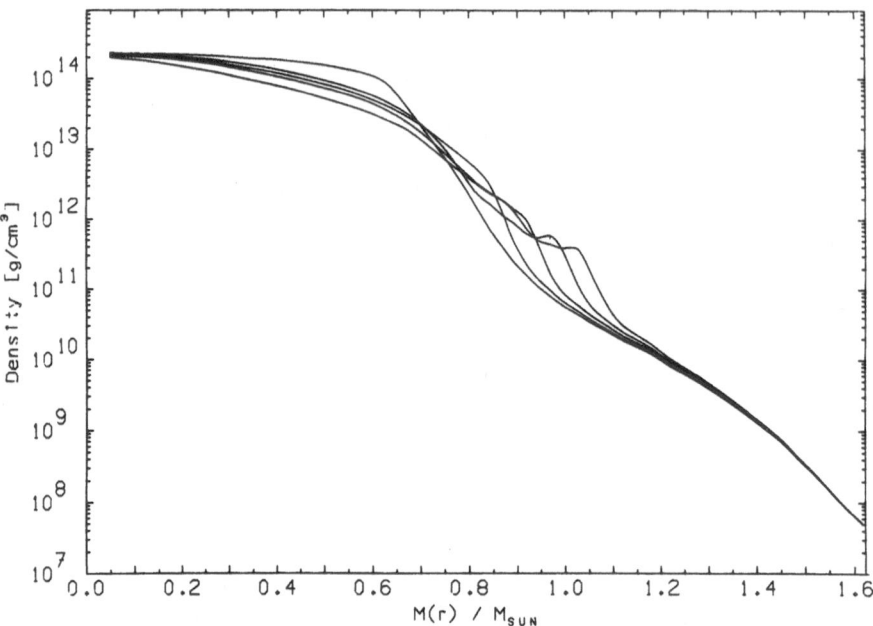

Figure 5. Density versus mass for different times. The snapshots
and models are the same as in Fig. 3.

First of all, in all computations the collapse proceeds up to
or slightly beyond nuclear matter density ($\sim$ 2.5 x 10^{14} to
5 x 10^{14} g cm^{-3}), the actual value depending on the stiffness of
the (nuclear matter) equation of state and on whether or not
relativistic effects were included. Generally speaking, stiffer
equations of state tend to decrease the central density at bounce
somewhat whereas general relativity tends to increase it, but the
differences are certainly minor if realistic equations of state
are used (see also Hillebrandt and Müller 1981 and Van Riper
1981). Also the mass of the homologously collapsing core just
before the innermost parts of the star start to decelerate ('last
good homology') are not so different in the various computations
and vary only around 0.75 M$_\odot$ with a spread of about 0.15 M$_\odot$ in
both directions depending mostly on the exact lepton concentration at neutrino trapping but also somewhat on whether or not
general relativity was included. As could be expected both
Newtonian hydrodynamics and increasing lepton concentration do
also increase this mass, which is almost exactly equal to the
fraction of the core mass that remains unshocked and approaches
hydrostatic equilibrium about a millisecond after bounce. With
the exception of Arnett (1982) and Hillebrandt (1981) who find,

probably because of differences in the initial stellar model, rather low infall velocities at bounce ($\sim 4 \times 10^9$ cm sec^{-1}), the computations based on the Weaver et al. (1978) stellar models all give $6-7 \times 10^9$ cm sec^{-1} infall velocities, again in very good agreement. Similarly good agreement is found for the central temperature ($T \sim 10-12$ MeV) (with the exception of Wilson 1980, who obtains ~ 20 MeV) and entropy ($S \sim 1.3-1.5$ k$_B$/nucleon) at bounce, and the shock forms between a radius of 15 and 20 km. From all these numbers we can conclude that:

1. During core collapse and throughout core bounce the results are rather sensitive to the details of the input physics if these are varied within reasonable limits. This gives us some hope that the general physics of the stellar collapse is sufficiently well understood.

2. The energy initially available to drive the shock and ultimately to eject matter in a supernova explosion should be approximately the same in all computations and should be around 4 to 7×10^{51} ergs.

3. The main difference between (successful) adiabatic models and more detailed and realistic computations seems to be the difference in the mass of the unshocked core and in general, as was discussed earlier, having more matter unshocked favours explosions.

b) Shock propagation

In all recent calculations the infall velocity at the edge of the homologous core near bounce is sufficiently supersonic that an outward propagating shock front forms, but apart from Wilson's (1980) computation all fail to give supernova explosions. There are several reasons for this failure, among which the most important one seems to be the energy needed to dissociate the matter passing through the shock front. In those computations based on the Weaver et al. (1978) stellar models the pressure and velocity just behind the shock front are not sufficiently large to replace the energy from the pdV work done once the shock has reached densities of about 10^{11} g cm^{-3}, a radius of about 100 km and a mass of about 1.2 $M_\odot$ (see Fig. 3). At that stage Van Riper (1981) finds neutrino luminosities of about 2×10^{50} erg/ms only (although his neutrino leakage scheme may overestimate the neutrino losses) which has to be compared with the rate at which energy is used to dissociate the nuclei, which is about ten times larger. In the computations of Mazurek et al. (1980) the neutrino luminosity turns out to be a factor of about 5 larger than in Van Riper's models, and therefore becomes comparable to the energy losses due to nuclear dissociation. Since their transport scheme may again overestimate the neutrino losses they

have performed also a computation in which the neutrinos were
kept (Mazurek et al. 1981) but did not find significant
differences, in agreement with the results of Hillebrandt and
Müller (1981) who came to the same conclusion. Van Riper's (1981)
statement that his adiabatic models did explode and therefore
deleptonization behind the shock front caused the shock to die is
misleading because he compared models with significantly
different core mass at bounce, which in our opinion was the main
reason for the differences. In the analytic investigations of
shock propagation of Bethe et al. (1980) and Brown et al. (1981)
it is assumed that the temperature distribution behind the shock
is uniform and that the density behind the shock increases
proportional to r^{-2}. Both assumptions do not agree with the
recent numerical studies, the only exception again being Wilson's
(1980) computation, who indeed obtained an almost uniform
temperature distribution behind the shock front, which was
probably caused by the fact that he also obtained unreasonably
fast deleptonization of the inner shocked core. We therefore
think that the disagreement between numerical and analytic
studies should not be taken too seriously at present, but more as
an indication where detailed numerical studies have to be very
precise.

In the computations discussed so far the shock would have to
propagate at least an additional 200 km (or 0.3 $M_\odot$ in mass)
before nuclear dissociation becomes unimportant and further
damping may not be expected. Stellar models with less mass of the
iron-nickel cores will therefore be more favourable for
explosions. Smaller core-masses, on the other hand, are certainly
not out of the range of possibilities. Arnett (1982) has computed
a collapse based on his 4 $M_\odot$ He-core stellar model. This differs
from those used by other authors by: 1) a significantly lower
initial entropy (~ 0.7 k_B/nucleon in the centre) and 2) somewhat
smaller core-mass (~ 1.45 $M_\odot$). Due to the lower entropy, electron
captures are slightly reduced and consequently less pressure
deficit is obtained during collapse. Therefore the infall
velocities are also significantly lower than in all other
computations and the shock forms at somewhat larger mass
(0.75 $M_\odot$), comparable to what one obtains for the Weaver et al.
(1978) initial models if electron captures on heavy nuclei is
neglected (Mazurek et al. 1980: Hillebrandt and Müller (1981).
By the time the shock front has reached about 1.1 $M_\odot$, the
maximum infall velocity is less than 2 x 10^9 cm s^{-1}, compared to
about 3 x 10^9 cm s^{-1} for the other models. Moreover, because of
the different structure of Arnett's model, the shock is then
already at a radius of 140 km, compared to about 70 km in the
more massive, higher entropy stellar cores. From the analysis of
Brown et al. (1981) one can estimate that due to the lower rate
at which nuclei are dissociated in Arnett's computation, this
effect alone cannot stop the shock. He finds, however, that after

the shock has passed the neutrino photosphere the outward velocity has dropped to about 10^9 cm s^{-1}, which, after all, was not sufficient for an explosion. Out at a radius of about 500 km the rest of the shock energy was going into the rest mass of electron-positron pairs and could not be replenished by pdV work. In general, at low densities ($\rho \lesssim 10^{10}$ g cm^3) and rather high temperatures, pair creation lowers the adiabatic index considerably, an effect that has also been discussed by Hillebrandt and Muller (1981), who found that their shocks would propagate significantly further out if pairs were suppressed. Hillebrandt (1981) has also run a computation in which he started from Arnett's 4 $M_\odot$ He-star, but kept the neutrinos once the shock propagated until the density behind the shock had dropped below 10^{10} g cm^{-3}. In this computation the outward velocity by the time the shock had reached a radius of about 300 km (at 1.3 $M_\odot$) was still about 2 x 10^9 cm sec^{-1}, twice the value obtained by Arnett, and equal to the escape velocity. It is very likely that this model will lead to an explosion, but the computations had to be stopped because they were becoming too expensive. Since the same procedure applied to the 25 $M_\odot$ stellar model of Weaver et al. (1978) failed to give an explosion, and the differences in the models were not very large, future stellar evolution calculations should try to predict the properties of the cores at the onset of collapse more precisely.

c) Summary and conclusions

From all recent stellar collapse computations one comes to the following conclusions:

1) Models based on the more massive initial iron-nickel cores are unlikely to give explosions generated directly from core-bounce. This failure of the models is caused by the fact that energy losses from the shock due to nuclear dissociation and, to a certain extent, due to neutrino emission are too severe and turn it into an accretion shock at about 1.3 $M_\odot$, which is still deep inside the original iron-nickel core. This result seems to be independent of differences in the input physics, such as in the equation of state, electron capture rates, and neutrino transport models. Nevertheless all models are sufficiently close to an explosion that additional effects, such as rotation, magnetic fields, etc., may also qualitatively change the picture.

2) Smaller iron-nickel cores and lower entropy favours explosions, although for these models details of neutrino-transport may become decisive. Since there is a general tendency in stellar evolution calculations to relate smaller iron-nickel core masses to lower main sequence masses, it may well be that only stars with originally between 8 and 12 $M_\odot$, say,

undergo core-collapse type II supernova explosions and most of
the very massive ones form black holes. Such a possibility can
certainly not be ruled out from observations. Moreover, even if
one takes into account that the stellar birth rates are rather
uncertain for very massive (M $\gtrsim$ 15 M$_\odot$) stars, the number of type
II supernovae makes it rather unlikely that these are the typical
progenitor stars (Tammann 1982). The failure of most of the
recent models to predict supernova explosions should therefore
not be considered as a disaster, but maybe the predictions are
indeed correct.

3. TWO-DIMENSIONAL (2-d) MODELS

3.1 Rotating models

Because at present core-collapse calculations based on rather
massive stars (~ 20 M$_\odot$) do not lead to supernova explosions, but
on the other hand at least some of these stars should explode in
order to explain several of the heavy element yields in the
galaxy (see e.g. the contribution of Woosley and Weaver (1981) to
this conference), one should be open-minded and investigate
generalizations of the simple core-collapse picture.

The most simple and straightforward extension is to include
rotation, since we know that stars do rotate and it is very
unlikely that they lose much of their angular momentum during
their evolution. As the ratio of centrifugal force to gravity
increases as r^{-1} with decreasing radius it is obvious that
rotation becomes more and more important as the cores of the
stars get more and more compact. It may even be that initially
not very rapidly rotating configurations become unstable against
triaxial deformations during collapse or may even fragment. Hoyle
(1946) has suggested that such instabilities may cause supernova
explosions.

With the numerical codes available at present these extreme
(3-d!) situations cannot be simulated, but it is feasible to
investigate more moderate cases in which the cores remain stable
to triaxial deformations, at least on dynamical timescales and
which therefore can be treated with 2-d hydrocodes. Muller et
al. (1980) have computed rotating models following essentially
the spirit of the spherically symmetric adiabatic collapse
calculations e.g. by assuming a n = 3 polytropic stellar model of
1.4 M$_\odot$, a parameterized equation of state and adiabatic
hydrodynamics. Of course in rotating models a certain amount of
angular momentum as well as the angular momentum distribution
has to be chosen, because these parameters are not known from
stellar evolution models. Müller et al. (1980) assumed rigid
rotation and $(\Omega/\Omega_{crit}) = 0.3$ at the outer boundary of the core,
where $\Omega_{crit} = (GM/R^3)^{-1/2}$, G is the gravitational constant, M the

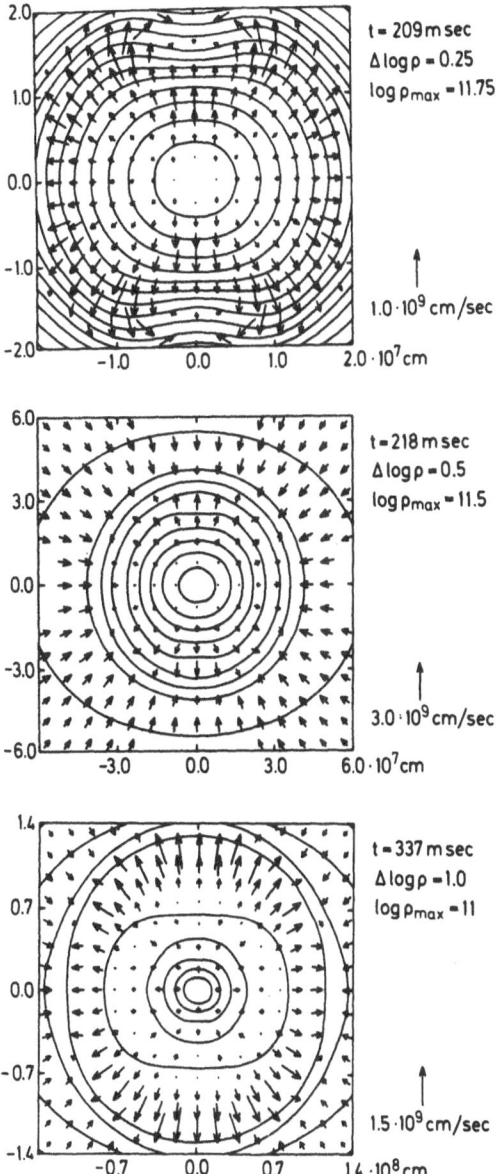

Figure 6. Rebound and shock propagation in a rotating adiabatic core-collapse model (from Müller et al. 1980). The bounce occurred at a central density of 10^{12} g cm^{-3}. Isopycnic contours as well as velocities are shown. The outward velocities are significantly greater at the poles and are beyond escape velocity. About 2 ms after core-bounce large scale circulation flows develop, which disappear some 8 ms later.

mass and R the radius. The parameters of the equation of state were taken from Van Riper (1978). They found significant qualitative and quantitative differences between rotating and non-rotating models (see Fig. 6):

1) The central density at bounce was two orders of magnitude lower for the rotating cores ($\rho_c \approx 10^{12}$ g cm^{-3}) and the bounce was caused by centrifugal forces rather than by the stiffness of the equation of state.

2) Large scale circulations developed, which were set up by a rather complicated interplay between centrifugal forces and gravity.

3) In cases where they found explosions, the expansion velocities were significantly larger along the polar axis than at the equator. Therefore _jets_ can be expected from those models.

Recently Tohline _et al_ (1981) have repeated the computations of Muller _et al_. (1980) and arrived at the same conclusions.

In these earlier models effects from rotation were certainly enlarged by the fact that the adiabatic index below nuclear matter density was chosen to be very close to its critical value ($\gamma = 1.33$). Consequently the effective adiabatic index, including rotation, exceeded the critical value already at low densities ($\rho \approx 10^{12}$ g cm^{-3}) and therefore low-density bounces resulted. Recent realistic equations of state, however, predict significantly lower values of γ and it could be expected that effects from rotation were less important for more realistic models. Müller and Hillebrandt (1981) therefore computed the collapse of rotating stellar cores, starting from a realistic stellar model (the 25 M$_\odot$ star of Weaver _et al_. 1978), incorporating a recent equation of state (El Eid and Hillebrandt 1980) and allowed for changes in the electron concentration, which was taken from the spherical computations of Hillebrandt and Müller (1981). In addition they investigated different amounts and distributions of angular momentum in the initial model. Their results are shown in Table 4 and Figs. 7-11 and can be summarized as follows:

During collapse the dynamics was not much affected by rotation, due to the limitation to 'slow' rotation in the sense discussed earlier, and due to the low adiabatic index of the equation of state. As in the non-rotating models the collapse stopped at nuclear matter density due to the stiffness of the equation of state and not due to centrifugal forces.

Only the more rapidly rotating cores ($E_{rot}/E_{grav} \approx 0.15$ at bounce) showed large deviations from spherical symmetry in the

Table 4

Properties of a rotating stellar core
(from Müller and Hillebrandt 1981)

Initial model	25 $M_\odot$ (Weaver et al. 1978)
Initial law of rotation	$\Omega(r) = $ const $\quad r < 10^8$ cm ($\lesssim 1.2$ $M_\odot$) $\Omega(r) \sim r^{-2} \quad r \gtrsim 10^8$ cm
Initial E_{rot}	6.2×10^{49} erg (for inner 1.4 $M_\odot$)
$(\Omega/\Omega_{crit})^2$	0.063 (at 1.4 $M_\odot$)
ρ_c at bounce (g cm^{-3})	2×10^{14}
E_{rot} at bounce (g cm^{-3})	1 for inner 0.8 $M_\odot$
$E_{rot}/|E_{grav}|$	0.15 for inner 0.8 $M_\odot$
maximum infall velocity at the pole at the equator	7×10^9 cm s^{-1} 3.5×10^9 cm s^{-1}
maximum shock radius (end of calculation) at the pole at the equator	180 km 150 km

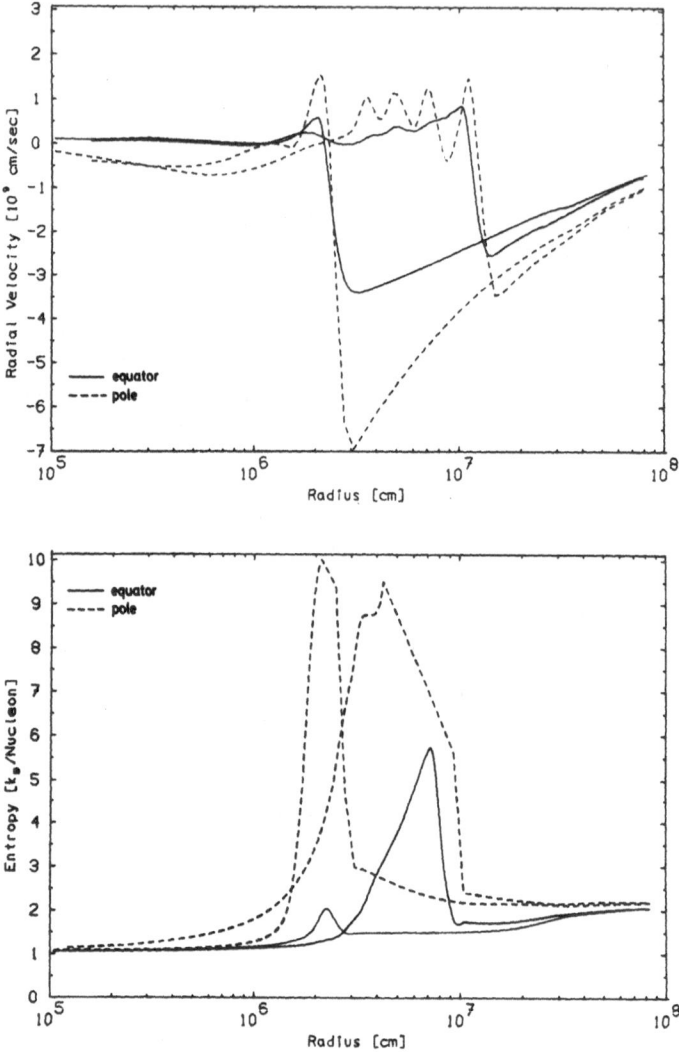

Figure 7a,b.   Velocity and entropy profiles in a rotating
non-adiabatic model after core-bounce (from Müller and
Hillebrandt 1981). The properties of the model are given in Table
4. Shown are the velocities and entropies at the equator (solid
lines) and at the pole (dashed lines) about 1 ms and 7 ms after
bounce.

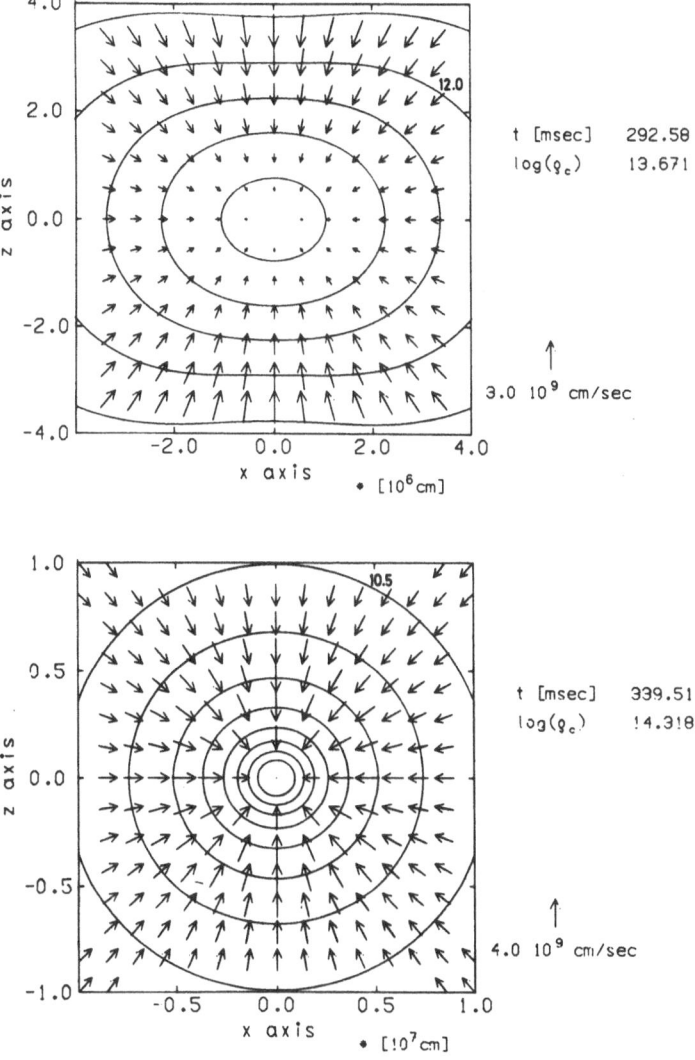

Figure 8a,b. Isopycnic contours and flow pattern for the model of Table 4 and a slowly rotating model ($\Omega/\Omega_{crit} \sim 0.1$, Fig. 8b) just before core-bounce. The axis of rotation is identical with the z-axis. The numbers on the contours give the logarithm of the density, the spacing is 0.5 on a logarithmic scale. Given are also velocity scales, central density and time after the onset of collapse.

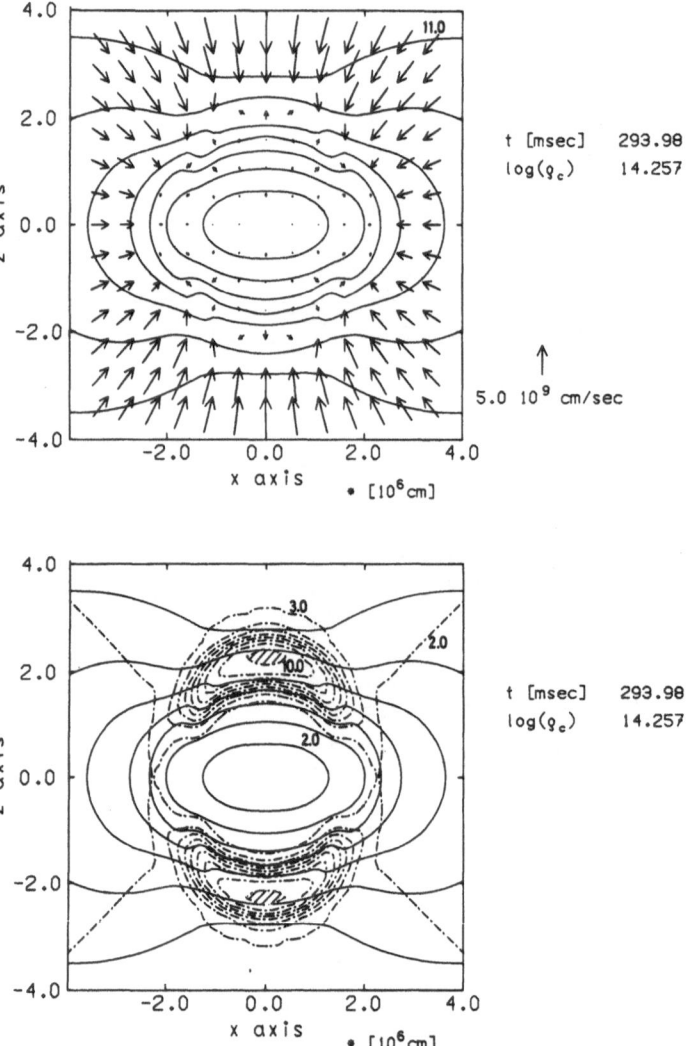

Figure 9a,b. Isopycnic and isentropic contours and flow pattern for the model of Table 4 just after bounce. The spacing of the density contours (solid lines) is $\Delta(\log \rho) = 0.1$, the spacing of the isentropic contours (dashed dotted lines) is 1 k_B/nucleon.

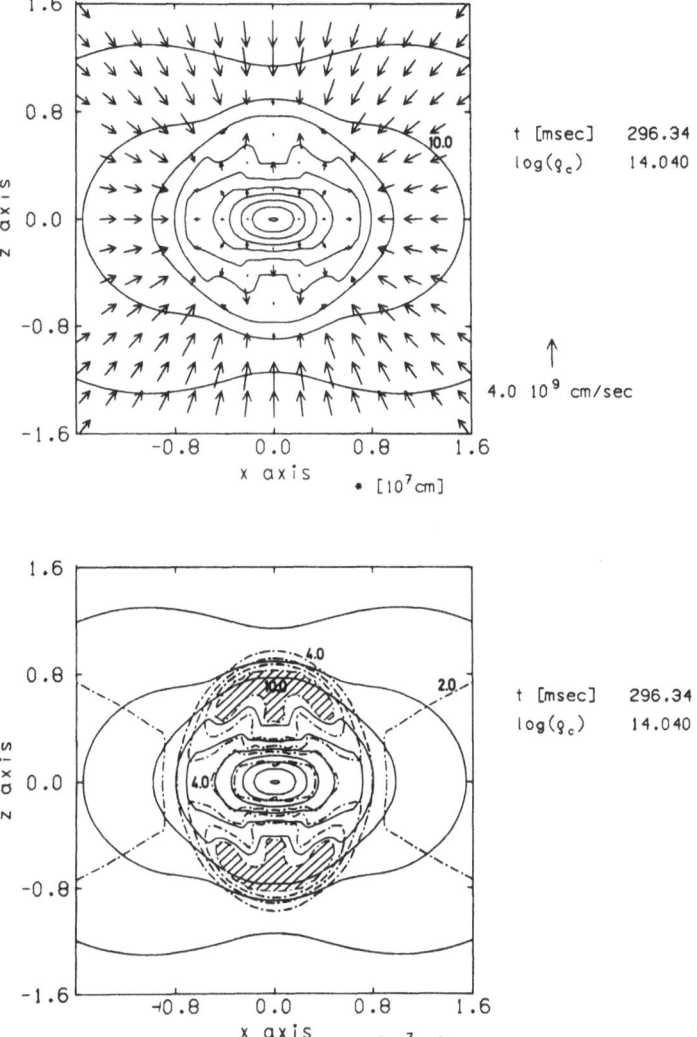

Figure 10a,b.   Same as Fig. 9a,b, but 3 ms after bounce. The spacing of the contours is $\Delta(\log \rho) = 0.5$ and $\Delta S = 2$ k_B/nucleon, respectively.

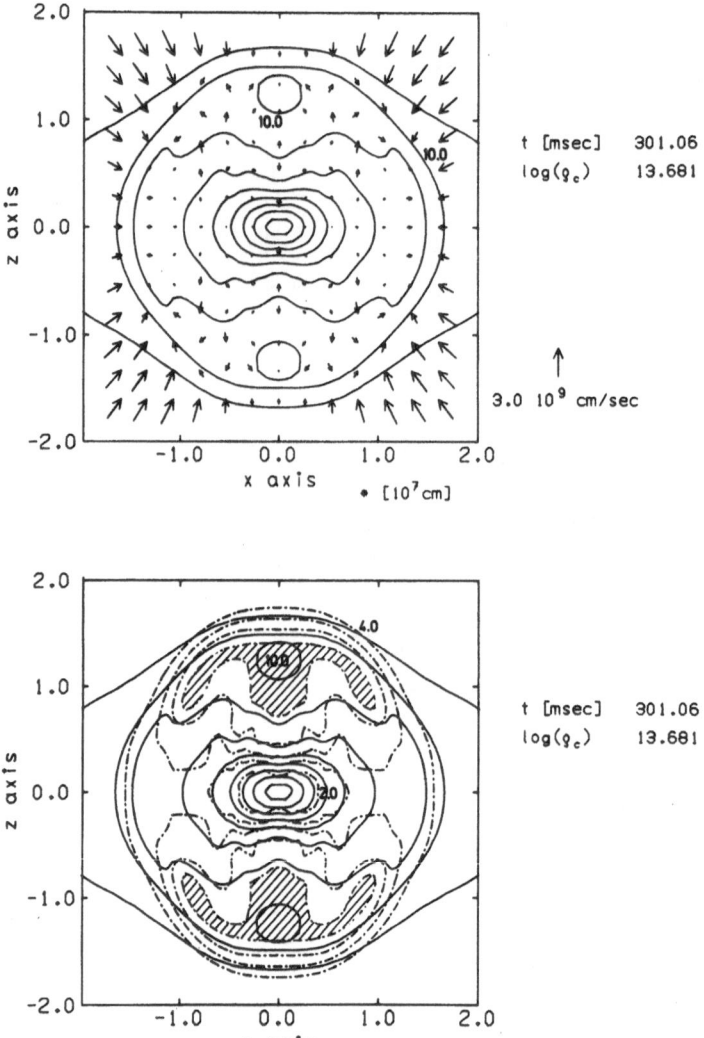

Figure 11a,b.   Same as Fig. 9a,b, but 7 ms after bounce. Spacing
as in Fig. 10a,b. Very hot (high entropy) matter forms at the
poles and gives rise to large scale circulation flows.

velocity field, the density and entropy distribution. In these models the collapse was stopped first on the axis of rotation because there larger pressure gradients had built up. Because of the very high infall velocity of matter at the poles a blob of high entropy matter formed there (S $\lesssim$ 10 k_B/nucleon), which tried to rise and expand and thereby drove large scale circulations. Those circulation flows with typical velocities of about 10^9 cm sec^{-1} first penetrated deep into the hydrostatic core, but decayed into higher modes about 4 msec after bounce. The further evolution of the shock was then very similar to the non-rotating models and in none of the cases considered was an explosion found.

There are further aspects of the rapidly rotating models of Muller and Hillebrandt (1981) that are very interesting and should be investigated in the future. The circulation flows found by them will transport neutrinos much more effectively from the hydrostatic core to the mass-zones behind the shock front than ordinary diffusion does. Rough estimates show that indeed an energy flow of about 10^{51} erg/ms can be obtained in this way, which may be sufficient to balance nuclear dissociation losses. In fact, for rapidly rotating cores the situation is very similar to what Bruenn et al. (1979) thought large scale convection, driven by lepton gradients, could do, namely mixing the leptons from the hydrostatic parts of the core to the shock front in a few milliseconds.

But even if this mixing should not supply enough energy for a supernova, rapidly rotating cores may still find another possibility to explode. As was pointed out by Ostriker (1981), some of the models of Muller and Hillebrandt form central condensations that are unstable against triaxial deformations on secular timescales. Those inner cores have typically 0.8 to 1.0 $M_\odot$ and rotation energies of about 10^{52} erg. Any non-axisymmetric perturbation may therefore grow and transform the inner core into a triaxial configuration that rotates differentially inside an axisymmetric infalling stellar mantle. It is very likely that a significant fraction of the rotational energy of the inner core will then be transferred to the mantle by either sound waves or shock waves and thus may trigger a supernova explosion. A computer simulation of this process, however, requires much more advanced numerical codes than are available to date.

3.2 Supernovae triggered by rotation?

Once we are convinced that we have to include rotation into our computer simulations of stellar collapse and supernova explosions, several additional aspects of the problem should

also be considered, among which we will discuss only those which
are the least speculative ones.

a) Rotating Magnetized Stellar Cores

From the magnetic fields observed in stars, where the ratio
of magnetic energy to gravitational energy is found to be less
than 10^{-10} even in the most extreme cases, one would suspect that
effects from magnetic fields can hardly ever become important
during stellar collapse. Even magnetic fields of the order of
10^{12} Gauss, which are observed in neutron stars (Kendziorra et
al. 1977), will have a negligible effect if the collapse is
spherically symmetric. In rotating models, however, differential
rotation may lead to large amplification of the magnetic field
locally in some regions of the star, which then may even dominate
the dynamics. In general the collapse of a rotating magnetized
stellar core will be a three-dimensional problem unless one
restricts the initial field configuration to be both axial and
equatorial symmetric (an aligned dipole field for example or a
homogeneously magnetized core). With these assumptions the
problem is still a two-dimensional one, but is feasible with
present day computers. Le Blanc and Wilson (1970) have performed
the only complete 2-d magneto-hydrodynamical (MHD) computation
and found supernova explosions if an initial ratio of magnetic to
gravitational energy of about 10^{-4} to 10^{-3} was assumed, which,
however, is too large to be realistic. The mechanism that caused
the MHD-explosion was a complicated interplay of rotation and
magnetic fields. As was discussed in the preceding section in
collapsing rotating cores large scale circulations develop.
Because of the very high electric conductivity of the matter, the
magnetic field-lines are frozen in and a certain fraction of the
kinetic and rotational energy is transformed into magnetic energy
in regions of significant differential rotation. The magnetic
pressure then can either lead to a direct explosion or to a
buoyancy instability followed by a jet of matter being ejected
along the polar axis. A crucial problem these MHD-supernova
models have to face is whether or not the timescale for the field
amplification is smaller or at least equal to the infall
timescale of the stellar mantle, because only in that case one
can expect large MHD-effects, as was discussed by Meier et al.
(1976). Unfortunately the initial stellar model chosen by Le
Blanc and Wilson (1970), a 7 $M_\odot$ iron-nickel core, was by far too
unrealistic to answer this question.

More recent investigations of the same problem replace the
forming neutron star by an inner boundary condition and solve the
MHD-equations only for the infalling mantle. Moreover either
cylindrical symmetry (Bisnovatyi-Kogan et al. 1976:
Bisnovatyi-Kogan 1980) is assumed, or the dynamics in the
equatorial plane only is considered (Muller and Hillebrandt

1979). Whereas the latter assumption may be not too bad for slowly rotating cores, cylindrical symmetry largely misrepresents the gravitational potential and therefore the results should be taken with some care. In both types of computations the winding of magnetic field lines in the equatorial plane due to differential rotation caused mass ejection when the azimuthal component of the magnetic field approached 10^{15} Gauss. Since in these models energy is fed into the magnetic field from the rotational energy of the inner core, this energy has to be at least of the order of 10^{51} erg and comparable to the gravitational binding energy of the stellar mantle. This condition can probably be fulfilled by realistic initial models. The second important parameter is the ratio of magnetic to rotational energy of the core at bounce, because the amplifiction timescale of the magnetic field scales approximately as the inverse square-root of this ratio (Bisnovatyi-Kogan 1980). In order to have an amplification timescale of the order of a second, ratios of about 10^{-4} are required, which is a rather high value. Infall timescales of the mantle, on the other hand, probably do not exceed a few seconds unless the initial model is already close to critical rotation (Bisnovatyi-Kogan et al. 1976) or the hydrodynamical shock from core-rebound has almost been successful (Muller and Hillebrandt 1979). Also other models of this type (Bodenheimer and Ostriker 1974: Kundt 1976) have to face the same difficulties. Finally it is highly questionable whether such high magnetic flux concentrations as required by the models of Bisnovatyi-Kogan et al. (1976) or Muller and Hillebrandt (1979) can really be contained in more realistic full 2-d computations. Due to their symmetry assumptions the field cannot expand perpendicular to the equatorial plane and therefore is strongly confined. More realistic models allowing for this degree of freedom are certainly feasible and should be investigated. One may, however, suspect that the magnetic flux concentrations will drive a Rayleigh-Taylor instability similar to what the high entropy blobs did in the rotating models discussed in section 3.1. In this case most of the energy would be dissipated in circulation flows and turbulent motions rather than in driving a explosion.

b) Rotation and Explosive Nuclear Burning

The unshocked mantle of a massive star after the shock from core-bounce has turned into an accretion shock still contains a lot of unburnt nuclear fuel, which is a potential source of energy for a supernova explosion. In particular the 3 to 6 $M_\odot$ oxygen found in those stars may liberate about 10^{51} ergs in a few seconds in explosive oxygen burning. The crucial question is, however, whether or not the oxygen shell is falling in supersonically by the time explosive burning is ignited. If the infall were indeed supersonic, no outward propagating detonation

front could develop and all the energy and mass would be
swallowed by a central black hole. Spherically symmetric
computations indicate (Woosley and Weaver 1981) that this is in
fact the case.

For rotating stellar models, on the other hand, the situation
can be considerably different, as was discussed already by Fowler
and Hoyle (1964) and recently by Bodenheimer and Woosley (1980),
since centrifugal forces may slow-down the infall sufficiently to
make it subsonic. The recent models are discussed elsewhere in
this volume (Woosley and Weaver 1982), so we do not have to go
into details, but only want to mention some of the computational
difficulties. Explosive oxygen burning occurs typically about a
second after core-bounce. All numerical codes applied so far to
the problem of rotating stellar cores are explicit ones, which
means that the timesteps in the numerics are limited by the
Courant condition, e.g. $\Delta t < \min(\Delta r/v_s)$, where Δr is the spatial
finite thickness of discrete mass-zones and v_s is the local
velocity of sound. This timestep is determined by the spatial
resolution in the inner central condensation, the proto-neutron-
star, and is typically of the order of or less than 10^{-6} sec! A
computation of explosive oxygen burning requires therefore
several million timesteps, which are by far too expensive. With
the code of Müller and Hillebrandt (1981), for example, such a
computation would cost at least a thousand hours of CPU-time on a
Cray-1. Bodenheimer and Woosley (1980) circumvented this
difficulty by imposing inner boundary conditions at the edge of
the central core, but the choice of those boundary conditions
will always be somewhat artificial and the results may depend
even qualitatively upon the assumptions made. In fact they do
find that with some boundary conditions oxygen burning leads to
supernova explosions, whereas with some others it does not. In
any case it seems very likely at present that for certain models
of rotating massive stars a supernova explosion can be initiated
by explosive oxygen burning. In these models a rapidly spinning
black hole rather than a neutron star would be left over.

c) Rotation and Lepto-convection

As was first realized by Epstein (1979), the matter behind
the shock front in the core-bounce supernova models is depleted
in leptons, giving rise to a negative lepton to baryon number
gradient there. This lepton number gradient may be sufficiently
steep that the energy transport from the inner core to the shock
front is no longer due to neutrino diffusion but rather due to
convection. Wilson (1980) incorporated Epstein's ideas into his
1-d code by assuming that lepton driven convection can be treated
like ordinary (semi-) convection but found little effect. Colgate
and Petschek (1980), Bruenn et al. (1979), and Livio et al.
(1980), on the other hand, thought that rather than generating

microconvection, as was assumed by Wilson (1980), the unstable
lepton number gradient might favour the large eddies and thus
lead to an overturn of the whole core on timescales of typically
a few milliseconds. This picture has been criticised by Lattimer
and Mazurek (1981) and Smarr et al. (1981). They point out that
the unshocked core as well as the inner parts of the dissociated
matter are stabilized by positive entropy gradients and thus only
a small fraction (about 0.2 to 0.3 $M_\odot$) of the shocked matter just
behind the shock front will be mixed by large scale macro-
convection. This may still affect the dynamics of the propagating
shock front in a 2-d computation (Smarr et al. 1981), but
certainly much less than what was envisaged by Bruenn et al.
(1979). It should be emphasized, however, that the stabilizing
effect of entropy gradients in a 1-d model can disappear in a 2-d
computation if rotation is also considered (Müller and
Hillebrandt 1981). In their models the primary cause of the
circulation flows was rotation rather than lepton number
gradients. But all that is required for the core-overturn
supernova models to work is fast mixing of matter on timscales of
about a few milliseconds independent of what mechanism is
responsible for this mixing.

4.   CONCLUDING REMARKS

Although by now we have not been able to present a mechanism
which beyond doubt can tell us what makes massive stars explode
as supernova, several of the models presented look very
promising.

The simple and most straightforward one-dimensional core-
bounce models may work after all, at least for certain classes of
initial models. In particular the computer simulations indicate
that explosion can result from core-bounce and reflected shock
waves if either the rebounding core has more than about 1 $M_\odot$,
which corresponds to an electron concentration at neutrino
trapping of at least 0.4 (or an adiabatic index almost exactly
equal to 4/3) or if the initial model has a low mass iron-nickel
core of about 1.3 to 1.4 $M_\odot$. This latter result indicates that we
have to pay more attention to stellar evolution in order to
predict this mass more precisely. Moreover it may mean that only
massive stars with main sequence masses of around 10 $M_\odot$ finally
undergo type II sypernova explosions unless some other explosion
mechanism is also effective.

Here rotating stellar models open a variety of different
possibilities. Many of these models are still rather speculative
at present, but the computations do clearly show that
two-dimensional effects help rather than hurt, which was not
clear before the computations were done. Because from
nucleosynthesis arguments it is desirable that also some rather

massive stars (M ~ 20 M$_\bullet$) do explode, these two-dimensional
models may provide the mechanisms needed, but clearly much
additional work is required. In particular stellar evolution
calculations including rotation from the beginning and predicting
the law of rotation and the angular momentum inside highly
evolved massive stars at the onset of collapse deserve a high
priority. Several of the different potential supernova mechanisms
mentioned, and one may add a few more to this list, work on
timescales much longer than the dynamical timescale of the
central neutron star. A proper treatment of these processes
requires implicit 2-d hydrocodes rather than the explicit ones
used so far in the field, but again these problems are out of our
range.

REFERENCES

Arnett, W.D. 1977a. Astrophys.J.Suppl., 35, 145.
Arnett, W.D. 1977b. Astrophys.J., 218, 815.
Arnett, W.D. 1982. This volume.
Bethe, H.A. 1982. This volume.
Bethe, H.A., Applegate, J. and Brown, G.E. 1980. Astrophys.J.,
    241, 343.
Bethe, H.A., Brown,G.E., Applegate, J. and Lattimer, J.M. 1979.
    Nucl.Phys., A324, 487.
Bisnovatyi-Kogan, G.S. 1980. Ann.N.Y.Acad.Sci., 336, 389.
Bisnovatyi-Kogan, G.S., Popov, Yu.P. and Samochin, A.A. 1976.
    Astrophys.Space Sci., 41, 287.
Bludman, S.A. and Van Riper, K.A. 1977. Astrophys.J., 212,
    859.
Bodenheimer, P. and Ostriker, J.P. 1974. Astrophys.J., 191,
    465.
Bodenheimer, P. and Woosley, S.E. 1980. Bull.Am.Astr.Soc.,
    12, 833.
Bonche, P. and Vautherin, D. 1980. Preprint IPNO/TH 80-66.
Brown, G.E. 1982. This volume.
Brown, G.E., Bethe, H.A. and Baym, G. 1981. NORDITA preprint,
    81/17.
Bruenn, S.W., Buchler, R.J. and Livio, M. 1979. Astrophys.J.
    Letters, 234, L183.
Buchler, R.J. and Epstein, R.J. 1980. Astrophys.J.Letters,
    235, L91.
Colgate, S.A. and Petschek, A.G. 1980. Astrophys.J., 238, 139.
El Eid, M.F. and Hillebrandt, W. 1980. Astr.Astrophys.Suppl.
    42, 215.
Epstein, R.J. 1979. Mon.Not.R.astr.Soc.; 188, 305.
Fowler, W.A. and Hoyle, F. 1964. Astrophys.J.Suppl.; 91, 201.
Fuller, G.M. 1981. Preprint, OAP-622.
Fuller, G.M., Fowler, W.A. and Newman, M.J. 1981. Preprint,
    OAP-620 and 621.

Hillebrandt, W. 1981. In preparation.
Hillebrandt, W. and Müller, E. 1981. Astr.Astrophys., 103, 147.
Hoyle, F. 1946. Mon.Not.R.astr.Soc., 106, 343.
Kendziorra, E., Staubert, R., Pietsch, W., Reppin, C., Sacco, B. and Trumper, J. 1977. Astrophys.J.Letters, 217, L93.
Kundt, W. 1976. Nature 261, 673.
Lamb, D.Q., Lattimer, J.M., Pethick, C.J. and Ravenhall, D.G. 1978. Phys.Rev.Letters, 41, 1623.
Lattimer, J.M. 1981. Ann.Rev.Nucl.and Part.Sci., 31, 337.
Lattimer, J.M. and Mazurek, T.J. 1981. Astrophys.J., 246, 995.
Le Blanc, J.M. and Wilson, J.R. 1970. Astrophys.J., 161, 541.
Lichtenstadt, I., Sack, N. and Bludman, S. 1980. Astrophys.J., 237, 903.
Livio, M., Buchler, J.R. and Colgate, S.A. 1980. Astrophys.J. 238, 139.
Mazurek, T.J. 1981. Paper presented at this conference.
Mazurek, T.J., Cooperstein, J. and Kahana, S. 1980. Proc. Dumand Workshop, ed. V.J. Stenger (Honolulu: University of Hawaii), p. 142.
Mazurek, T.J., Lattimer, J.M. and Brown, G.E. 1979. Astrophys. J., 229, 713.
Meier, D.L., Epstein, R.I., Arnett, W.D. and Schramm, D.N. 1976. Astrophys.J., 204, 869.
Müller, E. and Hillebrandt, W. 1979. Astr.Astrophys., 80, 147.
Müller, E. and Hillebrandt, W. 1981. Astr.Astrophys., 103, 358.
Müller, E., Rozyczka, M. and Hillebrandt, W. 1980. Astr. Astrophys., 81, 288.
Nomoto, K., Miyaji, S., Yokai, K. and Sugimoto, D. 1979. IAU Coll. no. 53, ed. Van Horn and Weidemann. p. 56.
Ostriker, J.P. 1981. Paper presented at this conference.
Smarr, L.L., Wilson, J.R., Barton, R. and Bowers, R. 1981. Astrophys.J., 246, 515.
Sugimoto, D. and Nomoto, K. 1980. Space Sci.Rev., 25, 155.
Takahashi, K., El Eid, M.F. and Hillebrandt, W. 1978. Astr. Astrophys., 67, 185.
Tammann, G.A. 1982. This volume.
Tohline, J.E., Schombert, J.M. and Boss, A.P. 1981. Los Alamos preprint, 4R80-2870.
Van Riper, K.A. 1978. Astrophys.J., 221, 304.
Van Riper, K.A. 1981. Preprint, Univ.Illinois at Urbana-Champaign.
Van Riper, K.A. and Arnett, W.D. 1978. Astrophys.J.Letters, 225, L129.
Van Riper, K.A. and Lattimer, J.M. 1981. Astrophys.J., 249, 270.
Weaver, T.A., Zimmerman, G.B. and Woosley, S.E. 1978. Astrophys.J., 225, 1021.
Wilson, J.R. 1978. Proc.Int.Sch.Phys.'Enrico Fermi', 65, 644.
Wilson, J.R. 1980. Ann.N.Y.Acad.Sci., 336, 358.
Woosley, S.E. and Weaver, T.A. 1982. This volume.

GRAVITATIONAL RADIATION FROM COLLAPSING ROTATING STELLAR CORES

E. Müller

Max-Planck-Institut für Physik und Astrophysik
Institut für Astrophysik

1. INTRODUCTION

The current and near-future efforts to detect gravitational waves (articles of Epstein and Weiss in Smarr 1979: Weber 1979: Douglass and Braginsky 1979) have increased the demand for reliable theoretical calculations of gravitational radiation from all possible energetic astrophysical sources. Among all potential observed sources, supernovae, possessing highly asymmetric cores and being situated in our own galaxy (see the review by Thorne 1980), have been thought to be the most promising candidates, probably already within the realm of second generation gravitational wave detectors.

Therefore, in the last few years, several authors have tried to calculate more or less idealized models of stellar collapse leading to the production of gravitational radiation. These investigations can be divided into two somewhat opposite approaches. One group of authors (Thuan and Ostriker 1974: Novikov 1975: Shapiro 1977: Saenz and Shapiro 1978, 1979, 1981) considered the aspherical collapse of uniform density, uniformly rotating cores with increasing but still crude attention paid to the thermodynamics, i.e. the equation of state. None of these one-shell models was able to study the influence of density gradients and angular velocity gradients in arbitrary axisymmetric configurations.

In the second approach, one tries to calculate the thermodynamics and hydrodynamics of spherical collapse as accurately as possible drawing certain conclusions about probable gravitational radiation losses (articles of Arnett, Wilson,

157

M. J. Rees and R. J. Stoneham (eds.), Supernovae: A Survey of Current Research, 157–166.
Copyright © 1982 by D. Reidel Publishing Company.

Kanzanas and Schramm in Smarr 1979). In this context Turner and
Wagoner (1979) have calculated the gravitational radiation
produced by rotationally-induced perturbations of spherically
symmetric stellar collapse models of Van Riper (1978) and Wilson
(1977).

The above-mentioned calculations predict that the energy
carried off by gravitational waves produced in supernovae should
lie in the range 10^{-2} to 10^{-5} $M_\odot$ c^2, where $M_\odot$ is the solar mass
and c is the speed of light, the only exception being Turner and
Wagoner (1979) who found due to their method (limited to slowly
rotating cores) values much less than 10^{-10} $M_\odot$ c^2. Considering
these numbers, one should bear in mind that the maximum energy
available from the collapse is of the order 0.1 $M_\odot$ c^2 (the
difference in binding energy of a degenerate dwarf and a neutron
star).

In this paper we calculate the gravitational radiation
produced by the axisymmetric, Newtonian stellar collapse models
of Müller and Hillebrandt (1981) - cited henceforth as MH
- within a post-Newtonian multipole formalism developed by
Wagoner (1977). The collapse models of MH combine the efforts of
the two approaches mentioned above by treating the relevant
microphysics and the propagation of the shock generated through
core bounce as accurately as possible as well as considering
general axisymmetric configurations.

2. MODEL

MH calculated the axisymmetric collapse of slowly and
rapidly, rigidly and differentially rotating stellar cores using
a two-dimensional explicit hydrodynamic Newtonian code including
a (tabulated) realistic equation of state (for further details
see MH). Although improving many of the weak points of the
earlier approaches, two shortcomings remain even in the quite
elaborate collapse models of MH. Instead of solving the full,
non-linear Einstein field equations in axial symmetry, they
employ Newtonian equations of motion, requiring slow velocities
$v^2 \ll c^2$ and weak gravitational potentials $\phi \ll c^2$. Both
requirements are fulfilled to about 10% accuracy in the models of
MH. Considering the enormous additional expense of the
calculation when going beyond the Newtonian limit and other
present uncertainties (equation of state, initial model), the
Newtonian treatment of the collapse in MH is justified.

The second shortcoming from which the model of MH (as well as
all other approaches to calculate the collapse of rotating
stellar cores) suffers is related to the unknown initial model.
No evolutionary calculations for (massive) rotating stars up to
the stage of core collapse exist. Therefore, the initial model,

especially the amount and the distribution of angular momentum at the onset of collapse, is somewhat arbitrary. MH have tried to overcome this lack of information by doing a parameter study. They evolved four different models which are divided into two groups (A and B) each consisting of a slow and a fast rotating model.

We have analysed the slowly rotating model A1 and the two rapidly rotating models A1 and B2. All models bounced around nuclear matter density and led to the formation of an outgoing shock front (being strongly asymmetric in case of models A2 and B2), which did not result in any explosive mass ejection (for further details see MH).

Following Wagoner (1977) we describe the emitted gravitational radiation by a multipole expansion of the transverse-traceless (TT) part of the gravitation-wave field tensor h_{jk}^{TT}. This symmetric spatial tensor is (with appropriate choice of gauge) the metric perturbation associated with the waves. The tensor h_{jk}^{TT} is given by (in natural units with $G = c = 1$)

$$\underline{h}^{TT} = \frac{1}{r} \sum_{L>2} \sum_{M=-L}^{+L} \{A_{LM}^{E2} (t-r)\underset{\sim}{T}_{LM}^{E2} (\theta,\phi) + A_{LM}^{B2} (t-r)\underset{\sim}{T}_{LM}^{B2}(\theta,\phi) \}, \quad (1)$$

at a distance r from the source much greater than the characteristic wavelength of the radiation. The 'electric' and 'magnetic' tensor spherical harmonics T_{LM}^{E2} and T_{LM}^{B2} are defined as in Thorne (1980). A_{LM}^{E2} and A_{LM}^{B2} are the 'electric' and 'magnetic' multipole amplitudes containing all the information carried by a wave.

Knowing the multipole amplitudes in the expansion above, the radiated energy and the energy spectrum of the gravitational wave can be calculated by

$$\Delta E_{Gw} = \frac{1}{32\pi} \sum_{L,M} \int \{|\partial A_{LM}^{E2} / \partial t|^2 + |\partial A_{LM}^{B2} / \partial t|^2 \}dt \quad (2)$$

and

$$\frac{d\Delta E_{Gw}}{d\omega} = \frac{1}{16\pi} \sum_{L,M} \{ |\tilde{A}_{LM}^{E2}(\omega)|^2 + |\tilde{A}_{LM}^{B2}(\omega)|^2 \} \omega^2, \quad (3)$$

respectively, where ω is the frequency of the radiation and $\tilde{A}_{LM}^{E2}$ and $\tilde{A}_{LM}^{B2}$ are the Fourier transformed amplitudes.

The post-Newtonian expansion of Wagoner (1977) with the expansion parameter being $\varepsilon := v/c$ also leads to a multipole expansion exactly analogous to that given by equation (1). The lowest order nonvanishing multipole amplitudes are A_{20}^{E2} and A_{30}^{B2}, which are of order $\varepsilon^2 \chi$ and $\varepsilon \sqrt[3]{\chi}$, respectively, where χ is the ratio of centrifugal to gravitational forces. The detailed expressions for these amplitudes read:

$$A_{20}^{E2} = \frac{d^2}{dt^2} M_{20}^{E2} = \frac{d^2}{dt^2} \frac{32\pi^{3/2}}{\sqrt{15}} \int_0^1 \int_0^\infty \rho(r,x,t)(\frac{3}{2}x^2-1)r^4 dr dx \qquad (4)$$

$$A_{30}^{B2} = \frac{d^3}{dt^3} M_{30}^{B2} \qquad (5)$$

$$= \frac{d^3}{dt^3} 8\pi \sqrt{\frac{\pi}{105}} \int_0^1 \int_0^\infty v^\phi \rho(r,x,t) \sqrt{1-x^2}(5x^2-1)r^5 dr dx,$$

where $x := \cos\theta$, v^ϕ is the ϕ – component of the velocity vector, and ρ is the density.

3. RESULTS

In their hydrodynamical calculations, MH had stored the quantities M_{20}^{E2} and M_{30}^{B2}, which are defined by equations (4) and (5), for each time-step in a time interval of about 10 msec size around core bounce. Due to the explicit difference scheme used in their calculations, these raw-data are non-equidistantly distributed in time. In a first step we constructed an equi-distantly spaced set of values M_{20}^{E2} and M_{30}^{B2} by applying linear in-terpolation to evaluate the quantities at a prescribed time. The calculation of an equidistant table was necessary for the (discrete) Fourier transformation employed later in our analysis.

In a second step we calculated the amplitudes A_{20}^{E2} and A_{30}^{B2} by means of numerical differentiation. Subsequently we calculated the amplitude $\tilde{A}_{20}^{E2}(\omega)$ and the energy spectrum by means of a dis-crete Fourier transformation. The octupole term was not analysed further because of its smallness compared to the quadrupole term. Applying the discrete Fourier transformation, one must be one must be careful because as is well known a time domain truncation of the signal (data only available in a certain time interval) introduces additional frequency components, which are termed leakage, in the frequency domain (see e.g. Brigham 1974). To reduce the leakage, we employed a Gaussian time domain truncation function which has side-lobe characteristics which are smaller in magnitude than those of the $\sin(\omega)/\omega$ function resulting from the usage of a rectangular truncation function. We have analysed the raw-data in a second way to check for noise

introduced in the signal by the numerical differentiation. We
first Fourier transformed the amplitude $M_{20}^{E2}(t)$ and subsequently
multiplied the Fourier amplitudes by the frequency squared, which
is equivalent to calculating the second derivative in 'ordinary'
space.

The waveforms, that is, the amplitudes A_{20}^{E2} as a function of
time, and the energy spectra for the models A2, B2 and A1 are
given in Figures 1 to 3. The waveforms for the different models
are qualitatively very similar, showing a large amplitude just at
core bounce and slowly decreasing oscillations on a timescale of
milliseconds afterwards. Additional high frequency structures can
be seen in all three models. The maximum amplitude for model B2
is about 50% larger than for model A2 and more than a factor of 5
larger than for model A1.

The energy spectrum for model A2 (Fig. 1b) after leakage
reduction shows a double maximum at around 1 kHz and a second
broad maximum of nearly the same magnitude at around 8 kHz. The
region between 2 kHz and 5 kHz is reduced in strength by about a
factor of 100. There are only tiny changes in the spectrum if we
analyse the data in the second way described above. To estimate
the amount of noise present in the signal (due to the
inaccuracies of the hydro-code) we, instead of analysing the
whole waveform, picked out only the first two milliseconds of the
signal and Fourier transformed it. That part before bounce should
contain nearly no signal strength – being mostly noise. The
corresponding Fourier spectrum is given by the dashed curve in
Figure 1b. It shows a steep increase of the noise contribution
at frequencies greater than 6 kHz and a flat part at lower
frequencies. At around 10 kHz the magnitude of the noise exceeds
that of the signal indicating that all frequency structures
beyond 10 kHz are hidden in the noise of the hydro-code.
Integrating the energy spectrum up to 10kHz leads to an emitted
energy in the form of gravitational waves of 5×10^{47} erg or 3×10^{-7}
$M_\odot$ c^2.

The energy spectrum for model B2 (Fig. 2b) shows in contrast
to model A2 a broad single maximum around 1 kHz and another
maximum, a factor of 5 larger, at 8 kHz. In the intermediate
region (2 kHz to 7 kHz) the spectrum exhibits a much smaller dip.
The noise contribution which is given by the dashed curve becomes
larger than the signal at around 10 kHz. The integration up to
this frequency gives rise to a radiated energy of 8×10^{48} erg or
$4 \times 10^{-6} M_\odot c^2$ which is a factor of 12 more energy than in the case
of model A2.

Figure 3b displays the energy spectrum for model A1. It is
relatively flat and contains a number of peaks with almost equal
spectral energy density in the range 1 kHz to 50 kHz. The noise

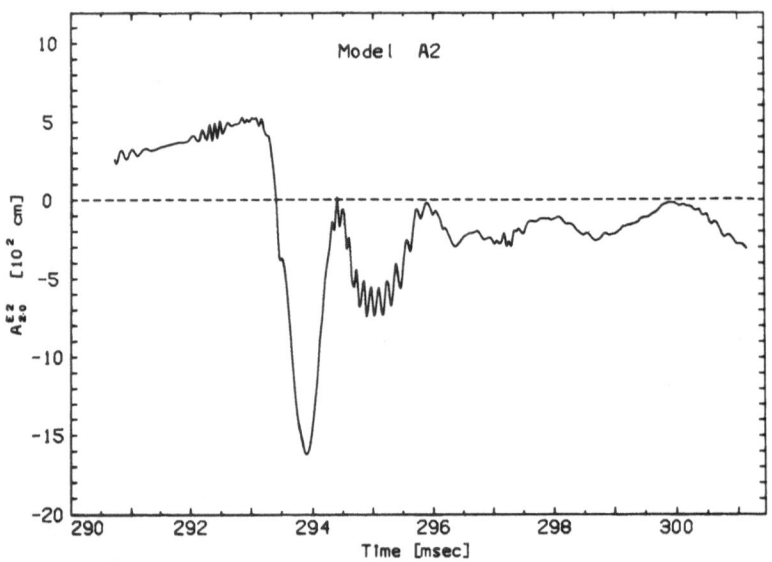

Figure 1a.   Quadrupole wave form for model A2.

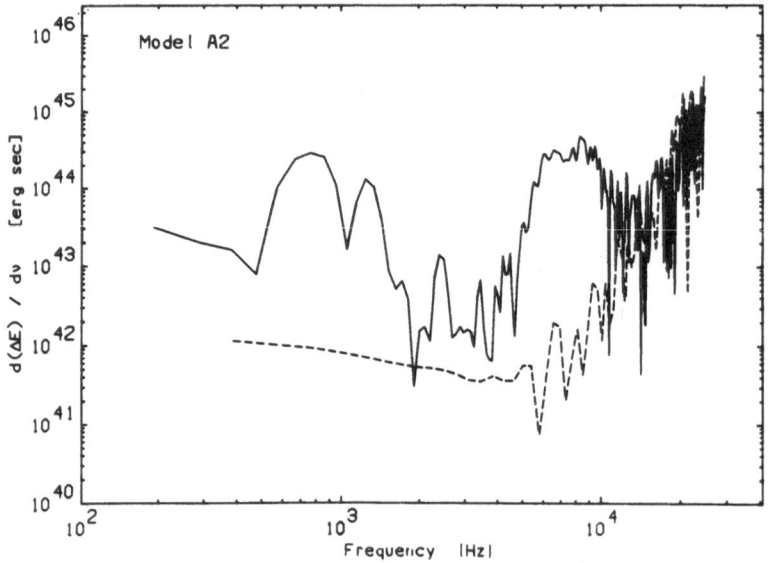

Figure 1b.   Energy spectrum and noise contribution (dashed curve) corresponding to the quadrupole amplitude of model A2.

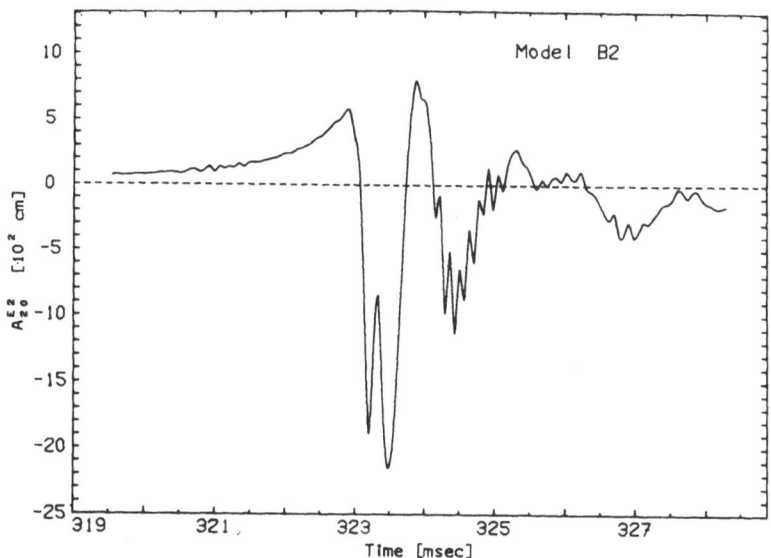

Figure 2a.   Quadrupole wave form for model B2.

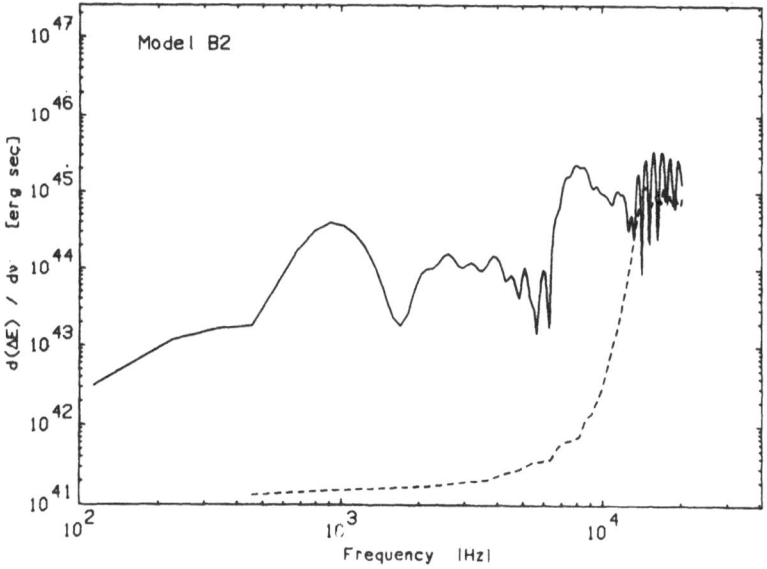

Figure 2b.   Energy spectrum and noise contribution (dashed curve) corresponding to the quadrupole amplitude of model B2.

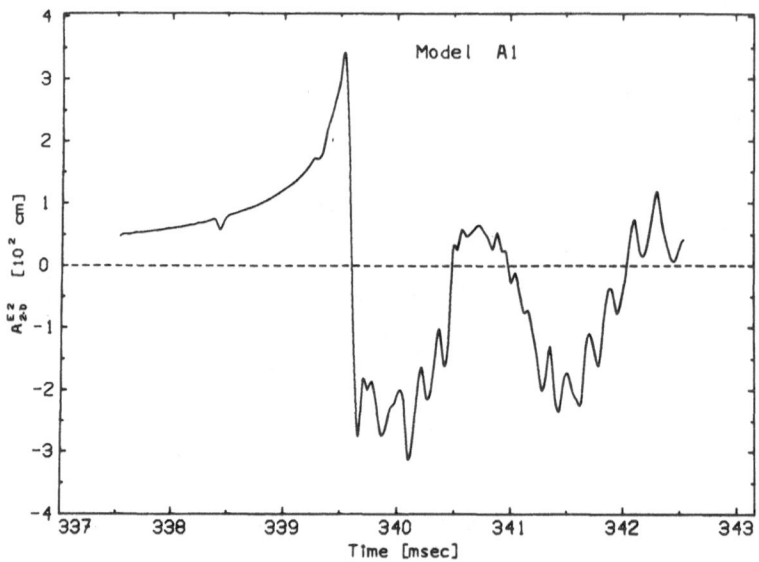

Figure 3a.   Quadrupole wave form for model A1.

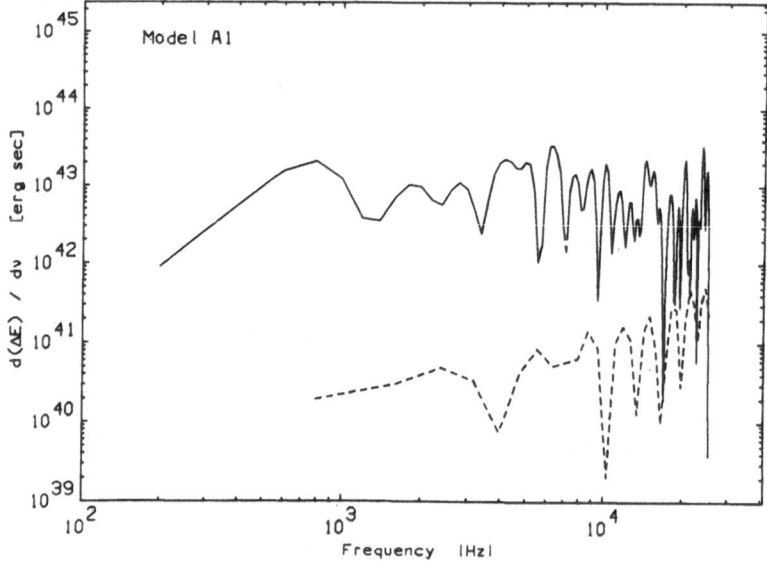

Figure 3b.   Energy spectrum and noise contribution (dashed curve) corresponding to the quadrupole amplitude of model A1.

contribution is relatively small and flat up to 50 kHz; but this is only due to our procedure to estimate the noise. Because model A1 remained spherically symmetric to a high degree up to core bounce, the noise is underestimated. As the model becomes more and more nonspherical, more expansion coefficients are involved and a larger noise contribution is expected. For this reason we have adopted a similar frequency cutoff in the spectrum of model A1 as in the models A2 and B2. This gives rise to a radiated energy of 10^{47}erg or $5\times10^{-8}M_\odot c^2$.

The octupole amplitude is in all three models about a factor of 50 smaller in magnitude than the quadrupole amplitude, which is consistent with an order-of-magnitude estimate. In the collapse calculations of MH, the expansion parameter $\varepsilon:=v/c$ was about 0.1 and $\chi < 0.5$. As the ratio of quadrupole to octupole amplitude $|A_{20}^{E2}|$ / $|A_{30}^{B2}|$ scales as $\sqrt{\chi}\,\varepsilon^{-2}$, one finds, in agreement with our detailed analysis, a ratio of about 50.

4. DISCUSSION

Our results can be summarized as follows: the radiated energies in the form of gravitational waves lie in the range $10^{-7}M_\odot c^2$ to $4\times10^{-6}M_\odot c^2$, the efficiency being larger if one increases the total amount of the angular momentum but keeps the angular distribution fixed (model A1 and A2). Secondly, there is a significant dependence of the radiation efficiency on the angular momentum distribution. The initially differentially rotating model B2 has ten times higher efficiency than the initially rigidly rotating model A2, although the angular momentum inside the innermost 0.8 $M_\odot$, where most of the radiation comes from, is 20% smaller in model B2. Comparing these results with the results of Saenz and Shapiro (1981), one finds that our efficiencies are more than a factor of ten smaller, probably mostly due to their one-shell-model approach and differences in the equation of state.

The dimensionless wave amplitudes for a source located at a distance of 10 Mpc are 6×10^{-23}, 8×10^{-23} and 1×10^{-23} for models A2, B2 and A1, respectively. Our amplitudes are more than a factor of ten smaller than those thought to be measurable with third generation wave detectors (Thorne 1980) which will be able to measure amplitudes larger than 10^{-21}. Therefore our predictions seem to be discouraging to gravitational wave experimentalists, but one should always keep in mind that these are predictions of only one very special model.

Clearly the energy radiated in the form of gravitational waves will greatly increase if the star or its core becomes triaxial. After collapse the rapidly rotating models of MH are just above the limit where they are secularly unstable against triaxial configurations. On the other hand, if the star does not explode as a supernova it will collapse to a black hole due to its large mass. Then one might expect an even larger signal later on in the evolution. Finally, if there are changes in the equation of state, leading to different densities at bounce (see MH), the radiation efficiency will also change. For these reasons, the aim of this work can only be to supply the gravitational wave experimentalist with further theoretical predictions from another, probably more realistic, model calculation.

REFERENCES

Brigham, E.O. 1974. The Fast Fourier Transform, Prentice
    Hall, Inc. Englewood Cliffs, New Jersey, USA.
Douglass, D.H. and Braginsky, V.B. 1979. In General
    Relativity: an Einstein Centenary Survey, ed. S.W.
    Hawking and W. Israel (Cambridge University Press).
Müller, E. and Hillebrandt, W. 1981. Submitted to Astr.
    Astrophys.
Novikov, I.D. 1975. Sov.Astr., 19, 398.
Saenz, R.A. and Shapiro, S.L. 1978. Astrophys.J., 221, 286.
Saenz, R.A. and Shapiro, S.L. 1979. Astrophys.J., 229, 1107.
Saenz, R.A. and Shapiro, S.L. 1981. Astrophys.J., 244, 1033.
Shapiro, S.L. 1977. Astrophys.J., 214, 566.
Smarr, L. (Ed.) 1979. Sources of Gravitational Radiation,
    Cambridge University Press.
Thorne, K.S. 1980. Rev.Mod.Phys., 52, 185.
Thuan, T.X. and Ostriker, J.P. 1974. Astrophys.J.(Letters),
    191, L105.
Turner, M.S. and Wagoner, R.V. 1979. In Sources of
    Gravitational Radiation, ed. L. Smarr (Cambridge University
    Press).
Van Riper, K.A. 1978. Astrophys.J., 221, 304.
Wagoner, R.V. 1977. In International Meeting on Experimental
    Gravitation, ed. B. Bertotti (Pavia, Italy).
Weber, J. 1979. In General Relativity and Gravitation,
    Einstein commemorative volume, ed. A. Held (Plenum: New
    York).
Wilson, J.R. 1977. Unpublished calculations.

THEORETICAL MODELS OF TYPE I SUPERNOVAE

J. Craig Wheeler

Department of Astronomy
University of Texas at Austin

1. OBSERVATIONAL CONSTRAINTS

Before discussing the recent models of Type I supernovae (SNI), the observational constraints should be summarized. None of these constraints are as unambiguous as we would desire. Indeed the discussion at the meeting showed that there could be considerable disagreement as to which constraints were the strongest and which could be assigned lower weight. Here the arguments and their various caveats will be presented and readers will be invited to assign their own weights to the individual factors. A similar discussion with some more detail is given by Wheeler, Branch, and Falk (1980).

1.1 Population

The population of SNI is a point of hot contention. The events are not concentrated in spiral arms to the degree evidenced by Type II supernovae (SNII) (Maza and van den Bergh 1976). On the other hand, they suffer an inclination effect, displaying a lower rate in inclined than in face-on spirals (Tammann 1977). This suggests that they are not a true halo population, but are to be associated with the disk. This statement is already somewhat at odds with the classical notion that since SNI are the only type to appear in ellipticals they must be of an old population, formally population II. The similar but conflicting fact is that SNI are also the only type to be observed in irregular galaxies of type IO. These galaxies are distinguished by their evidence for recent star formation overlain on an otherwise ordinary old red stellar component, and by their high rate of supernova production particularly when normalized to unit galactic luminosity. The high supernova rate

167

M. J. Rees and R. J. Stoneham (eds.), Supernovae: A Survey of Current Research, 167–203.
Copyright © 1982 by D. Reidel Publishing Company.

in IO galaxies and related arguments have been used by a number
of people over the years , most recently and forcibly by Oemler
and Tinsley (1979), to argue that SNI must be associated with
active star formation.   The implication is that SNI are
relatively young and occur in stars of at least a few solar
masses.

    The statistics on which the arguments for youthful SNI are
based are suggestive but not as rigorous as one would prefer.
There is probably reason to remain somewhat sceptical.   In this
vein, Tammann argued here that proper correction for internal
absorption in IO galaxies gives a higher galactic luminosity and
hence a lower supernova rate per unit luminosity. His feeling
that this mitigates the problem also does not seem sufficiently
reassuring .

    If one entertains the notion of young SNI, just how young
and how massive can they be? The only constraint is that they
should not concentrate in spiral arms to the same degree as
SNII.   If all stars are born in spiral arms, a limit can be set
on the mass of SNI progenitors such that they live long enough
to lose memory of their origin.   This limit is about six solar
masses.   Recent arguments (eg Bash and Visser 1981) have
suggested that primarily O stars are born in spiral arms and
that most B stars may be born in the interarm region.   In this
case there is very little constraint on the mass of SNI
progenitors, only that they not be O stars.   In principle they
might be as massive as fifteen solar masses.   SNI show a
remarkable degree of spectral homogeniety.   That there should be
very different classes of progenitor stars is difficult to
accept.   The implication is that if SNI are relatively youthful
and massive, then elliptical galaxies must be the sites of
ongoing star formation   (see Gunn, Stryker, and Tinsley 1981).

1.2 Composition

    The subject of the composition of the ejecta of SNI has
been one of recent rapid development.   Axelrod(1980a,b) has
argued that the late-time spectra show evidence not only for
iron but also for cobalt as would be predicted by models for the
light curve (see below).   Branch et al. (1981) have analyzed the
spectrum of the recent bright SNI in NGC 4536 at maximum light
(see the review by Kirshner for a sample of the spectra).   They
find evidence for a variety of intermediate mass elements:
oxygen, magnesium, sulfur, silicon, and calcium.   There is
tentative evidence that the ratios of these elements compared to
silicon is approximately solar.   Silicon is a convenient
reference element since it has one strong unblended line for
which the blue-shifted absorption is at about 6125 Angstroms.
Iron is not detected at maximum light, the upper limit being

consistent with a solar abundance with respect to silicon. There is no detectable hydrogen. There is also no cobalt in the concentrations which would be predicted by some models. The helium abundance is particularly uncertain because the expected lines are of high excitation and would be strongly affected by a temperature gradient above the photosphere. Similar arguments apply to carbon.

Axelrod reported here his conclusion from study of the late time spectra that the material mixed in with the calcium seen in absorption at late times could not be more than 0.5 helium by mass. More helium would imply a greater trapping of gamma rays in the expanding envelope and excessive heating. An important question is then whether the calcium seen at late times is from the same material layer as that seen near maximum light. Similar velocities suggest this may be the case.

These observations show that one does not have to account for excessive amounts of iron at maximum accompanied by none of the cobalt expected for fresh synthesis, a possibility raised by Branch (1980a,b) in his analysis of post-maximum spectra of SN 1972e. There continues to be an apparent need for some "envelope" to shield the nickel, cobalt, and iron produced by the explosion in the core from direct observation at maximum light. The elements observed at maximum light are similar to those detected in the remnants of putative SNI historical remnants by the solid state spectrometer on the Einstein Observatory (eg Becker et al. 1980a,b; Szymkowiak 1980), namely the products of oxygen burning. A crucial question is whether these elements are freshly synthesized or unprocessed.

Preliminary estimates of the amount of calcium based on the early spectrum by Branch and on the late spectra by Axelrod suggest about 10^{-3} $M_{\odot}$. This is considerably more than one would expect if the calcium were only that present when the star was born. The X-ray spectra also suggest real variations from solar abundances, but of course there is no assurance that the spectra at maximum light refer to the same matter as the X-ray spectra. Seward discussed here the mass estimate for SN1572 based on X-ray observations. The estimate is high, 8 $M_{\odot}$, if H or He are abundant and considerably lower if heavier elements dominate the X-ray emission.

One development we can look forward to is the rationalization of the early spectra due to resonance scattering on a thermal continuum as analyzed by Branch with the late spectra due primarily to forbidden emission lines as calculated by Axelrod. There is evidence for net emission contaminating the spectra within a few weeks, and for resonance absorption, particularly by calcium, after many hundreds of days so the

present division of techniques, while understandable, is
artificial. Progress will come when Branch uses his technique
to calculate line profiles which arise from scattering the
quasi-continuum of emission lines given by Axelrod. There are
several problems which such a calculation could help clarify.
One is that despite the impressive fit of his synthetic spectra
to certain features of the data, and the overall qualitative
reproduction of the spectra, there are features which Axelrod
does not fit particularly well. Although amplitudes are not so
crucial, the failure of Axelrod to fit some key features in
terms of wavelengths is bothersome. Adding some scattering may
improve the situation, but at the expense of making the fits
more *ad hoc*. Likewise, there is an obvious redistribution of
the ultraviolet flux at maximum light which results in an
optical continuum which does not correspond precisely to a black
body despite the assumption of the resonant scattering model for
the early spectra. Understanding the nature of this continuum
and the nature of the net emission which comes in several weeks
after maximum would obviously be a great step in the right
direction.

1.3 Kinematics

The velocity of the material at the photosphere near
maximum light is nearly a constant for a given event.
Unfortunately, disentangling the basic kinematic effects from
those due to opacity is very difficult. The material may be
moving in a manner such that appreciable mass is moving with the
same velocity. This requires a steep density gradient in the
homologous expansion. Alternatively, the opacity could conspire
to lock the photosphere into a particular lagrangian point which
perforce moves out with a constant velocity. Branch (1981)
following up the studies of Pskovskii(1977) concludes that there
is a real variation between the photospheric velocity of
different SNI, the range being from 10,000 to 12,000 km/s.
Their work also shows that there are indications that the
photospheric velocity is correlated with the light curve
classification of "fast" versus "slow" (see also Barbon et al.
1973). This correlation is opposite to that one might expect in
the sense that the events with slower light curves, which might
be thought to be thinning out more slowly, are, in fact,
expanding more rapidly. Again the differential effects of
opacity may play a role here and be difficult to isolate.

There are other constraints from the velocities besides the
necessity to have some silicon, calcium, etc. moving with a
characteristic bulk velocity of about 10,000 km/s. The blue
edge of the absorption profile is difficult to pick out from the
continuum with great accuracy , but there is clearly material of
the same composition moving as fast as 20,000 km/s or more.
Care must be taken that models with various envelope tampers do

not do their job too thoroughly and prevent any appreciable matter from moving at such velocities. As a lower bound, the calcium which still appears in absorption at late times moves at about 8000 km/s. If this is the same calcium-bearing layer as appears at maximim light, which seems reasonable since the velocity of the absorption minimum is seen to drift downward gradually from maximum light, then no material from this layer is ejected at less than 8000 km/s. This might imply a composition change at the lagrangian point corresponding to this velocity.

1.4 Compact Remnant

Nomoto and Tsuruta (1981) have given the most up-to-date discussion of the failure to detect the thermal X-radiation from a cooling neutron star surface in the historical "SNI" remnants (see also Glen and Sutherland 1980; Van Riper and Lamb 1981). There are still uncertainties concerning possible exotic cooling mechanisms and the proper emissivity, but the statement can safely be made that there is no direct evidence in favor of neutron stars in these remnants. Helfand (this conference) has reached the same conclusion by noting the lack of synchrotron-radiation X-ray nebula in the "SNI" remnants which are seen around several known pulsars and which should be detectable independent of any beaming of the pulsar radiation.

As a caveat to the growing circumstantial evidence that SNI do not leave neutron star remnants, Ostriker pointed out that there may be two classes of pulsars, one of which occurs at high latitudes and hence may be due to SNI. Such an argument may be resolved by other explanations for pulsar classes, and in any case runs headlong into the previous question of whether SNI are actually an old, high latitude, population as has so long been supposed.

1.5 Light Curve

The light curve is a crucial diagnostic tool. The last two years have seen great progress toward understanding the late time light curve in terms of radioactive decay. The same period has seen a transition in attitude toward differences in the initial peaks from event to event. In the past there was a strong tendency to regard these differences as dominated by observational error. The view is gaining support that these variations are real and that the physical reason must be sought. The initial peak contains some important clues to the nature of SNI, but its features are fairly model independent (see also the discussion here by Arnett), since white dwarfs and red giants can give reasonable looking shapes. Interpretation of the light curve can then be quite model dependent.

One important number derivable from the light curve is the bolometric luminosity at maximum light. From the black body which best fits the optical continuum at maximum and a modest contribution from the truncated UV component, Wheeler, Branch and Falk (1980) estimate the minimum luminosity for an SNI at maximum light to be 0.5×10^{43} erg/s. This estimate assumes $H_0 = 100$ km/s/Mpc or the distance of SN1972e to be 2 Mpc or the distance of the new SNI in NGC 4536 to be 10 Mpc. The estimated luminosity could be appreciably larger since it varies as the square of these quantities. If the progenitor star is sufficiently compact that all the shock energy is dissipated in adiabatic expansion, then all the light at maximum may be presumed to arise from radioactive decay of Ni to Co and Fe. In this case, one can simply estimate the amount of Ni ejected. For the above minimum luminosity one gets about 0.2 $M_\odot$. This estimate also increases as the square of the mean or specific distance. For the late time spectrum of SN 1972e one can make another set of minimizing assumptions and obtain a different lower limit to the amount of ejected Ni. If one assumes all the positrons are trapped, neglects reddening, assumes no correction for non-optical emission, and again adopts a minimal distance of 2 Mpc, one estimates the Ni ejecta to exceed 0.08 $M_\odot$.

The analysis of "fast" versus "slow" light curves by Pskovskii and Branch have raised tantalizing questions in this context as well. They argue not only that the "fast" decaying light curves are associated with the slowest velocities at the photosphere but also with the smallest luminosities. Branch (1981) estimates that the range in peak brightness could be as much as two magnitudes. Branch argues that rather than account for the apparent variation, observational errors, which are clearly present, mask the true variation. As shown below (see also Arnett in this volume and Chevalier 1981) the correlation of greater peak luminosity with "slow" light curves is counter to the expectation for the simple class of models in which various amounts of fuel are burned to Ni in a star of fixed total mass, e.g., the Chandrasekhar mass. In any case, if some SNI (i.e., the "slow" ones) are one to two magnitudes brighter than the dimmest, then their peak luminosity would be of order 2×10^{43} erg/s, and, by the arguments just given, they would require the ejection of 0.8 $M_\odot$ of Ni. One should bear in mind that if the progenitor star has a very extended envelope, $R > 2 \times 10^{13}$ cm as occurs in some realistic models of helium stars, then there can be some shock contribution to the peak luminosity and the amount of nickel ejected is no longer a direct function of the peak luminosity.

Dopita reported here his study of collisionless shocks in supernova remnants in the Large Magellanic Cloud, which he identifies as remnants of SNI. Dopita deduces the association

of a large burst of ionizing radiation with the supernova
explosion. Since extended envelopes are required to give the
amount of ionizing radiation required (Klein and Chevalier 1978;
Falk 1978) Dopita concludes that the progenitors of SNI have
extended envelopes.

1.6 Galactic Iron Abundance

Since many current models for the light curves and spectra
of SNI predict the ejection of appreciable Ni which decays to Fe
there is a concern that such models will overproduce iron in the
Galaxy. The results of a recent galactic evolution model which
gives a constraint on the allowable amount of iron which may be
ejected per SNI are given in section 4.1. A possibly related
issue is the failure of the Einstein Observatory to detect the
predicted overabundance of iron in the young "SNI" remnants of
1006, 1572, and 1604. To rationalize the predictions with the
observations, the ejected iron must not be heated to X-ray
temperatures. Fabian argued here that a solar mass of iron
could be present in SN 1006 with the corresponding emission at
less than 1/4 keV.

1.7 Summary

The population of SNI is an open question, and exploration
of both young and old models is legitimate. The post-explosion
structure seems to consist of an inner incinerated region
composed of Ni surrounded by a masking envelope which is
optically thick at maximum light. There is some evidence that
SNI do not produce neutron stars. The light curves and spectra
give conflicting evidence concerning the progenitor
configuration and the nature of the envelope. Solar ratios of
intermediate mass elements and the possible requirement for a
large burst of ionizing radiation suggest an extended envelope.
The seemingly large amount of Ca and a low He abundance are, at
face value, not consistent with an extended envelope. Also, a
CNO processed helium envelope should be devoid of oxygen.
Possible relations among the peak luminosity, light curve speed,
and material velocity are a challenge to theorists to find consistent models which will elucidate the requirements for a compact or extended progenitor.

2. REVIEW OF MODELS FOR SNI

2.1 Light Curve Developments

The notion that the radioactive decay of Ni (6.1 day
halflife) to Co (77 day halflife) to Fe could play an important
role in the generation of the light curve of an SNI was first
made by Pankey (1962). Later Colgate and McKee (1969)
independently investigated the idea in some detail. They found
considerable difficulty in reproducing the light curves because
the observed exponential tail falls more rapidly than the

natural 77 day timescale of the Co.

Great progress has been made on this problem in the last two years. Meyerott (1980) addressed a number of the important atomic physics processes which determine the formation of the forbidden emission line spectrum at late times in the expanding radioactively heated ejecta. Arnett (1979) drew attention to the possible role of positrons and spurred Colgate, Petschek, and Kriese (1980) to do a detailed Monte Carlo calculation of the deposition of the gamma rays which are the dominant decay product and to estimate the behavior of the associated positrons.

Axelrod (1980a,b) has independently brought all these components together to calculate self-consistent light curves and spectra for the nebular phase at times exceeding one hundred days. Axelrod argues that the spectrum gives tighter constraints on models than the associated light curves. His synthetic spectra give evidence for a particular feature at about 5900 Angstroms being due to CoIII. The change in the strength of this feature with respect to blends of iron lines corresponds to that predicted by the radioactive decay model. As mentioned previously the precise wavelength correspondence of some features is not as good as one would like but Axelrod's models have the obvious strength that they are calculated, with certain simplifying assumptions, from first principles. They give strong encouragement that the radioactive decay picture contains elements of the true behavior of SNI. One might bear in mind, however, that the large Doppler broadening leads to a bothersome increase in wavelength coincidences. The region at 5900 Angstroms also corresponds to lines of He, Na and Si. What appears to the eye in the spectra of the supernova in NGC 4536 to be a single feature is identified by Branch as Na twenty days after maximum and by Axelrod as Co ninety days after maximum. Both these claims must be closely examined in the future.

The picture which has emerged from this effort is one in which the progressive leakage of the gamma rays from the expanding nebula causes the efficiency of deposition of energy to decrease with time. Thus the rate of deposition of energy and hence the radiated luminosity fall faster than the natural decay time of the Co and the observed quasi-exponential decay with a half-life of about 55 to 60 days can be reproduced.

Maintaining the proper slope of the light curve for periods of several hundred days is still not a completely solved problem. Axelrod finds that as the nebula cools a condition is reached when the emission is dominated by fine structure lines of iron. After a certain model-dependent time of order 500 days most of the luminosity in the models comes in the far infrared.

The temperature and the optical luminosity then plummet in what Axelrod terms the "infrared catastrophe." Axelrod's initial models were of constant density, and he is optimistic that inclusion of density gradients will alleviate this problem.

The role of the positrons which represent about twenty percent of the Co decay energy is also still debated. Following Arnett's suggestion Colgate et al. (1980) argued that any pre-existing magnetic field could be combed out in the explosion allowing the progressive leakage of the positrons as well as the gamma rays, and a continued drop in the efficiency of energy deposition after several hundred days when the nebula is virtually transparent to the gamma rays, and none of their energy is deposited. Axelrod has assumed that the ambient magnetic field will give the positrons a small Larmor radius and serve to trap them. At late times his light curve thus follows the 77 day halflife curve of the Co until the luminosity in the infrared becomes appreciable and the optical luminosity drops more rapidly. Colgate argued in this meeting that the well-defined exponential light curves of SN 1937c and SN 1972e are more consistent with his picture of a smooth decrease in the positron deposition than are the "kinky" curves of Axelrod which are flatter at first, but then drop off when the infrared catastrophe occurs. Axelrod predicts that accurate photometry at very late times will show that there is a range in the behavior of the light curves. He is also concerned that if the positron energy is only partially deposited the nebula will be even cooler than he predicts at a given time and that this will hasten the onset of the infrared catastrophe. The far infrared luminosity at late times represents a potentially important discriminent of these opposing points of view. Unfortunately, observations of the event in NGC 4536 seem to be just beyond the capability of present equipment.

While the great interest and concern with the radioactive decay picture is well-justified, one should also bear in mind that there are as yet unresolved alternatives in terms of the generation of the initial peak in the light curves of SNI. For any progenitor configuration with an initial radius much less than 10^{13} cm the initial shock energy is dissipated in adiabatic expansion and the initial hump as well as the exponential tail must be generated entirely by an added source of energy, presumably radioactive decay. As Lasher (1975,1980) showed, however, many properties of the initial peak in the light curve can be well reproduced by an explosion in an extended helium envelope without any contribution from radioactive decay. Such envelopes could affect the dynamics, light curves, and spectra in a number of ways, and are not completely unexpected in terms of possible progenitor evolution. The behavior of such envelopes is discussed further below.

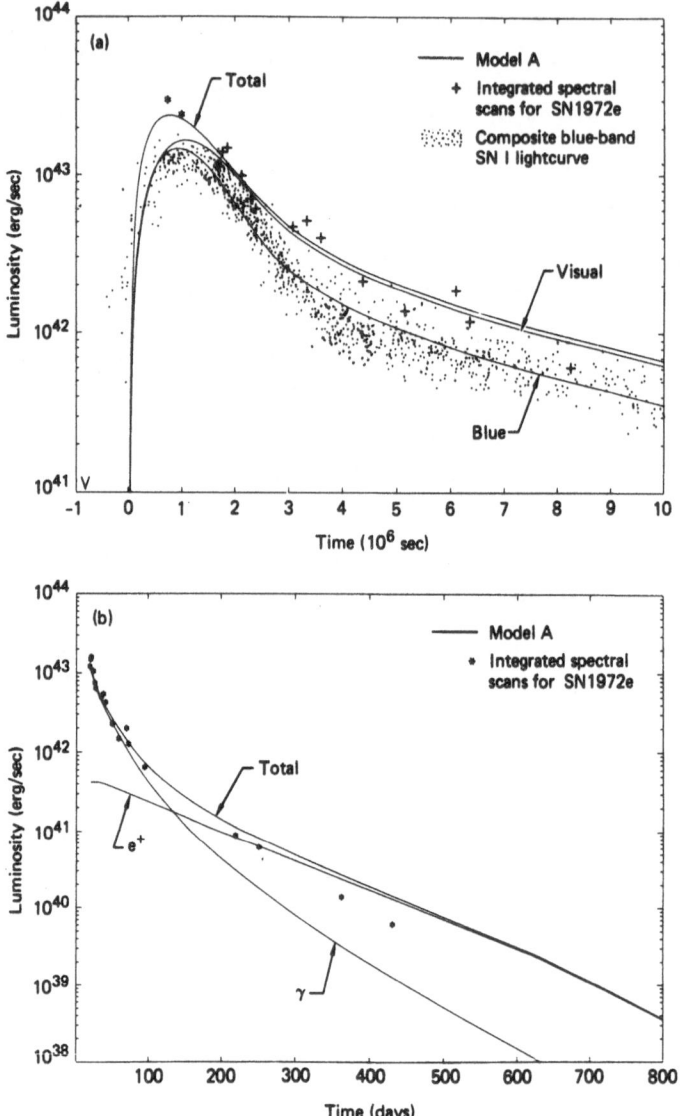

Figure 1. Theoretical light curve for the "double detonation" model from Weaver et al. (1980). Note that with the positrons trapped the late-time light curve tends to fall above the data points.

Models of SNI have recently been discussed by Chevalier (1981), Arnett(1981, and this volume), Nomoto (1981), Woosley, Weaver, and Taam (1980), and Weaver, Axelrod, and Woosley (1980). The state of the art of SNI model calculations is represented in figures 1 and 2 taken from Weaver et al.(1980). Figure 1 is based on the "double detonation" model (Woosley et al., 1980; Nomoto 1980,1981) which occurs when hydrogen is accreted at moderate rates onto a carbon/oxygen white dwarf. The hydrogen burns stably and helium accumulates in a thick degenerate shell. Eventually helium ignites explosively at the base of the shell sending a detonation outward through the helium and another inward through the core. The star is almost completely incinerated to Ni and is completely disrupted at relatively high velocities. This model gives a fair approximation to the light curve, although with the positrons trapped it does not fall rapidly enough at the latest times. Figure 2 shows the light curve for a model in which a 1.34 $M_\odot$ core explodes inside an extended envelope of mass 0.3 $M_\odot$ and initial radius 2.4×10^{11} cm. This model gives conditions which better match the requirements of the spectral synthesis, but it is subject to the infrared catastrophe, as shown in the lower illustration. These models and others are compared to various observational constraints in the next section.

2.2 Constraints on Specific Models

A variety of models have been proposed for the progenitor stars and mechanism of explosion of SNI. Many of these are based on particular evolutionary scenarios. The purpose of this section is to confront some of these models with the perceived observational constraints again recognizing that different people will want to apply the constraints with different weights, so that some of the conclusions reached here will be controversial.

2.2.1 White Dwarf Models

Most specific models for SNI continue to be based on white dwarfs despite the recent suggestions that SNI are associated with young stars. A young age is one constraint which might be downplayed, and in any case white dwarfs are not necessarily old, though many of the models must be in practice.

Detonation of helium white dwarfs triggered by accretion in a binary system has been suggested as an SNI mechanism by Mazurek (1973) and by Nomoto and Sugimoto (1977). Helium has a high specific energy and tends to give too large a velocity to correspond to observations of SNI. High mass helium dwarfs, of order a solar mass, incinerate entirely into nickel, in violation of the early spectrum. Lower mass stars might only burn partially, leaving a helium blanket (Mazurek 1980), but

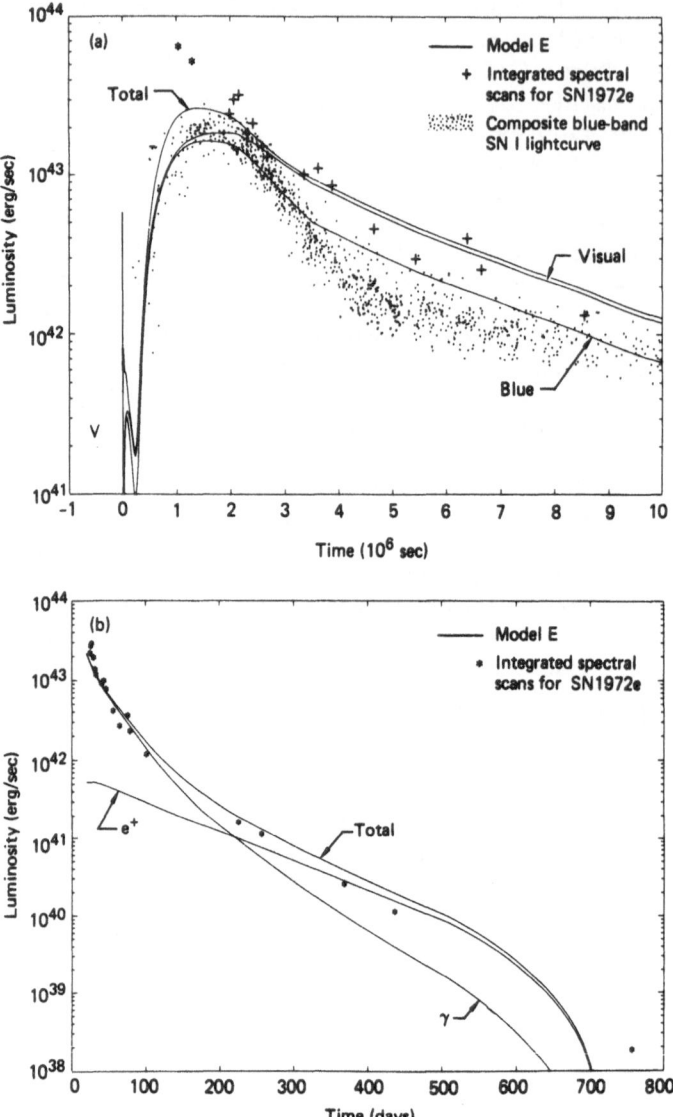

Figure 2. Theoretical light curve for the model with an incinerated Ne/O core inside a helium envelope from Weaver et al. (1980). Note that this model undergoes the "infrared catastrophe" at late times.

such cases would perhaps eject too little Ni, and off-center ignition in the pre-white dwarf state could prevent such a star from forming. A low mass helium star must come from a low mass progenitor, and so must be old and not associated with recent star formation. All things considered, this model does not seem a good candidate for the progenitors of SNI.

Nomoto (1980) has discussed the evolution of a binary white dwarf composed of O, Mg, and Ne. This type of dwarf would be the core of a star of a bit over $7M_\odot$ which would have burned carbon non-violently. With this composition and a mass intrinsically near the Chandrasekhar limit, the dwarf will collapse due to electron capture. The fuels will be consumed in collapse without generating an explosion and a neutron star will form. The outcome of the collapse of this kind of model is not known, but rather than Ni any ejecta is apt to be odd, neutron-rich material. This model also does not seem to meet the needs for SNI.

There is a variety of models based on accretion onto carbon/oxygen white dwarfs. The outcome depends on the mass of the dwarf and the mass accretion rate (Taam 1980; Fujimoto 1980; Nomoto 1980; Woosley, Weaver and Taam 1980).

For accretion rates somewhat in excess of 4×10^{-8} $M_\odot$/yr, hydrogen and helium both burn rapidly on the surface of the star and neither accumulates appreciably. The dwarf grows to the Chandrasekhar limit and undergoes central carbon ignition. The carbon probably does not detonate, but burns in a complicated subsonic, semi-dynamical combustion (Mazurek and Wheeler 1980; Sugimoto and Nomoto 1980),which is commonly, if not strictly correctly, referred to as deflagration. The outcome of this process might be the incineration of the central portion into Ni and the partial burning and mixing of an outer layer into the unburned outermost parts. The result would give a composition in qualitative agreement with the early spectra, and predicts an enrichment in elements of moderate atomic mass. One half to one solar mass of Ni might plausibly be ejected, in accordance with the requirements of the light curve and late time spectrum. There are conflicting results concerning the velocity of the unburned material (see section 3.2). Some calculations give rather low, constant velocity profiles, others give rising profiles very similar to completely incinerated white dwarfs. Even in the latter case an appropriate photospheric velocity might result from variations in the opacity. This progenitor system could be young or old depending on the mass of the mass-losing star. This suggestion seems to be the only one which can naturally provide this flexibility of age without changing the nature of the explosion in some substantial way. The crucial issue is to determine whether the early spectra are consistent with the net enrichment of elements predicted. In the meantime this model seems worth keeping in the list of reasonable possibilities.

For accretion rates between about 1×10^{-9} and 4×10^{-8} $M_\odot/yr$, hydrogen burns but helium accumulates in a cold, dense, degenerate shell. Eventually, however, helium ignites at the base of the shell, and in the numerical models sends a detonation inward through the core, and outward through the helium shell (Woosley, Weaver and Taam 1980; Nomoto 1980). This process results in virtually total incineration of the star to Ni and excessive velocities. It does not account for intermediate mass elements moving at about 12,000 km/s, as indicated by the early spectra. Although this model arises from rather natural conditions, it does not, as it stands, correspond to the observations of SNI. As a physical process, double-detonation may play a role in SNI if, for instance, the accretion rate increased toward the end, leaving a non-degenerate helium envelope which was not consumed in the explosion, as was suggested by Weaver at this conference. Such a possibility is worth checking, but one might find that hydrogen, not helium, was left by accretion at the necessary rate. Another possibility is that nova explosions eject sufficient matter that the conditions for the double detonation do not arise. Still another alternative is that the assumption of ignition in a spherical shell is misleading, and that in the realistic case of ignition at an off-center point the outcome is quite different.

For accretion rates less than about 10^{-9} $M_\odot/yr$, the outcome depends on the mass of the white dwarf (Fujimoto 1980; Nomoto 1980). For a mass in excess of about 1.2 $M_\odot$ a degenerate helium shell forms, but carbon ignites in the center before helium ignites in the base of the shell. The outcome again depends on the nature of the subsequent carbon burning, but this model will lead to an outcome similar to the high accretion rate, central carbon ignition case. One difference is that there will be unburned helium as well as intermediate mass elements in the ejecta. This type of model could give a range of ages, but they will tend to the long side, constraining the mass of the companion, because of the low accretion rate required. Again the relative and absolute abundances implied by the early spectra will provide an important constraint on this type of model. For the time being it remains a possibility worth considering.

If the mass of the white dwarf is less than about 1.2 $M_\odot$ the degenerate helium shell will form, ignite, and detonate, but the carbon will not detonate. The reason is that the shock is weak compared to the high Fermi energy of the very dense carbon

(Nomoto private communication). A model has been calculated (in collaboration with P.G. Sutherland, see section 3.2) to explore this case. The model consisted of a C/O core of 1.09 $M_\odot$ with a helium shell of 0.31 $M_\odot$. The helium was flashed instantaneously, and the subsequent dynamics followed. The Ni produced by flashing the helium was ejected with very high velocities. The inward propagating shock produced no burning except in the central zone where shock convergence caused the flashing of that single zone, a singularly untrustworthy result. The star was completely disrupted, with the C/O meekly following the Ni at low velocity. This model explores a legitimate limit of the helium shell detonation phenomenon, but it provides such a rotten representation of any SNI characteristics that it deserves a "Dos Equis" label for being excessively poor as a model for SNI.

The models discussed above involving central carbon ignition have central temperatures of a few hundred million kelvins and ignite at a central density of about 3×10^9 g/cm^3. Another class of models involving central carbon ignition are those for which the temperature is effectively zero. These represent dwarfs which have cooled for a long time, and hence may be regarded as a limiting case of very low mean accretion rate. The carbon is ignited by pycnonuclear reactions driven by the zero point energy, rather than by thermonuclear reactions. The density of ignition may approach 10^{10} g/cm^3 (see Mazurek and Wheeler 1980, Sugimoto and Nomoto 1980, or Wheeler 1981, for a discussion of the literature on this model).

Canal, Isern, and Labay (1982) have discussed a variation of these very low temperature models in which solidification leads to a separation of the carbon and oxygen, with the latter falling like "snowflakes" to the center. The oxygen has a high ignition threshold density and will first undergo electron capture. Canal et al. consider the formation and burning of ^{16}C which provides insufficient energy to explode the star but rather triggers collapse. This is an interesting suggestion, but like the O/Mg/Ne dwarfs, the resulting collapse of a bare white dwarf is difficult to relate to the observations of SNI. Canal et al. consider a high accretion rate regime which might ensue after a long cooling period. Nomoto argued in this meeting that his calculations of this case show that the heat of accretion melts the core and prevents or removes the chemical separation. In his discussion of this work at this meeting Canal raised the possibility of off-center carbon ignition. As he noted, deflagrative carbon burning in such a situation might lead to a more viable SNI model. It would probably yield a layer of Ni sandwiched between the oxygen core and an outer layer of intermediate mass elements. This is a strange configuration, but it is perhaps not excluded at the present time.

If the ignition is at densities less than 10^{10} g/cm^3 and there is no chemical separation, electron captures will ensue, although not at a rate sufficient to cause the star to collapse after a detonation (Bruenn 1972, Mazurek and Wheeler 1980). As there is some question concerning the outcome of subsonic burning in such a model, we calculated the dynamics following ignition of a 1.4 $M_\odot$ C/O core with an initial central density of 10^{10} g/cm^3. The prescription for the rate of burning was taken from Nomoto, Sugimoto, and Neo (1976). This model resulted in total disruption, but the central regions suffered enough electron capture to alter their composition appreciably as suspected by Nomoto (private communication). The mean number of electrons per baryon was Y_e=0.484, and in the central 0.2 $M_\odot$ it was 0.45 or less. This probably represents too much contamination by neutron-rich species (eg ^{54}Fe - Woosley, private communication) to be ejected by each SNI. Thus such a possibility can probably be ruled out even though it is otherwise similar to other models with central carbon ignition.

2.2.2 Models With Extended Helium Envelopes

Some models have been proposed for SNI in which the exploding core is surrounded by an extended helium envelope. The only models of this class which are associated with a particular evolutionary scenario are those in which stars of mass in the range of about 8 to 12 $M_\odot$ are envisaged to lose their hydrogen envelopes. The envelope loss is *ad hoc* but this class of models is conservative in the sense that these stars are known to form helium cores, and the bare helium cores are known to develop extended helium envelopes. Thus these models represent the smallest leap from known evolution to plausible helium star models for SNI.

Nomoto (1980) has discussed the evolution of the core of a star of about 8 to 10 $M_\odot$ after an unspecified process of mass loss. The remaining core will burn carbon non-violently and undergo electron capture and collapse. This process will leave a neutron star, which may be deemed undesirable, and ejects little or no Ni. Although it is an apparently unsatisfactory model for an SNI, this model may represent an interesting progenitor for the Crab nebula (Nomoto, these proceedings).

Woosley, Weaver and Taam (1980) have discussed the evolution of a star of about 10 $M_\odot$ as a possible SNI progenitor. For somewhat more massive stars they find that violent Ne shell flashes eject both the hydrogen and the helium shells, leaving a bare core, a process they think may be relevant to a class of SNII. They conjecture that for 10 $M_\odot$ the hydrogen envelope could be ejected, but the helium retained. While this case has not actually been calculated , they can point to a physical process of envelope loss which is not totally *ad hoc*.   The remaining core would, after a fairly complex evolution, undergo core collapse.   In the meantime the helium envelope swells in size. An important feature of these massive star models for SNI is that they are unambiguously young.   There may even be a problem of having them   so young as to be tightly confined to spiral arms, in contradiction to the evidence. This may not be the case, however, if such stars are not, in fact, born in spiral arms (Bash and Visser 1981).   This model also predicts the formation of a neutron star.   Unlike the model of Nomoto just discussed, this model has a layer of Si surrounding the collapsing core.   Presuming the core collapse to result in an explosion, a shock will propagate out through this Si layer, burning some of it to produce Ni.   Such a process probably produces less than 0.3 $M_\odot$ of Ni.   There are reasons to think that this is insufficient Ni to explain the light curve, but they are not without possible rejoinders. If one ignores the possible spread in SNI luminosities and invokes distances on the extreme small end of the range of possibilities, then this amount of Ni may be sufficient to account for the peak light of SNI.   Since these models have extended envelopes, there is a possibility that there could be a significant contribution to the maximum light from the deposited shock energy.   The luminosity of the exponential tail alone provides a less stringent constraint on the amount of Ni required so this relatively small amount could suffice for that.   This type of model for SNI pushes a lot of parameters to their extreme limits and the idea that all SNI are very young is still uncomfortable to many, but this model seems worth retaining as a possibility until more rigorous arguments for or against it are marshalled.

2.3 Summary

Having surveyed these various models, and ruled out some of them with varying degrees of confidence, some preliminary conclusions can be reached. To provide an ample amount of Ni and still not violate the restrictions on the composition and velocity from the early spectra, models involving central carbon ignition and deflagration seem generally promising, and worth exploring.   Likewise, the effect of extended helium envelopes seems worthy of further consideration.   Such envelopes could play a number of roles, some indirect.   The presence of an envelope will send a reverse shock back into the expanding core ejecta, reheating it after the initial phase of adiabatic cooling. Thus one might observe the radiation of shock energy from the core at peak light even if the original envelope had become optically thin.   Any envelope will also moderate the velocities and flatten the velocity profile of the core ejecta. Between the effect on the velocities and the possible addition of shock heating at maximum light, envelopes provide a way to

understand the small but apparently real variations among SNI.
If any envelope is optically thick at maximum light, there would
be a natural explanation for the apparent "solar" distribution
of intermediate mass elements observed at that phase. Again the
absolute abundances, or even more precise relative abundances,
will represent a crucial test of this possibility.

There is a temptation to consider a model which combines
both aspects, -a deflagrating core and an extended helium
envelope. Unfortunately, there is presently no known way to
evolve such a configuration. Nature may be giving hints in the
form of such helium stars as the R Cr Bor stars which apparently
have a degenerate core within their extended helium envelopes.
Whether the core contains carbon, as assumed in most models of
these stars (Paczynski 1971), can not be determined directly,
nor is there any evidence that they retain their envelopes long
enough to reach an explosive endpoint. Nevertheless the
exploration of models with exploding C/O cores in extended
helium envelopes in a parametrized way would seem to be a useful
exercise to place limits on possible models for SNI. Weaver,
Axelrod, and Woosley (1980) have calculated models of this
nature but the radii of the envelopes were relatively small, of
order 10^{11} cm, so shock energy could play no role. Lasher
(1975,1980) examined the behavior of larger envelopes, but he
did not include any effects of radioactive decay and neglected
the effect of the exploding core. In the next section new
models which investigate more thoroughly the behavior of
deflagrating cores and of extended envelopes are discussed.

3. CURRENT DEVELOPMENTS

Nomoto reported at this meeting his recent calculations of
the evolution of helium stars. He points out that normal stars
could leave behind evolving helium cores if they undergo mass
loss at the rate of about 10-6 $M_\odot$/yr during core helium burning.
Loss rates this high or higher are known during later thin shell
burning phases, but in those phases very small amounts of helium
are left sandwiched between the thin shells. Whether rather
large mass loss rates occur at the relatively early phase of
core helium burning when substantial helium remains is a crucial
question. If this mass loss occurs, stars of initial mass
greater than about 8 $M_\odot$ leave helium cores in excess of 2 $M_\odot$
which burn carbon and then undergo electron capture and
collapse. The arguments against this type of event being
assocated with SNI were given in the previous section. Stars of
initial mass less than about 8 $M_\odot$ would leave helium cores of
less than 2 $M_\odot$ which form, in turn, degenerate C/O cores and
evolve to explosive carbon ignition if they retain their
envelopes. Nomoto described his calculation of a 2 $M_\odot$ helium
star to the point of carbon ignition, and his initial
exploration of the subsequent dynamics. At carbon ignition the

model has 0.6 $M_\odot$ in an envelope of approximately constant density which extended to a radius of 2×10^{13} cm. This is large enough to make the contribution of some shock energy to peak light a realistic possibility. We are engaged in a collaborative study to explore carbon burning and the subsequent dynamics and light curve in this model.

3.1 Rayleigh Taylor Instability

A central complexity in all partially burned deflagrating models is that they are intrinsically Rayleigh Taylor unstable. Indeed, some form of overturn instability is at the heart of the process by which the subsonic burning is presumed to propagate. As discussed below, several prescriptions have been proposed for the rate of burning, but whatever the detailed physics assumed, this burning releases energy which causes the incinerated matter to heat and expand. The result is hot, low density material overlain by the cold, dense, unburned material. The Rayleigh Taylor instability of this material has not been explicitly examined previously. It is complicated numerically because the hot matter is in nuclear statistical equilibrium and cannot simply be mixed with the unburned matter. In addition, the mixing results in the heating and burning of the cold matter, so the mixing requires more than a simple homogenizing of the composition. In the models to be discussed below a prescription for calculating the Rayleigh Taylor overturn of the material has been employed. It is based loosely on the prescriptions of Falk and Arnett (1977) and Chevalier (1976) which were invoked for the core-envelope instability. It will be outlined here, and the details published elsewhere.

The growth time for the instability is taken to be

$$\tau_{RT} \sim \left(\frac{\Delta\rho}{\rho} \frac{\nabla P}{\rho} \frac{2\pi}{\lambda} \right)^{-1/2} \qquad (1)$$

where the pressure gradient is approximated by $\nabla P \sim \Delta P/\Delta R$. The characteristic wavelength of the perturbation is approximated by the zone size, $\lambda \sim \Delta R$. While this may seem quite artificial it does have the interesting feature that the Rayleigh Taylor timescale becomes proportional to the distance over which the mixing occurs. With these approximations we write

$$\tau_{RT} = \alpha \frac{\rho \, \Delta R}{(2\pi \, \Delta\rho \, \Delta P)^{1/2}} \qquad (2)$$

where the dimensionless factor α is a scale factor expected to

be of order unity. Note that in equation 1 for a strong
instability, $\Delta\rho \sim \rho$, and $\nabla P/\rho \sim g$, the Rayleigh Taylor timescale is
comparable to the dynamical time. The burning propagated in
such circumstances can be quite rapid. To conserve momentum an
acceleration

$$a_{RT} = \frac{\Delta v}{\tau_{RT}} \qquad\qquad (3)$$

is added into the force equation. The difference in the
internal energy with and without this acceleration represents
the turbulent energy of the mixing. This energy is assumed to
dissipate rapidly into thermal energy, the quantity of which was
calculated but always found to be negligible. In the code,
zones were checked sequentially for instability such that the
density and pressure gradients were oppositely directed. If so,
the composition and internal energy of the zones were mixed on
the timescale of equation 2. This technique was used simply to
mix the compositions and reduce density irregularities in models
where a portion of the core was assumed to be instantaneously
incinerated, and was used in other models as the actual process
by which the burning was propagated to compare this recipe to
others in the literature.

3.2 Models

A survey of the properties of deflagrating models with and
without envelopes is underway in collaboration with Peter
Sutherland. A list of the models which have been calculated to
date is given below:

- 1.4 $M_\odot$ C/O white dwarf, total instantaneous
  incineration to NSE

- 1.4 $M_\odot$ C/O white dwarf, instantaneous incineration of
  central $M_\odot$ to NSE

- 0.71 $M_\odot$ C/O white dwarf, instantaneous incineration of
  central 0.5 $M_\odot$ to NSE

- 1.4 $M_\odot$ C/O core totally incinerated inside 0.5 $M_\odot$
  envelope, constant density, $R=2 \times 10^{13}$ cm

- 1.4 $M_\odot$ C/O core totally incinerated inside 0.5 $M_\odot$
  envelope, constant density, $R=2/3 \times 10^{13}$ cm

- 1.4 $M_\odot$ C/O core totally incinerated inside 1.5 $M_\odot$
  envelope, constant density, $R=1 \times 10^{13}$ cm

- 1.09 $M_\odot$ C/O core, 0.31 $M_\odot$ He shell incinerated instantaneously

- 1.4 $M_\odot$ C/O white dwarf, central density 1×10^{10} g/cm^3, Nomoto et al.deflagration plus electron capture

- Rayleigh Taylor-driven deflagraton in 1.4 $M_\odot$ C/O white dwarf

The light curves for these models were calculated using a flux-limited diffusion scheme. Within the single scattering approximation, the gamma ray deposition was treated rigorously for an arbitrary density distribution. Details of the deposition function will be given elsewhere. The positrons were assumed to be completely trapped. The opacity was assumed to be constant, κ=0.1 cm^2/g. At present, only bolometric light curves have been calculated.

One of the central questions we wish to explore is the mechanism by which the velocity at the photosphere remains so nearly constant near peak light. Understanding the effects of opacity will be difficult, but one can look for intrinsic features of the dynamics of various models which will promote constant velocity at the photosphere in the simplest case, namely constant opacity. The main criterion is a steep density gradient as this will serve to restrict the photosphere from receding rapidly in a lagrangian sense.

Of particular interest in this regard is the further discussion by Lasher (1980) of his extended models. In his figure 1, he shows that the combined effect of the initial shock and subsequent rarefaction wave on an extended envelope of initially constant density is to make a shell for which the density profile is steeply decreasing at radii beyond the point of maximum density. The result is a relatively flat velocity profile, as shown in his figure 2. Similar behavior is found in some of the extended envelope models for SNII calculated by Falk and Arnett (1977). This result is not completely general for extended envelopes, however. Our calculations show that when the envelope has less than the mass of the core, the core determines the density profile of the material in the envelope, even as the presence of the envelope affects the final density distribution of the core. This is illustrated in figure 3 which shows the final density profile of the totally incinerated white dwarf and the result of putting on an envelope of 0.5 $M_\odot$. One can see that a reverse shock propagates into the core, enhancing its relative density, but that the envelope assumes the same density profile as the the outer portions of the bare-core model. The presence of the relatively massive core completely alters the density profile of the envelope, an effect ignored by Lasher and minimized by the massive envelopes of Falk and

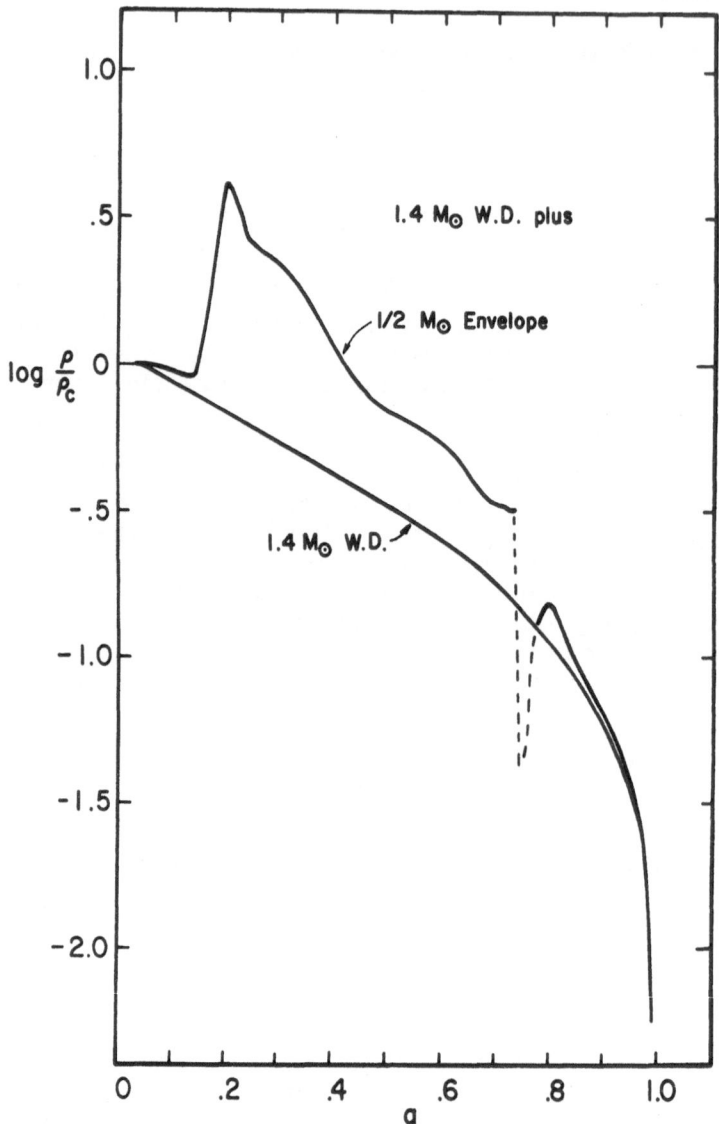

Figure 3. The post-explosion density profiles as a function of mass fraction q are shown for a bare white dwarf and for the inclusion of a 1/2 M$_\odot$ envelope. Note the similar outer density structure. The dashed line represents a numerical fitting problem.

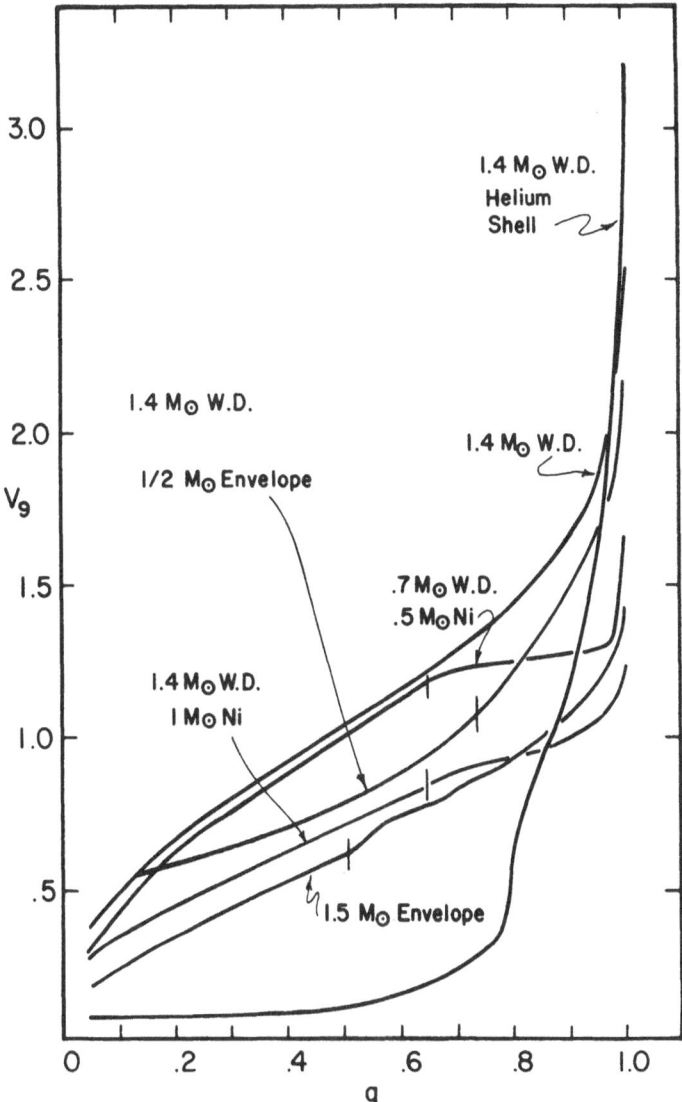

Figure 4. Velocity profiles are given as a function of mass fraction q for a number of models. See text for discussion.

Arnett.

Figure 4 shows the velocity profiles for most of the models listed previously. The totally incinerated white dwarf shows the steeply rising profile characteristic of such models. The region with velocity of order 10,000 to 12,000 km/s is buried deeply within the star. There are no elements of intermediate mass to be found there, and the photosphere will recede rapidly through this mass if the opacity is sensibly constant. The model for which the outer helium shell burns but the inner C/O core does not shows the most bizarre profile. The Ni, formerly helium, expands with great velocity while the core follows at a sedate 1000 km/s, or so. The model with the 1.5 $M_\odot$ envelope moves very slowly, despite the release of a maximum amount of energy from its C/O core. The opacity would have to be such as to trap the photosphere in the outer one percent of the envelope to give the proper velocity, and this seems very unlikely.

The other models have properties which are of some interest for SNI. The two models with 0.5 $M_\odot$ envelopes have identical velocity profiles, dominated as they are by the core kinematics. The initial radius is unimportant for these models, but not, perhaps, for others with a smaller energy release. The velocity profile is flattened in the very inner regions of the core, but is relatively steep in the outer portions of the core and throughout the envelope material. This will promote rapid recession of the photosphere, counter to the observations. In the two partially incinerated models the burned portions seem to squeeze up the outer unburned layers, giving a steep density profile and a relatively flat velocity profile. Such a profile would naturally lead to a constant velocity at the photosphere as the photosphere receeded through the unburned layers, but the result must currently be regarded as suspect. Although it seems to be a general characteristic of the present set of calculations, with and without the Rayleigh Taylor feature, the velocity profile given in figure 6 of Nomoto (1980) for a partially deflagrated model shows the steep profile characterizing the totally incinerated models. The cause of this discrepancy is under investigation.

Figure 5 shows the bolometric light curves of the bare, totally incinerated core, and that core surrounded by envelopes of various characteristics. The 1.5 $M_\odot$ envelope yields too slow a light curve to match any observations. The full width at half maximum is in excess of forty days, a factor of about two too large. The large envelope mass affects the speed of the light curve in two ways. The larger mass in the same volume lengthens the diffusion time. In addition, with a fixed input of thermonuclear energy, the greater mass has a lower velocity. The epoch when the diffusion time becomes comparable to the

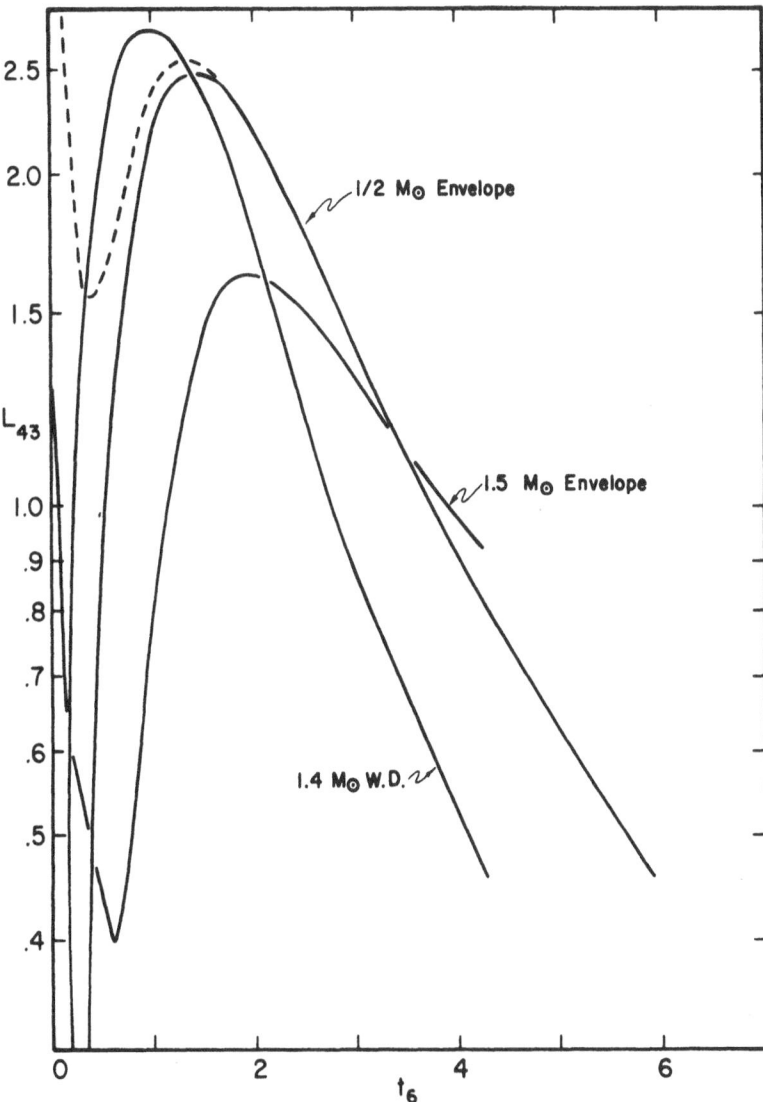

Figure 5. Light curves are given for a bare, totally incinerated white dwarf and three envelope models. The 1.5 M$_\odot$ envelope is too massive and slow. The initial radius of the 0.5 M$_\odot$ envelope is unimportant for these particular models (the solid versus dashed lines).

dynamic time, the condition that determines peak light, is then delayed.

The light curves of the two models with 0.5 $M_\odot$ merge just after maximum light. The model of initially larger radius has a stronger initial shock "spike", and a less pronounced minimum prior to the diffusion wave peak, but these features are rapid, hot, and directly observable only in the most fortunate of circumstances. Dopita argued here that there is indirect evidence in favor of such spikes, and hence large envelopes, for SNI. In the second, optical, maximum the shock energy initially deposited in the larger envelope supplies less than 10 percent of the luminosity. Further investigation of this class of models is encouraged because they closely resemble the detailed evolutionary model of a 2 $M_\odot$ helium star by Nomoto. With a slightly larger initial radius and a partially deflagrated core supplying less decay energy, the shock energy and the decay energy could play roughly equal roles in providing the luminosity at maximum light. This would make using the peak light to deduce the amount of Ni more difficult.

In terms of understanding the possible variations of the light curves among different SNI, the light curves in figure 5 may be considered as a sequence. The sequence is one in which envelopes of increasing mass (beginning with zero in the limit of the bare core) are added to a core which provides a fixed explosion energy. Examining the curves shows that as mass is added the peak gets dimmer and wider, or "slower". The range of variation is of an interesting magnitude, but is in the opposite sense to the observations, if Pskovskii and Branch are to be believed. Recall they argue that the "slow" light curves seem to correlate with the brighter events. In addition, the "slow" SNI seem to have the higher velocities at the photosphere, whereas for the sequence of models in figure 5 the slowness of the light curve is intrinsically related to the lower material velocity of the larger masses. As usual the subject of the velocity at the photosphere is confused because of the uncertain nature of the opacities which will set the location of the photosphere. Nevertheless, this series of models is a quite natural one for producing differences among SNI and the fact that the variations are correlated in an opposite manner to that suggested by the observations illustrates the depth of the problem which Pskovskii and Branch have raised.

Another rather natural way to introduce differences among SNI is to fix the mass of the exploding star and to vary the fraction which explodes. In particular, one could envisage the deflagration of differing amounts of a white dwarf with Chandrasekhar mass. Physically, such a white dwarf model has a fairly unique structure, and one's first guess would be that the fraction burned in a deflagration would be a constant, to first order, for central carbon ignition. If driven in this direction, however, one can imagine invoking variations in the composition or temperature structure which might have the desired effect of changing the fraction incinerated. Unfortunately, this scheme also runs counter to the interpretation of Pskovskii and Branch. The release of a smaller energy by deflagrating a smaller fraction of the white dwarf gives smaller material velocities, promoting "slower" light curves, and less Ni giving a dimmer maximum. This can be seen by comparing the bare core model of figure 5 with the middle light curve of figure 6 which represents the burning of the central 1 $M_\odot$ in a white dwarf of 1.4 $M_\odot$. The peak of the latter is dimmer by about 50 percent, peaks somewhat later, and is broader, hence "slower" (see also the models of Arnett in these proceedings).

The sequence of models in figure 6 represents a possible, albeit *ad hoc*, way to reproduce the type of variation advocated by Pskovskii and Branch. The procedure is to vary both the total mass, to get a longer diffusion time and greater trapping of the decay energy, and the amount of Ni produced in order to move the greater mass faster and to provide a greater luminosity. The bottom curve is the light curve for the burning of 0.5 $M_\odot$ in a white dwarf of 0.7 $M_\odot$, the middle curve represents doubling both these quantities, and the top one roughly tripling them by dint of adding a half solar mass envelope. This series does accomplish the bare minimum goal of giving light curves for which the "slower" are the brighter. As seen from figure 4 the envelope model also has larger material, and hence presumably photospheric, velocities than the 1.4 $M_\odot$ model. The small white dwarf moves faster than the larger one due to its lower binding energy, but presumably this could be adjusted by appropriate changes in the total and/or incinerated mass.

The sequence of models in figure 6 has manifest problems, however. There is no known astrophysical way by which a white dwarf of appreciably less than the Chandrasekhar mass can undergo central carbon ignition as induced here in the 0.7 $M_\odot$ model. In addition, the probability that the series of models in figure 6 could reproduce the spectral uniformity observed in SNI near maximum light seems very small.

Nomoto has pointed out that the "double detonation" models hold some promise for producing the "slow"/bright correlation. Models which have larger total mass and larger fractional C/O core mass will produce more Ni, for greater brightness, but with a lower energy release due to the lower specific energy of the C/O than the helium. This suggestion would still seem to result

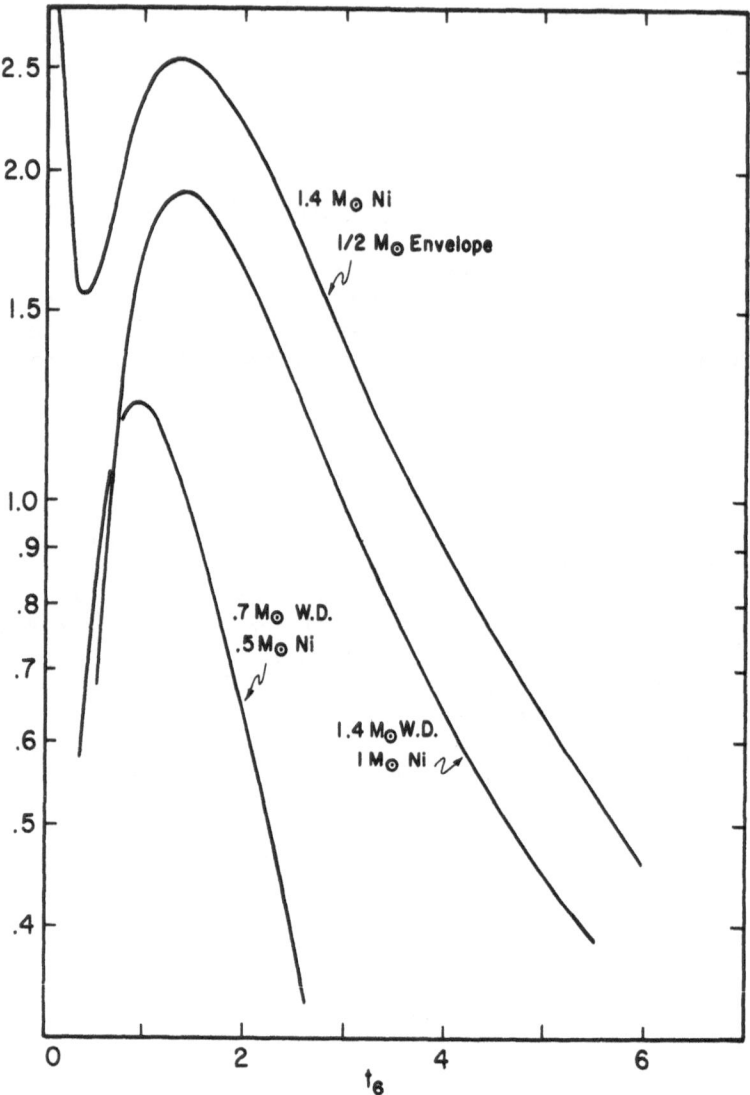

Figure 6. A sequence of light curves is given for which the "slower" are the brighter, the correlation suggested by Pskovskii (1977) and Branch (1981). Simultaneous increase of both the total mass and the incinerated mass seems necessary, a difficult requirement in practice.

in intrinsically slower material and photospheric velocities for the "slow" events.

The subject of variations in the properties of different SNI events is one which requires much further study. The claim that there are real variations among events is surely not universally accepted, and the specific correlations advocated by Pskovskii and Branch require much more refinement. Nevertheless, there is a current need for further theoretical study to put limits on the kinds of variations which are physically possible and plausible.

Another issue which has been brought strongly to the fore with the new observations of the SNI in NGC 4536 is the composition at peak light. The observations suggest that the abundances of a variety of intermediate mass elements, O, Na, Mg, Si, S, and Ca, are present in roughly solar proportions in the material at the photosphere. The simplest explanation for such ratios is that one is looking at an envelope in which the hydrogen has been transformed to helium, leaving the heavier elements intact. That is one motivation for considering models involving helium envelopes. In such a picture the absolute abundances of the heavy elements should be low, just those amounts originally contained in the star. The absolute abundances are quite uncertain, although both Branch and Axelrod spoke at this meeting of the possibility that about 10^{-3} $M_\odot$ of Ca is present. This is ten thousand times the amount expected to be originally present in a star of roughly a solar mass, and so would have to be virtually entirely freshly synthesized. If this were the case, the implication is that the other intermediate mass elements observed must also be freshly synthesized, but in a way that roughly reproduces the solar distribution. From this point of view the question is what models would be able to produce freshly synthesized elements in the proper amounts. For this type of model even the abundance ratios could be a stringent discriminent. Since the absolute abundances are not known well they may not constrain this type of model as much.

One possibility for producing freshly synthesized elements in roughly solar ratios is the explosion associated with core collapse. The 3 or 4 $M_\odot$ helium core of a star of original mass of 15 $M_\odot$ would give a reasonable match (Weaver and Woosley 1980). In such a picture the intermediate mass elements are formed primarily in the quasi-static evolution and then ejected in the explosion. There are problems with associating such an object with the progenitors of SNI, as described in section 2.2.2, but the element production is a positive feature.

Nomoto (1980, and in this meeting) has raised another

possibility in this regard. He suggests that the desired
composition might arise naturally in the deflagrating white
dwarf models. In his deflagrating models, Nomoto finds that as
the deflagration dies, so that incineration completely to NSE is
no longer possible, overturn continues to transfer heat into the
unburned C/O. The result is the burning of a few tenths of a
solar mass of C/O into Si peak products. These models are not
yet sufficiently refined to accurately predict the abundance
ratios. One might suspect that the unburned oxygen might be
enhanced compared to the Si peak elements. In addition, freshly
synthesized Ni is mixed into the Si peak layer which may result
in an unacceptable enhancement in the Fe/Si ratio and in the
Co/Fe ratio since the Fe would be predicted to have arisen in
the decay of Ni via Co. Nevertheless this possibility is
sufficiently interesting to merit further attention.

To follow up on this idea a model was calculated in which
the first zone was artificially incinerated in a C/O white dwarf
of 1.4 $M_\odot$. Subsequently, the Rayleigh Taylor prescription
described in section 3.1 was used to calculate the propagation
of the burning. For a value of the scale parameter $\alpha = 1$ in
equation 3 the mixing time was comparable to the dynamical time.
The result was the propagation of the burning at about the local
sound speed. The burning soon became self-propagating,
effectively a detonation, and proceeded even if the mixing were
turned off. The result was the total incineration and
disruption of the star.

With the value $\alpha = 10$, the slower mixing time gave results
which more closely resembled those of Nomoto. One $M_\odot$ of the
star was incinerated to NSE and then the deflagration died, as
the Si burning timescale became longer than the dynamical and
mixing timescales. The spontaneous overturn caused the burning
of one more zone to Si, but the mixing was too slow to cause any
appreciable burning in   the next zone. The result was 1 $M_\odot$
burned to NSE (eventually to become Ni) and the remainder
ejected as unburned C/O.

In Nomoto's calculations mixing continues to propagate heat
at a fair rate from the incinerated regions causing the burning
up to the bottleneck at Si. This result might have been
achieved in the present models by artificially increasing the
rate of mixing once the deflagration to NSE ceased, but such a
step would have been totally *ad hoc*. At the very least one
might conclude that the amount of Si peak elements that can be
produced through partial deflagration is rather model dependent.
Nomoto used the prescription for mixing time presented in
Nomoto, Neo, and Sugimoto (1976), which is based on a convective
velocity and which differs quantitatively from the prescription
adopted here. Buchler and Mazurek (1975) give yet another

prescription for the rate of deflagration. Effort is underway to reproduce Nomoto's results using his propagation velocity and to understand the differences in the two methods of calculation.

4. POTPOURRI: Other Aspects of SNI

4.1 Iron Production by SNI

A galactic chemical evolution model has been calculated with Bruce Twarog. The central purpose of this model was to explore the constraints set by new observations of the CNO abundances of old disk stars. As a by-product the calculation gives various limits on the allowed iron production by SNI.

The model invokes infall to solve the metal poor dwarf problem. The star formation rate in the solar neighborhood is taken to be 5 $M_\odot/pc^2/10^9$yr, or roughly 5 $M_\odot$ per year in the Galaxy. The return fraction is R = 0.35, the initial iron abundance in the disk is one-tenth solar, and the initial oxygen abundance one-third solar.

If no Fe is produced by massive stars, and all the Fe in the galaxy derives from SNI then SNI could eject about 0.6 $M_\odot$ per event if they explode at a rate of one per 100 years in the Galaxy. If, on the other hand, the yield of Fe/O from massive stars is as estimated by Arnett (1978) or Chiosi and Caimmi (1979) then, allowing for possible variations in the mass function, SNI must eject less than 0.2 $M_\odot$ per event at a rate of one per 100 years. Note that these estimates are uncertain to the extent that the chemical evolution model puts limits on the Fe in the solar neighborhood while the SNI rate is known, and only crudely, for the Galaxy as a whole. Given these caveats, the ejection of a fair amount of Fe per SNI seems difficult to rule out. A possibility for relaxing the limit even further was discussed at the meeting in the context of the various interesting "fountain" models. Supernova driven winds or fountains might expel any extra Fe. Whether this outflow can be rationalized with the infall assumption which is important for the present model is an interesting question.

4.2 The Effect on the Upper Atmosphere of Gamma Rays from SNI

Rood et al. (1979) have raised the possibility that gamma rays from the decay of Ni in galactic SNI events could have induced the excess NO_3^- found in Antarctic ice cores at times roughly corresponding to 1572 and 1604. There are various problems with this suggestion as discussed by Rood in a memo he has circulated privately. We were concerned that one of the weaker links was the assumed efficiency of production of NO_3^-, one molecule of which was assumed to be produced for every induced ionization. With David Slavsky an appropriate set of chemical rate equations have been integrated to follow the chemistry in some detail. We find that even neglecting photodissociation there are numerous back reactions which hinder the formation of NO_3^-. The amount of NO_3^- which can reasonably be produced is several hundred times less than assumed by Rood et al. Unless there is some very selective process which focusses all the freshly produced NO_3^- in a small area of the Antarctic, we are pessimistic that SNI can represent the source of the enhancements in the ice cores.

4.3 Light Element Synthesis in Helium Envelopes

The question of the production of elements like deuterium and lithium by shocks in supernova envelopes has been discussed at length by Epstein et al (1976) and by Weaver (1976). There is a small range of parameter space not completely shut down by these studies which still holds some promise as a site for light element production. With Susan Chandler a study has been done of element synthesis in extended helium envelopes, patterned after that of Epstein et al. Some observed helium stars are enriched in ^{12}C and might plausibly be in ^{13}C as well. We have found that a fairly moderate ion precursor shock in a helium envelope doped with some ^{13}C could produce an interesting amount of ^7Li and maybe even a bit of deuterium.

5. CONCLUSIONS

In surveying the properties of various models and comparing them to the observations, several conclusions are indicated. No proposed configuration involving detonating helium seems to work very well. Helium is very volatile and tends to produce excessive velocities and total incineration. At least there is a strong challange to discover ways to utilize helium detonation in a manner that is astrophysically reasonable and consistent with the observations of SNI. Models based on massive stars or other configurations which leave neutron stars are also difficult to reconcile with a variety of observational constraints. The very act of leaving a neutron star may prove to be the downfall of such models. There are also potential difficulties with the massive star models in terms of the population and kinematics of SNI. Certain massive star models may give the right sort of composition to match the observations at maximum light, but the observations and their interpretation must be refined in this regard.

Rapid accretion onto a C/O white dwarf could give nearly identical explosive events with a wide range of ages for the progenitor systems. To give a distribution of elements which is roughly solar, such models must produce fresh Si, etc., to match the pre-existing unburned oxygen. They would also seem to predict fresh Fe and associated Co which is not observed at maximum light. Whether sufficient Si can be made without mixing in an excessive amount of Fe and Co remains to be seen. Slow accretion onto a C/O dwarf will result in central carbon ignition, as for the rapid accretion, but with a layer of degenerate helium on the outside. The progenitor systems in this case must be fairly old, a problem if one thinks SNI are associated with recent star formation. There are indications that the helium envelope could be ejected at moderate velocities by a process of partial deflagration of the C/O core. If the photosphere resides in this layer at maximum light, one might account for the solar ratios of heavy elements. If the heavy elements prove to have absolute abundances in excess of solar, this type of model might still fit, but only if the helium layer were completely transparent at maximum light, and then it would be susceptible to the problems of the rapidly accreting model just discussed.

Models with extended helium envelopes are still worth exploring. Such envelopes can provide kinematical and light curve variations among different SNI. Further study is needed to determine under what conditions the photosphere will be in the envelope near maximum light and whether the presence of the envelope can result in the addition of shock energy to the luminosity at peak light, directly from the shock passage through the envelope, or indirectly due to the reflected shock which penetrates and reheats the expanding core.

There is some empirical evidence suggesting that a deflagrating carbon core surrounded by a non-degenerate helium envelope represents the most reasonable picture for an SNI. Such a model produces the amount of Ni needed for the luminosity and late time spectra fairly naturally. It also gives a shroud with which to reproduce the spectra at maximum light and account for some of the variations among SNI. If helium stars with deflagrating cores prove to play a role in SNI, explaining the origin of such helium stars will remain a challenge. One requires a total mass of the helium star to be less than about 2 $M_{\odot}$ to generate a degenerate C/O core, but the core itself must be at the Chandrasekhar limit to initiate central carbon deflagration. There is currently no known way to evolve such a configuration.

Tinsley (1979) suggested that one should look for the origin of SNI in stars of mass 4-6 $M_{\odot}$. Current models for the evolution of stars in that mass range do not give reasonable SNI progenitors. Perhaps it is time to take Tinsley's suggestion seriously and look for a way to lose an extensive amount of mass from such stars before they reach the double shell burning phase. In any case, recent theoretical and observational developments leave us with much fertile territory to be

explored.   There are many intriguing questions and in addition
to the necessary model building the subject of Type I supernovae
gives ample room for, and probably even demands, bold new ideas.

REFERENCES

Arnett, W. D. 1978, Ap. J., 219, 1008.

Arnett, W. D. 1979, Ap. J.(Letters), 230, L37.

Arnett, W. D. 1981, preprint.

Axelrod, T. S. 1980a, Ph.D. Thesis, University of California,
     Santa Cruz, unpublished.

Axelrod, T. S. 1980b, Proceedings of the Texas Workshop
     on Type I Supernovae, ed. J. C. Wheeler (Austin:
     University of Texas) p.80

Barbon, R., Ciatti, F., and Rosino, L. 1973, Astr. Ap., 25, 24.

Bash, F. N., and Visser, H. C. D. 1981, Ap. J., 247, 488.

Becker, R. H., Boldt, E. A., Holt, S. S., Serlemitsos, P. J.,
     and White, N. E. 1980b, Ap. J.(Letters), 237, L77.

Becker, R. H., Holt, S. S., Smith, B. W., White, N. E.,
     Boldt, E. A., Mushotzky, R. F., and Serlemitsos, P. J.
     1980a, Ap. J.(Letters), 235, L5.

Branch, D. 1980a in Supernovae Spectra ed. R. Meyerott
     and G. H. Gillespie (New York: American Institute of
     Physics) p. 39.

Branch, D. 1980b, in Proceedings of the Texas Workshop
     on Type I Supernovae, ed. J. C. Wheeler (Austin:
     University of Texas) p. 66.

Branch, D. 1981, Ap. J., in press.

Branch, D., Buta, R., Falk, S. W., McCall, M. L.,
     Sutherland, P. G., Uomoto, A., Wheeler, J. C.,
     and Wills, B. J. 1981, Ap. J.(Letters), in press.

Bruenn, S. W. 1972, Ap. J., 177, 459.

Buchler, J. R., and Mazurek, T.J. 1975,
     Mem. Soc. Roy. Sci. Liege, 8, 453.

Canal, R., Isern, J., and Labay, J. 1982, this volume.

Chevalier, R. A. 1976, Ap. J., 207, 872.

Chevalier, R. A. 1981, Ap. J., 246, 267.

Chiosi, C., and Caimmi, C. 1979, Ast. & Ap., 80, 234.

Colgate, S. A., and McKee, C. 1969, Ap. J., 157, 623.

Colgate, S. A., Petschek, A. G., and Kriese, J. T. 1980,
    Ap. J.(Letters), 237, L81.

Epstein, R. I., Arnett, W. D., and Schramm, D. N. 1976,
    Ap. J. Suppl., 31, 111.

Falk, S. W., Jr. 1978, Ap. J.(Letters), 225, L133.

Falk, S. W., Jr. and Arnett, W. D. 1977, Ap. J. Suppl., 33,
    515.

Fujimoto, M. Y. 1980, in Proceedings of the Texas Workshop
    on Type I Supernovae, ed. J. C. Wheeler (Austin:
    University of Texas) p. 155.

Glen, G., and Sutherland, P. 1980, Ap. J., 239, 671.

Gunn, J. E., Stryker, L. L., and Tinsley, B. M. 1981,
    Ap. J., 249, 48.

Klein, R. I., and Chevalier, R. A. 1978, Ap. J.(Letters),
    223, L109.

Lasher, G. 1975, Ap. J., 201, 194.

Lasher, G. 1980 in Supernova Spectra ed. R. Meyerott
    and G. H. Gillespie (New York: American Institute of
    Physics) p. 1.

Maza, J., and van den Bergh, S. 1976, Ap. J., 204, 519.

Mazurek, T. J. 1973, Ap. and Space Sci., 23, 365.

Mazurek, T. J. 1980 in Proceedings of the Texas Workshop
    on Type I Supernovae, ed. J. C. Wheeler (Austin:
    University of Texas) p. 182.

Mazurek, T. J., and Wheeler, J. C. 1980, Fundamentals
    of Cosmic Physics, 5, 193.

Meyerott, R. E. 1980, Ap. J., 239, 257.

Nomoto, K. 1980, in Proceedings of the Texas Workshop
    on Type I Supernovae, ed. J. C. Wheeler (Austin:
    University of Texas) p.164.

Nomoto, K. 1981, Ap. J., in press.

Nomoto, K., and Sugimoto, D. 1977, Pub. Astr. Soc.
    Japan, 29, 765.

Nomoto, K., Sugimoto, D., and Neo, S. 1976, Ap. and
    Space Sci., 39, L37.

Nomoto, K., and Tsuruta, S. 1981, Ap. J. (Letters), 250, L19.

Oemler, A. Jr., and Tinsley, B. M. 1979, A. J., 84, 985.

Paczynski, B. E. 1971, Acta Astron., 21, 1.

Pankey, T. Jr. 1962, Ph.D. Thesis, Howard University
    (Ann Arbor: University Microfilms).

Pskovskii, Y. P. 1977, Sov. Astron., 21, 675.

Rood, R. T., Sarazin, C. L., Zeller, E. J., and
    Parker, B. C. 1979, Nature, 282, 701.

Sugimoto, D., and Nomoto, K. 1980, Space Science Review,
    25, 155.

Szymkowiak, A. E. 1980 in Proceedings of the Texas
    Workshop on Type I Supernovae ed. J. C. Wheeler
    (Austin: University of Texas) p. 32.

Taam, R. E. 1980, Ap. J., 237, 142.

Tammann, G. A. 1977, in Supernovae, ed. D. N. Schramm
    (Dordrecht: Reidel), p. 95.

Tinsley, B. M. 1979, Ap. J., 229, 1046.

Van Riper, K. A., and Lamb, D. Q. 1981, Ap. J.(Letters),
    244, L13.

Weaver, T. A. 1976, Ap. J. Suppl., 32, 233.

Weaver, T. A., Axelrod, T. S., and Woosley, S. E. 1980,
    in Proceedings of the Texas Workshop on Type I

Supernovae, ed. J. C. Wheeler (Austin: University of Texas) p. 113.

Weaver, T. A., and Woosley, S. E. 1980, Ann. N. Y. Acad. Sci., 336, 335.

Wheeler, J. C. 1981, Rep. on Prog. in Phys., 44, 85.

Wheeler, J. C., Branch, D., and Falk, S. W.,Jr. 1980 in Proceedings of the Texas Workshop on Type I Supernovae ed. J. C. Wheeler (Austin: University of Texas) p. 199.

Woosley, S. E., Weaver, T. A., and Taam, R. E. 1980, in Proceedings of the Texas Workshop on Type I Supernovae, ed. J. C. Wheeler, (Austin: University of Texas) p. 96.

THE ORIGIN OF THE CRAB NEBULA AND ELECTRON CAPTURE SUPERNOVA OF 8-10 M⊙ STARS

Ken'ichi Nomoto

NASA-Goddard Space Flight Center, Greenbelt, USA
and
Department of Physics, Ibaraki University, Mito, Japan

The chemical composition of the Crab Nebula is compared with several presupernova models. The small carbon and oxygen abundances in the helium-rich nebula are only consistent with a presupernova model of the star whose main-sequence mass was M_{ms} = 8-9.5 $M_{\odot}$. More massive stars contain too much carbon in the helium layer and smaller mass stars do not leave neutron stars. The progenitor star of the Crab Nebula lost an appreciable part of the hydrogen-rich envelope before the hydrogen-rich and helium layers were mixed by convection. Finally it exploded as an electron capture supernova: the O+Ne+Mg core collapsed to form a neutron star and only the extended helium-rich envelope was ejected by the weak shock wave.

1. INTRODUCTION

The Crab Nebula is the remnant of the supernova in 1054 (SN 1054). It contains a pulsar which has provided solid evidence for neutron star formation through a supernova explosion. Therefore, it is quite important to answer the questions: i) what was the progenitor star of SN 1054? and ii) what was the mechanism of explosion? However, the origin of the Crab Nebula has been obscure despite a lot of observations.

In this paper, I discuss these questions from the point of view of the chemical composition of the Crab Nebula. Based on the recent IUE observations (Davidson et al. 1981) and the new theoretical work on the evolution of 8-10 $M_{\odot}$ stars (Nomoto 1981), I argue that SN 1054 was an electron capture supernova in a star which had a mass of 8-9.5 $M_{\odot}$ on the zero-age main-sequence.

205

M. J. Rees and R. J. Stoneham (eds.), Supernovae: A Survey of Current Research, 205–213.
Copyright © 1982 by D. Reidel Publishing Company.

2. CHEMICAL COMPOSITION OF THE CRAB NEBULA

Many optical observations (Davidson 1979) and the recent UV observations with the IUE (Davidson et al. 1981) have determined the chemical composition of the Crab Nebula. The abundances of hydrogen, helium, carbon, and oxygen and also the mass of the nebula are summarized as follows (X_H, X_{He}, X_C, and X_O denote their mass fractions):

i) The Crab Nebula is helium-rich, i.e. $1.6 \lesssim X_{He}/X_H \lesssim 8$ (Henry and MacAlpine 1981): hydrogen and helium are distributed through the nebula and certainly are not well segregated (Davidson et al. 1981).

ii) Oxygen abundance is less than solar, i.e. $X_O \sim 0.003$, and oxygen-to-hydrogen ratio is roughly solar (Davidson et al. 1981).

iii) Carbon-to-oxygen ratio is $0.4 \lesssim X_C/X_O \lesssim 1.1$ (Davidson et al. 1981).

iv) The mass of the Crab Nebula is larger than $1.2 \; M_\odot$ (Henry and MacAlpine 1981) and probably around 2–3 $M_\odot$ (Davidson et al. 1981).

Since the Crab Nebula consists mostly of the ejecta of SN 1054 rather than interstellar material, its composition provides the important clue to determine the presupernova model of SN 1054 as will be discussed in the following sections.

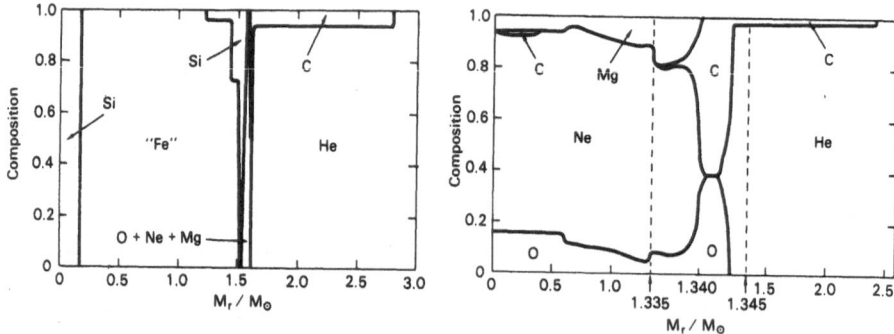

Figure 1. Composition of ~ 12 $M_\odot$ star ($M_H = 3 \; M_\odot$) at presupernova stage (onset of silicon flash).

Figure 2. Composition of ~ 9.5 $M_\odot$ star ($M_H - 2.6 \; M_\odot$) at the onset of the dredging up of helium layer.

3. COMPARISON WITH MASSIVE STAR MODELS OF M $>$ 10 M$_\odot$

High helium abundance and low oxygen abundance of the Crab
Nebula imply that only the hydrogen (H)-rich envelope and helium
(He) layer of the progenitor star were ejected to form the nebula
(Arnett 1975) because there is an oxygen-rich layer just below
the He-burning shell. This also implies that the materials
contained below the He-burning shell should collapse into the
neutron star. (Hereafter, M_H, M_{He} and M_C denote the mass
contained interior to the burning shell of hydrogen, helium, and
carbon, respectively, and M_{ms} denotes the stellar mass at the
zero-age main-sequence.) The values of M_{He} at the presupernova
stage for stars of $M_{ms} \gtrsim 10$ M$_\odot$ are $M_{He} = 2.6 - 1.8$ M$_\odot$ for $M_{ms} =$
14 - 15 M$_\odot$ (Arnett 1975: Sparks and Endal 1980: Weaver et al.
1978): 1.6 M$_\odot$ for $M_{ms} \sim 12$ M$_\odot$ (Nomoto 1980b): 1.55 M$_\odot$ for $M_{ms} =$
10 M$_\odot$ (Woosley et al. 1980). Composition of a $M_{ms} \sim 12$ M$_\odot$ star
($M_H = 3$ M$_\odot$) is shown in Figure 1.

If we assume the gravitational mass of the neutron star to be
around 1.4 M$_\odot$, the progenitor models with $M_{ms} > 15$ M$_\odot$ may be
inadequate because oxygen-rich material would be ejected, and
probably $M_{ms} \lesssim 12$ M$_\odot$ is required for the progenitor star.

The small carbon abundance gives a more strict constraint on
M_{ms} of the progenitor star. During the late stages of evolution
of stars with $M_{ms} \gtrsim 10$ M$_\odot$, helium shell burning is so active that
it forms a convective zone and produce appreciable amount of
carbon in the He-layer. For stars of $M_{ms} = 12$ M$_\odot$ (Figure 1) and

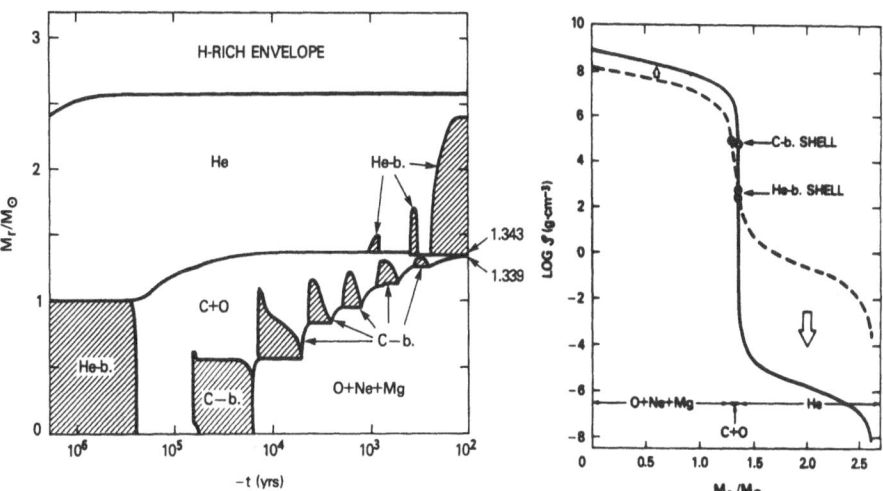

Figure 3. Chemical evolution   Figure 4. Expansion of the He-
of the core of $M_H = 2.6$ M$_\odot$.   layer at $-t = 10^2$ yrs
                                        in Figure 3.

10 $M_\odot$ (Woosley et al. 1980), X_C is 0.06 and 0.04, respectively, in the convective shell of the He-layer (0.05 and 0.03 if averaged over the He-layer). Although X_C could be reduced at most by a factor of 2 due to the mixing with H-rich matter, it is still too large to be compatible with such a low carbon abundance as $X_C \lesssim 0.003$ in the Crab. Therefore we conclude that the progenitor star of SN 1054 should have M_{ms} smaller than 10 $M_\odot$.

4. COMPARISON WITH 8-10 $M_\odot$ STARS

The evolution of M_{ms} = 8-10 $M_\odot$ stars is essentially different from more massive stars because strong electron-degeneracy sets in when O+Ne+Mg core is formed. The final state of these stars is electron capture supernovae (Miyaji et al. 1980: Nomoto 1981), i.e. the O+Ne+Mg core collapses due to electron captures and a neutron star will be left after the explosion. Therefore, the star in this mass range is a potential candidate for the Crab's progenitor.

The composition of the supernova ejecta, in particular the carbon abundance in the He-layer, depends on M_{ms}. I have computed the evolution of the helium core of 8-10 $M_\odot$ stars systematically for M_H = 2.4, 2.6 and 2.8 $M_\odot$ (Nomoto 1980a, 1981). The evolution of the core of M_H = 2.6 $M_\odot$, which corresponds to $M_{ms} \sim 9.5$-10 $M_\odot$, is as follows (Figure 3): Carbon burning proceeds under non-degenerate condition. The resultant O+Ne+Mg core of mass M_C grows through the phases of several carbon-shell flashes. These shell flashes are too mild to induce any dynamical events. After a maximum temperature of 1.1×10^9 K is attained in the outer shell, the temperature begins to decrease because of the neutrino losses. Neon ignition does not occur because M_{He} = 1.343 $M_\odot$ is smaller than the critical mass of 1.37 $M_\odot$ for the neon ignition. The electron degeneracy becomes stronger as M_C increases. When M_C reaches 1.339 $M_\odot$, which is very close to M_{He} (= 1.343 $M_\odot$), the He-layer expands greatly as seen in Figure 4. This implies that the surface convection zone begins to penetrate into the He-layer and dredges up most of the material in the He-layer.

As for the carbon abundance, the He-shell burning becomes active enough before the penetration of the surface convection zone so that carbon of X_C = 0.03 is produced in the convective shell of He-layer (shaded region in Figure 3: see also Figure 2 for the composition of this core at the onset of the dredging up of the He-layer). Therefore the star of M_H = 2.6 $M_\odot$ also contains too much carbon in the H-He envelope to be compatible with the small carbon abundance in the Crab Nebula.

However, such a carbon production in the He-layer is prevented for stars with slightly smaller M_{ms}, i.e. $M_{ms} \lesssim 9.5$ $M_\odot$

($M_H < 2.6$ $M_\odot$). The preliminary results for the core of $M_H = 2.4$ $M_\odot$ shows that the penetration of the surface convection zone into the He-layer begins earlier than for the core of $M_H = 2.6$ $M_\odot$, i.e. at the early stages of carbon shell burnings (see also Weaver et al. 1980) because of the stronger electron-degeneracy (see Sugimoto and Nomoto 1980). Most of the He-layer is dredged up before the He-shell burning becomes so active as to produce appreciable carbon. For stars with 8 $M_\odot \lesssim M_{ms} \lesssim 9.5$ $M_\odot$, therefore, no carbon enrichment takes place in the envelope at this stage.

5. EVOLUTION OF THE PROGENITOR STAR OF THE CRAB NEBULA

According to the discussions in the preceding sections, the following scenario of the evolution of the Crab's progenitor star is most plausible and consistent with the observations. On the zero-age main-sequence, the star had a mass of $M_{ms} = 8$-9.5 $M_\odot$. The star spent 3×10^7 yrs, 2×10^6 yrs, and 6×10^4 yrs for the H, He, and C burning phases, respectively. During the blue and red supergiant stages, the star lost ~ 5-6 $M_\odot$ of its envelope by mass loss to reduce the H-rich envelope mass down to ~ 0.5-1 $M_\odot$.

When the O+Ne+Mg core formed after the exhaustion of carbon in the central region, the surface convection zone penetrated into the He-layer and most of the materials in the He-layer (~ 1.2 $M_\odot$) were dredged up into the H-rich envelope. At this stage, the H-He envelope had a mass of ~ 2 $M_\odot$ and had a composition of $X_H \sim$ 0.2-0.3, $X_{He} \sim$ 0.8-0.7, and a solar ratio of X_C/X_H and X_O/X_H. Since the He-layer had contained $\sim 1\%$ ^{14}N produced by the CNO cycle, ^{14}N abundance in the envelope was enhanced to about $X_N \sim$ 0.005 by the mixing. (X_{He}/X_H and X_N depend on the mass ratio between the He-layer and H-rich envelope, i.e. on M_{ms} and mass loss rate. Also X_N is proportional to the initial CNO abundances.)

Afterwards the mass of the O+Ne+Mg core ($M_H = M_{He} = M_C$) increased from ~ 1.3 $M_\odot$ through 1.38 $M_\odot$ due to triple shell-burning of H, He, and C. During this phase, the carbon abundance in the envelope would be somewhat enhanced by the recurrence of the He-shell flashes in the thin He-zone and the associated dredging up of the processed material into the H-He envelope. Therefore X_C could be close to X_O in the presupernova envelope.

Finally the star became an electron capture supernova (Miyaji et al. 1980). At the stage with $M_C = 1.38$ $M_\odot$, the degenerate O+Ne+Mg core collapsed because the Chandrasekhar limiting mass was reduced by electron captures on ^{24}Mg, ^{20}Ne, etc. During the collapse, the oxygen deflagration was ignited and incinerated the core materials into nuclear statistical equilibrium composition. Since the effects of electron capture dominated over the oxygen deflagration, the core continued to collapse.

Although the hydrodynamic behaviour of the bounce of this core has not been investigated, from the recent computations of the collapse and bounce of the iron core we would expect that the reflecting shock wave would not be strong enough to eject the core material. However, such a weak shock wave could eject the extended H—He envelope rather easily because the shock wave would be strengthened due to the very steep density gradient round the core-envelope interface as seen in Figure 4 and because the binding energy of the extended envelope is as small as 10^{46} erg.

Then the supernova ejecta consisted of the material of the He-rich envelope so that the composition was consistent with the Crab Nebula. This scenario predicts the enrichment of ^{14}N in the nebula: in fact, recent observations (Fesen and Kirschner 1981) suggest that ^{14}N may be overabundant. The interaction between the ejected He-rich envelope and the circumstellar H-rich material lost before the explosion might cause the suggested abundance variations among filaments (Fesen and Kirshner 1981).

Such a weak shock model is consistent with the small kinetic energy of expansion of the Crab Nebula. Also even the weak shock wave could produce the maximum luminosity of SN 1054 because the radius of the supernova star would be as large as $\sim 10^{14}$ cm (Chevalier 1981). The light curve at late times would be powered by the pulsar (Ostriker and Gunn 1971).

It should be noted that the progenitor model involving the star with $M_{ms} \lesssim 8\ M_\odot$ ($M_H \lesssim 2\ M_\odot$) is not adequate for the Crab because the star explodes completely as a carbon deflagration supernova (Nomoto et al. 1976) and does not leave a neutron star. Instead, helium stars of 1.5-2 $M_\odot$ or C+O white dwarfs originating from stars of $M_{ms} \lesssim 8\ M_\odot$ may become Type I supernovae as discussed in the Appendix.

It is a pleasure to thank Drs W.M. Sparks, R.A. Fesen, T.R. Gull, S.P. Maran and T.P. Stecher for useful discussions and H. Nomoto for preparation of the manuscript. This work has been supported by NRC-NASA research associateships in 1979-1981.

APPENDIX

CARBON DEFLAGRATION IN HELIUM STARS AND TYPE I SUPERNOVAE

Observations of Type I supernovae (SN I) have shown that their progenitor stars should be hydrogen-deficient and produce a large amount of ^{56}Ni through the explosion (Wheeler and

Sutherland 1981). It has been shown that supernovae in accreting white dwarfs satisfy these constraints (Nomoto 1980a, 1981: Weaver et al. 1980: Woosley et al. 1980). Here I demonstrate that helium stars of 1.5-2.0 $M_\odot$ also evolve into supernovae to produce a large amount of ^{56}Ni. Such low mass helium stars may form from the stars of M_{ms} = 6-8 $M_\odot$ if they lose the H-rich envelope by mass loss before the surface convection zone dredges up the He-layer.

The evolution of 2 $M_\odot$ helium star is computed from the beginning of the helium burning. It forms a C+O core whose mass, M_{He}, grows by the helium shell burning. When M_{He} reaches ~ 1 $M_\odot$ and electrons become strongly degenerate, the helium envelope expands greatly as seen from the drastic change in the density distribution in Figure 5 (see also Paczynski 1971). Afterwards the star moves upward along the Hayashi line in the HR diagram as M_{He} increases.

Since M_{He} grows at such a high rate as 4-6 x 10^{-6} $M_\odot$ yr^{-1}, the resultant compressional heating ignites several carbon-shell flashes. Thus the O+Ne+Mg zone forms in the outer layers of the core (i.e. $M_r \gtrsim 1.1$ $M_\odot$). Also the luminosity due to the gravitational energy released by the core compression is as high as the luminosity due to the He-shell burning.

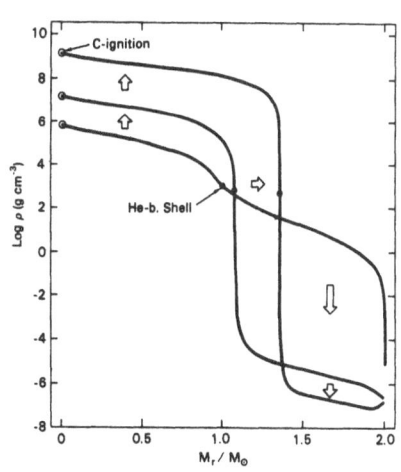
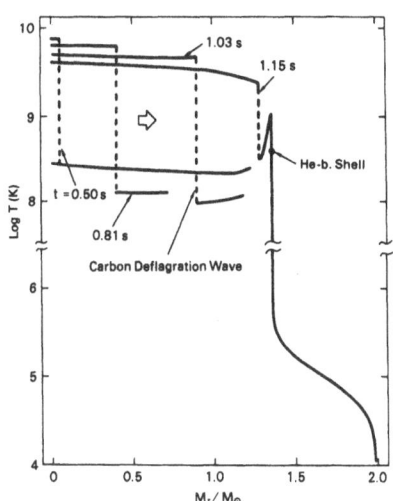

Figure 5. Evolutionary change in the density distribution. As the C+O core grows, the He-layer expands to form a red-giant-like envelope.

Figure 6. Propagation of the carbon deflagration wave and the associated change in the temperature profile.

Because of such a rapid growth of the core, carbon is ignited in the centre at the stage with M_{He} = 1.36 $M_\odot$ and the central density of ρ_c = 1.5 x 10^9 g cm^{-3}: this is earlier than the degenerate carbon ignition in normal stars (Sugimoto and Nomoto 1980). The carbon flash grows into the deflagration.

The carbon deflagration wave (DFW) propagates due to convective heat transport. Its hydrodynamical behaviour is computed by the same method as in Nomoto et al. (1976) for several parameters involved in the propagation velocity of the DFW, v_{def}. For the case the $v_{def} \sim$ 0.3 v_s (where v_s denoted the sound velocity behind the DFW), the propagation of the DFW is shown by the change in the temperature distribution in Figure 6.

As the DFW propagates, it becomes weaker, i.e. the temperature and density at the DFW decrease because of the core expansion. In the outer layer, moreover, the carbon deflagration changes into neon deflagration which is weaker because of the smaller nuclear energy release.

Accordingly, the material in the inner layer of 0.90 $M_\odot$ is incinerated into ^{56}Ni while 0.38 $M_\odot$ Ca-Si-Mg-O is synthesized by the partial burning of Si, O, and Ne in the decaying DFW. The rest, 0.08 $M_\odot$ O+Ne+Mg, remains unburned.

The mass of ^{56}Ni, M_{Ni}, synthesized in the DFW depends on v_{def}. For $v_{def} \sim$ 0.5 v_s, the core material is almost completely incinerated into ^{56}Ni, i.e. M_{Ni} = 1.36 $M_\odot$, while M_{Ni} = 0.7 $M_\odot$ for $v_{def} \sim$ 0.25 v_s.

We note: i) the carbon/neon deflagration in the helium star produces a sufficient amount of ^{56}Ni to power the light curve of SN I by the radioactive decay of ^{56}Ni $\to$ ^{56}Co $\to$ ^{56}Fe (Wheeler and Sutherland 1981). ii) Since this star has an extended helium envelope with the radius of $\sim 10^{13}$ cm (Figure 5), the rapidly expanding core hits the envelope and forms a very strong shock wave near the core-envelope interface. Such a shock wave may contribute somewhat to the SN I light curve (Wheeler and Sutherland 1981). iii) The synthesis of Ca-Si-Mg-O in the outer layer leads to a surface composition consistent with the recent observation of SN I 1981 in NGC 4536 (Branch 1982): the spectrum of this SN I at maximum light is well interpreted by the presence of Ca, Si, S, Mg, and O.

REFERENCES

Arnett, W.D. 1975. Astrophys.J., _195_, 727.
Branch, D. 1982. This volume.
Chevalier, R.A. 1981. Fund. of Cosmic Phys., _7_, 1.
Davidson, K. 1979. Astrophys.J., _228_, 179.

Davidson, K., Gull, T.R., Maran, S.P., Stecher, T.P., Fesen, R.A., Parise, R.A., Harvel, C.A., Kafatos, M. and Trimble, V.L. 1981. Astrophys.J. in press.

Fesen, R.A. and Kirschner, R.P. 1981. Astrophys.J. submitted.

Henry, R.C. and MacAlpine, G.M. 1981. Paper presented at this conference.

Miyaji, S., Nomoto, K., Yokoi, K. and Sugimoto, D. 1980. Publ. Astr.Soc. Jpan, 32, 303.

Nomoto, K. 1980a. In Proc.Texas Workshop on Type I Supernovae ed. J.C. Wheeler (Austin: University of Texas), p. 164.

Nomoto, K. 1980b. In preparation (see Sugimoto and Nomoto 1980).

Nomoto, K. 1981. In IAU Symp. No.93, Fundamental Problems in the Theory of Stellar Evolution, ed. D. Sugimoto, D.Q. Lamb and D.N. Schramm (Reidel: Dordrecht), p. 295.

Nomoto, K., Sugimoto, D. and Neo, S. 1976. Ap.Space Sci., 39, L37.

Ostriker, J.P.and Gunn, J.E. 1971. Astrophys.J.(Letters), 164 L95.

Paczynski, B. 1971. Acta Astr. 21, 1.

Sparks, W.M. and Endal, A.S. 1980. Astrophys.J., 237, 130.

Sugimoto, D. and Nomoto, K. 1980. Space Sci Rev., 25, 155.

Weaver, T.A., Zimmerman, G.B. and Woosley, S.E. 1978. Astrophys.J., 255, 1021.

Weaver, T.A., Axelrod, T.S. and Woosley, S.E. 1980. In Proc. Texas Workshop on Type I Supernovae, ed. J.C. Wheeler (Austin: University of Texas), p. 113.

Wheeler, J.C. and Sutherland, P. 1981. Paper presented at this conference.

Woosley, S.E., Weaver, T.A. and Taam, R.E. 1980. In Proc.Texas Workshop on Type I Supernovae, ed. J.C. Wheeler (Austin: University of Texas), p. 96.

CARBON-OXYGEN WHITE DWARF PRESUPERNOVA MODELS

R. Canal, J. Isern and J. Labay

Departamento de Fisica de la Tierra y del Cosmos
Universidad de Barcelona, Spain

ABSTRACT

The long-term evolution of accreting C-O white dwarfs as a possible source of neutron stars and/or SNI events is considered. The effects of solidification of the star's core are included, in particular carbon-oxygen separation. Different degrees of chemical differentiation and a wide range of accretion rates are studied. We determine the pre-flash configurations and we show that a variety of results are possible, ranging from total collapse to partial disruption due to off-centre ignition of carbon.

1. INTRODUCTION

We will study the long-term evolution of accreting C-O white dwarfs and we will try to determine the conditions for these objects to be a source of neutron stars (low-mass binary X-ray sources, X-ray bursters, pulsars perhaps?) and/or Type I supernova events. Our scenario will be a binary system consisting of a massive C-O white dwarf plus a low-mass main-sequence companion. The chemical composition is assumed to be a standard $X_C = X_O = 0.50$. It should be noted that the detached phase of such a system may last very long: $t > 10^9$ yr, and so the white dwarf may cool down to low temperatures: $T < 5 \times 10^7$ K ($t_{cool} = 9 \times 10^8$ yr for a 1 $M_\odot$ star). At such low temperatures the star's core becomes solid. This may have, as we will show, two important consequences. The first one is to initially exclude convection as a mode for burning propagation. The second one is the possible separation of oxygen and carbon (Stevenson 1980),

215

M. J. Rees and R. J. Stoneham (eds.), Supernovae: A Survey of Current Research, 215–220.

with the solid oxygen accumulating in the central regions of the
star. We will consider both effects combined with different
accretion rates and we will show how they can lead to a variety
of results ranging from total collapse to partial disruption of
the star.

2. RESULTS AND DISCUSSION

The effects of mass accretion on the outer layers of the star
will be provisionally excluded. There are essentially two regimes:
low rates ($\dot{M} < 10^{-10}$ $M_{\odot}$ yr^{-1}), corresponding to accretion from
the companion's stellar wind or to decay of the orbit by
gravitational radiation, and high rates ($\dot{M} > 10^{-8}$ $M_{\odot}$ yr^{-1}),
corresponding to Roche lobe overflow by evolution of the
secondary away from the main sequence.

Solidification

If the ignition line for ^{12}C in the (ρ, T) plane is attained
when the central layers of the star are still in the fluid phase,
we are faced with the same problem as for carbon ignition in the
degenerate cores of intermediate-mass ($4M_{\odot} < M < 8 M_{\odot}$) stars. The
result from one-dimensional approaches is unclear (Buchler,
Colgate and Mazurek 1980) and the hydrodynamical problem awaits a
two-dimensional treatment (Arnett 1981).

When solidification takes place, we have the following
situation:

a) Thermal stability in front of the pycnonuclear reactions
for $\rho_c \lesssim 6 \times 10^9$ g cm^{-3}, corresponding to $M \lesssim 1.357$ $M_{\odot}$.

b) Dynamical instability due to the e^{-}-captures on ^{16}O for
$\rho_c > 1.92 \times 10^{10}$ g cm^{-3}, corresponding to $M > 1.365$ $M_{\odot}$.

c) The burning can only propagate by conduction (if we
exclude a detonation being initially formed: see Mazurek, Meier
and Wheeler 1977).

Fast accretion

$\dot{M} \sim 10^{-6}$ $M_{\odot}$ yr^{-1} (of the order of the Eddington limit). This
case has been considered by Canal and Schatzman (1976), Duncan
et al. (1976), Ergma and Tutukov (1976) and Canal and Isern
(1979). It leads to ^{12}C ignition at $\rho_c = 1.37 \times 10^{10}$ g cm^{-3}. But,
assuming deflagration to be the mode of burning propagation, we
have that conductive propagation is $\sim 10^4$ times slower than the
convective propagation assumed by Nomoto, Sugimoto and Neo
1976). The density is high enough for fast e^{-}-captures to occur
on the incinerated material. Thus, the central layers were

starting to collapse in our above-referred calculation (Canal and Isern 1979).

Slow accretion

$\dot{M} < 10^{-10}$ $M_\odot$ yr^{-1}. It leads to ignition at $\rho_c \sim 6 \times 10^9 g$ cm^{-3}. The outcome is uncertain.

We turn now to another important aspect of solidification: its being a process which may induce chemical differentiation. Stevenson (1980) has indicated the presence of a pronounced eutectic in the phase diagram of ^{12}C-^{16}O mixtures. The eutectic composition (the composition corresponding to the temperature minimum in the phase diagram) is 66.8% ^{12}C and 33.2% ^{16}O, by number. With our assumed chemical composition ($X_C = X_O = 0.50$), a likely sequence of events is the following:

1) ^{16}O 'snowflakes' settle in the centre of the star.

2)   Rehomogenization of the oxygen-depleted fluid by a 'salt-finger' instability.

3)   Formation of solid ^{12}C when the fluid composition becomes eutectic.

4)   Solid ^{12}C rises and re-dissolves higher up.

5)   A completely differentiated body is formed.

It must be noted, however, that the cooling process is slowed down by the chemical differentiation itself, since it involves a release of gravitational potential energy. This effect can easily be estimated:

The relative change in gravitational potential energy is: $\Delta\Omega/\Omega \sim 10^{-3}$ (Pollock and Hansen 1973). Typical values for a 1 $M_\odot$ star at the beginning of crystallization are: $-\Omega = 4.72 \times 10^{50}$ erg and $L = 1.6 \times 10^{-3}$ $L_\odot$. So: $\Delta t = \Delta\Omega/L = 2.5 \times 10^9$ yr, which means a non-dramatic increase of the time scale of cooling.

In order to calculate the pre-collapse and/or pre-supernova evolution of chemically differentiated C-O white dwarfs, we have combined two models for the chemical separation:

A)   Complete carbon-oxygen separation (~ 0.7 $M_\odot$ of ^{16}O surrounded by ~ 0.7 $M_\odot$ of ^{12}C).

B)   Central oxygen surrounded by an eutectic carbon-oxygen mixture (~ 0.25 $M_\odot$ of ^{16}O surrounded by ~ 1.15 $M_\odot$ of ^{12}C-^{16}O).

with two extreme values of the accretion rates:

1) $\dot{M} = 10^{-6}$ $M_\odot$ yr^{-1} (of the order of the Eddington limit).

2) $\dot{M} = 10^{-12}$ $M_\odot$ yr^{-1} (accretion time of the order of the Hubble time).

In the A1, A2, and B1 cases, the evolution follows the sequence:

Heating of the star's centre by the e^--captures on ^{16}O from $\rho_c = 1.92 \times 10^{10}$ g cm^{-3}.

^{16}O melts and a superadiabatic temperature gradient is built up, but convection is stopped by the μ_e-gradient.

The ^{16}C abundance grows through $^{16}O + 2e^- \longrightarrow {}^{16}C + 2\nu_e$.

^{16}C ignites when its abundance reaches $\sim$ 10% by number.

The nuclear reaction network, involving neutronization plus n- and α-captures has been simplified to:

$^{16}C + {}^{16}C \longrightarrow {}^{28}Mg + 4n$ (Q = 10.13 MeV)

$^{16}O + n \longrightarrow {}^{17}O + \gamma$ (Q = 4.14 MeV).

So, the total energy released by the chain is: Q_{tot} = 26.7 MeV.

The stellar configuration at ^{16}C flash is shown in Figure 1 for the A1 case. Temperature and density profiles are displayed against stellar mass fraction. Also shown are the energy releases by the nuclear (fusion) reactions and the e^--captures, together with the thermal neutrino losses. The A2 and B1 cases are similar. In the B2 case (partial carbon-oxygen separation plus low accretion rate) an off-centre ignition seems likely. The overpressures in the star's centre (cases A1, A2, B1), of the order of 20%, are not high enough to start a detonation (the zone involved being extremely small, spherical damping strongly adds to this). In those cases burning propagates outwards by conduction and the electron captures on the incinerated material are fast, so favouring collapse. We see that different results are possible, depending on the degree of carbon-oxygen separation.

The behaviour of the accreted layers sets additional constraints on our scheme. High accretion rates: $\dot{M} > 4 \times 10^{-8}$ $M_\odot$ yr^{-1} or low accretion rates: $\dot{M} < 1 \times 10^{-9}$ $M_\odot$ yr^{-1} plus high initial masses: $M_{init} > 1.2$ $M_\odot$ are needed to avoid helium detonation in the outer layers (Nomoto 1980: Taam 1980).

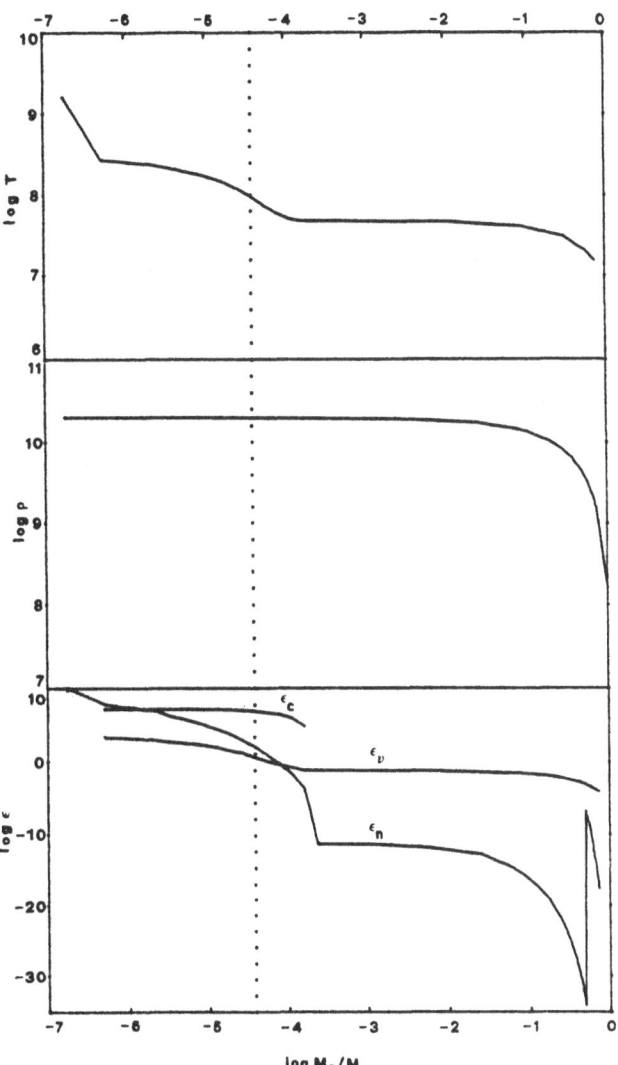

Figure 1. Stellar structure at the time of ^{16}C flash for the
Al case. Vertical dotted line separates the melted (left) and the
solid (right) zones.

3. CONCLUSIONS

Carbon-oxygen white dwarfs appear as viable progenitors of
neutron stars, either through a quiet collapse or through a
partial explosion of variable strength.

The possibility of an off-centre ignition of ^{12}C, either in the quasi-hydrostatic phase or during the hydrodynamic phase of the evolution is worth exploring as a mechanism for mass ejection (in variable amounts and composition) and so perhaps for SNI displays.

The model parameters (initial masses and temperatures, accretion rates, degree of carbon-oxygen separation and its time-dependence) should be further explored as should the hydrodynamical phase. The effects of rotation, magnetic fields and the non-sphericity of the mass-accretion process should also be included in a more realistic model.

Anyway, the physics of dense matter in the white dwarf range clearly offers new possibilities for neutron star formation and maybe also for Type I supernova production.

REFERENCES

Arnett, W.D. 1981. Private communication.
Buchler, J.R., Colgate, S.A. and Mazurek, T.J. 1980. J. de
    Phys.Suppl. No.3, 41, C2-159.
Canal, R. and Isern, J. 1979. In IAU Colloquium 53, White
    Dwarfs and Variable Degenerate Stars, ed. H.M. van Horn and
    V. Wiedemann (University of Rochester), p. 52.
Canal, R. and Schatzman, E. 1976. Astr.Astrophys., 46, 229.
Duncan, M.J., Snell, R.L., Mazurek, T.J. and Wheeler, J.C. 1976.
    Astrophys.Lett., 17, 19.
Ergma, E.V. and Tutukov, A.V. 1976. Acta Astr., 26, 69.
Mazurek, T.J., Meier, D.C. and Wheeler, J.C. 1977. Astrophys.J.
    213, 518.
Nomoto, K. 1980. In Proc.Texas Workshop on Type I Supernovae,
    ed. J.C. Wheeler (Austin: University of Texas), p. 164.
Nomoto, K., Sugimoto, D. and Neo, S. 1976. Astrophys.Space Sci.
    39, L37.
Pollock, E.L. and Hansen, J.P. 1973. Phys.Rev.A., 8, 3110.
Stevenson, D.J. 1980. J. de Phys.Suppl., No. 3, 41, C2-53.
Taam, R.E. 1980. Astrophys.J., 237, 42.

THE NATURE OF SUPERNOVAE AS DETERMINED FROM THEIR LIGHT CURVES

W. David Arnett*

Max-Planck-Institut für Astrophysik, Garching bei
München, West Germany; Institute of Astronomy,
Cambridge, England; and Enrico Fermi Institute,
University of Chicago

* A. von Humboldt Senior Awardee.

What are supernovae? An analysis of the probable nature of
supernovae is carried through systematically using analytic
models which are sufficiently general to include both types I and
II. Many aspects of the models are quantitatively compared with
observations. An improved version of the distance and mass
independent relation between gamma ray flux and observed apparent
magnitude is given. Preliminary results for cosmological distance
determination using supernovae are presented and expressed as a
Hubble ratio. Implications for the energy, mass, composition and
pre-explosion structure are summarized. Questions regarding
"silent" (dim) supernovae and the pulsar and supernova rate, as
well as iron production in supernovae, are discussed.

I. INTRODUCTION

    This review will attempt to work backward from observations
of supernova explosions to infer the physical nature of the
events. For clarity analytical models (Arnett 1980, 1982a) will
be used in the discussion, although numerical work played a
leading role in the history of this topic.

M. J. Rees and R. J. Stoneham (eds.), Supernovae: A Survey of Current Research, 221–251.

II. SOME GENERAL CONSIDERATIONS

A. Nickel-56 Production

Pankey (1962) first suggested that ^{56}Ni might explain the characteristic behavior of the light curve of type I supernovae. Truran, Arnett and Cameron (1967) found that ^{56}Ni would be a likely product of the explosive conditions expected in a stellar explosion, and Bodansky, Clayton and Fowler (1968) showed that this result was a general feature of advanced thermonuclear burning. Colgate and McKee (1969) examined the effects of ^{56}Ni in gravitational collapse models; Arnett (1969) suggested ^{56}Ni would be produced by degenerate ignition models.

With 28 neutrons and 28 protons, ^{56}Ni is a doubly magic nucleus, and is therefore relatively tightly bound. For Z > 40 (calcium) the valley of beta stability curves away from a Z = N line toward more neutron rich isobars. Thus ^{56}Ni decays by electron capture (or positron emission) to ^{56}Co which in turn decays to ^{56}Fe. About 0.1% of all the known mass in the universe seems to be in the form of ^{56}Fe; this nucleus is the fifth most abundant, following ^{1}H, ^{4}He, ^{16}O and ^{12}C. Nickel-56 has the highest binding energy per nucleon of any nucleus with equal neutron and proton number. Thus thermonuclear reactions in hot Z = N matter will tend to form ^{56}Ni. Such matter is common in evolved stars because the first and most common burning stages make it:

> Hydrogen burning makes ^{4}He (Z = N = 2), and
> Helium burning makes ^{12}C (Z = N = 6) and ^{16}O (Z = N = 8)

These ashes, if heated enough (T $\geq$ 4 x 10^{9}°K), will burn to ^{56}Ni (even on an explosive time scale).

The ^{56}Ni will be destroyed if allowed to decay during the burning process. The half-life is 6.1 days. It appears that burning to iron peak nuclei can occur on such long time scales only by hydrostatic evolution in massive stars (M $\geq$ 15 M$_{\odot}$), and even for these only in the <u>central</u> regions. Thus explosive burning of surrounding material can produce ^{56}Ni even in such objects.

Explosive burning involves time scales of the order of a second or less. At high densities electron capture speeds the rate of ^{56}Ni destruction. For all but very high densities, $\rho \lesssim 10^{9}$ g/cc, electron capture is too slow to destroy much explosively formed ^{56}Ni; expansion reduces the electron fermi energy before much damage is done.

Given these considerations it comes as no surprise that ^{56}Ni

is expected to be a common product of supernova explosions.
Virtually all quantitative models of supernova explosions imply
the production of some ^{56}Ni; perhaps the real question is not
whether it is made, but how much is ejected.

B. Homologous Expansion

A general property of explosive disruptions of stars is a
tendency toward homologous expansion, that is an expansion in
which at any instant of time the velocity at a point is proportional to its distance from the center of explosion. This
property holds for a variety of distributions of density with
radius. Simply put, the inner zones must either go slower than
the outer zones or push them. This push then causes acceleration
(and deceleration) so that the velocities tend toward homologous
flow. Even if outer regions move especially fast, as for example
due to shock acceleration in a density gradient, at later times
we have $r \propto vt$, so homology is again approached. Note that there
is still freedom here involving the distribution of mass with
radius (the density structure). Our strategy is to assume
homologous flow and examine the (weak) dependence of the light
curves upon the choice of various mass distributions.

C. Opacity

In an ordinary star the average opacity is taken as the
Rosseland mean, that is, regions of photon frequency having small
cross sections dominate the energy flow (a "leaky bucket"). If
there is a significant velocity gradient over a mean free path,
this motion tends to red or blue shift the photons out of these
"holes" and on the average reduce the flow of energy. The
effective opacity is increased. Supernovae have larger velocities
and longer mean free paths than ordinary stars, so this effect
must not be neglected. A preliminary estimate based on work by
Karp et al. (1977) suggests that the effect (for our conditions)
is important but not dominant, but further investigation is
obviously needed. As a matter of simplicity rather than conviction
we will ignore such effects below.

Stellar opacity is dominated by Thomson scattering on free
electrons and by absorption processes on electrons in the field
of an ion. The relative importance of the two depends upon the
relative abundance of bound states (or quasi-bound states) to free
states, and on the photon distribution in energy. At high
temperature and low density, ionization is so extensive that
Thomsom scattering dominates. At lower temperatures the bound-
free and free-free processes become more important, especially
for higher densities. However at still lower temperatures,
recombination removes the free electrons and shifts the Planck
distribution of photon energies so low that many bound states

cannot be excited, and the opacity plummets. These patterns of behavior can be seen in Figure 3-16 of Clayton (1968, see also Cox 1965) which shows the opacities for a Population I mixture.

To estimate the temperature and density conditions appropriate to supernovae, consider the following crude argument. Supernovae remain near maximum brightness for $\tau_{SN} \simeq 0.1$ years. This is roughly the radiative diffusion time for the supernovae. The radiative diffusion time for the sun is $\tau_\odot \simeq 10^4$ years. The mean opacity for the sun is found to be about ten times the Thomson opacity. The diffusion time is $\tau \sim R^2/\lambda \sim M\kappa/R$ if we choose a given mass M and note $M \sim \rho R^3$. Thus $\tau \simeq 0.1$ year if $R \simeq 10^4 R_\odot \simeq 7 \times 10^{14}$ cm, and scales directly with the mass. The mean density of the sun is $\rho_\odot \simeq 1$ g cm^{-3} so for the supernova, $\rho_{SN} \simeq 10^{-12} (M_{SN}/M_\odot)^{-2}$ g cm^{-3}. Also the temperature is lowest at the surface because of radiative diffusion, so $T_{SN} \geq T_e$ where the observed effective temperatures are in the range $T_e \simeq (5 \text{ to } 10)10^3 °K$ or higher. For such conditions the opacity is essentially the Thomson cross section times the number of free electrons; that is, almost constant with respect to temperature density variation.

It should be noted that these conditions also imply that radiation pressure is large compared to gas pressure.

III. ANALYTIC SOLUTIONS

A. Types of Models

We begin by examining a very simple model of type I supernovae. We imagine that thermonuclear ignition of fuel (^{12}C or perhaps ^4He) in an electron-degenerate star ("white-dwarf") of mass $M \simeq 1.4 M_\odot$ causes an explosion, converting about half the mass to ^{56}Ni. In this model the energy of the explosion, E_{SN}, comes from the burning of fuel:

$$E_{SN} = M_a q_a \qquad\qquad (1)$$

where M_a is the mass of ashes formed and q_a the average energy release per unit mass of ash formed. If the burning goes mostly to ^{56}Ni, $M_a \simeq M_{Ni}$ and $q_a \simeq 7 \times 10^{17}$ erg g^{-1}.

As the matter expands thermal energy is lost by escape of photons and by cooling through expansion. Usually the total loss by radiation is only a few percent. If most of the original supernova energy E_{SN} is eventually converted to kinetic energy, $E_{SN} = 1/2 M <v^2>$, so that for thermonuclear models,

$$M_a/M = <v^2>/2q_a, \qquad\qquad (2)$$

which relates the r.m.s. velocity (a quantity which can in principle be observed) to the fraction of fuel burned.

Alternatively the energy of explosion might come from
gravitational collapse of a stellar core, so that the supernova
explosion energy is not necessarily related in a simple way to the
amount of ^{56}Ni formed.

By choosing various values of ^{56}Ni mass, total mass, explosion
energy and initial radius we can reproduce many observed characteristics of both type I and type II supernovae.

B. An Illustrative Case: Simple Ni Decay

The equations to be solved are (a) the first law of thermodynamics,

$$\dot{E} + P\dot{V} = - \frac{\partial L}{\partial m} + \varepsilon, \tag{3}$$

(b) radiative diffusion,

$$L/4\pi r^2 = - (\lambda_c/3) \frac{\partial(aT^4)}{\partial r}, \tag{4}$$

(c) radiative decay,

$$\varepsilon/\varepsilon_{Ni} = e^{-t/\tau_{Ni}}, \tag{5}$$

where $\varepsilon_{Ni} = q_{Ni}/\tau_{Ni} = 4.78 \times 10^{10}$ erg g^{-1} s^{-1}, and (d) hydrodynamic expansion. After the (explosion) shock hits the surface
(radius R) of the presupernova star (at t = 0), the motion tends
toward a homologous coasting phase, so

$$R(t) \simeq R(0) + v_{sc}t \tag{6}$$

where v_{sc} sets the velocity scale for the model. For a given
density structure v_{sc} is uniquely related to $<v^2>$ above. The
form for energy input ε given in (5) implies that gamma rays and
positrons are trapped but neutrinos escape. As we shall see, this
method of solution can be immediately generalized to more complex
assumptions (in particular ^{56}Co decay and partial gamma escape)
with no difficulty.

The mathematical details have been given in Arnett (1980,
1982a). Here we will discuss the strategy and the results. From
(6) and the homology assumption, we find that the specific volume
satisfies

$$\dot{V} = 3V v_{sc}/R(t). \tag{7}$$

Then (3) and (4) can be combined, and the time and space variables
separated, if the space distribution of energy density $E = aT^4V$

is proportional to that of energy input ε. This is not strictly true but it is a surprisingly good approximation. Most of the energy is released as gamma rays of 1-2 MeV. These diffuse from the Ni (or Co) with a mean free path λ_{KN} determined by the Klein-Nishina cross section, until they escape or degrade to energies $E_\gamma \simeq 100$ KeV. Below this energy the photo-electric absorption gives a quick thermalization. Now λ_{KN} is about the same size as the mean free path for Thomson scattering of thermal photons, which causes the two distributions to have similar shapes and justifies the approximation.

The solution for surface luminosity is

$$L(t) = \varepsilon_{Ni} \, M_{Ni} \, \Lambda(x,y) \tag{8}$$

where M_{Ni} is the mass of ^{56}Ni at t = 0, a characteristic time is

$$\tau_m \equiv \sqrt{2\tau_d \, \tau_h} \,, \tag{9}$$

the diffusion time at t = 0 is

$$\tau_d = 3R^2\kappa/V\alpha c \big|_{t=0}, \tag{10}$$

α is the eigenvalue for the spatial distribution of thermal energy (π^2 for uniform density), and the hydrodynamic time at t = 0 is

$$\tau_h = R(0)/v_{sc}. \tag{11}$$

The dimensionless variable x is a time coordinate,

$$x = t/\tau_m \tag{12}$$

while the dimensionless parameter y is

$$y = \tau_m/(2\tau_{Ni}). \tag{13}$$

The dimensionless integral $\Lambda(x,y)$ is

$$\Lambda(x,y) = e^{-x^2} \int_0^x (\varepsilon/\varepsilon_{Ni}) \, e^{z^2} \, 2z \, dz \tag{14}$$

where from (5)

$$\varepsilon/\varepsilon_{Ni} = e^{-2zx}. \tag{15}$$

Note that we obtain (14) for arbitrary time dependence of the energy source ε. In fact $\Lambda(x,y)$ is essentially a Green's function for the diffusion equation with homologous expansion.

The behavior of $\Lambda(x,y)$ is shown in Figure 1 as a function of $xy = t/2\tau_{Ni}$. For small y, $\tau_d \, \tau_h/\tau_{Ni}^2$ is small. If we examine

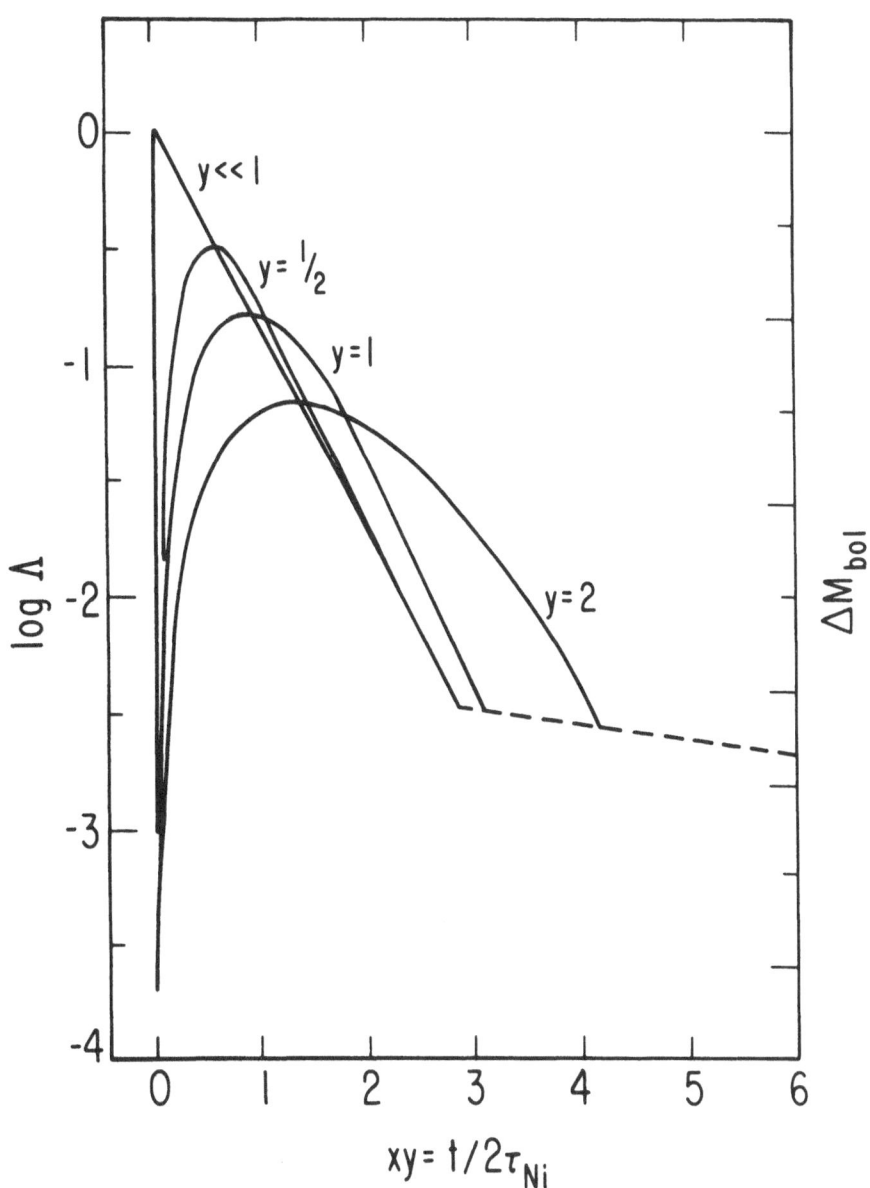

Figure 1.

times xy which are not small, we consider conditions for $\tau_d \ll \tau_{Ni}$.
The solution relaxes quickly so that the diffusion loss of energy
follows the radioactive production. For larger y, Λ "lags" behind
the radioactive source, broadening the luminosity peak. Note that
even for y = 2 the broadening is pronounced. For this simple
model, a limit on this broadening of the luminosity peak is a
limit on

$$(2\tau_{Ni} \, y)^2 = 2\kappa M/\beta c v_{sc} \qquad (16)$$

where $\beta \simeq 13.7$. If the velocity scale is constrained by observations of Doppler shifted features, then the product κM is constrained. In fact, it appears that $M/M_\odot \simeq 1.5 \pm 0.5$ or so for
Thomson opacity.

Using $L = 4\pi R^2 \sigma T_e^4$ gives

$$T_e(t) = 15,493°K \, \left(\frac{M_{Ni}}{M_\odot}\right)^{1/4} \, \left(\frac{10^9 \, \text{cms}^{-1}}{v_{sc}}\right)^{1/2} \, \left(\frac{\Lambda}{x^2 y^2}\right)^{1/4}, \qquad (17)$$

the effective temperature.

C. Large Initial Radius (SNII)

In the previous discussion we implicitly assumed that the
thermal energy E_{th} all resulted from radioactive heating. Suppose
the thermal energy from shock heating is not small. This implies
that cooling by adiabatic expansion is not exceedingly large, so
that the initial radius is not small compared to the radii at times
of interest. Alternatively this corresponds to a limit in which
the radioactive heating (or mass of ^{56}Ni) is small.

Using the same procedures we can obtain solutions for this
case also. The transient effects which occur as the shock emerges
at the surface will not be discussed here (see Arnett 1980); we
consider only the lowest order spatial eigenfunction.

The surface luminosity is

$$L(x,y,w) = \frac{E_{th}(0)}{\tau_d} \, [e^{-u(x)}] + \varepsilon_{Ni} \, M_{Ni} \, \Omega(x,y,w) \qquad (18)$$

where

$$u(x) = wx + x^2, \qquad (19)$$

$$w = (2\tau_h/\tau_d)^{1/2}, \qquad (20)$$

and

$$\Omega(x,y,w) = e^{-u(x)} \int_0^x (w + 2z) e^{-2yz + u(z)} \, dz. \qquad (21)$$

For $w = 0$ and $E_{th}(0)/\tau_d \ll \varepsilon_{Ni} M_{Ni}$ we recover the previous solutions. For $E_{th}(0)/\tau_d \ll \varepsilon_{Ni} M_{Ni}$ we have "type I" solutions (radioactive heating dominant) while in the opposite limit we have "type II" solutions (shock heating dominant).

If $M_{Ni} = 0$, for example,

$$L = \frac{E_{th}(0)}{\tau_d} e^{-(wx + x^2)}, \tag{22}$$

which is just the analytic solution for type II supernovae (Arnett 1980). If at time $t = 0$ the shock emerges at the stellar surface, we expect that the thermal energy and kinetic energy are related by

$$E_{th}(0) \simeq E_{KE}(0) \simeq 1/2 \, E_{SN}. \tag{23}$$

Now $E_{SN} = 1/2 \, M \langle v^2 \rangle$, so we have

$$\frac{E_{th}(0)}{\tau_d} = \frac{c\beta}{4\kappa} R(0) \langle v^2 \rangle \tag{24}$$

and the peak luminosity (ignoring transients) is proportional to the initial radius, the expansion velocity squared and the reciprocal of the opacity:

$$L \sim R(0) \langle v^2 \rangle / \kappa. \tag{25}$$

For a typical type II model, $w \ll 1$, so

$$L \sim e^{-x^2} = e^{-(t/\tau_m)^2}. \tag{26}$$

Now the width in time of the bolometric light curve is determined by $\tau_m = \sqrt{2\tau_h \tau_d}$, so

$$\tau_m^2 = 2\kappa M / c\beta v_{sc}. \tag{27}$$

If $\kappa \simeq \kappa_{Thomson}$ and v_{sc} is measured, we get an estimate of the mass (a "diffusion mass", M_d) which is independent of distance. For a typical type II, $M_d \simeq 8M_\odot$. The actual mass may be larger ($M \geq M_d$) because this "volume average" weights regions of low density most heavily.

Can a large initial radius affect type I light curves? Yes, as is shown in Figure 2. The bolometric luminosity rises more sharply at early times. The B-band luminosity is much less affected; the extra luminosity is in the ultraviolet. To get a large effect in B requires $R(0) \simeq 10^{14}$ cm and gives $T_e \simeq 50,000°K$. There is no indication (yet) of such high effective temperatures, but early premaximum data is rare. Stellar structure theory suggests that it might be difficult to get such a large radius for a

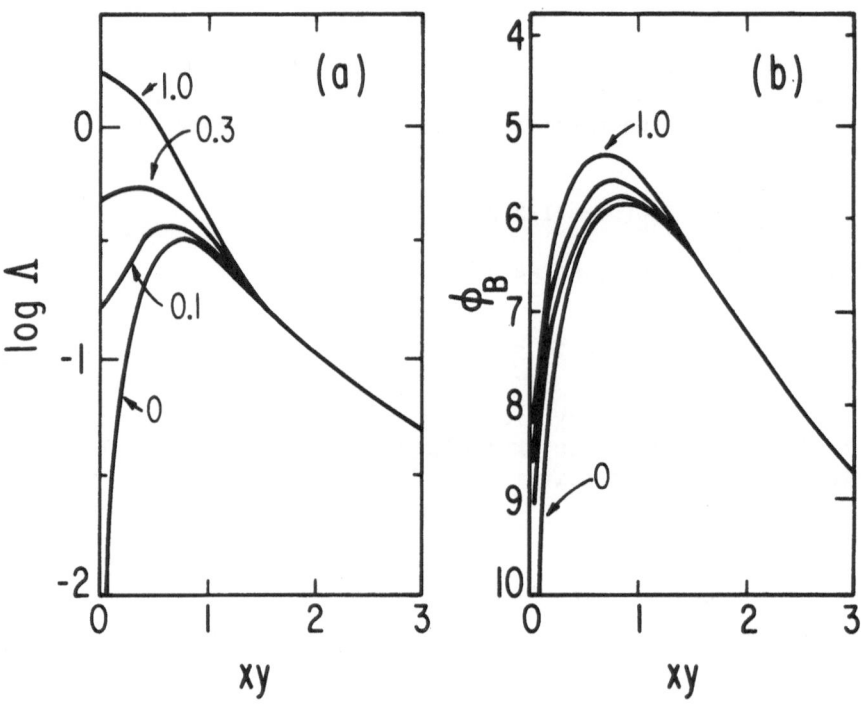

Figure 2.

type I presupernova (a low mass, hydrogen poor star; see Arnett 1982a).

D. Gamma Ray Escape

As mentioned above, not all of the gamma rays produced by ^{56}Ni and especially ^{56}Co decay are trapped and thermalized. In a detailed numerical investigation Colgate, Petschek and Kriese (1980) have derived a "deposition function" which allows us to generalize (5) to include gamma escape as well as ^{56}Co decay. Figure 3 compares the gamma ray luminosity to the thermal photon luminosity and the energy release from the ^{56}Ni-^{56}Co-^{56}Fe decay chain. At maximum light the gamma luminosity is 10% of the optical. About a month (28 days) after maximum the gamma luminosity exceeds the optical and rises slowly to a maximum which is about a third of the optical maximum. This allows us to improve the distance and mass independent relation between gamma flux and observed apparent magnitude (Arnett, 1979). If our apparatus can detect one gamma per decay, the flux is

$$\log_{10}(F/10^{-4} \text{ cm}^{-2}\text{s}^{-1}) \simeq 0.4(11.4 - m_B(\text{max})) \qquad (28)$$

where m_B(max) is the apparent B-band magnitude at maximum light. This flux is considerably smaller than previously estimated because the more accurate solutions show the supernovae to be significantly more opaque.

A type I supernova as bright as $m_B = 11.4$ was detected and identified roughly once every 2.5 years during the interval 1965-75. On the same basis, a flux of $F \geq 10^{-5}$ cm^{-2}s^{-1} would be expected from an identified SNI every 9 months on average ($m_B <$ 13.9).

The New Hampshire group (Chupp et al., 1974) have upper limits on the gamma flux from the very bright type I supernova 1972e. On May 3 they find the 0.847 MeV Co line to be $F < 6 \times 10^{-3}$cm^{-2}s^{-1}; the theoretical prediction is $F = 5 \times 10^{-4}$ cm^{-2}s^{-1}. On May 23 they find $F < 3 \times 10^{-2}$ cm^{-2}s^{-1} for all lines of energy greater than 0.65 MeV; theory predicts 9×10^{-4} cm^{-2}s^{-1} (per decay), and a maximum of 1.6×10^{-3} cm^{-2}s^{-1} during the second month after maximum.

As gamma ray detectors gain greater sensitivity, type I supernovae should become desirable targets: they are bright in gamma rays, the spectra contains a signature of characteristic lines (^{56}Co), and there is an optical precursor (the visual peak) which tells when and where to look. The great breadth of the gamma ray peak (about 4 times that of the optical peak) should allow repeated observations to improve the signal to noise ratio.

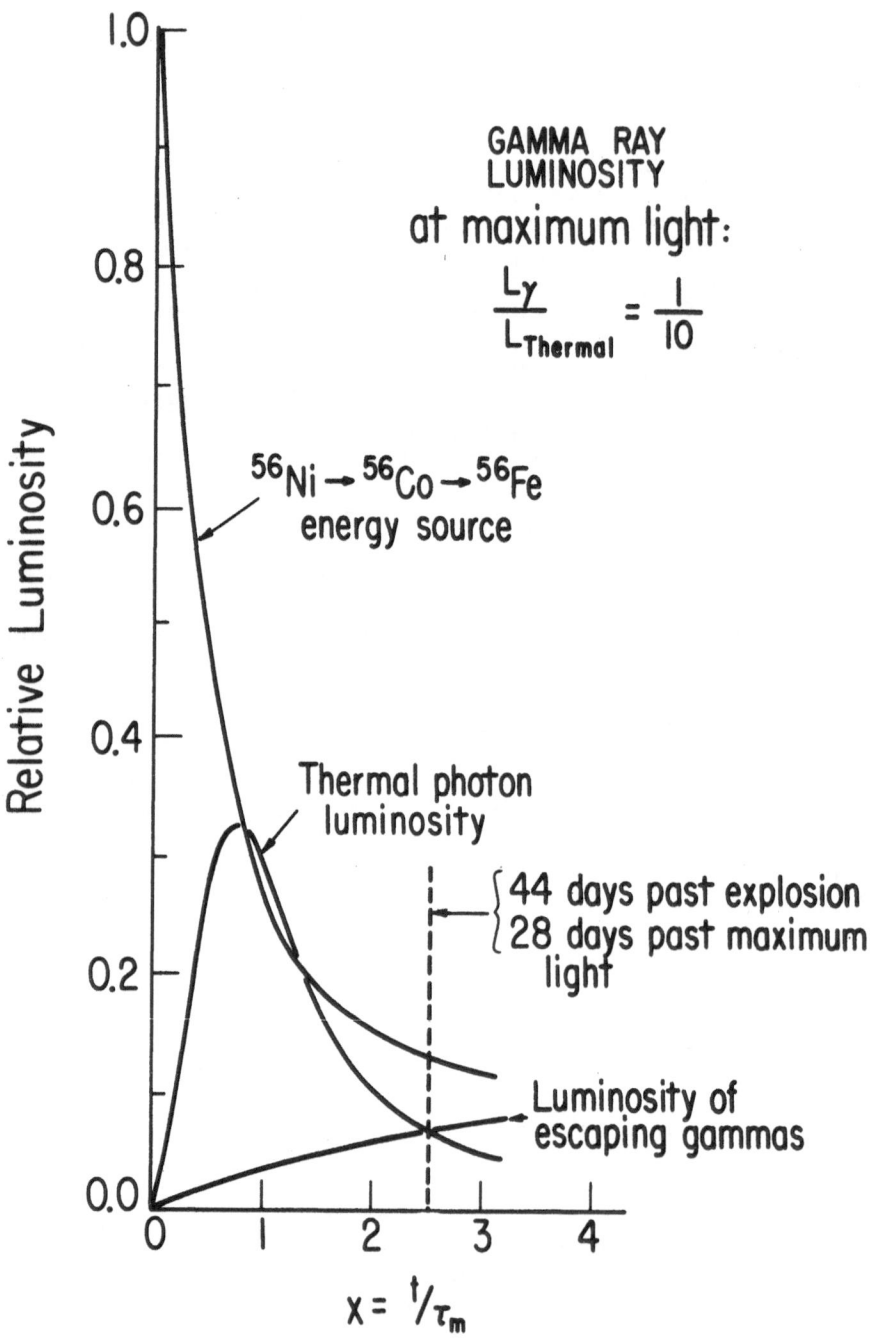

Figure 3.

IV.   COMPARISON WITH OBSERVATIONS

A. Model Parameters

For specific comparison with observation we will choose a
subset of possible models. This choice is not unique but is
representative in many respects. We choose a thermonuclear model.
We assume all burned fuel is burned to ^{56}Ni. We choose a total
mass of M = 1.45 $M_\odot$; the width of the light curve about maximum
suggests $1 \lesssim M/M_\odot \lesssim 2$. If we now choose a value of M_{Ni}, this
specifies the explosion energy and expansion velocity from (1) and
(2). The mean opacity is chosen to be $\kappa = 0.08$ cm^{2}g^{-1} for both
the ^{56}Ni "core" and the unburned "mantle."

This restricted set of models has one adjustable parameter:
the mass of ^{56}Ni. Models A, B and C have M_{Ni} = 0.5, 0.7 and 1.0
solar masses of ^{56}Ni, respectively. Detailed characteristics of
these models can be found in Arnett (1982a).

We emphasize that, on the basis of the physics described by
these solutions, it is not possible to distinguish between thermonuclear and collapse models. Whether both can give the correct
range of parameters is as yet unclear; current collapse models do
not seem to produce so much ^{56}Ni as even model A.

B. Bolometric Luminosity

The bright type I supernova 1972e was observed shortly after
peak, and for about 700 days thereafter. In the post-peak epoch,
most of the energy radiated was in the B and V bands (Kirshner,
et al. 1973; Holm, Wu and Caldwell, 1974; Ardeburg and deGroot 1973).
It is interesting therefore, to compare the shape of the bolometric
luminosity curve with that observed in the V and B bands.

In Figure 4a the dimensionless luminosity Λ is compared with
observed B and V magnitudes for the type I supernova 1970j in
NGC 7619. Model A lies along the B points in the early part of
the peak when they carry most of the energy, and along the V points
when they dominate. Model C falls too quickly after peak light.
However in Figure 4b, which has data from the bright type I supernova 1972e in NGC 5253, the situation is reversed. Model A falls
too slowly while model C has the correct slope.

The steepness of the decline in Λ is determined by the
parameter y, which for the thermonuclear models depends upon M_{Ni},
the amount of ^{56}Ni synthesized. For the collapse models y may
vary with both the strength of the explosion and the amount of
^{56}Ni ejected. The difference in 1970j and 1972e corresponds (in
both cases) to a factor of two variation in explosion energy.
The behavior seems to be related to the "fast" and "slow" subclasses

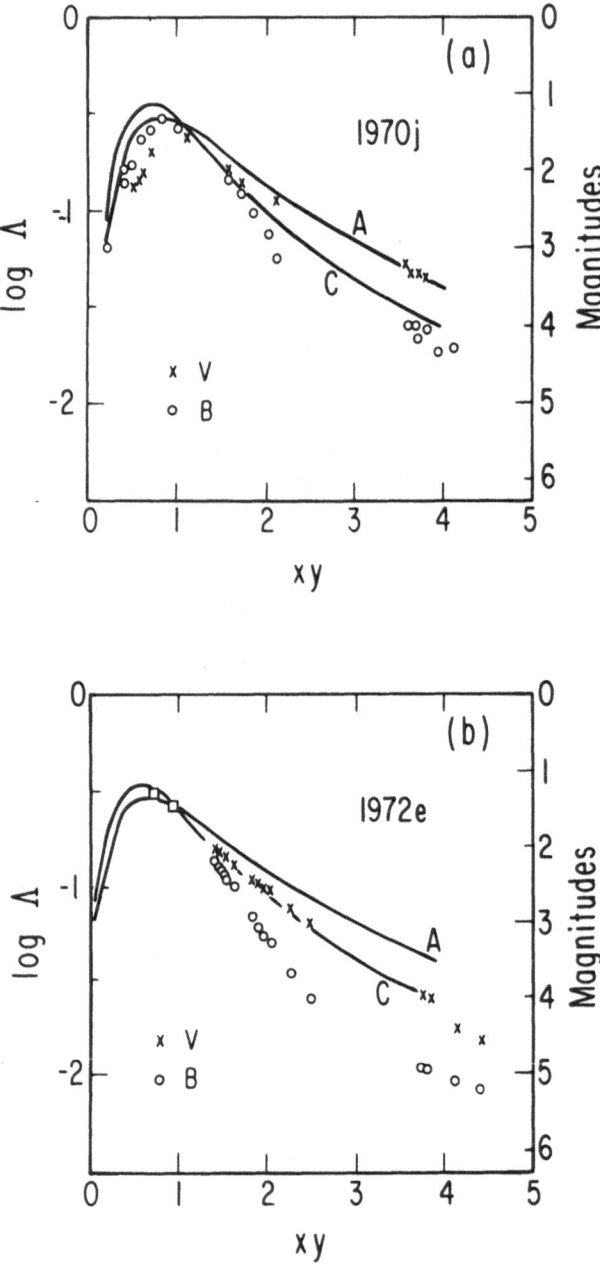

Figure 4.

of type I supernovae defined by Barbon, Ciatti and Rosino (1973). Although their discussion concerned the B magnitudes, the same behavior is apparent here also. Roughly speaking a bigger explosion gives a bigger peak, but also faster expansion and hence quicker decline due to transparency for these models. Note however that y also varies with the total mass M and average opacity κ which we have arbitrarily taken to be constant.

C. The Blackbody Supernova

In an attempt to obtain a more detailed understanding of type I supernova, we introduce the hypothetical construct of a "blackbody supernova." In this construct the radiation escaping has a blackbody spectrum corresponding to the temperature of the photosphere. The difficulty then lies in deciding where the photosphere is.

Not only does the expanding supernova become transparent to gamma rays, but also to thermal photons. This occurs only slightly later than gamma transparency because the Thomson cross section for electron scattering is only slightly larger than the Klein-Nishina cross section at the appropriate gamma ray energies. This transparency to thermal photons has two effects of fundamental importance:

1) it keeps the effective temperature of the photosphere higher, and

2) it causes much of the radiation to be "nebular" rather than "photospheric."

This epoch of the SNI evolution provides a fascinating problem in radiative transfer. Unfortunately it is also a formidable one, and quite beyond the scope of this discussion.

We will use simple approximations to explore the onset of transparency. First we will determine an effective temperature from

$$\sigma T_e^4 = L/4\pi R_e^2 \tag{29}$$

where $R_e = R - 2/3\lambda$, and λ is the transport mean free path for thermal photons. This is the Eddingtion solution to the Milne problem, and appropriate to planar geometry. This rapidly becomes inaccurate as $\lambda \to R$, due to a number of effects (spherical geometry, absorption versus scattering, energy source outside the photosphere, variable opacity, "nebular" line emission, Doppler effects, etc.). Although imperfect, this approximation does allow us to follow the evolution of the type I supernova well past maximum light before it breaks down disastrously. Further details of the "blackbody supernova" are given in Arnett (1982b).

The quantities ϕ_B and ϕ_V are calculated which represent U,
B and V magnitudes of our "blackbody supernova" at D = 1 Mpc if
we add some corrections for the zero scale of the UBV system.
The corrections were obtained from the absolute flux calibration
given by Johnson (1966). The ϕ magnitudes are useful for several
reasons. They are functionally similar to UBV magnitudes (the
shapes of theoretical and observational graphs can be compared
easily) and they represent an absolute scale. The ϕ's are roughly
equal for temperatures near $10^4°$K. At $T_4 = 1.31$, $\phi_U - \phi_B = 0$
and $\phi_B - \phi_V = -0.11$ while at $T_4 = 1.05$, $\phi_U - \phi_B \simeq 0.14$ and
$\phi_B - \phi_V \simeq 0$. The scales of the UBV system were chosen so that for
a real spectrum (an AO dwarf), U = B = V. Because real stars have
an ultraviolet deficiency relative to a blackbody, the correction
is particularly significant. However real supernovae also have
a large deficiency so that these corrections tend to cancel when
we compare observed UBV with the ϕ magnitudes of our blackbody
supernova.

The behavior of the ϕ magnitudes for a blackbody supernova
(model C) is shown in Figure 5 Before maximum light the shapes
of all three of the curves are similar. The peak occurs first in
ϕ_U, then ϕ_B, then ϕ_V; the interval is $\Delta xy \simeq 0.1$ or about 2 days.
This is in excellent agreement with the estimates of Ardeberg and
deGroot (1973) for 1972e and the results of Bertola (1964) for
the peculiar type I supernova in NGC 1073 discovered in 1962.
The decline from peak becomes linear on the magnitude scale
(exponential in flux), with a different slope for each band. In
our previous discussion of the bolometric curve (Λ) and the V
light curve of 1972e we identified the time JD 2441455 with xy =
1.4. This synchronized the time scales. Table 1 gives the
observed slopes of the UBV light curves and the ϕ magnitudes at
this time. The agreement may be misleadingly good; the effective
temperature of the model is $T_e \simeq 7,800°$K while Kirshner et al.
(1973) found $T_e \simeq 10,000°$K. Their temperature was chosen to fit
the shape of the visual spectrum; it is simply not clear whether
or not our effective temperature would imply a similar shape when
processed through a realistic atmosphere.

Table 1. Slopes of Light Curves, in magnitudes/day,
for 1972e at JD2441455 and Model C.

Slope	1972e	Model C
dV/dt	0.053	0.060
dB/dt	0.080	0.088
dU/dt	0.114	0.114

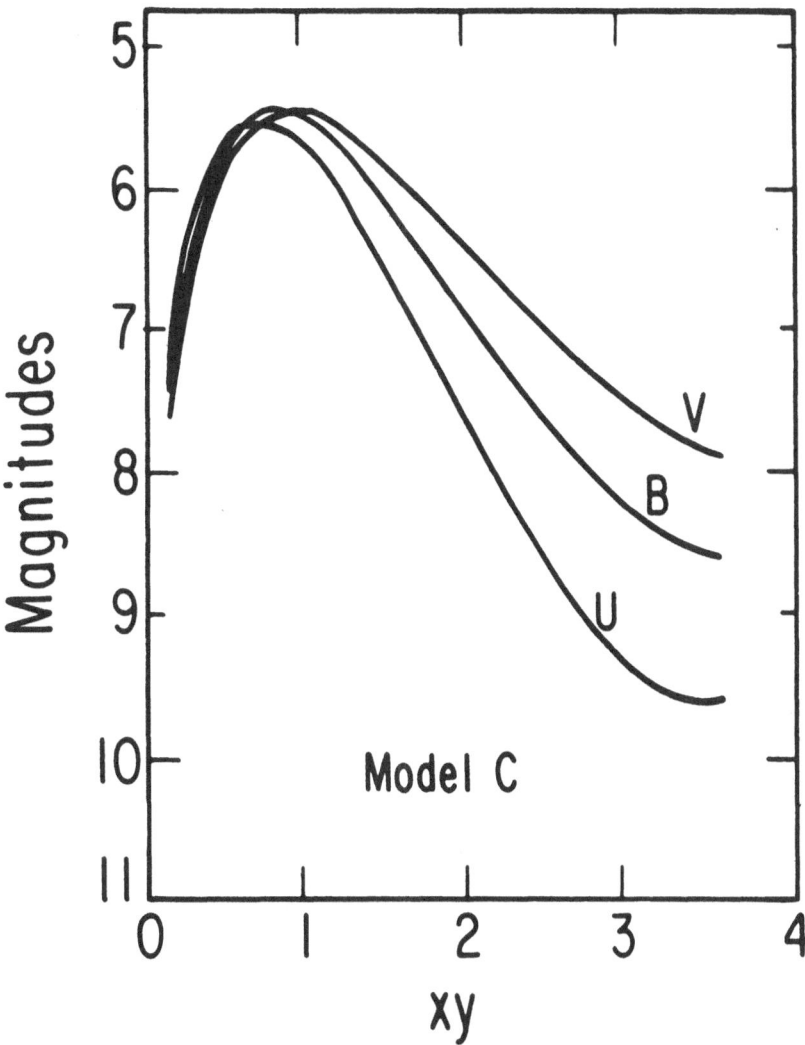

Figure 5.

In Figure 6 the ϕ_B and ϕ_V magnitudes from model A are compared
with the B and V magnitudes of Barbon, Ciatti and Rosino (1973)
for 1970j in NGC 7619. Three adjustable parameters were picked:
the time axis and the assumed distance were shifted until the B
curve fitted the ϕ_B curve. The V curve was then shifted upward
(slightly) to the data. The observational errors are estimated
to be about ±0.15 magnitudes. The agreement is excellent until
time xy ≃ 2.2; then V is brighter and B dimmer than the model.
At this time the photosphere is at 88% of the radius, so about
one-third of the volume is "transparent." Our neglect of "nebular"
effects begins to be serious, so a discrepancy is to be expected.
Prior to this time the blackbody model represents the observed B
and V magnitudes quite well; for later times the V curve follows
the bolometric luminosity as noted before (see Figure 4a).

D. Near Maximum Light

Kirshner, Arp and Dunlap (1976) observed type I supernova
1975a in NGC 2207, beginning about 7 days before maximum light
and continuing until shortly after maximum. At 5 days before
maximum they found a radius of 1.0×10^{15} cm, at temperature of
12,000°K and a fluid velocity of 11,600 km/sec. The radius
estimate depends on the distance attributed to NGC 2207. Table
2 gives the corresponding quantities for the models, all taken at
xy = 0.6 which is about 5 days before maximum light. Within the
uncertainties imposed by our ignorance of supernova atmospheres,
the agreement is good for all three models. There is a weak indication that 1975a was neither "fast" nor "slow," but intermediate
between 1972e and 1970 because model B fits the velocity better
than A or C.

Table 2. Comparison of Theory with Observations
of 1975a.

Model	$v_e/10^4 \mathrm{kms}^{-1}$	$R_e/10^{15}$cm	$T_e/10^4$°K	ϕ_B
A	0.93	0.85	1.25	6.55
B	1.10	1.00	1.29	6.11
C	1.30	1.18	1.32	5.66
SNI(1975a)*	1.16	1.0±0.2	1.2±0.2	

*Kirshner, Arp and Dunlap (1976).

Far more important is the variation in the ϕ_B magnitude
shown. For models A and C the difference of 0.89 magnitudes corresponds roughly to the fact that model C had twice as much ^{56}Ni.
If actual type I supernova are like these models, then they are

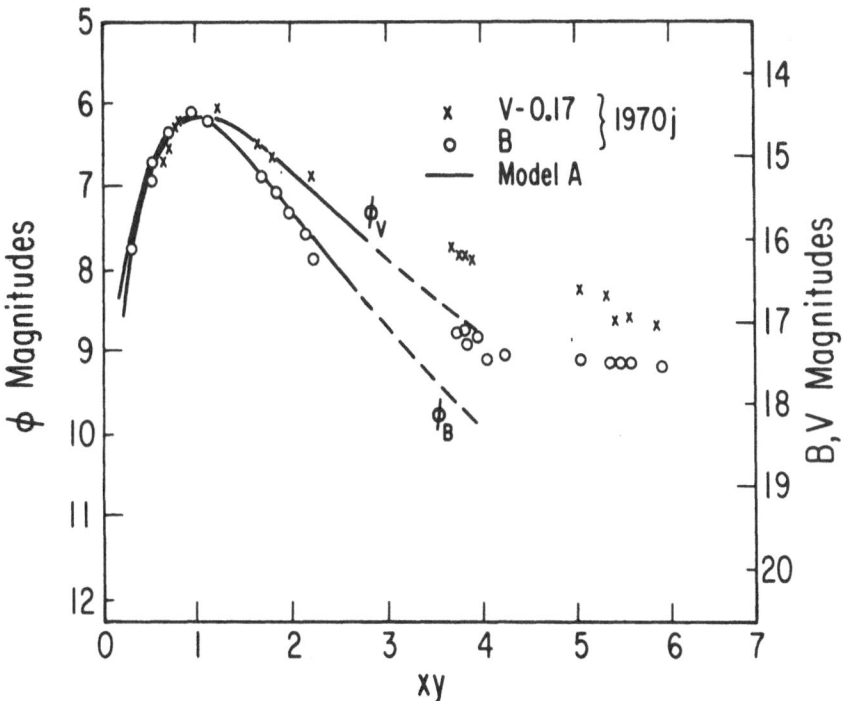

Figure 6.

<u>not</u> an especially uniform set of standard candles. Branch and
Bettis (1978) estimate the standard deviation in intrinsic luminosity of type I supernovae to be $\sigma < 0.5$ magnitudes, corresponding
to a range of variation of about $2\sigma \lesssim 1$ magnitude. This is slightly
above the 0.89 magnitude difference in models A and C. For these
thermonuclear models, (1) the shape of the light curve, and
(2) the velocity scale may be used to infer the intrinsic luminosity more accurately. The problem of distance determination
from supernovae is itself complex, and will be discussed separately.

E. Homogeneity of Spectral Evolution.

    It has long been known that type I supernovae form a very
homogeneous group with regard to the nature and evolution of their
spectra (Minkowski, 1964), so much so that the estimated error on
the date of maximum light for 1972e was only ±2 days (Ardeberg and
deGroot, 1973). How does this striking regularity come about?
Stellar spectra are determined primarily by temperature and
composition of the photosphere, and to a lesser extent the photospheric density (gravity). To obtain regularity in spectra we
must assume regularity in composition (it would be of considerable
interest to know how much variation in abundance is allowed by the
observations). The requirement that the luminosity peak not be
too broad and that the velocity scale be of order 10^9cm/s con-
strains the already weak dependence on photospheric density.

    The remaining condition is that the temperature scale be
consistent from supernova to supernova. Despite complications
from shock effects and transparency (discussed above), this condition is essentially reduced to a consideration of eq. (17), or

$$T_e \propto (M_{Ni}^\circ/v_{sc}^2)^{1/4} \ (\Lambda/x^2y^2)^{1/4}. \tag{30}$$

The <u>shape</u> of the luminosity curve fixes the factor Λ/x^2y^2. Thus,
to keep the temperature scale consistent, we would require

$$M_{Ni}^\circ \propto v_{sc}^2. \tag{31}$$

This is essentially just eq. (2), a <u>necessary</u> condition for the
degenerate thermonuclear explosion. It is a <u>possible</u> condition
for a gravitational collapse supernova (a more violent explosion
might process and eject more ^{56}Ni), but is not a necessary one.
Further, the constant of proportionality is also given correctly
for $2\ ^{12}C + 2\ ^{16}O \rightarrow\ ^{56}$Ni. Figure 7 shows the evolution of effective temperature T_e for models A, B and C. While the precise
shape of these curves may be modified by more sophisticated treatment of the atmosphere/nebula, their similarity to one another
will probably not be affected.

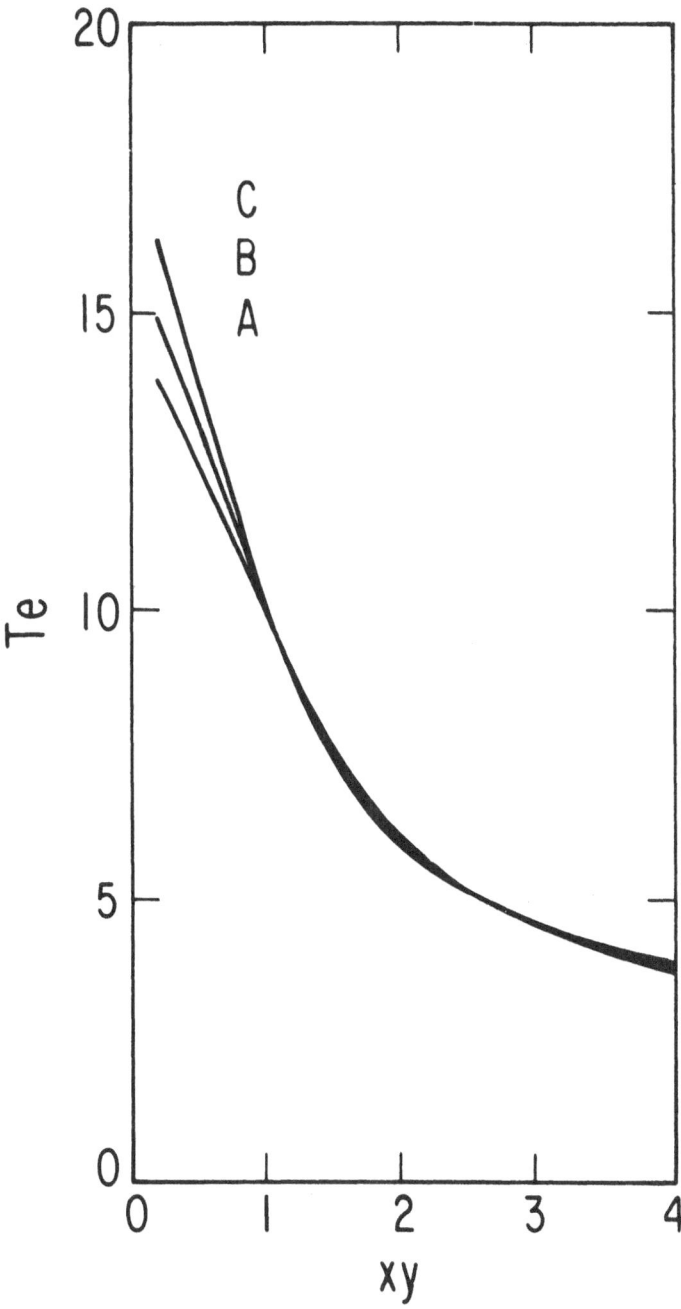

Figure 7.

V. DISTANCES OF SUPERNOVAE

A. Methods of Distance Determination

Supernovae are among the brightest events known in astronomy; their peak luminosity is an appreciable fraction of that of even a luminous galaxy. Therefore, they can be observed to relatively large redshifts. They are _events_ rather than _objects_; the time dependence of their behavior contains more information than would a steady source of light. If we can establish a quantitative understanding of the physics of the supernova phenomena, then the time behavior of all observed supernova characteristics can be used together to infer the intrinsic nature of the given event. In particular, the evolution of spectral shapes and line profiles can be used to infer the angular distance and the absolute luminosity of a given event, providing for extragalactic astronomy a primary distance indicator which is independent of other methods.

A complete description of a supernova outburst would involve computation of the trajectories and states of every particle and field associated with the event. This is obviously impractical so that we must resort to "simple" models. Because even the most complex models we might use are incomplete in this sense, the determination of distance by supernova methods will involve two stages: (1) _verification_ of the model and (2) its _utilization_ for actually inferring distances.

Supernova methods of distance determination may be divided into two classes: _kinematic_ and _dynamic_ methods. In classical mechanics the term _kinematics_ refers to a study of types of motions in themselves, leaving out of account the causes of that motion (Whittaker 1960, p. 1). By "supernova kinematical methods" we will mean those which involve the velocities and temperatures of the supernova photospheres without any statement as to the causes of those velocities and temperatures. The Baade (1926) method for variable stars, as revised for supernovae (Branch and Patchett 1973, Kirshner and Kwan 1974), is an example.

In the kinematical method one seeks to determine only a single characteristic of the event: its distance. A motivation for this approach is the hope that by minimizing the assumptions about the event one might also minimize systematic error resulting from an incorrect choice of theoretical model (Wagoner 1977). By the same token, the approach is more sensitive to observational errors because it does not make use of all available knowledge of the event for error correction.

By "supernova dynamical methods" we will mean those which involve not only the velocities and temperatures of the supernova photospheres but also the causes of those velocities and temperatures. In particular, hydrodynamics, thermodynamics and radiative diffusion will be involved. The method used by Schurmann, Arnett and Falk (1979) is one specific example.

In the dynamical method one attempts a physical description of the event, and as a result obtains its distance as well as other information concerning its nature (explosion energy, mass involved, etc.). One wishes to capture as many as possible of the observational phenomena in a model with as few physical parameters as possible. In a dynamical description variations in one quantity are correlated (in general) with those in another quantity. Hopefully a few observational data will serve to determine the solution, so that the remainder of the data may be used to reduce observational error, and make more plausible the uniqueness of the interpretation. The dynamical method is advantageous in dealing with poor or fragmentary data and in synthesizing disparate data. However dynamical models are more difficult to verify.

In practice numerical models often appear so complex as to make awkward the use of the dynamical method. The analytical solutions for supernovae of both. types I and II allow a clearer application of this method.

B.   The Hubble Ratio from Type I Supernovae

Following Branch and Bettis (1978) we see that as a consequence of the Hubble law, $v_0 = H_0 D$, a sample of supernovae of a given type is expected to satisfy

$$m_{pg}^{\circ} = C_{pg} + 5 \log v_0 \qquad (32)$$

where m_{pg}° is the apparent photographic magnitude at maximum brightness, corrected for interstellar extinction, v_0 is the radial velocity (of recession) of the parent galaxy as seen in the rest frame of our galaxy, and C_{pg} is a constant. The mean absolute magnitude at maximum light is

$$M_{pg} = M_{pg}(100) + 5 \log(H_0/100), \qquad (33)$$

where H_0 is the Hubble constant (kms^{-1}Mpc^{-1}) and

$$M_{pg}(100) = C_{pg} - 15. \qquad (34)$$

Recall that the quantity ϕ_B is related to the B-band magnitude for a hypothetical blackbody supernova (BBSN). Now, we choose

$$M_B(BBSN) = \phi_B - 25. \qquad (35)$$

Let us denote

$$\Delta B(BB,SN) = M_B(BBSN) - M_B(SN) \tag{36}$$

as the difference in B magnitudes of our hypothetical and a real supernova; we will later set this ΔB to zero for our atmosphere approximation. Further, let

$$\Delta M(pg,B) \equiv M_{pg} - M_B$$

$$= m_{pg}{}^\circ - B \tag{37}$$

$$\simeq -0.35 \pm 0.1$$

for a real supernova, as is observationally determined. Therefore, we have

$$M_{pg} = -25 + \phi_B + \Delta M(pg,B) - \Delta B(BB,SN). \tag{38}$$

Branch and Bettis (1978) find that type I supernovae occurring in spirals appear dimmer; this is very likely due to extinction in the parent galaxy which is difficult to correct for. We will confine our discussion to ellipticals to avoid this problem. Branch and Bettis (1978) find   ·

$$M_{pg} = -18.70 \pm 0.14 + 5 \log(H_0/100). \tag{39}$$

Taking model B as the mean type I supernova, we have

$$\phi_B - 25 = -19.12. \tag{40}$$

Therefore

$$M_B = -19.12 - \Delta B(BB,SN) \tag{41}$$

and

$$M_{pg} = -19.5 \pm 0.1 - \Delta B(BB,SN), \tag{42}$$

so from (31) we have

$$H_0 = (69 \pm 6 \text{ km s}^{-1} \text{ Mpc}^{-1}) \ 10^{-(\Delta B/5)}. \tag{43}$$

The error of ± 6 is statistical and <u>not</u> the dominant error. The systematic error which is partially represented by $10^{-(\Delta B/5)}$ is probably larger, <u>but is quite uncertain at present</u>. A crude estimate is $\Delta B = 0.0 \pm 0.3$ so

$$H_0 = 69 \pm 12 \text{ km s}^{-1} \text{ Mpc}^{-1}, \tag{44}$$

but the error estimate should not be taken too seriously. It is a measure of the probability that a new and comparable set of data, taken with this given model of explosion and atmosphere, will give a different value for H_0. A harder and more interesting question is whether a different model and/or atmosphere will give a better representation of the complete data set for all type I supernovae, and ultimately a different H_0. Experience may eventually answer this question.

C. The Distance to Virgo, Coma, and Pegasus

The Branch and Bettis list contains 10 type I supernovae in the Virgo cluster of galaxies and 5 in the Coma cluster. The apparent brightness of these supernovae of maximum light should show some dispersion due to differences in intrinsic brightness and due to improper corrections for extinction. The latter effect should be statistically more significant for Sb or Sc parent galaxies which have more matter to cause extinction. The observed distribution is consistent with this interpretation. We will drop all supernovae in Sb or Sc galaxies from our consideration. We note that only two supernovae, both in Virgo, seem to be sig-nificantly obscured. They are 1963i in the SB_C II galaxy NGC 4178 and 1971g in the Sc galaxy NGC 4165, with m_{pg}° of 12.64 and 13.02 respectively.

At this point we consider what at first glance appears to be a side issue. We require that our model reproduce the variation in shape of the B-band light curves which Barbon, Ciatti and Rosino (1973) describe by "fast" and "slow" supernovae. We must use models A, B and C, corresponding to an amount of ^{56}Ni of 0.5, 0.7 and 1.0 $M_{\odot}$. We interpret the two groups of Barbon et al. (1973) as one σ deviation from a gaussian distribution about model B. Therefore,

$$\phi_B - 25 = -19.12 \pm 0.4, \qquad (45)$$

where 0.4 is one standard deviation in the intrinsic B band mag-nitude to be expected from a set of our model supernovae.

Our reduced set of Virgo cluster supernovae contains 1919a, 1939b, 1957b, 1960r, 1961h, and 1965i. Their mean apparent photo-graphic magnitude is

$$\langle m_{pg}^{\circ} \rangle = 11.57 \pm 0.16 \qquad (46)$$

where one standard deviation of the distribution is 0.40. We have arrived at a startling result. By requiring that our paradigm reproduce the range of fast to slow light curve charac-teristics, we predict a dispersion in maximum intrinsic brightness which in fact is equal to that observed in this set of supernovae in the Virgo cluster of galaxies! This agreement is an unexpected bonus, and supports our choice of paradigm. It differs from the suggestion but not the numbers derived by Tammann (1978) on the dispersion in intrinsic brightness of type I supernovae. The distance modulus for Virgo is then

$$\langle m_{pg}^{\circ} \rangle - \langle M_{pg} \rangle = 31.1 \pm 0.4, \qquad (47)$$

or a distance of D = 16 ± 2.2 Mpc, where we use (34), (38) and ΔB = 0.0 ± 0.3. By doing so we have made the perhaps reasonable but unproven assumption that the average luminosity corresponds to the average shape between fast and slow light curves. Further examination of individual supernovae is needed to test this quantitatively. This estimate for the distance modulus is a bit smaller than some previous ones (see Peebles 1978 for a review and detailed references), but in agreement with the recent value of 30.98 ± 0.08 found by Mould, Aaronson and Huchra (1980).

For the Coma cluster of galaxies we have five supernovae (1961d, 1962a, 1963c, 1963m and 1973f), none of which occurred in Sb or Sc galaxies. Their mean apparent magnitude is

$$\langle m_{pg}^{\circ} \rangle = 15.47 \pm 0.11. \qquad (48)$$

The standard deviation is reduced from 0.40 for Virgo to 0.25 for Coma; this may simply be due to the small numbers in the samples, to observational error, to failure to detect dimmer supernovae in the more distant cluster, or perhaps other causes. The distance modulus for Coma is

$$\langle m_{pg}^{\circ} \rangle - \langle M_{pg} \rangle = 35.0 \pm 0.3. \qquad (49)$$

and the distance is D = 100 ± 14 Mpc. If dimmer (slow) supernovae have been lost from the sample then the cluster is farther away than this estimate. The largest contribution to the formal error is from ΔB, the correction between the flux in the B-band from a blackbody supernova and a real one. For relative measurements this error cancels, so that the relative modulus for the Coma and Virgo clusters is 3.9 ± 0.2 magnitudes. This is in essential agreement with Tammann (1978) who used supernovae as a relative measure of distance, requiring calibration by other methods. This differs from Aaronson et al. (1980) who obtained 3.50 ± 0.11. They have suggested that this discrepancy may be removed by a statistical treatment different from Tammann (1978), but require M_{pg} = -18.5. This absolute photographic magnitude is less than even the slowest supernova, model A, for which $M_{pg} \simeq -19.15 \pm 0.1$.

Aaronson et al. (1980) gave a related argument concerning the two supernovae (1970j in NGC 7619 and 1972j in NGC 7634) observed in the Pegasus I cluster of galaxies. SN 1970j was the canonical "slow" supernova used above and its light curve has been shown to best match model A. Although Barbon (1978) classified 1972j as "fast," sparseness of data near peak precludes a unique fit; its M_{pg} is uncertain. Using model A we find for 1970j,

$$m_{pg}{}^{\circ} - M_{pg} = 33.6 \pm 0.3 \qquad (50)$$

and a distance D = 52 ± 8 Mpc. The relative modulus to Virgo is 2.6 ± 0.3 which favors neither the conventional (redshift) estimate of 2.96 ± 0.08 nor the Aaronson et al. (1980) estimate of 2.35 ± 0.2. Note that this last use of the dynamical method involved an individual supernova rather than statistical samples, and is free from statistical problems inherent in the semi-empirical, calibrated methods (Tammann 1978).

Table 3 gives a summary of the results from this preliminary investigation; for more detail see Arnett (1982b). While there remains much to be learned about supernovae before this technique can be considered satisfactory, even these early results are of some interest because this method is independent of previously used ones.

Table 3.   Supernova Distances

Object	$m_{pg}{}^{\circ}$	distance modulus	distance (Mpc)	Hubble ratio
SN 1975a in NGC 2207			39.6±5.9	62±9
Virgo Cluster	11.57±0.40	31.1±0.4	16±2.2	62±11
Coma Cluster	15.47±0.11	35.0±.3	100±14	70±12
Pegasus I (NGC 7619)	15.2±0.1	33.6±.3	52±8	76±11
All SNI in E galaxies				70±15

VI. CONCLUSION

A. Type I and II

From the foregoing we can begin to define the characteristics of supernovae of both types in terms of their intrinsic properties. Our preliminary conclusions are as follows:

Type I supernovae have an explosion energy $E_{SN} \simeq 10^{51}$ ergs and a total ejected mass $M \simeq 1.5 \pm 0.5\ M_{\odot}$. They require a significant

amount of ^{56}Ni; probably $M_{Ni} = 0.75 \pm 0.25$ $M_{\odot}$. The spectra suggest that the presupernova had essentially no hydrogen left. The
presupernova radius need not have been large. Thermonuclear
models seem to produce the observed features in a natural way.

Type II supernovae are a more diverse set of events, so the statements below may be thought of as representative of many but perhaps
not all events so classified. The explosion energies are $E_{SN} \simeq$
10^{51} ergs again, and a diffusion mass $M_d \geq 8$ $M_{\odot}$. In some cases
the later part of the light curve shows a tail, indicating ^{56}Co
decay. The spectra show that the presupernova had a hydrogen rich
envelope. A large initial radius ($R > 10^{13}$ cm) is required, so
the presupernova was a red supergiant. Earlier in its life the
object was an OB star.

B. Dim Supernovae

It is sometimes suggested that the theoretical ideas concerning supernovae have a serious flaw because the rate of pulsar
formation exceeds the rate of supernovae. This gives rise to the
idea of "silent" supernovae (or more correctly, dim ones). First
we note that the uncertainties in both the pulsar and supernovae
rates are such to make it unclear whether any pulsars need be
formed without a supernova display. Perhaps a more weighty
argument is to note that so few supernova remnants are of plerionic
(Crab-like) type. So let us assume that some pulsars form without
making either a supernova (a visual outburst) or a supernova remnant (a radio shell). There are no problems with the first possibility; we simply need to say that the object ejected no ^{56}Ni
and was not a red supergiant. That would make its explosion quite
dim. For the second condition of no radio shell, we would have
to say that there was little mass ejected, or perhaps that it
expanded in a "hole" in the interstellar medium so that its surface
brightness was unusually low. It would appear that the problem
(if it is one) is why (or if) the two radioastronomical objects,
pulsars and supernova remnants, have different rates. It may well
be easier to form a dim supernova than a radiosilent supernova
remnant. Finally, a white dwarf nudged over the Chandrasekhar
limit might not form much in the way of visual outburst or radio-
supernova remnant (the total mass in the Crab is probably much
larger, perhaps as much as twice the Chandrasekhar mass).

C. Iron and SNI

We have argued that the type I phenomena are dominated by
^{56}Ni and ^{56}Co decay. The x-ray spectrum of Tycho's remnant (a
type I supernova) has been obtained by Einstein Observatory.
Becker et al. (1980) have argued that the abundances derived from
the x-ray features are evidence against type I light curves being
powered by ^{56}Ni and ^{56}Co decay. This has several possible flaws:

the abundance determinations (1) assume ionization equilibrium, (2) assume homogeneous composition, and (3) are inconsistent in that the abundances determined are not the same as the abundances used to model the spectra. It seems likely that Tycho's supernova did synthesize a lot of iron (0.5 to 1 $M_\odot$). It is not clear in what form such iron would be (gas, solid drops, dust?) or whether it would be in the matter currently heated by the shock formed from interaction with slower material. It is therefore of considerable importance to understand precisely why Becker et al. (1980) did not discover a strong iron line in Tycho's x-ray spectrum.

On the other hand, the work of Meyerott (1978, 1980) and Axelrod (1980, this conference) on the optical spectra at late times (t $\gtrsim$ 70 days or xy $\gtrsim$ 4) strongly supports the decay of ^{56}Co as the cause of the exponential tail in luminosity. Branch (1980) finds no indication of cobalt lines in the spectrum of 1972e at velocities $\gtrsim$ 8,000 km/s (Axelrod obtains a similar constraint from the late spectra). This and the Becker et al. (1980) result seem to rule out models which convert essentially all the star to ^{56}Ni. However this may not be awkward if the explosion burns to completion only the inner part of the star. This important clue cannot yet be further understood because our knowledge of the nature of the hydrodynamics associated with the thermonuclear burning is inadequate.

Thus we have an interesting confrontation: some observations seem to imply that there is (was) no ^{56}Ni in a type I supernova while others seem to imply that there is.

D. Other Opportunities

Two other large problems remain: (1) how are the optical spectra formed (the atmosphere/nebula problem), and (2) what evolutionary paths lead to this result? There are several good prospects for immediate progress. Although standard astronomical methods for radiative transfer seem inappropriate, scattering and Doppler shifts may actually make the atmosphere/nebula problem easier, or so we may hope! An extensive comparison of these models with observations will define the allowed range of parameters such as total mass, nickel mass and explosion energy. These models can be generalized to include nonuniform density distributions and more realistic treatment of gamma ray and thermal photon escape.

REFERENCES

Aaronson, M., Mould, J., Huchra, J., Sullivan, W.T. III, Schommer, R.A. and Bothun, G.B. 1980, Ap. J. 239, p. 12.
Ardeburg, A. and deGroot, M. 1973, Astron. and Ap. 28, p. 295.

Arnett, W.D. 1969, Ap. Space Sci. 5, p. 180.
Arnett, W.D. 1979, Ap. J. (Letters) 230, p. L37.
Arnett, W.D. 1980, Ap. J. 237, p. 541.
Arnett, W.D. 1982a, Ap. J., in press.
Arnett, W.D. 1982b, Ap. J., in press.
Axelrod, T. 1980, in Proc. Texas Workshop on Type I Supernova,
    ed. J.C. Wheeler (U. of Texas Press: Austin).
Baade, W. 1926, Astr. Nachr. 228, p. 359.
Barbon, R. 1978, Mem. Soc. Astr. Italy 49, p. 331.
Barbon, R., Ciatti, F., and Rosino, L. 1973, Astron. and Ap. 25,
    p. 241.
Becker, R.H., Holt, S., Smith, B.W., White, N.E., Boldt, E.A.,
    Mushotzky, R.F., and Serlemitsos, P.J. 1980, Ap. J. (Letters)
    235, p. L5.
Bertola, F. 1964, Astron. J. 69, p. 236.
Bodansky, D., Clayton, D.D. and Fowler, W.A. 1968, Ap. J. Suppl.
    16, p. 299.
Branch, D. 1980, in Proc. Texas Workshop on Type I Supernovae
    (U. of Texas Press: Austin).
Branch, D. and Bettis, C. 1978, Astron. J. 83, p. 224.
Branch, D. and Patchett, B. 1973, M.N.R.A.S. 161, p. 71.
Chupp, E.L., Forrest, D.J., Suri, A.N., Adams, R., and Tsai, C.
    1974, in Supernovae and Supernova Remnants, ed. C.B.Cosmovici
    (D. Reidel: Dordrecht), p. 311.
Clayton, D.D. 1968, Principles of Stellar Evolution and Nucleo-
    synthesis (McGraw Hill: New York), p. 224.
Colgate, S.A. and McKee, C. 1969, Ap. J. 157, p. 623.
Colgate, S.A., Petschek, A.G. and Kriese, J.T. 1980, Ap. J. 237,
    p. L81.
Cox, A.N. 1965, in Stellar Structure, ed. L.H. Aller and
    D.B. McLaughlin (U. of Chicago Press: Chicago).
Holm, A.V., Wu, C.-C. and Caldwell, J.J. 1974, Pub. Astron. Soc.
    Pacific 86, p. 296.
Johnson, H.L. 1966, Ann. Rev. Astron. Ap. 4, p. 193.
Karp, A.H., Chan, K.L., Lasher, G.J. and Salpeter, E.E., 1977,
    Ap. J. 214, p. 161.
Kirshner, R.P. and Kwan, J. 1974, Ap. J. 193, p. 27.
Kirshner, R.P., Arp, H.C., and Dunlap, J.R. 1976, Ap. J. 207,
    p. 44.
Kirshner, R.P., Willner, S.P., Becklin, E.E., Neugebauer, G.,
    and Oke, J.B. 1973, Ap. J. 180, p. L97.
Meyerott, R.E. 1978, Ap. J. 221, p. 975.
Meyerott, R.E. 1980, Ap. J. 239, p. 257.
Minkowski, R. 1964, Ann. Rev. Astron. Ap. 2, p. 247.
Mould, J., Aaronson, M. and Huchra, J. 1980, Ap. J. 238, p. 458.
Pankey, T., Jr. 1962, Possible Thermonuclear Activities in Natural
    Terrestrial Materials. Ph.D. Thesis, Howard University
    (University Microfilms: Ann Arbor).
Peebles, J. 1978, Comments Astrophys. 7, p. 197.

Schurmann, S.R., Arnett, W.D. and Falk, S.W. 1979, Ap. J. 230, p. 11.
Tammann, G.A. 1978, Mem. Soc. Astr. Italy 49, p. 315.
Truran, J.W., Arnett, W.D. and Cameron, A.G.W. 1967, Can.J. Phys. 45, p. 2315.
Wagoner, R.V. 1977, Ap. J. (Letters) 214, p. L5.
Whittaker, E.T. 1960, A Treatise on the Analytical Dynamics of Particles and Rigid Bodies, 4th Ed. (Cambridge University Press: Cambridge).

TYPE II SUPERNOVAE PHOTOSPHERES AND DISTANCES

Robert V. Wagoner

Institute of Theoretical Physics, Department of Physics
Stanford University, California

ABSTRACT

The kinematical method of determining distances to supernovae
is briefly reviewed. A program to compute the required continuum
flux from model atmospheres of Type II supernovae is outlined.
The results of a perturbed gray-body calculation are presented to
indicate qualitatively some of the important features to be
expected. It is found that if the contribution of scattering to
the opacity is greater than that of true absorption (as seems
likely), the flux is diluted, which tends to increase the derived
value of the Hubble constant. Fortunately, it appears that the
Balmer jump may be a sensitive indicator of this opacity ratio.

1. SUPERNOVAE AND COSMOLOGY

As a means of measuring cosmological distances, supernovae
may have an important advantage over other probes such as
galaxies, clusters of galaxies, or quasars. They may well be
simpler physical systems, whose characteristics we can discover
from a detailed analysis of the time-dependence as well as the
frequency-dependence of their radiation. Spectrophotometry is
possible from the ground out to redshifts $z < 0.1$, the limit
being due to the background light from the parent galaxy, not the
size of the telescope. The Space Telescope will make available
the important ultraviolet wavelengths, as well as allowing us to
reach larger redshifts.

So far three distinct methods have been proposed for using
supernovae to determine cosmological distances, and therefore H_0
and q_0. The first, advocated by Colgate (1979, 1982) and

M. J. Rees and R. J. Stoneham (eds.), Supernovae: A Survey of Current Research, 253–266.
Copyright © 1982 by D. Reidel Publishing Company.

Tammann (1979), involves the use of Type I supernovae as
'standard candles'. The difficulties inherent in this method
include calibration and selection effects in the determination of
H_0, and evolutionary effects in the determination of q_0. Is there
a simple physical reason why the maximum luminosities of all Type
I supernovae should be identical?

The second, proposed by Schurmann, Arnett and Falk (1979),
employs a hydrodynamical model of the supernova explosion. From
any particular model, which consists of a specification of the
initial energy input and density distribution, observables such
as the colour indices can be related to the luminosity. The
difficulty with this method involves the determination of the
correct model.

The third, we propose to call the kinematical method, in
contrast to the second, which we propose to call the dynamical
method. The idea was first proposed by Baade (1926) as a means of
measuring the distance to Cepheids, while Oke and Searle (1974)
first suggested that it be applied to supernovae. This method
has been reviewed by Wagoner (1980), and additional details have
been considered by Wagoner (1981). A direct measure of the
distance to a supernova can be determined if the velocity of the
matter at the photosphere can be obtained from the line profiles
and the angular size of the photosphere can be obtained from the
ratio of observed to intrinsic flux. The major difficulty at
present with this method is the determination of the intrinsic
flux from the shape of the observed continuum. A model atmosphere
must be constructed, and the parameters which specify the model
must be obtainable from the observed spectrum.

In the next section, we shall outline the full model
atmosphere problem that we are beginning to attack. In the third
section, some exploratory results are presented. These were
obtained in the perturbed gray-body approximation, which provides
only a qualitative understanding of certain new features of such
low-density photospheres. In the final section, the implications
of these qualitative results for the kinematical distance
determinations are discussed.

2. THE MODEL ATMOSPHERE PROBLEM

We are interested in understanding the continuum radiation
from Type II supernovae, especially during the first month of
their existence. During this period, the lines contribute very
little to the net flux. The model atmospheres that we intend to
compute may be applicable to other astronomical objects such as
very massive stars and accretion disks.

The problem is to calculate the emergent flux as a function of frequency from an atmosphere that satisfies the following assumptions:

a) Structure of the photosphere

We shall consider atmospheres in which the density ρ is a rapidly decreasing function of distance R. In particular, we shall assume that $\rho \propto R^{-n}(n \gg 1)$ in the neighbourhood of the photosphere. Initially, we shall consider the situation in which n is large enough to produce one-dimensional planar geometry, as in stars or accretion disks. Eventually, we shall consider the additional terms of order 1/n in the transfer equation for spherically-symmetric objects. Hydrodynamical models of supernova explosions, such as those computed by Weaver and Woosley (1980), indicate that $n > 10$ after the shock has passed.

The calculation is carried out in a coordinate system moving with the matter, so that the opacity and emissivity are iso-tropic. The exterior column mass m will be used as our comoving spatial coordinate, so that physical distance $R = R(m,t)$, with the matter velocity $v = \partial R/\partial t$. With the density dependence we have assumed, we then obtain

$$\rho = m/\Delta R, \tag{1}$$

where the scale height $\Delta R \approx R/n \approx$ constant in the neighbourhood of the photosphere at any time t, but may change with time. Equation (1) is employed in lieu of hydrostatic equilibrium.

b) Nonrelativistic motion

In supernovae, the observed expansion velocities at the photosphere satisfy $v \leqslant 0.03c$. In addition, the effects of acceleration and gravitation on the propagation of photons is negligible. Therefore it is sufficient to retain only terms through first order in v/c in the equation of transfer.

c) 'Eddington approximation'

We shall assume that the specific intensity $I_\nu(\nu,\mu,t)$ is related to the average intensity $J_\nu(\nu,m,t)$ and the Eddington flux $H_\nu(\nu,m,t)$ by the equation

$$I_\nu = J_\nu + 3\mu H_\nu. \tag{2}$$

The outward normal component of the unit propagation vector of the photon beam is μ. This relation is an especially good approximation when the intensity is nearly isotropic, so that $H_\nu \ll J_\nu$. Such a situation is expected at the low densities of supernova photospheres, since the relatively large scattering optical depth keeps the radiation fairly isotropic. The usual Eddington approximation, $J_\nu = 3K_\nu$ (where K_ν is the second moment of the intensity), is a consequence of equation (2).

With these three assumptions, the equations of transfer reduce to

$$\frac{1}{R'}\frac{\partial H_\nu}{\partial m} - \frac{\varepsilon_1}{R}\,\nu\,\frac{\partial J_\nu}{\partial \nu} = \eta_\nu - \chi_\nu J_\nu, \tag{3}$$

$$\frac{1}{R'}\frac{\partial J_\nu}{\partial m} - \frac{\varepsilon_2}{R}\,\nu\,\frac{\partial H_\nu}{\partial \nu} = -3\chi_\nu H_\nu, \tag{4}$$

with $R' \equiv \partial R/\partial m = -\rho^{-1}$. The functions $\varepsilon_1 \approx v/c$ and $\varepsilon_2 \approx 3v/c$, while χ_ν is the opacity and η_ν is the emissivity. Because of assumptions (a) and (b), the explicit time-derivatives and comparable terms in the transfer equation are negligible. Thus we are dealing with quasi-static photospheres, in which the photon propagation time scale is less than the hydrodynamical time scale. There is no explicit dependence of any quantity on time.

d) Opacities and Emissivities

We assume normal Population I abundances, as indicated by the lines in Type II supernova spectra. It will also be initially assumed that most of the free electrons come from hydrogen, which puts a lower limit of about 5000K on the temperature at the low densities of interest. The sources of opacity and emissivity that we shall explicitly include are electron scattering, bound-free transitions from hydrogen and helium, and free-free transitions.

As indicated by Wagoner (1981), an accurate calculation of the contribution of the many atomic transitions to an effective bound-bound continuum opacity represents a separate project of great complexity. The large Doppler shifts make this source of opacity much more important than for stellar continua. Giora Shaviv will collaborate with us on this part of the problem. Equations (3) and (4) will be solved by a two-dimensional computer calculation. Note that in our comoving frame, the contribution of each transition i to χ_ν and η_ν is proportional to $\delta(\nu - \nu_i)$, since thermal broadening is so much less than the velocity-gradient broadening. What is desired is the effective contribution of the bound-bound transitions to the transfer equation within a bandwidth $\Delta\nu \ll \nu$ which however is broad enough to contain many lines. The results of Karp et al. (1977) represent a first step toward this goal. We have shown that the effect of the many weaker transitions is to convert the

frequency-derivative terms in equations (3) and (4) into terms of
the same form as the right-hand sides of the equations. That is,
such transitions do contribute to the continuum opacity and
emissivity in the usual way, as Karp et al. (1977) assumed.
However, they also assumed local thermodynamic equilibrium (LTE),
which we shall not do. Because of the low densities ($10^9 \lesssim$ n
$\lesssim 10^{13}$ cm^{-3}) that we shall deal with, LTE is in general not a
good approximation. In fact, it appears as if most of the bound-
bound opacity contributes to the scattering rather than to the
absorption (Wagoner 1981).

Until the full effects of the bound-bound transitions have
been calculated, we will use the results of Karp et al. (1977) as
an approximate opacity, which we will add to the electron-
scattering opacity (they are comparable). Equations (3) and (4)
can then be written in the more familiar forms

$$\frac{\partial^2 J_\nu}{\partial \tau_\nu^2} = \frac{3}{\chi_\nu} \, (K_\nu J_\nu - \eta'_\nu) \tag{5}$$

$$H_\nu = \frac{1}{3} \frac{\partial J_\nu}{\partial \tau_\nu}, \tag{6}$$

where the differential optical depth $d\tau_\nu = \chi_\nu \rho^{-1} dm$. We have
separated out the contributions of electron scattering and
bound-bound effective scattering, writing

$$\chi_\nu = (1 + Q_\nu) \chi_{es} + K_\nu \tag{7}$$

$$\eta_\nu = (1 + Q_\nu) \chi_{es} J_\nu + \eta'_\nu . \tag{8}$$

e) _Radiative equilibrium_

We assume that the dominant mode of energy transport through
the photosphere is via photons. Under our quasi-static
conditions, we then obtain the usual form of the equations of
radiative equilibrium:

$$\int_0^\infty (\eta'_\nu - K_\nu J_\nu) d\nu = 0, \tag{9}$$

$$\int_0^\infty H_\nu d\nu = \text{constant} = \mathcal{F}/4\pi \equiv (\sigma/4\pi) T_e^4. \tag{10}$$

f) _Statistical equilibrium_

Under the quasi-static conditions we have specified, it is a
good approximation to assume that P_{ij}, the rate of radiative and

collisional transitions from atomic level i to j (including the
continuum), is much greater than the expansion rate of the
medium. We then obtain the usual form of the equations of
statistical equilibrium for each species:

$$n_i \sum_{j \neq i} P_{ij} - \sum_{j \neq i} n_j P_{ji} = 0. \tag{11}$$

Initially we shall explicitly consider the lowest four levels of
hydrogen, and later include other levels of hydrogen and helium.

The level populations n_i must be calculated from equation
(11) because LTE appears to be a poor approximation at these low
densities. We note that radiative detailed balance in the lines
is also probably not a good approximation, since the large
Doppler broadening should prevent the lines from being very
optically thick at the photosphere, as they are in stars. We
also note that in general the collisional excitation rates should
be smaller than the radiative excitation rates.

With the above set of approximations, we have generated a
complete set of equations which govern the fundamental variables
$J_\nu(m, \nu)$, matter temperature $T(m)$, electron density $n_e(m)$,
bound-state densities $n_i(m)$, and nucleus densities $n_p(m)$,
$n_\alpha(m), \ldots$ This set consists of equations (1), (5), (9), (11),
plus charge neutrality and the specification of the relative
abundances. In addition, the transfer equation (5) is subject to
the boundary conditions

$$\frac{\partial J_\nu}{\partial \tau_\nu} \rightarrow 3h_\nu J_\nu, \qquad (\tau_\nu \rightarrow 0) \tag{12}$$

$$\frac{\partial J_\nu}{\partial \tau_\nu} \rightarrow \frac{3\overline{\chi}_R T_e^4}{16\chi_\nu T^3} \left(\frac{\partial B_\nu}{\partial T}\right), \qquad (\tau_\nu \rightarrow \infty) \tag{13}$$

where $\overline{\chi}_R$ is the Rosseland mean opacity and B_ν is the Planck
function. It should be a good approximation to take the quantity
$h_\nu(\nu) = 1/\sqrt{3}$, the value for pure scattering. The free parameters
which characterize each model atmosphere are the effective
temperature T_e and scale height ΔR. In addition, the bound-bound
opacity will initially be treated as a parameter. Of course, it
will always depend upon the velocity gradient.

Since all the fundamental variables interact globally
throughout the atmosphere (especially under the non-LTE
conditions of this problem), the best way to solve these
equations appears to be the complete linearization method, as
described in Mihalas (1978, pp. 230-234). The equations are
discretized in depth m and frequency ν. The solution at each
depth point is represented by a vector whose components are the

fundamental variables. The first-order variation of our set of
fundamental equations then generates a linear set of equations
relating small changes in all the fundamental variables.
Beginning with some starting solution (e.g. LTE), the equations
are then iterated to convergence, which should be rapid. The
Feautrier elimination scheme seems to be the most efficient way
to solve the system on a computer. The author, a Stanford
graduate student (Stephen Hershkowitz), and a Princeton
undergraduate (Eric Linder) have begun writing the computer
program.

We plan to compute spectra $\mathcal{F}_\nu(\nu)$ for various choices of
$T_e \gtrsim 6000K$. The scale height ΔR will be varied over a wide enough
range so that we may check our calculation against the published
continuum spectra of ordinary stars.

3. PERTURBED GRAY-BODY APPROXIMATION

In order to begin to understand some of the important
physical effects in these low-density photospheres, exploratory
calculations have been carried out by making additional
simplifying assumptions. The first assumption is that the
opacity and emissivity can be represented by isotopic scattering
plus LTE emission-absorption. That is, equations (7) and (8) are
taken to be of the form

$$\chi_\nu = \chi_s + \chi_a \tag{14}$$

$$\eta_\nu = \chi_s J_\nu + \chi_a B_\nu. \tag{15}$$

The second assumption is that the scattering and absorption
opacities are of the form

$$\chi_s = \overline{\chi}_s(m) \tag{16}$$

$$\chi_a = \xi(m)\overline{\chi}_\nu(m)[1 + \delta_\nu(\nu,m)], \tag{17}$$

with $\delta_\nu \leqslant 1$, and its frequency average $\overline{\delta}_\nu(m) = 0$. This is a
perturbed gray-body approximation, in which we have only
considered the effects of frequency dependence in the absorptive
opacity because we will be interested in the effect of the
bound-free opacity discontinuities, in particular the resulting
Balmer jump. Adopting the optical-depth scale $d\tau = \chi_\nu \rho^{-1} dm$, the
relevant equations (transfer and radiative equilibrium) become

$$\frac{\partial H_\nu}{\partial \tau} = \xi(1 + \delta_\nu)(J_\nu - B_\nu), \tag{18}$$

$$\frac{\partial J_\nu}{\partial \tau} = 3(1 + \xi \delta_\nu) H_\nu, \tag{19}$$

$$\int_0^\infty (J_\nu - B_\nu)(1 + \delta_\nu) d\nu = 0. \tag{20}$$

The solution to these equations in the zero-order (gray-body) approximation $\delta_\nu = 0$ is given by Wagoner (1981) for a general power-law dependence of ξ on τ (or m). However, we shall here only consider the case ξ = constant, which however is a representative and realistic choice. The emergent flux is then

$$H_\nu(\nu,0) = \frac{\sqrt{\xi/3}}{1+\sqrt{\xi}} \int_0^\infty B_\nu(T) e^{-\tau^*} d\tau^*, \tag{21}$$

with

$$T^4 = \frac{3}{4} T_e^4 [\tau + \tau_0] \cong \frac{3}{4} T_e^4 [(3\xi)^{-1/2} \tau^* + \tau_0], \quad (\tau_0 \sim 2/3) \tag{22}$$

The effective absorption optical depth $\tau^* \sim 1$ at the photosphere. Recall that ξ is the ratio of absorptive to total opacity.

From these results, the following features emerge:
a) The spectral shape is close to that of a blackbody.
b) The flux is diluted from that of a blackbody by a factor $\sim 1/2$.
c) The colour temperature $T_c \sim \xi^{-1/8} T_e$. We have taken $T_c = h\nu_m/3k$, where ν_m is the frequency of maximum flux.

These features can be understood from the fact that the spectrum is formed at a large optical depth $\tau \sim \xi^{-1/2}$ if the absorption is a small fraction of the total opacity. At such depths, the flux $H_\nu \sim \partial J_\nu/\partial \tau \sim \partial B_\nu/\partial \tau \sim B_\nu/\tau \sim \xi^{1/2} B_\nu \ll B_\nu$. But the shape of the spectrum reflects the temperature $T \sim \xi^{-1/8} T_e$ which determines B_ν, not H_ν.

Let us try to estimate the order of magnitude of the critical factor ξ that has emerged from this gray-body analysis. In order to do that, we have presented the results of an LTE calculation of some (Planck) mean opacities in Figure 1. Explicitly shown are the regions in the electron density-temperature plane in which the bound-free and free-free opacities of hydrogen are less than that of electron scattering. Also shown is the region where electron excitation rates between atomic levels should be less than radiative excitation rates. One expects that this ratio R_e/R_γ should be roughly equal to the ratio of absorption to scattering by bound-bound transitions. The relation between n_e and T for photospheres with two different scale heights is also shown.

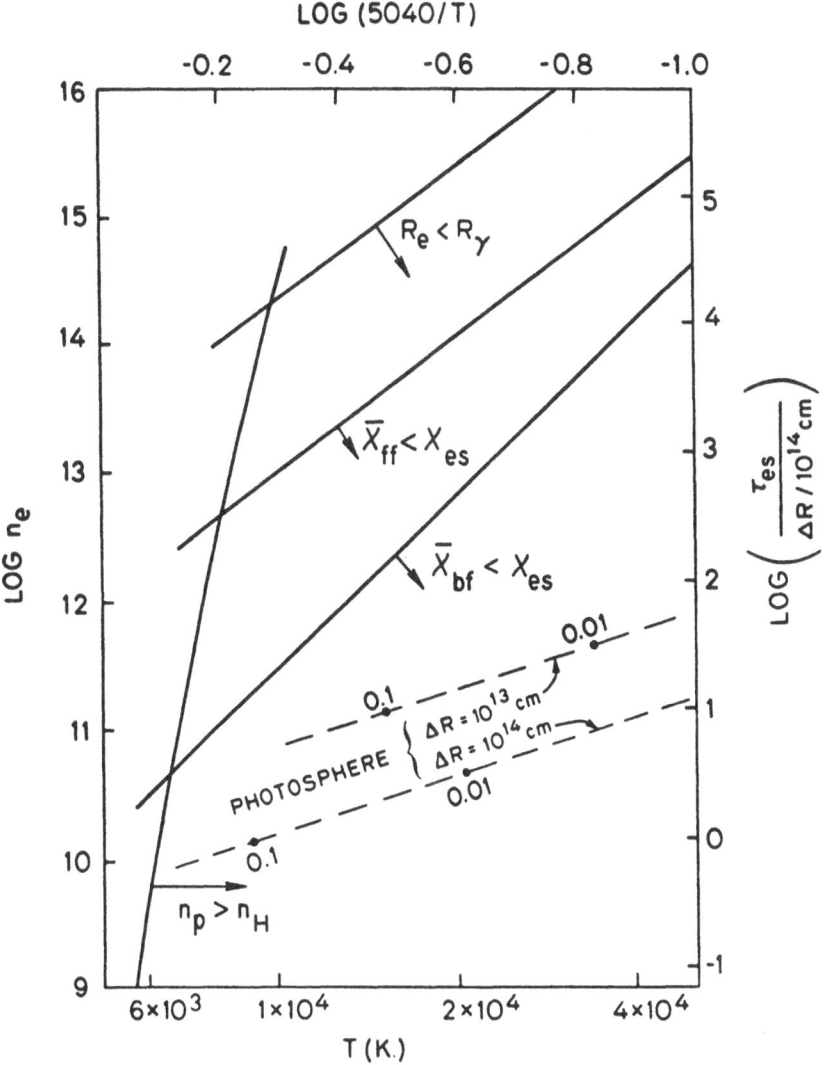

Figure 1. Various regions of interest in the electron density (or equivalently, the electron-scattering optical depth τ_{es}), temperature plane are indicated. Included are regions where hydrogen is ionized, the electron excitation rate is less than the radiative excitation rate for a typical atomic transition, and mean absorptive opacities are less than the electron-scattering opacity. Also shown is the location of the photosphere for two choices of scale height, assuming that electron scattering and bound-bound transitions contribute equally to the scattering opacity, while bound-free transitions dominate the absorptive opacity. Two values of ξ are indicated along each of these curves. LTE has been assumed.

The results of Karp et al. (1977) indicated that the total bound-bound opacity was comparable to that of electron scattering under the physical conditions expected within supernovae. In order to estimate ξ, we have therefore assumed that bound-free processes dominate the absorption, while electron scattering and bound-bound transitions contribute equally to the scattering. The resulting values of ξ are indicated at two points along each photospheric line in Figure 1. More details are given by Wagoner (1981).

This estimate of ξ may be grossly in error, however. The assumption of LTE and the use of Planck means are particularly suspect. What is clearly needed is a way to obtain ξ from observations. To indicate how this can be done, we now consider the second part of our approximation scheme, the first-order deviations from the gray-body results produced by a frequency dependence of the absorptive opacity. In particular, let us consider the effects of a jump in the opacity above some frequency due to bound-free transitions.

If $\Delta\chi_a = \overline{\chi}_a \Delta\delta_c = \xi\chi_\nu\Delta\delta_c$ is the increase in absorptive opacity at the critical frequency ν_c, the resulting first-order fractional difference in flux is given by

$$\frac{\Delta H_\nu(\nu_c,0)}{H_\nu(\nu_c,0)} = [\int_0^\infty B_\nu \, e^{-\tau*} d\tau*]^{-1} \int_0^\infty [B_\nu - J_\nu - \sqrt{3\xi} \, H_\nu] \, \Delta\delta_c e^{-\tau*} \, d\tau*. \quad (23)$$

The functions B_ν, J_ν, and H_ν are obtained from the zero-order (gray-body) solution. The absorptive opacity jump $\Delta\chi_a$ has been computed at the Balmer continuum threshold (still assuming LTE). The fractional absorptive opacity jump $\Delta\delta_c$ was computed by using the Planck mean of the total hydrogen bound-free opacity. The resulting Balmer jump

$$D_B = 2.5 \, \log_{10}[H_\nu(3647^+\text{Å})/H_\nu(3647^-\text{Å})].$$

In stellar spectra at the temperatures of interest (T ~ 10^4K), the Balmer jump is large. Thus a perturbed gray-body is not a good approximation, at least for $\xi = 1$. Therefore we have normalized our results to those of computed stellar models (without H^- absorption, which is negligible at supernova densities) for $\xi = 1$, but have retained the consistent first-order approximation $D_B \propto \Delta H_\nu/H_\nu$. The results, which should be considered as only a qualitative indication of the effect of scattering on the Balmer jump, are presented in Figure 2. They were computed by Stephen Hershkowitz and Eric Lander.

The decrease in the Balmer jump (even through zero) as the ratio of absorption to scattering is reduced can be roughly understood in the following way. At frequencies just below the

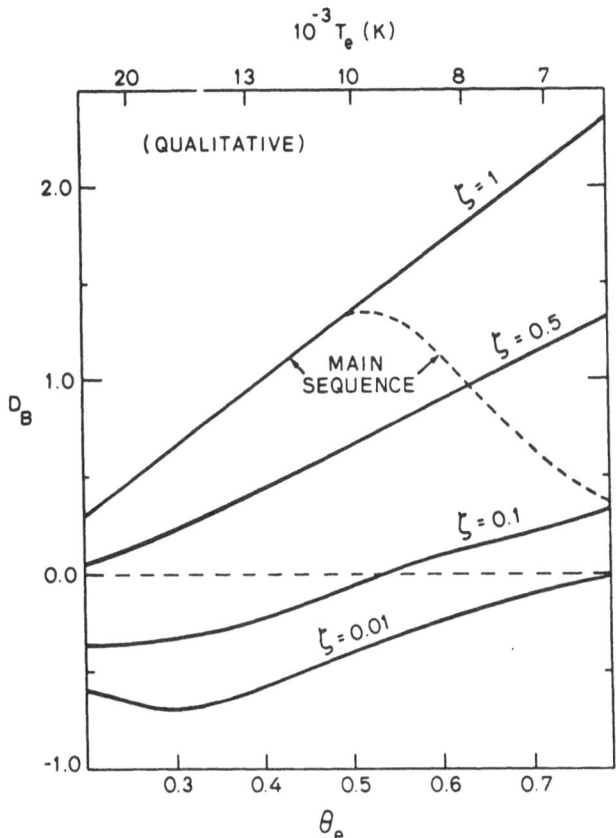

Figure 2. Schematic results of a perturbed gray-body calculation of the Balmer jump as a function of ζ and T_e (or $\theta_e \equiv 5040/T_e$).

Balmer continuum threshold, we look deeper into the star because the absorptive opacity is less. This means that we see a higher temperature than we see at higher frequencies, which normally implies a greater flux. But if scattering dominates absorption, the radiation becomes more isotropic as we look deeper, which reduces the flux compared to the mean intensity. It appears as if the second effect can become comparable to the first when scattering dominates.

Although these results are very approximate, they are encouraging because they indicate that a detailed fit of our computed spectra to observed spectra (particularly in the regions of the Balmer, Paschen, etc. jumps) may allow us to determine the parameters of our model atmosphere, thus determining the flux.

4. THE HUBBLE CONSTANT

Let us first briefly review the kinematical method of determining the distance to a supernova. The basic idea applies to any spherically-symmetric, expanding object. The additional requirement that the photosphere be very sharp (Wagoner 1980) can be dropped if the object is expanding freely [v = v(m)], which is the case well after the shock wave has passed. The (proper motion) cosmological distance $d_M = cH_0^{-1}z[1-0.5(1+q_0)z+0(z^2)]$ is then obtained from the relation

$$d_M(\theta_2/v_2-\theta_1/v_1) = t_2 - t_1. \tag{24}$$

Here v_i and θ_i are the matter velocity at and the angular size of the photosphere at time t_i, with t_i much later than the time the matter was last accelerated. It is assumed that the redshift z of the parent galaxy has been measured.

An analysis of the profiles of the spectra lines allows one to determine v(t) (Branch 1980), while the observed spectral flux

$$f_{\nu_0}(\nu_0) = \theta^2(1+z)^{-3}\mathcal{F}_\nu(\nu)\exp[-\tau_1(\nu_0)] \tag{25}$$

can be used to determine $\theta(t)$ if the intervening absorption optical depth τ_1 and the intrinsic spectral flux [as a function of emitted frequency $\nu = \nu_0(1+z)(1-v/c)$] are known. The 2200 Å dust feature should be a good indicator of τ_1. Note that the time independence of d_M as computed from equation (24) is an important check of the assumptions.

In the past, it has been assumed that $\mathcal{F}_\nu$ was the blackbody function because the observed continuum was claimed to have a 'thermal' shape. The resulting comparisons with the visible spectra of both Type I supernovae (Branch and Patchett 1973: Branch 1977, 1979) and Type II supernovae (Kirshner and Kwan 1974) have been used to determine a 'blackbody Hubble constant' $H_{BB} \sim 40-60$ km s^{-1} Mpc^{-1}.

For illustrative purposes, consider observations of a supernova sufficiently separated in time so that $\theta_2/v_2 \gg \theta_1/v_1$. Then from equations (24) and (25) we see that

$$H_0 \propto d_M^{-1} \propto \theta_2 \propto \mathcal{F}_\nu^{-1/2}(t_2). \tag{26}$$

But our gray-body approximation indicated that $\mathcal{F}_\nu \sim \xi^{1/2}\mathcal{F}_{BB}$. It then follows that the true value of the Hubble constant is

$$H_0 \sim \xi^{-1/4} H_{BB} > H_{BB}. \tag{27}$$

In any case, it is clear that the magnitude and the frequency dependences of the sources of opacity will have an important effect on the derived value of the Hubble constant.

In order to estimate how strong the effect of the scattering opacity may be, let us consider the Balmer jump. We have inspected all published spectra of Type II supernovae, although only ~ 6 appear to be of good quality at wavelengths as short as ~ 3500 Å. The colour temperatures assigned to these spectra range from T_c = 5,000–13,000 K. The only one in which a possible Balmer jump appears is one published by Panagia et al. (1980). However, it is possible that this is an instrumental effect (Panagia 1981). From Figure 2, it is obvious that the lack of a detectable Balmer jump means that the photospheres are certainly very different from those of stars with the same colour temperature. Moreover, one ready explanation (within our approximate models) is that scattering dominates absorption: $\xi \leqslant 1/4$. Equation (27) then indicates that H_0 could be significantly greater than H_{BB}.

These conclusions must be considered as suggestive at best until the detailed model atmosphere construction outlined in Section 2 has been completed.

The author would like to thank Eric Lander and Stephen Hershkowitz for their help with the computer calculations. This work was supported in part by a grant (PHY 79-20123) from the National Science Foundation.

REFERENCES

Baade, W. 1926. Astr.Nachr., 228, 359.
Branch, D. 1977. Mon.Not.R.astr.Soc., 179, 401.
Branch, D. 1979. Mon.Not.R.astr.Soc., 186, 609.
Branch, D. 1980. In Supernovae Spectra (AIP Conference Proceedings No 63), ed. R. Meyerott and G.H. Gillespie (New York: Amer.Inst.Physics).
Branch, D. and Patchett, B. 1973. Mon.Not.R.astr.Soc., 161, 71.
Colgate, S. 1979. Astrophys.J., 232, 404.
Colgate, S. 1982. This volume.
Karp, A.H., Lasher, G., Chan, K.L. and Salpeter, E.E. 1977. Astrophys.J., 214, 161.
Kirshner, R.P. and Kwan, J. 1974. Astrophys.J., 193, 27.
Mihalas, D. 1978. Stellar Atmospheres (2nd ed: San Francisco: Freeman).
Oke, J.B. and Searle, L. 1974. Ann.Rev.Astr.Astrophys., 12, 315.
Panagia, N. 1981. Private communication.
Panagia, N. et al. 1980. Mon.Not.R.astr.Soc., 192, 861.
Schurmann, S.R., Arnett, W.D. and Falk, S.W. 1979. Astrophys. J., 230, 11.

Tammann, G.A. 1979. In Scientific Research with the Space
    Telescope (IAU Colloquium No 54), ed.M.S. Longair and J.W.
    Warner (US Govt. Printing Office).
Wagoner, R.V. 1980. In Physical Cosmology, (Les Houches,
    Session XXXII), ed. R. Balian, J. Audouze and D.N. Schramm
    (Amsterdam: North Holland).
Wagoner, R.V. 1981. Astrophys.J.(Letters), 250, L65.
Weaver, T.A. and Woosley, S.E. 1980. In Supernovae Spectra
    (AIP Conference Proceedings No 63), ed. R. Meyerott and G.H.
    Gillespie (New York: Amer.Inst.Physics).

TYPE I SUPERNOVAE - OBSERVATIONAL CONSTRAINTS

David Branch

University of Oklahoma

ABSTRACT

Some of the constraints on models of Type I supernovae which can be inferred from the observations are discussed. Comparison of McDonald Observatory spectra of the SN I of 1981 in NGC 4536 establishes that elements other than nickel, iron, and cobalt are represented in the maximum-light spectrum, and that permitted Fe II lines are present 17 days later. For models which power the light-curve peak by radioactivity, the lower limits on the mass of ejected nickel are discussed. The rate of decline of the post-peak light curve appears to be correlated with expansion velocity and absolute magnitude. If these correlations can be firmly established they will provide stringent tests of SN I models.

INTRODUCTION

The level of research activity on the explosive phases of Type I supernovae (SN I) has increased sharply in the last few years, owing primarily to the successes of the nickel-56 radio-activity model (Colgate and McKee 1969, Arnett 1979). The delayed release of energy by nickel-56 and its decay product, cobalt-56, leads to an explanation for the early-time SN I light curve as well as the late-time exponential (Weaver, Axelrod, and Woosley 1980, Chevalier 1981). The demonstration that many of the features in the late-time spectra can be interpreted as forbidden emission lines of iron (Kirshner and Oke 1975, Meyerott 1980) and cobalt (Axelrod 1980a,b) adds weight to the arguments that nickel decay plays a crucial role.

M. J. Rees and R. J. Stoneham (eds.), Supernovae: A Survey of Current Research, 267–279.
Copyright © 1982 by D. Reidel Publishing Company.

The requirement that substantial quantities of nickel be
ejected narrows the field of detailed models for SN I, and raises
the hope that the identification of the SN I progenitor and its
explosion mechanism is almost within reach. However, there are
a variety of scenarios which lead to nickel production. For example, accreting white dwarfs can explode in very different ways,
depending on their initial masses and accretion rates, while
producing large amounts of nickel (Fujimoto 1980, Nomoto 1980),
and white dwarfs disrupting inside extended envelopes can do the
same thing. If only a small fraction of a solar mass of nickel
is required, core-collapse models which leave behind neutron
stars are also viable.

To make progress towards the identification of the correct
model, we need to make inferences about the physical conditions
and abundances in the ejected matter from the observations.
The difficulty here is that observations, even good ones, do not
readily translate into reliable constraints. If all of the
inferences one can make from the observations were taken seriously,
all models - at least all of those which have been discussed in
the literature so far - would have to be ruled out. Model-makers
need to know which of the constraints are firm, and which are
only indications that might change as observations accumulate
and interpretations improve.

I will attempt to summarize here some of the constraints
which are inferred from observations of SN I, and to indicate
which seem to be the firm ones. Constraints based on other
observations, such as the types of the parent galaxies, the
possible lack of neutron stars in galactic remnants of SN I, and
limits from galactic evolution models and from X-ray observations
of remnants on the amount of iron produced per SN I, will not be
discussed.

The next section is concerned with the interpretation of the
early-time spectra. The mere identification of spectral features
in the early-time spectra is enough to usefully constrain the
models, by establishing that the identified elements are present
in a specified velocity range. Quantitative information on
abundance ratios would be better, but the current estimates are
very uncertain. The third section contains some comments on the
direct determination of the distances to SN I. This is relevant
to the models because, barring a contribution to the peak luminosity from shock energy deposited in an initially extended envelope
(Lasher 1975), the peak luminosity depends to first order just on
the mass of ejected nickel. The final section discusses the
degree to which SN I differ from one, another. If the small observed differences among SN I can be shown to be real, and especially if correlations between observed properties can be established, valuable constraints will emerge.

EARLY-TIME SPECTRA

Attempts to identify lines in the early-time spectra of SN I, usually as P Cygni-type features due to permitted lines of singly ionized metals superimposed on a thermal continuum, have been based primarily on wavelength coincidence (e.g., Pskovskii 1969, Branch and Tull 1979, Gordon 1981). Because of the Doppler broadening and the blending, the identifications, although reasonable, have always been less than convincing. Synthetic spectra such as I have been computing recently can help establish line identifications by allowing for blends. The assumptions and approximations on which the calculations are based have been discussed elsewhere (Branch 1980). The basic assumption is that lines are formed by resonant scattering above a sharp photosphere. An individual line profile has an emission peak at the rest wavelength and an absorption component blueshifted by the velocity at the photosphere. Examination of calculated blends shows that it is best to refer to lines by their absorption components, because the positions of the absorption are less sensitive than the emissions to blends. The optical depth at the photosphere of one line of each ion is a fitting parameter, and the strengths of the other lines are fixed by assuming Boltzmann excitation. The temperature is chosen to give the best fit to the slope of the continuum, using a Planck function. The synthetic spectra can be used to maximum advantage only if fully calibrated spectra having a resolution of ≤ 10 Å are available. An excellent series of such spectra has recently been obtained at the McDonald Observatory for the SN I of 1981 in NGC 4536. Detailed interpretation of these spectra will be presented elsewhere in collaboration with the Texas astronomers. Some of the results from the two earliest spectra are discussed here.

Spectra obtained on the nights of March 6, 7, 8, and 9 have been combined by Alan Uomoto to make the first high-quality spectrum for an SN I at maximum light (Figure 1). The synthetic spectrum to which it is compared has T = 17,000 K and V = 12,000 km/sec. The synthetic spectrum accounts for all of the major observed features, and the agreement is probably as good as could be hoped given the simplicity of the calculations. There is some doubt about the O I identifications, which are new, because both features involving O I are blended in the synthetic spectra. The identification of He I $\lambda 5876$ is also very uncertain because it is blended with $\lambda 5979$ of Si II in the synthetic spectrum. However, the fit is improved when the O I and He I lines are included.

A similar comparison for March 24, about 17 days after maximum light, is shown in Figure 2. The synthetic spectrum has T = 8000 K and V = 11,000 km/sec. Lines of Ca II, Si II, and perhaps O I are still present, but Mg II and S II have faded, and

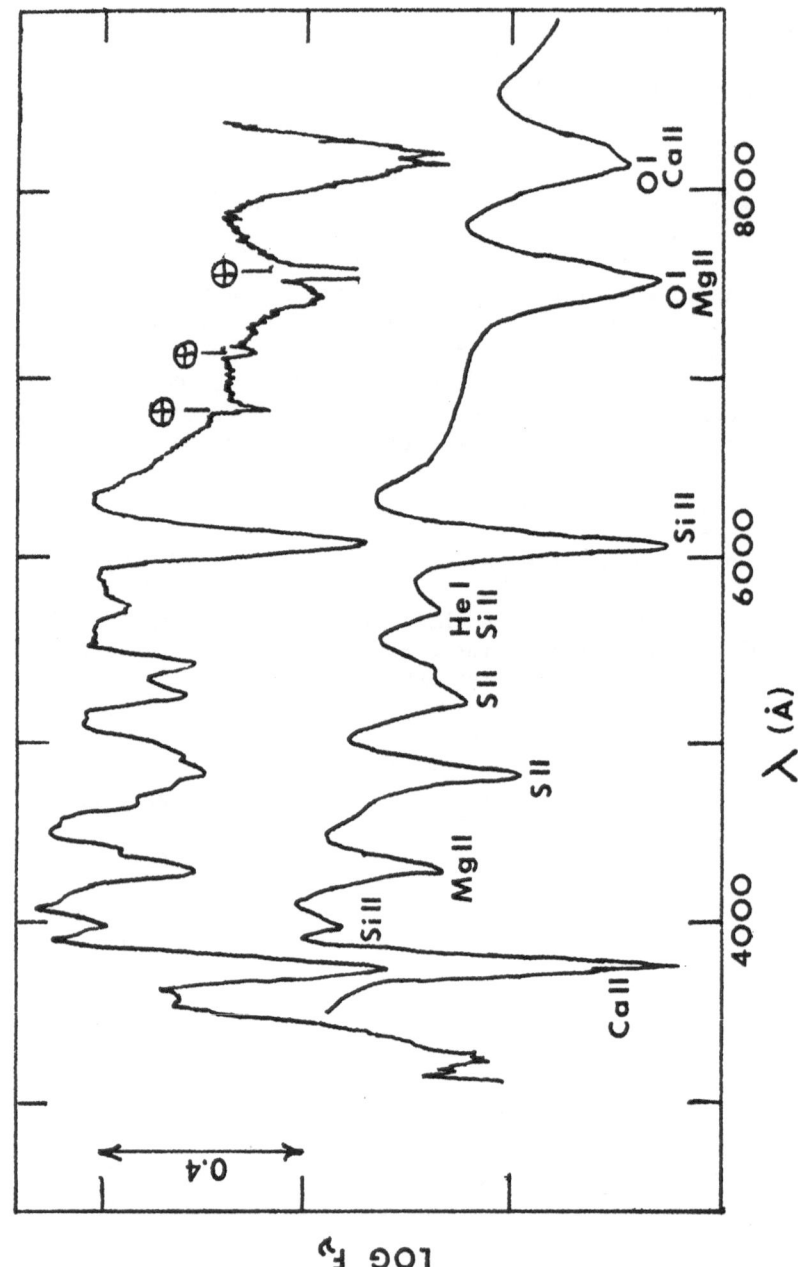

Figure 1. Upper: a McDonald Observatory spectrum of the SN I of 1981 in NGC 4536 at maximum light. Lower: A synthetic spectrum having $T = 17,000$ K, $V = 12,000$ km/sec, and $A_V = 0.3$.

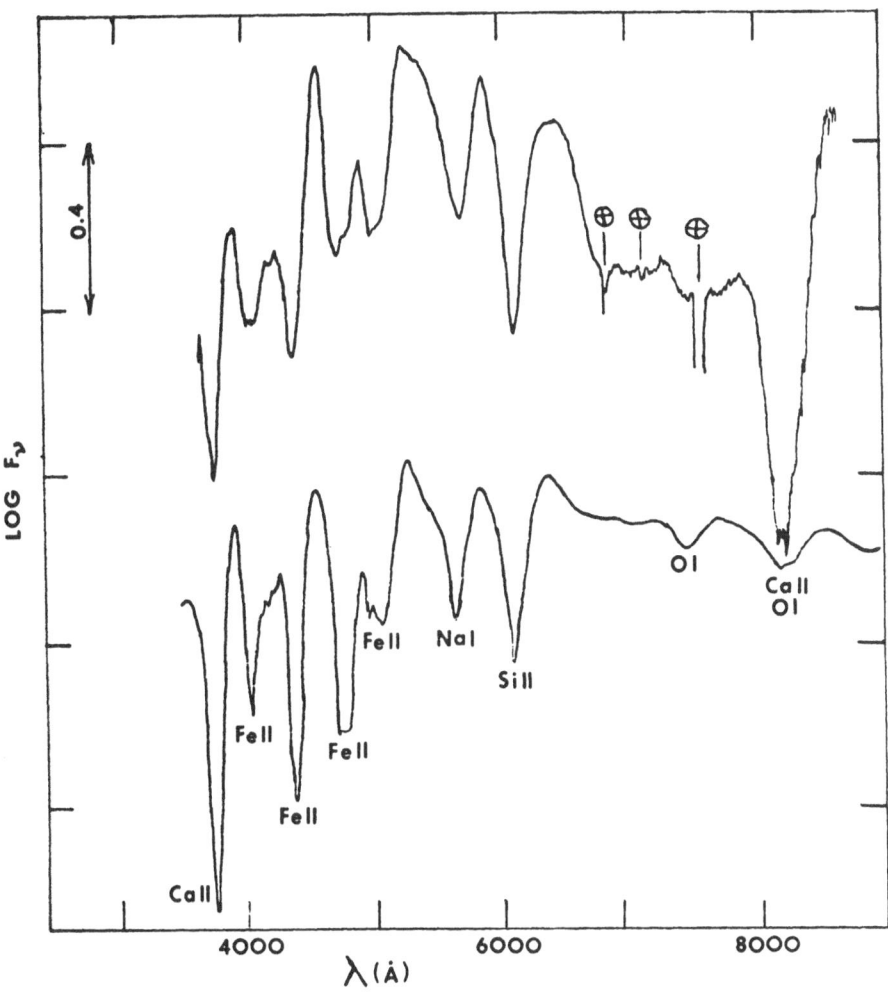

Figure 2. Upper: McDonald Observatory spectrum of the SN I of 1981 in NGC 4536, 17 days after maximum light. Lower: a synthetic spectrum having T = 8000 K, V = 11,000 km/sec, and A_V = 0.3.

much of the blue-green is interpreted as blends of (permitted)
Fe II. The Na I D lines have been introduced to account for the
absorption at 5715 Å. Although there are obvious discrepancies,
especially between the observed and computed strengths of the
infrared Ca II triplet, the most important point is that the
wavelengths of the observed and computed absorption features are
in good agreement. The evidence for Fe II lines at this phase
is convincing.

The presence of elements other than nickel, cobalt, and iron
in the material ejected at $\sim$10,000 km/sec is also a firm constraint, and it rules out at least one model - the bare "edge-
lit" or "double-detonation" accreting carbon-oxygen white dwarf
which ignites degenerate helium at the bottom of the accreted
layer and incinerates itself almost entirely to nickel-56
(Woosley, Weaver, and Taam 1980, Nomoto 1980). To confront other
models, estimates of abundance ratios are needed.

To translate the optical depths used for the synthetic spectra into abundance ratios, one can carry out non-LTE calculations
in the escape-probability approximation, as Feldt (1980) has done
for Si II and hydrogen in schematic supernova envelopes, or assume
LTE. The density at the photosphere is low and radiative transitions dominate collisional transitions, but detailed balancing
will keep the level populations not too far from LTE if the radiation field at the photosphere is nearly Planckian. This is borne
out by Feldt's calculations. The emergent spectrum of an SN I
obviously is not Planckian; the spectrum from 0.4 to 2.2 microns
can be approximated by a Planck function (Kirshner et al. 1973a)
but shortward of 0.4 microns the spectrum falls far beneath
the extrapolated blackbody curve. However, this does not necessarily mean that the spectrum is non-Planckian at the optical
photosphere. The UV-deficiency is due to an enhanced opacity in
the UV relative to longer wavelengths, so the UV spectrum is
formed in a higher, cooler layer than the optical spectrum. The
optical photosphere is beneath the UV photosphere and the UV
radiation field, which drives the level populations, may be
nearly Planckian at the optical photosphere.

The assumption of LTE leads to an interesting result. The
relative abundances derived from the optical depths of the lines
of Ca II, Si II, S II, Mg II, O I, and Fe II are not orders of
magnitude different from solar. These relative abundances are
not too temperature-sensitive, and it is tempting to attach
some significance to this result. However, a note of caution is
introduced by the fact that in LTE at 17,000 K, lines of Si III
and O II should be stronger than those of Si II and O I. This
may be an indication of departures from LTE, but another possibility which needs to be considered is that a strong temperature
gradient near the photosphere causes the optical depth of these

high-excitation (~20 eV) lines to fall off very rapidly above the
photosphere. This could produce weak lines. He I λ5876 is also
a 20-volt line, so the weakness of the observed feature is no
guarantee that the helium abundance is low.

If the iron seen in the March 24 spectrum was synthesized via
nickel-56 at the time of the explosion, the Co/Fe ratio on March
24 would have been about 4. With the Co II oscillator strengths
given by Kurucz and Peytremann (1975), Co II lines are then predicted to dominate the blue part of the spectrum. The absolute
levels of the Kurucz and Peytremann oscillator strengths for
other ions of iron group elements are approximately correct, and
there is no particular reason to think that the absolute level
for Co II is badly wrong. Some doubt must remain until accurate
measurements of the Co II oscillator strengths have been made,
but it is unlikely that the iron seen in the March 24 spectrum
had been freshly synthesized.

Apparently the supernova "atmosphere" had not become optically thin by 17 days after maximum light. This may be reasonable in terms of nickel-56 models. However, most of the features
in the March 24 spectrum, especially those attributed to Fe II,
persist with little change in blueshift until later times, beyond
the transition to the exponential phase of the light curve. The
nickel-56 models for the light curve require that the Ni-Co-Fe
core be revealed by the beginning of the exponential phase. It
may be that for some time after the transition, when the atmosphere has become optically thin in the continuum, most of the
spectral features are formed in the atmosphere but are superimposed on the emergent spectrum of the core. Tim Axelrod and I
plan to collaborate on this problem by combining our two approaches
to spectral synthesis.

DISTANCES

From the apparent magnitudes of SN I outside the local Super-
cluster and the radial velocities of their parent galaxies one
can estimate that the absolute blue magnitude of SN I is about
-20.0 if the Hubble constant if 50 km/sec/Mpc, or about -18.5 if
the Hubble constant is 100. Estimates of the Hubble constant can
be based on the local distance scales and the properties of
galaxies (de Vaucouleurs and Bollinger 1979, Aaronson et al. 1980,
Tammann, Sandage, and Yahil 1979), or on the supernovae themselves by means of the Baade method. Three independent estimates
of the distance to the Type II SN 1979c in M100, a Virgo-cluster
member, are in excellent agreement: 24 Mpc (Panagia et al. 1980),
23 Mpc (Branch et al. 1981), and 22 Mpc (Kirshner, 1982).
This shows that the internal errors are under control. If
the component of our galaxy's peculiar motion towards Virgo

is about 300 km/sec, this distance corresponds to H_o ~60 km/
sec/Mpc.

Attempts to derive the Hubble constant from composite data
for SN I, on the assumption that they are standard candles, have
given $40 \leq H_o \leq 60$ km/sec/Mpc (Branch and Patchett 1973, Branch
1979). One problem with the use of SN I is that the time base-
line for the Baade method is uncomfortably short; before maximum
light the colors are blue and not sufficiently sensitive to the
temperature, and by two weeks after maximum the determination of
the color temperature is too uncertain because the lines begin
to develop net emission. Arnett (1981) suggests that rather
than derive the radius at maximum light by the Baade method, it
is better to simply estimate the radius from the velocity at
the photosphere and a rise time to maximum light based on pre-
maximum observations or theory. For the NGC 4536 supernova we
have V = 12,000 $\pm$ 1000 km/sec. The rise time for this supernova
is not determined but pre-maximum observations of other SN I
(Barbon et al. 1973, Pskovskii 1977) suggest 15 $\pm$ 2 days. With
an optical color temperature of 17,000 $\pm$ 3000 K, this leads to
M_B =-20.5 $\pm$ 0.4, and H_o = 40 $\pm$ 10. It is worth pointing out
that although at first glance the determination of the optical
color temperature from the spectrum would seem to be quite uncertain, the temperature is actually fairly well determined when
a synthetic spectrum is fitted. As for SN II, the internal errors
are tolerable. To go beyond the simple approximation of blackbody emissivity at the optical color temperature we will need to
know the degree of atmospheric extension (Branch 1979), the wavelength dependence of the opacity, and the ratio of scattering to
absorption (Wagoner 1981). The latter two depend on the chemical
composition, and bound-bound transitions must be taken into
account.

Another way to use SN I to estimate distances is to fit hydrodynamical models to the observations (Arnett 1981). A model
calculation gives values of the luminosity and radius at maximum
light. The problem then is to go from the bolometric luminosity
to something which is related to observation, such as the absolute blue luminosity. The effective temperature can be calculated from the luminosity and the radius, but if the supernova
is assumed to radiate like a blackbody at the effective temperature the blue luminosity will be underestimated. Supernovae,
like early type stars, are ultraviolet-deficient and their
effective temperatures are lower than their optical color temperatures. The assumption of blackbody emission at the effective
temperature "wastes" much of the luminosity in the UV, where it
does not actually emerge. A simple way to use the model luminosity to predict the blue luminosity would be to represent the
spectrum as a Planck function truncated at 4000 Å. This seems
reasonable because observations from space indicate that only a

small fraction of the luminosity emerges at shorter wavelengths (Holm, Wu, and Caldwell 1974, Benvenuti, 1981). For $M_B = -20$, a truncated 17,000 K blackbody would emit 2.0×10^{43} ergs/sec. This luminosity is not sensitive to the adopted value of the optical color temperature. Thus a first-order prescription for going from model luminosity to Hubble constant is

$$H_o = 50 \ (L/2.0 \times 10^{43} \ \text{ergs/sec})^{-\frac{1}{2}} \ \text{km/sec/Mpc}. \tag{1}$$

For models which power the peak by radioactivity, an approximate relation between Hubble constant and mass of ejected nickel can then be obtained by assuming a rise time of 15 days and that the luminosity is equal to the rate of energy deposition by decay of nickel and cobalt at that time. This leads to

$$H_o = 40 \ (M_{Ni}/M_\odot)^{-\frac{1}{2}} \text{km/sec/Mpc}. \tag{2}$$

The limits on the mass of ejected nickel from both the late-time light curve and the late-time spectra are less restrictive than this. The lower limit shown by Axelrod (1980b) from the late-time spectra of SN 1972e is for a particular choice of the thickness of the shell of nickel. As explained in Axelrod (1980a), other choices of shell thickness give other lower limits. The lower limit shown by Axelrod (1980a) from the late-time light curve is for a distance to NGC 5253 of 2 Mpc and an extinction in front of SN 1972e of $A_B = 0.9$ mag. This combination corresponds to a peak absolute magnitude $M_B = -19.1$, which is brighter than the $H_o = 100$ value.

White dwarf models which disrupt completely and produce up to a Chandrasekhar mass of nickel-56 imply low values of H_o. On the other hand, disregarding the possibility of a shock contribution to the peak luminosity, core-collapse models, if they can eject as much as 0.2 solar masses of nickel, correspond to high values of H_o.

Compact models which eject less than 0.2 solar masses of nickel can be ruled out.

STATISTICAL PROPERTIES

The most intriguing property of Type I supernovae is the uniformity of their optical spectra. Apart from a few obvious exceptions, they show almost identical spectra at a given phase. This impression is reinforced by comparison of the new spectra of the NGC 4536 supernova with those of SN 1972e (Kirshner et al. 1973b, Branch and Tull 1979). As calibrated, digitized spectra of SN I accumulate we will be able to express the spectral homogeneity and departures therefrom in a quantitative way.

Photometrically, SN I are also rather homogeneous, but the observed differences apparently can not be attributed completely to observational error. Barbon, Ciatti, and Rosino (1973) have divided SN I into "fast" and "slow" subclasses, based on several characteristics of their light curves, and Pskovskii (1977) has made a finer subclassification according to a parameter β which measures the rate of decline of the blue light curve between the peak and the transition point, in magnitudes per 100 days. Pskovskii finds that β correlates with both expansion velocity, inferred from spectroscopy, and absolute magnitude. Rust (1974) also found the light-curve decay rate to correlate with absolute magnitude. I have recently tried to check on these correlations (Branch 1981). From a survey of the literature on SN I spectra I compiled a list of apprently reliable measurements of the red Si II absorption feature, and found them to correlate with Pskovskii's measurements of β (Figure 3). The correlation is in the same sense found by Pskovskii - the faster the decline of the light curve, the slower the expansion velocity at the photosphere. I used distances to SN I parent galaxies on the scale of de Vaucouleurs (1979) to investigate the correlation between β and absolute magnitude. The correlation appeared, again in the sense found by Pskovskii - the faster the decline of the light curve, the fainter the supernova.

If the correlation between β and the absolute magnitude is real, it shows, of course, that SN I are not standard candles. It follows that SN I are subject to the Malmquist effect; because of observational selection, supernovae found at large distances will be intrinsically brighter, on the average, than those found relatively nearby. This is borne out by the fact that β correlates with parent-galaxy radial velocity (distance) in the expected sense. de Vaucouleurs (1979) has shown that SN I apparent magnitudes are consistent with his distance scale, assuming that they are standard candles. However, when account is taken of the fact that the more distant supernovae have smaller values of β and are brighter on average than the nearer ones, it turns out that SN I imply that the distance scale is non-linear, in the sense of being too compressed. An iterative procedure leads to a stretching of the de Vaucouleurs scale by a factor of about two (corresponding to a global H_o nearer to 50 than to 100 km/sec/Mpc) and the absolute magnitude $-\beta$ correlation shown in Figure 4. Linear regression gives

$$M_{pg} + 20.7 = 0.22 \ (\beta - 10.7) \atop \underline{+}\ 0.06 \eqno(3)$$

with a standard deviation of 0.69 magnitudes. According to Pskovskii, SN I have values of β ranging from 6 to 14, so the implication is that the total range in absolute magnitude approaches 2 magnitudes.

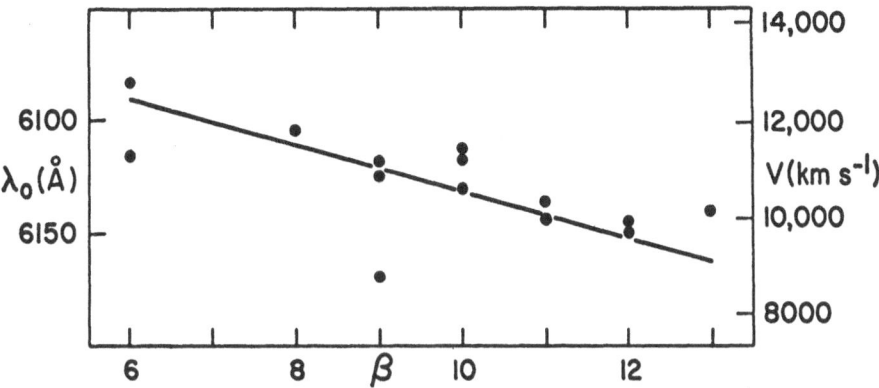

Figure 3. The wavelength and blueshift of the red Si II absorption are plotted against Pskovskii's light-curve parameter β. The $\lambda_0(\beta)$ regression line is shown

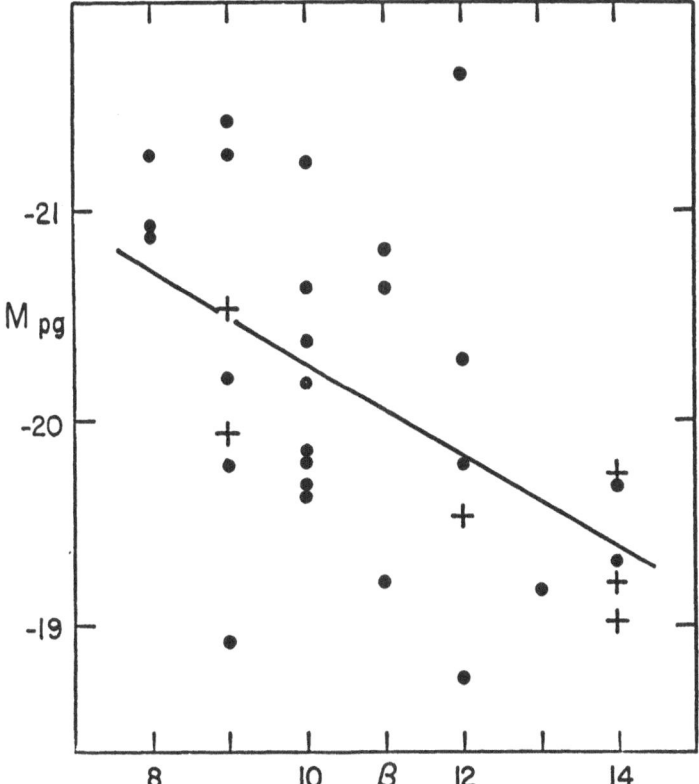

Figure 4. Absolute magnitude is plotted against Pskovskii's light-curve parameter. Crosses denote Virgo cluster supernovae. The $M_{pg}(\beta)$ regression line is shown.

The correlations of β with expansion velocity and absolute magnitude are both opposite in sense to what is expected from first-order considerations of exploding white dwarfs (Chevalier 1980, Arnett 1981). One reason fŏr caution, however, is that the values of β are given by Pskovskii without error estimates, and some of them may be poorly determined. It seems unlikely that errors would conspire to produce the three correlations, with expansion velocity, with distance, and with absolute magnitude, but they perhaps should not be considered firm until the uncertainties in β have been investigated. Bruce Patchett and I have begun to review the literature on light curves for this purpose. It is clearly important to measure many more accurate light curves for SN I!

Discussions with Tim Axelrod, Ken Nomoto, Peter Sutherland, Craig Wheeler, and others have been helpful. My research on supernovae is supported by the National Science Foundation through grant AST 7808672.

REFERENCES

Aaronson, M., Mould, J., Huchra, J., Sullivan, W., Schommer, R., and Bothum, G. 1980, Astrophys. J. 239, pp. 12-37.
Arnett, W.D. 1979, Astrophys. J. (Letters), 230, pp. L37-L40.
Arnett, W.D. 1981, Astrophys. J., in press.
Axelrod, T. 1980a, Ph.D. Thesis, Univ. Calif. Santa Cruz.
Axelrod, T. 1980b, in Proc. Texas Workshop Type I Supernovae, ed. J.C. Wheeler (Austin: Univ. Texas Press), pp. 80-95.
Barbon, R., Ciatti, F., and Rosino, L. 1973, Astron. Astrophys. 25, pp. 241-248.
Benvenuti, P. 1981, paper presented at this conference.
Branch, D. 1979, M.N.R.A.S., 186, pp. 609-616.
Branch, D. 1980, in Supernova Spectra, ed. R. Meyerott, and G.H. Gillespie (New York: American Institute of Physics), pp. 39-48.
Branch, D. 1981, Astrophys. J., in press.
Branch, D., Falk, S.W., McCall, M.M., Rybski, P., Uomoto, A.K., and Wills, B.J. 1981, Astrophys. J., 244, pp. 780-804.
Branch, D., and Patchett, B. 1973, M.N.R.A.S., 161, pp. 71-83.
Branch, D., and Tull, R.G. 1979, Astron. J., 84, pp. 1837-1839.
Chevalier, R.A. 1980, in Proc. Texas Workshop Type I Supernovae, ed. J.C. Wheeler (Austin: Univ. Texas Press), pp. 53-56.
Chevalier, R.A. 1981, Astrophys. J., 246, 267-277.
Colgate, S.A., and McKee, C. 1969, Astrophys. J., 157, pp. 623-643.
de Vaucouleurs, G. 1979, Astrophys. J., 227, pp. 729-755.
de Vaucouleurs, G., and Bollinger, G. 1979, Astrophys. J., 233, pp. 433-452.

Feldt, A. 1980, Ph.D. Thesis, Univ. Oklahoma.

Fujimoto, M.Y. 1980, in Proc. Texas Workshop Type I Supernovae, ed. J.C. Wheeler (Austin: Univ. Texas Press), pp. 155-163.

Gordon, C. 1981, Astron. Astrophys., 81, pp. 43-49.

Holm, A.V., Wu, C.C., and Caldwell, J. 1974, Publ. Astr. Soc. Pacific, 86, pp. 296-303.

Kirshner, R.P. 1982. This volume.

Kirshner, R.P., and Oke, J.B. 1975, Astrophys. J., 200, 574-581.

Kirshner, R.P., Oke, J.B., Penston, M.V., and Searle, L. 1973b, Astrophys. J., 185, 303-322.

Kirshner, R.P., Willner, S.P., Becklin, E.E., Neugebauer, G., and Oke, J.B. 1973a, Astrophys. J. (Letters) 180, pp. L97-L100.

Kurucz, R.L., and Peytremann, E. 1975, Smithsonian Astrophys. Obs. Spec. Rep. No. 362.

Lasher, G. 1975, Astrophys. J., 201, pp. 194-201.

Meyerott, R.E. 1980, Astrophys. J., 239, pp. 257-270.

Nomoto, K. 1980, in Proc. Texas Workshop Type I Supernovae, ed. J.C. Wheeler (Austin: Univ. Texas Press), pp. 164-181.

Panagia, N. et al. 1980, M.N.R.A.S., 192, pp. 861-879.

Pskovskii, Y.P. 1969, Sov. Astr., 12, pp. 750-756.

Pskovskii, Y.P. 1970, Sov. Astr., 21, pp. 675-682.

Rust, B. 1974, Ph.D. Thesis, Univ. Illinois.

Tammann, G.A., Sandage, A., and Yahil, A. 1979, Univ. Basel Preprint No. 1.

Wagoner, R. 1981, Astrophys. J. (Letters), in press.

Weaver, T.A., Axelrod, T.S., and Woosley, S.E. 1980, in Proc. Texas Workshop Type I Supernovae, ed. J.C. Wheeler (Austin: Univ. Texas Press), pp. 113-154.

Woosley, S.E., Weaver, T.A., and Taam, R.E. 1980, in Proc. Texas Workshop Type I Supernovae, ed. J.C. Wheeler (Austin: Univ. Texas Press), pp. 96-112.

RADIO SUPERNOVAE

K. W. Weiler[1], R. A. Sramek[2], J. M. van der Hulst[3],
N. Panagia[4]

[1]National Science Foundation, Washington, D. C. 20550
[2]NRAO - VLA, P. O. Box 0, Socorro, NM 87801
[3]Dept. of Astron., Univ. of Minn., Minneapolis, MN 55455
[4]Inst. Radioastron., Via Irnerio 46, 40126 Bologna, Italy

Three supernovae have so far been detected in the radio range shortly after their optical outburst. All are Type IIs. A fourth supernova, a Type I, is being monitored for radio emission but, at an age of approximately four months, has not yet been detected. For two of the supernovae, extensive data are presented on their "light curves" and spectra and a preliminary discussion is given of possible mechanisms. An interesting implication of this work is that no similarly strong radio supernova is likely to have occured in our Galaxy within the past 20 to 30 years and, if these supernovae are typical, the next galactic supernova will probably be discovered in the radio range.

I. INTRODUCTION

Although there has been a traditional split in the study of supernovae and their remnants between optical studies concentrating mainly on extragalactic supernovae and radio studies concentrating mainly on galactic supernova remnants, the increasing overlap between the two areas of study is profitable to all. The optical studies of remnants in the Galaxy and the radio observations of extragalactic remnants promise new insights into their properties and mechanisms. Now, with the increased sensitivity provided by the VLA[1], it is additionally possible to study the development of extragalactic supernovae in the radio range.

There are four examples which yield essentially all of our knowledge concerning the radio properties of supernovae. These are listed in Table 1. The first three (SN1970g, SN1979c, and SN1980k) are all Type II and are the only supernovae which have ever been

281

M. J. Rees and R. J. Stoneham (eds.), Supernovae: A Survey of Current Research, 281–291.

measured in the radio range. The last, the Type I supernova in NGC4536, is very recent and, although optically bright, has not yet been detected in the radio range to a limit of less than 0.3 mJy at an age of approximately 4 months. This lack of detection is, of course, not yet definitive (SN1979c was not found until almost one year after the optical maximum) so that the search for radio emission will continue.

Table 1: Radio Supernovae

Object	Galaxy	SN Type	Optical Max.
SN1970g	M101	II	$\sim$ 1 Aug. 1970
SN1979c	M100 (NGC 4321)	II	$\sim$23 Apr. 1979
SN1980k	NGC 6946	II	$\sim$ 3 Nov. 1980
SN1981?	NGC 4536	I	$\sim$ 5 Mar. 1981

In order to put the properties of radio supernovae in perspective, they are compared with other types of radio sources in Table 2. The first column gives the source name, while the second and third columns give the 6 cm flux density and approximate distance. From actual radio measurements or VLBI limits, an approximate angular size is given in column 4 and this is converted into a physical size in column 5. For the 4 young supernovae, an approximate size is calculated from the assumption of an average expansion velocity of 10^4 km s^{-1} from the time of optical maximum until the time of apparent radio maximum. A spherical volume, a spectral power emitted, and a spectral emissivity are given in columns 6, 7, and 8, respectively. Column 9 gives an example of how strong each type of source would be if it were placed at the standard distance of the galactic center (10 kpc) and column 10 gives the ratio of the 6 cm flux density of each source to that of Cassiopeia A, one of the most important supernova remnants and the strongest source in the sky at centimeter wavelengths. It should be noted that what is given as an extragalactic source is not an actual object but is representative of the small, bright, VLBI sources such as BL Lacertae objects and compact quasars. Between the parentheses on the line above are extreme values thought to occur in the most compact of extragalactic radio sources.

Examination of Table 2 shows that the young supernovae are likely to be very bright and compact objects emitting a spectral power (column 7) between that of old supernova remnants and extragalactic sources and a spectral emissivity (column 8) approaching the 10^{-30} W Hz^{-1} cm^{-3} seen in the most compact of extragalactic

Table 2: Comparisons

SOURCE	S_{6cm} Jy	DISTANCE kpc	SIZE arcsec	SIZE pc	VOLUME V cm^3	F $W\ Hz^{-1}$	F/V $W\ Hz^{-1}\ cm^{-3}$	S_{6cm} 10kpc Jy	RATIO To Cas A
VLBI Source	~ 1	$z \sim 0.5$ $\sim 3*10^6$	$\sim 10^{-3}$	(~ 1) ~ 10	$(\sim 10^{55})$ $\sim 10^{58}$	$\sim 10^{25}$	$(\sim 10^{-30})$ $\sim 10^{-32}$	$\sim 10^{11}$	$\sim 10^9$
Cas A	900	2.8	300	4	$9*10^{56}$	$8*10^{17}$	$9*10^{-40}$	70	1
3C10	30	5	600	14	$4*10^{58}$	$8*10^{16}$	$2*10^{-42}$	8	0.1
Crab	660	2	360	3.5	$6*10^{56}$	$3*10^{17}$	$5*10^{-40}$	26	0.4
3C58	29	8	480	18.5	$9*10^{58}$	$2*10^{17}$	$2*10^{-42}$	19	0.3
Peak SN1970g	$\sim.005$	$7*10^3$	---	$6*10^{-3}$(a)	$2*10^{49}$	$3*10^{19}$	$2*10^{-30}$	$2*10^3$	35
Peak SN1979c	.008	$16*10^3$	---	$1.3*10^{-2}$(a)	$3*10^{49}$	$2*10^{20}$	$7*10^{-30}$	$21*10^3$	300
Peak SN1980k	.003	$5*10^3$	---	$3.8*10^{-3}$(a)	$8*10^{47}$	$7*10^{18}$	$9*10^{-30}$	650	9
SN1981? NGC4536	$<.0003$	$20*10^3$	---	---	---	$<10^{19}$	---	<1200	<15

[a] Assumes an average expansion velocity of 10^4 km s^{-1}

VLBI sources. From column 10 it is apparent that the detected radio supernovae are 1-2 orders of magnitude stronger than the brightest supernova remnant known, Cassiopeia A. It is interesting to examine column 9 displaying the flux density of each type of object at a standard distance of 10 kpc. A BL Lacertae object would, of course, be extremely bright having about 10^{11} Jy, and all of the bright galactic supernova remnants would be relatively ordinary with flux densities less than 100 Jy. However, a supernova like SN1979c would be a spectacular radio source with flux density greater than 20,000 Jy, which would influence any modern radio telescope even when present in the distant sidelobes. From this, it is apparent that a supernova such as SN1979c must be relatively short lived in the radio range and it is unlikely that such an outburst has occured in our Galaxy since the advent of powerful radio telescopes at high frequencies -- roughly the past 20 to 30 years. However, it is also apparent that the next galactic supernova is likely to be first detected in the radio range. It is often said that we are "overdue" to observe an optical galactic supernova since, even with extinction, one should be close enough to be visible every 200 years or so and the last was seen in 1604 (Kepler's supernova). Since a supernova should occur somewhere in the Galaxy every 20 years or so, we may be similarly "overdue" for a galactic radio supernova.

Why do we call these objects "radio supernovae" rather than "supernova remnants," which are know radio emitters? There are at least two good reasons:

1) They are observed to be 1-2 orders of magnitude more luminous than any of the known supernova remnants and are thus likely to generate their radio emission by a different mechanism, and

2) like a supernova, they apparently flare up for only a brief period of months or years and then fade, leaving true remnant formation for the distant future.

The evidence for this latter is three-fold:

1) SN1970g appeared relatively quickly in the radio after maximum optical light and approximately 4 years later had again weakened below detection limits (Gottesman et al., 1972; Allen et al., 1976).

2) Searches for radio emission from several tens of extragalactic historical supernovae ranging in age from a few years to almost 100 years have never detected a single object down to a limit of less than a Cas A (Brown and Marscher, 1978). Thus, even if some supernovae are very bright in the radio immediately after outburst, they must fade away rather quickly.

3) We do not see several thousand Jansky point sources anywhere in the Galaxy. Since the supernova rate for our Galaxy is about one every 20 years and these should be roughly equally divided between Type I and Type II (Tammann, 1982), even if one assumes that only Type IIs emit significant radio emission a radio lifetime significantly shorter than about 40 years is implied.

SN1970g had only five significant detections (Marscher and Brown, 1978) so that little information is available on its radio light curve and spectrum. Also, the Type I supernova in NGC4536 has not yet been detected. Thus, we will concentrate on SN1979c in M100 (NGC4321) and SN1980k in NGC6946.

SN1979c IN NGC 4321 (M100)

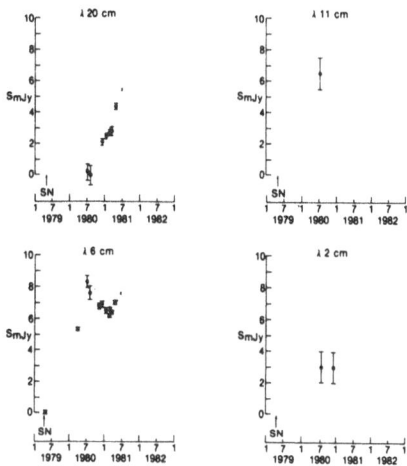

Figure 1:   The radio "light curves" for SN1979c in NGC 4321 (M100) are shown for 20, 11, 6, and 2 cm. All points are taken with the VLA except the 11 cm point by Swinbank taken with the Cambridge 5 km interferometer. The date of maximum optical light of the supernova on 23 April 1979 is marked. The last points at 6 and 20 cm shown without error bars are preliminary values.

II. RESULTS

The available information on the radio observations of SN1979c is shown in Figure 1. The observations at 20 cm (1.5 GHz), 6 cm (5 GHz) and 2 cm (15 GHz) are all taken with the VLA while the point

at 11 cm (2.7 GHz) is taken with the Cambridge 5 km interferometer
(E. Swinbank, private communication). The supernova was discovered on 19 April 1979 (Johnson, 1979) and reached maximum optical
brightness on approximately 23 April 1979. A radio observation
taken on 27 April 1979 showed no detectable emission stronger than
about 0.1 mJy at 6 cm wavelength. The next measurement in the
radio was approximately a year later on 6 April 1980 when the
supernova was detected with a flux density at 6 cm of 5.3 mJy
(Weiler and Sramek, 1980; Weiler et al., 1981). After that time,
regular monitoring was begun. The errors shown in Figure 1 are
formal measurement errors and may contain an additional systematic
calibration error which is estimated to not exceed 3 per cent. The
last values shown on the 6 and 20 cm plots without error bars are
preliminary values obtained on 6 May 1981. After the detection at
6 cm in April 1980, the source remained optically thick and
undetectable at 20 cm until 8 months later when 2.1 mJy was
measured on 4 December 1980. The radio spectrum of the supernova
has remained optically thick between 20 and 6 cm and optically thin
from 6 to 2 cm. However, while the flux density has remained
relatively constant at 6 and 2 cm, it has been increasing rapidly
at 20 cm so that the spectral index between the two frequencies has
been continually flattening (Figure 3). It is interesting to note
that a small increase in flux density seen in the last two points
at 6 cm has been accompanied by a larger increase at 20 cm. This
implies a changing optical depth along the line of sight to the
source.

An attempt has been made to detect linear polarization in the
radio radiation from SN1979c but the source has been found to be
essentially unpolarized (<1%). Also, scintillation of the radio
emission has been sought on a time scale of a few hours and no
fluctuations as strong as 10% have been seen.

The light curves for SN1980k are shown in Figure 2. As above,
the 20, 6, and 2 cm data taken with the VLA may be affected by a
small additional calibration error. The 2.8 cm (10.3 GHz) point
was taken with the 100m telescope of the Max Planck Institute for
Radioastronomy in Bonn, Germany (W. Sieber, private communication).
The optical supernova was discovered on 28 October 1980 (Wild,
1980) and reached maximum light on approximately 3 November 1980.
A radio measurement taken near maximum light on 3 November 1980
with the VLA at 6 cm showed no radio emission stronger than about
0.1 mJy. However, only a month later on 4 December 1980 a 6 cm
detection was obtained with a flux density of 0.7 mJy (Sramek et
al., 1980) and there followed a rapid flux density increase over
the next several months. The 20 cm radio emission was also detected very early showing 1.3 mJy on 23 March 1981. The 2 cm flux density is quite high (4.9 mJy) giving a spectrum which appears optically thick at radio wavelengths. As was the case for SN1979c, the
spectrum is flattening with time between 20 and 6 cm (Figure 3).

SN1980k IN NGC 6946

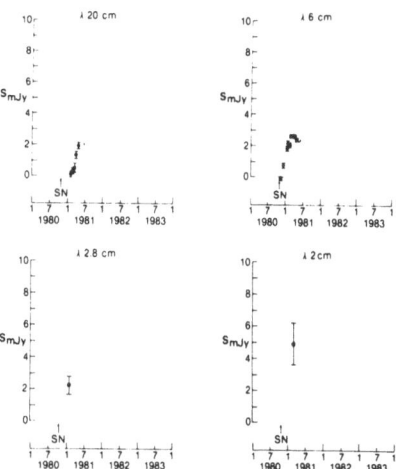

Figure 2: The radio "light curves" for SN1980k in NGC 6946 are shown for 20, 6, 2.8, and 2 cm. All points are taken with the VLA except the 2.8 cm point by Sieber with the MPIfR 100m telescope. The date of optical maximum on 3 November 1980 is marked.

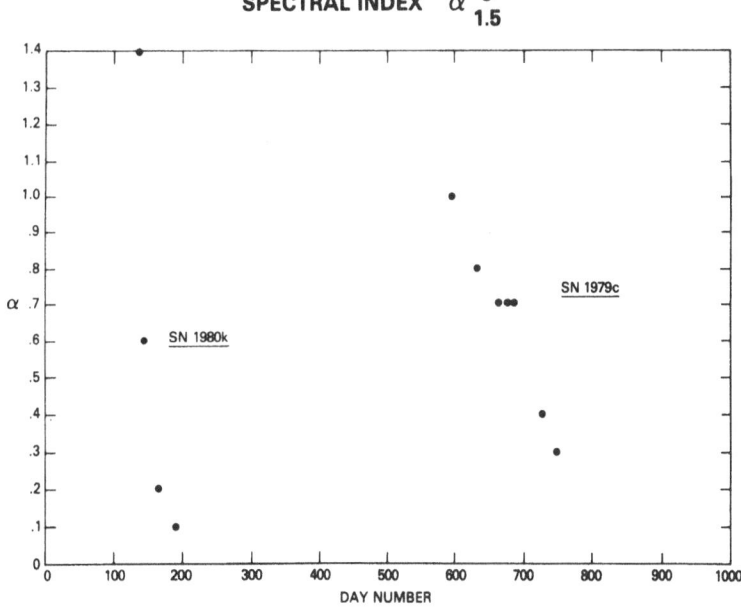

Figure 3: Spectral index ($S \propto \nu^{\alpha}$) between 20 cm (1.5 GHz) and 6 cm (5 GHz) plotted as a function of time.

III. DISCUSSION

For an initial interpretation of the data, the measurements
for both SN at the best studied frequencies of 20 and 6 cm are
plotted in Figure 4 in the form of traditional "light curves" with
the log of the flux density vs. the number of days since maximum
optical light. The points plotted on the 0.1 mJy line are upper
limits and not actual detections. Hand fitted curves are traced
through the data points to illustrate their main features. Some
minor deviations in the curves such as the dip at 6 cm at about 100
days for SN1980k and the bump in the curve at about 600 days for
SN1979c at 6 cm may not be real since they are dependent on single
data points. However, the low flux density for the first detection
of SN1979c at 6 cm and the increase in the flux density for the
last two points for SN1979c at 6 and 20 cm are both considered to
be reliable.

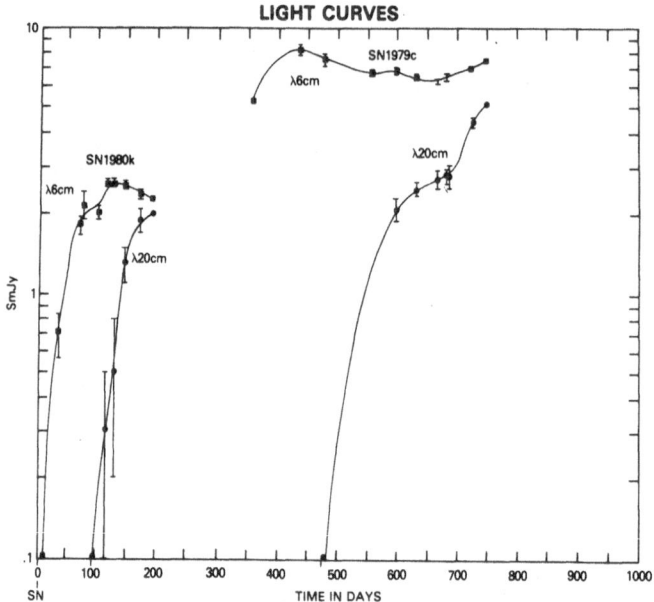

Figure 4:   Radio "light curves" for SN1979c and SN1980k at 20 and 6
             cm plotted as a function of time after maximum optical
             light. See Figures 1 and 2 for details. The points
             plotted on the 0.1 mJy line are only upper limits and
             not actual detections. The smooth curves are hand drawn
             to pass roughly through the measured points.

The main features of the curves are their extremely sharp
turn-on, the delays between 6 and 20 cm, the differing delays

before turn-on for SN1979c and SN1980k, and the resumption, after a
period of relative stability, of flux density increases at 6 cm and
even more strongly at 20 cm for SN1979c. Considering the simple
model of a uniformly filled sphere of fully ionized hydrogen
surrounding a central radio emitter which is constant in flux
density, if the sphere expands at 10^4 km s^{-1} and has a temperature
of 10^4 K the only free parameters are the maximum flux density at
each frequency and the total amount of hydrogen contained in the
sphere. For a hydrogen mass of $\sim 8*10^{-3}$ $M_\odot$, the model provides a
reasonable fit to the sharp turn-on of the radio emission in
SN1979c and to the time delay between 6 and 20 cm. Perhaps
fortuitously, this mass agrees in magnitude with the $\sim 10^{-2}$ $M_\odot$
estimated from IUE observations to be contained in the UV emitting
shell surrounding the supernova (Panagia et al., 1980).

The slight increase in the 6 cm flux density for SN1979c after
day 700 and the much larger increase in the 20 cm flux density at
the same time strengthen the case for an optical depth effect being
responsible for at least part of the flux density variation and
delay observed.

Such a simple picture does not fit SN1980k well. Although the
steep slope of the turn-on can be reproduced by an optical depth
variation model, a single mass for the HII sphere does not fit the
time delay observed between 6 and 20 cm. It can be said, however,
that because of its early turn-on, a much smaller mass of only
$\sim 10^{-4}$ $M_\odot$ is implied for SN1980k. Again perhaps fortuitously, this
agrees with the observed absence of a UV emitting shell in SN1980k
(Panagia, private communication).

Another model for which quantitative comparisons are possible
is that by Marscher and Brown (1978). It uses a co-mixed medium of
relativistic particles, magnetic fields, and thermal gas all
expanding at high velocity. The relativistic particles and
magnetic field are supplied by a central pulsar. It predicts, with
reasonable input parameters, a turn-on which goes as t^5 and a
decline after maximum as $t^{-0.25}$, both of which fit the data fairly
well. The observed delay between 6 and 20 cm can also be predicted. The model derives from earlier work on the Crab Nebula by
Pacini and Salvati (1973) and would be in agreement with a recent
discussion for SN1979c by Pacini and Salvati (1981) and the
suggestion that Type II supernovae are responsible for the creation
of extremely active pulsars (Weiler, 1978; Weiler and Panagia,
1980).

The impulsive model for radio supernovae and supernova remnants,
where the relativistic electrons are created in a single outburst
and then expand adiabatically (e.g. van der Laan, 1966), is, in any
case, not consistent with the data. It predicts a slow turn on at
any frequency and significantly higher maxima at higher frequen-

cies, neither of which are observed. Marsher and Brown (1978) reached this same conclusion for SN1970g.

The difficulty with any model proposing an internal source of radio emission is propagating the radiation out through the large amount of matter which is assumed to be present in the outer regions of a Type II supernova. This is especially true since the radio light curves show evidence for only a small fraction of a solar mass of ionized hydrogen. Solutions might be to coalesce the supernova envelope into small massive filaments, to postulate an asymmetric explosion where very little mass is ejected in some directions, or to maintain essentially all of the hydrogen envelope in neutral form. To avoid these problems, other workers such as Chevalier (1981) have suggested that the radio radiation is produced in the outer regions of the supernova where it is generated by a shock wave propagating through the mass lost in the later stages of stellar evolution. Unfortunately, no specific predictions yet exist for this model for comparison with the observations.

In conclusion, these new observations give us an unprecedented opportunity for studying supernovae and the interstellar medium. However, great uncertainties remain as to what are the mechanisms and how they operate. In order to extend the available data, monitoring of the presently known supernovae is continuing and attempts will be made to discover new objects for study.

---

[1] The VLA is operated by the National Radio Astronomy Observatory of Associated Universities Inc. under contract to the National Science Foundation.

REFERENCES

Allen, R.J., Goss, W.M., Ekers, R.D., Bruyn, A.G. de 1976, Astron. Astrophys. 48, 253.

Brown, R.L., Marscher, A.P. 1978, Astrophys. J. 220, 467.

Chevalier, R. 1981, paper presented at this conference.

Gottesman, S.T., Broderick, J.J., Brown, R.L., Balick, B. 1972, Astrophys. J. 174, 383.

Johnson, G. 1979, IAU Circular No. 3348.

Laan, H. van der 1966, Nature 211, 1131.

Marscher, A.P., Brown, R.L. 1978, Astrophys. J. 220, 474.

Pacini, F., Salvati, M. 1973, Astrophys. J. 186, 249.

Pacini, F., Salvati, M. 1981, Astrophys. J. Lett., 245, L107.

Panagia, et al. 1980, Monthly Notices Roy. Astron. Soc. 192, 861.

Sramek, R.A., Hulst, J.M. van der, Weiler, K. 1980, IAU Circular 3557.

Tammann, G. 1982, this volume.

Weiler, K.W. 1978, Mem. Soc. Ital. Astron. 49, 545.

Weiler, K.W., Sramek, R.A. 1980, IAU Circular No. 3485.

Weiler, K.W., Panagia, N. 1980, Astron. Astrophys. 90, 269.

Weiler, K.W., Hulst, J.M. van der, Sramek, R.A., Panagia, N. 1981, Astrophys. J. Lett., 243, L151.

Wild, P. 1980, IAU Circular No. 3532.

EVOLUTION AND NUCLEOSYNTHESIS IN POPULATION III STARS

W.W. Ober[1], M.F. El Eid[2] and K.J. Fricke[2]

1. INTRODUCTION

In recent years many arguments have been gathered in favour of the existence of a pregalactic generation of very massive objects (VMOs). An early generation of massive stars, which become pair unstable at the end of their evolution (M > 80 $M_\odot$) may manifest itself by its influence on the galactic chemical abundances and on the intensity and spectral distribution of the cosmic microwave background radiation. In the following we list some of these points.

(1) The well-known G-dwarf problem (Audouze and Tinsley 1976: Truran and Cameron 1971) may be resolved by prompt enrichment with the debris of exploded pregalactic (z > 10) massive stars.

(2) The high O/Fe ratios in metal-deficient stars (Sneden et al. 1979) suggest that oxygen has reached the solar abundance in the interstellar gas earlier than other heavy elements (see also Lambert et al. 1974). While most of the Fe in the disk could have been produced by the disk stars during the life of the Galaxy, a large fraction of the oxygen must have originated differently (Twarog 1980), possibly in a pregalactic phase.

(3) ^{14}N is significantly underabundant in high velocity stars (Clegg and Bell 1975) suggesting pregalactic processes.

1. Max-Planck-Institut fur Physik und Astrophysik, Karl-Schwarzschild-Strasse 1, 8046 Garching.
2. Universitats-Sternwarte Gottingen, Geismarlandstrasse 11, 3400, Gottingen.

M. J. Rees and R. J. Stoneham (eds.), Supernovae: A Survey of Current Research, 293–301.

The nuclei between A = 31 and 56 are underproduced in Type II supernovae (Woosley and Weaver 1982) and therefore call for an independent nucleosynthetic site. On the other hand the α-particle nuclei in the mass range are found to be enhanced in metal-poor stars (Peimbert 1974) pointing to a pregalactic origin of these nuclei by explosive oxygen burning.

In the solar neighbourhood r- and s-process elements have been detected in metal-poor low mass stars (Holweger 1980). Therefore these stars may have been formed out of gas which already was enriched with heavy elements (Truran 1980). Further points and arguments of a more cosmological nature for pregalactic massive objects have been outlined by Carr et al. (1982). The latter concern, for example, the missing mass problem and the distortion (Puget and Heyvaerts 1980) and the origin (Carr 1977) of the microwave background.

In this contribution we report on evolution and nucleo-synthesis calculations for massive primordial stars having no metals initially in the mass range $80 < M/M_\odot < 500$. The aim of this work is (i) to determine the mass range in which the evolution terminates explosively and (ii) to assess the kind of enrichment caused by the stars in this mass range.

2. THE HYDROSTATIC EVOLUTION OF VMOs

The equations of stellar structure for high mass stars were solved together with an equation of state which includes the effect of e^- e^+ pair creation, electron scattering opacity and expressions for nuclear energy generation in the various thermonuclear burning stages. Neutrino losses due to photo- and pair-annihilation-processes (Beaudet et al. 1967) have been taken into account. Initially the stars were assumed to consist only of H and He with X = 0.739 and Y = 0.261, respectively. In the absence of heavy elements, stars are much more compact than usual since the pp-chains are not sufficient to supply the luminosity and therefore the star has to contract in order to ignite the 3α-process at an enhanced central temperature. Thereby, very quickly, a tiny fraction of CNO nuclei is produced in the core and the CNO cycle takes over at a central temperature $\sim 10^8$ K. Convective energy transport is treated using the mixing length theory and the boundaries of the convection zones are determined with the Schwarzschild criterion. Mass loss was taken into account using the semi-empirical relation (Chiosi 1981)

$$M = (\frac{t_D}{t_{KH}})^{1/4} \frac{LR}{GM}$$

where t_D, t_{KH} are the dynamical and the Kelvin-Helmholtz time scales, respectively, and L, R, M, G have their usual meanings.

The mass loss rates derived this way are consistent with those obtained from theoretical calculations of pulsationally induced mass loss (Appenzeller 1970: Papaloizou 1973 and others) as well as with observational results.

The left part of Fig. 1 shows evolutionary tracks for stars of initial masses 80, 100, 150, 300 and 500 $M_\odot$. At the main sequence the stars have their minimum luminosity and $X_{CNO} \sim 10^{-9}$ depending slightly on mass (see Table 1). The tracks may be understood qualitatively by means of the heuristic 'mirror principle' taking into account the effect of mass loss on the internal structure of the stars. According to this principle the direction of the track must reverse itself each time an energy burning site emerges or disappears. The first turn occurs when the hydrogen shell source establishes itself, the second when He is ignited at the centre, the third when the shell source is killed as a result of the mass loss. Thereafter the evolution proceeds very quickly towards the helium main sequence. For comparison we have plotted on the right hand side of Fig. 1 some evolutionary tracks without mass loss which show the normal behaviour known from Pop I stars without mass loss (e.g. Maeder 1981). Up to central helium ignition our stars have lost about 30 to 50% of their original masses. The instantaneous mass values are indicated along the tracks for some stages. The stars initially develop large convective cores which shrink with progressive hydrogen depletion. More detailed information is presented in Table 1. We note that none of the models displays the density inversion reported by Ezer and Cameron (1971).

At the end of the He burning phase large C/O cores are present. The evolution towards carbon burning proceeds exceedingly fast due to effective neutrino losses. The e^- e^+ pairs copiously produced in this phase trigger the dynamical instability of the core. This dynamical phase as well as the preceding fast contraction stage have been treated employing an implicit hydrodynamical code.

3. THE FINAL PHASE OF VMOs

Here we describe as an example the dynamical evolution of a 116 $M_\odot$ core originating from a 200 $M_\odot$ primordial VMO by quasistatic evolution with mass loss. Fig. 2 is a T_c-ρ_c diagram for the whole evolutionary track of the central region of this object. After an initial homologous contraction with slope d log T_c/d log ρ_c $\approx$ 1/3 the core reaches the instability domain $\Gamma < 4/3$ and collapses to peak central values log $T_c = 9.53$ and log $\rho_c = 6.12$. The liberated energy by explosive oxygen burning in this phase was in this case sufficient to disrupt the star completely ($\sim 10^{52}$ erg). The explosion, like the collapse, proceeds nearly homologously (Fig. 2).

Table 1

Parameters of evolving VMOs between 80 and 500 $M_\odot$*

Phase	$M/M_\odot$	$\tau/10^6$ yr	$\log \frac{L}{L_\odot}$	$\log \frac{T_e}{K}$	$\log \frac{R}{cm}$	$\log \frac{T_c}{K}$	$\log \frac{\rho_c}{gcm^{-3}}$	x_c	x_s	y_c	x_{12}	q_c	$M/M_\odot$	$\dot{M}/M_\odot/$yr
1	80	0.031	5.943	4.959	11.421	8.081	1.650	0.736	0.739	0.264	7.5(-10)	0.795	80	4.13(-6)
	100	0.009	6.108	4.983	11.455	8.109	1.653	0.738	0.739	0.262	5.2(-10)	0.841	100	5.15(-6)
	150	0.028	6.392	4.996	11.571	8.109	1.519	0.734	0.739	0.266	9.0(-10)	0.884	150	9.48(-6)
	200	0.019	6.578	5.006	11.643	8.114	1.450	0.731	0.739	0.269	1.0(-9)	0.910	200	1.32(-5)
	300	0.034	6.821	5.154	11.747	8.121	1.353	0.732	0.739	0.268	1.3(-9)	0.940	299	2.04(-5)
	500	0.015	7.108	5.030	11.862	8.140	1.271	0.736	0.739	0.264	1.3(-9)	0.966	499	3.10(-5)
2	80	1.429	5.977	4.919	11.517	8.063	1.591	0.502	0.739	0.498	2.5(-9)	0.688	72	7.49(-6)
	100	1.280	6.142	4.932	11.573	8.069	1.534	0.502	0.739	0.498	2.8(-9)	0.725	90	1.03(-5)
	150	1.076	6.420	4.951	11.674	8.080	1.436	0.502	0.739	0.498	3.5(-9)	0.784	136	1.70(-5)
	200	0.966	6.605	4.964	11.742	8.087	1.370	0.500	0.732	0.501	4.0(-9)	0.810	183	2.30(-5)
	300	0.833	6.845	4.977	11.835	8.097	1.283	0.514	0.718	0.486	4.5(-9)	0.847	278	3.30(-5)
	500	0.747	7.133	4.986	11.961	8.107	1.175	0.501	0.697	0.499	5.4(-9)	0.871	468	5.15(-5)
3	80	3.370	6.047	4.946	11.497	8.164	1.950	0.000	0.338	1.000	1.0(-6)	0.691	42	2.00(-5)
	100	3.070	6.214	4.981	11.511	8.196	1.968	0.000	0.307	1.000	1.0(-5)	0.724	55	2.30(-5)
	150	2.656	6.494	5.020	11.575	8.233	1.942	0.000	0.268	1.000	7.6(-5)	0.767	88	3.13(-5)
	200	2.38	6.680	5.004	11.698	8.240	1.871	0.000	0.297	1.000	1.0(-5)	0.756	127	4.38(-5)
	300	2.23	6.917	5.057	11.712	8.322	2.000	0.000	0.263	1.000	2.0(-4)	0.789	196	5.26(-5)
	500	1.947	7.197	5.008	11.949	8.317	1.844	0.000	0.308	1.000	1.0(-4)	0.768	358	9.32(-5)
4	80	3.755	6.114	5.217	10.990	8.600	3.270	0.000	0.106	1.(-4)	1.65(-1)	0.896	34	4.86(-6)
	100	3.400	6.295	5.215	11.085	8.619	3.243	0.000	0.096	1.(-6)	1.30(-1)	0.903	46	6.93(-6)
	150	2.953	6.558	5.191	11.200	8.500	2.744	0.000	0.065	7.(-3)	1.22(-1)	0.937	78	1.35(-5)
	200	2.651	6.761	5.210	11.327	8.656	3.126	0.000	0.069	4.(-4)	3.01(-2)	0.923	116	1.55(-5)
	300	2.489	6.994	5.218	11.427	8.716	3.188	0.000	0.064	4.(-3)	1.00(-6)	0.930	179	1.96(-5)

*x_c, x_s are the fractions of hydrogen at the center, and at the surface; y_c, x_{12} are the He and C mass fractions at the center; q_c is the mass fraction of the convective core.

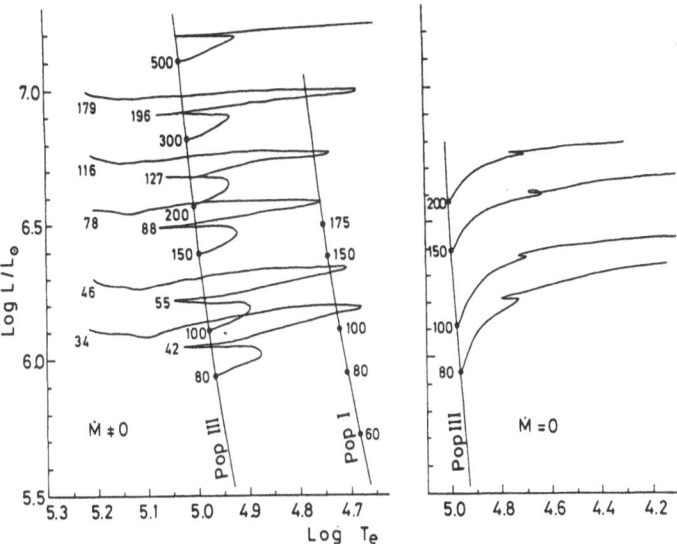

Fig. 1. Evolutionary tracks of VMOs with mass loss (lhs) and without (rhs). The numbers along the tracks are the instantaneous masses in solar units. The main sequences denoted Pop III and Pop I correspond to CNO mass fractions of $\sim 10^{-9}$ and 2×10^{-2}, respectively.

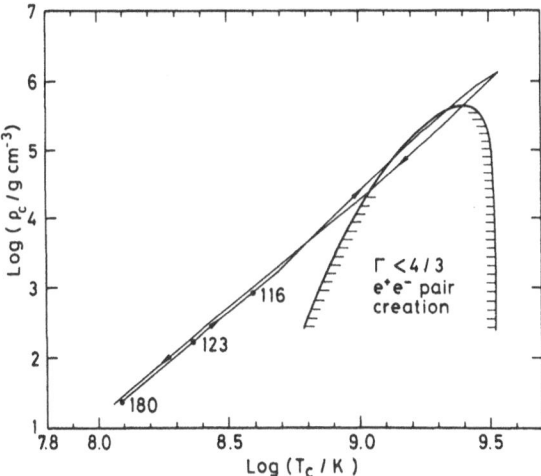

Fig. 2. The T_c-ρ_c diagram for a 200 $M_\odot$ VMO with mass loss. Instantaneous total masses are indicated. The hatched curve encloses the domain of pair instability where the adiabatic index $\Gamma < 4/3$.

Fig. 3 displays the distribution of the various nuclear species as functions of fractional mass in the exploded and processed material of 116 $M_\odot$. Most of the matter is in the form of oxygen. Only the inner 10 percent of mass had been transformed to Si and Ca nuclei. The helium shell has been processed explosively to some extent during the implosion. In Table 2 the enhancement factors over solar abundances of various species are listed and compared to the Pop III calculations of Woosley and Weaver (1982) for a 200 $M_\odot$ (without mass loss) and the calculations (also without mass loss) of a helium star of mass M_α = 100 $M_\odot$ by Arnett and Schramm (1973). The enhancement factor of ~ 80 for [16]O indicates that only a small fraction of primordial matter could have experienced conditions reached by a 200 $M_\odot$ primordial star.

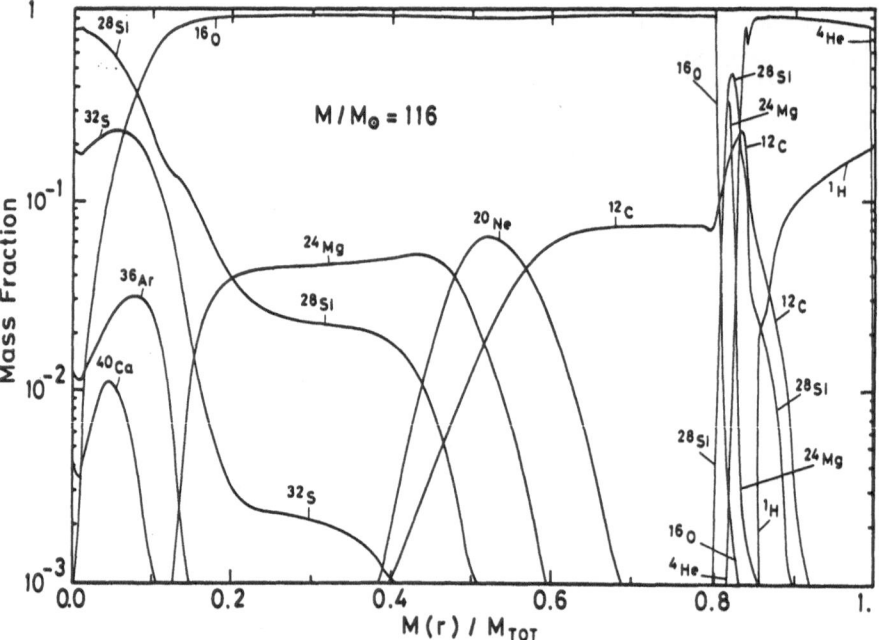

Fig. 3. The mass fractions of the nucleosynthesis products after explosion of a 116 $M_\odot$ core originating from a 200 $M_\odot$ primordial star as functions of fractional mass.

Table 2

Enhancement factors of products of explosive oxygen burning
relative to the sun for three model calculations

Species	This work	Woosley and Weaver 1982 $X_*/X_\odot$	Arnett and Schramm 1973
^{1}H	0.02	–	–
^{4}He	0.66	1.55	0.34
^{12}C	7.74	24.90	5.2
^{14}N	3.(-4)	0.14	–
^{16}O	80.28	51.80	54.2
^{20}Ne	5.46	18.70	2.2
^{24}Mg	48.19	77.7	40.7
^{28}Si	138.95	207.2	
^{32}S	74.08	145.1	517.6*
^{36}Ar	32.66	72.5	
^{40}Ca	9.83	98.4	
ejected ^{16}O mass fraction	0.663	0.43	0.45
$M_\alpha/M_\odot$	96	140	100
Population	III	III	I

*enhancement of Si through Ca nuclei.

Our results in Table 2 are much closer to Arnett and Schramm's than to those of Woosley and Weaver. This may be explained by the almost identical He core masses M_α in the first two cases (~ 100 $M_\odot$) while the neglect of mass loss by Woosley and Weaver leads to a quite different structure of the star and a He-core mass of ~ 140 $M_\odot$.

We have carried out dynamical calculations for the stars which we have evolved quasistatically (Fig. 1). Details of these calculations will be published elsewhere. We find that explosions are not possible for masses beyond $M = 300$ $M_\odot$ ($M_0 < 160$ $M_\odot$). The lower mass limits for the operation of the pair instability is found to be around $M = 100$ $M_\odot$ ($M_{CO} \sim 40$ $M_\odot$).

4. CONCLUSIONS

The results in Table 2 show that at least in some respects VMOs are indeed appropriate to account for the abundance anomalies described in Section 1: (i) oxygen is strongly enhanced, (ii) the G dwarf problem may find an explanation by VMOs, (iii) ^{14}N and the neutron rich nuclei are underabundant and (iv) the nuclei between Si and Ca are enhanced.

A more quantitative and detailed determination of primordial enrichment by VMOs will have to take account of a mass function of VMOs in the mass range $100 < M/M_\odot < 300$ and also of the steady enrichment (particularly of He) taking place during the hydrostatic evolution. This will be the subject of a paper in preparation.

ACKNOWLEDGEMENTS

This work has been supported in part under grants Fr 325/9 and Fr 325/14 by the Deutsche Forschungsgemeinschaft.

REFERENCES

Appenzeller, I. 1970. Astr.Astrophys., 5, 355.
Arnett, W.D. and Schramm, D.N. 1973. Astrophys.J., 184, L47.
Audouze, J.M. and Tinsley, B.M. 1976. Ann.Rev.Astr.Astrophys., 14, 43.
Beaudet, G., Petrosian, V. and Salpeter, E.E. 1967. Astrophys. J., 150, 979.
Carr, B.J. 1977. Astron.Astrophys., 60, 13.
Carr, B.J., Arnett, W.D. and Bond, J.R. 1982. This volume.
Chiosi, C. 1981. Astron.Astrophys., 93, 163.
Clegg, R.E.S. and Bell, R.A. 1975. Bull.Am.Astron.Soc., 7, 272.
Ezer, D. and Cameron, A.G.W. 1971. Astrophys.Space Sci., 14, 399.
Holweger, H. 1980. Mit.Astron.Ges.

Lambert, D.L., Sneden, G. and Ries, L.M.   1974.   Astrophys.J.,
    188, 97.
Maeder, A.   1980.   Astron.Astrophys., 92, 101.
Papaloizou, J.C.B.   1973.   Mon.Not.R.astr.Soc., 162, 169.
Peimbert, M.   1964.   In The Formation and Dynamics of Galaxies
    IAU Symposium No. 58.
Puget, J.L. and Heyvaerts, J.   1980.   Astron.Astrophys., 83, L10.
Sneden, C., Lambert, D.L. and Whitaker, R.W.   1979. Astrophys.J.,
    234, 964.
Truran, J.W. and Cameron, A.G.W.   1971.   Astrophys.Space Sci.,
    14, 179.
Truran, J.W.   1980.   Preprint.
Twarog, B.A.   1980.   Thesis, Yale University.
Woosley, S.E. and Weaver, T.A. 1982.   This volume.

THE BOUNDARY BETWEEN EXPLOSION AND COLLAPSE IN VERY MASSIVE OBJECTS

J.R. Bond[1,4], W.D. Arnett[2,5] and B.J. Carr[3]

ABSTRACT

A simple model emphasizing an entropic view of VMO evolution in the oxygen core phase is developed. Calculations of the effects of the pair instability, oxygen and silicon burning, and alpha quenching on a global effective potential allow us to predict the critical oxygen core mass for black hole formation, M_{0c}, and the abundance ratio of oxygen burning products to oxygen in those VMOs that explode. We find $M_{0c} \sim 10^2 M_\odot$, corresponding to an initial star mass $> 220\ M_\odot$.

1. INTRODUCTION

Very Massive Objects (VMOs) are defined to be stars in the mass range 10^2–$10^5\ M_\odot$. They are excellent candidates for Population III stars, those of nearly zero metallicity, since the fragmentation spectrum of metal-free clouds is likely shifted upward in mass from that of clouds with normal metallicity (Silk 1977; Tohline 1980). If little or no fragmentation occurs, Supermassive Objects are expected; SMOs are dynamically unstable due to general relativistic effects during the hydrogen burning phase (if $M > 10^5\ M_\odot$ for Pop III abundances; Fricke 1973).

1. Department of Astronomy, University of California, Berkeley.
2. Astronomy and Astrophysics Center, University of Chicago.
3. Institute of Astronomy, Cambridge University.
4. Present address: Department of Physics, Stanford University.
5. Max Planck Institute for Astrophysics, Munich.

303

M. J. Rees and R. J. Stoneham (eds.), Supernovae: A Survey of Current Research, 303–311.
Copyright © 1982 by D. Reidel Publishing Company.

If moderate fragmentation occurs, stars in the next mass range,
VMOs, would be expected. These stars are pair unstable during
the oxygen core phase, but are dynamically stable (though
pulsationally unstable) during hydrogen burning. If VMOs were
created in large numbers in the pregalactic era, they could have
had important cosmological effects. In order to understand these
cosmological effects, we must understand the stellar evolution of
VMOs. In this paper, we outline a simple model to calculate the
critical mass M_c: if $M > M_c$, the heavy element cores of VMOs
collapse completely to the black hole state; if $M < M_c$, complete
thermonuclear disruption occurs. Details of this work are
reported in Arnett, Bond, and Carr (1981a), ABC1. Cosmological
implications of VMOs are explored in Arnett, Bond, and Carr
(1981b), ABC2, and are outlined in this volume (Carr, Arnett and
Bond 1982).

2. EVOLUTION THROUGH CORE HYDROGEN AND HELIUM BURNING

Population III stars are pulsationally unstable if $M > 84M_\odot$
(Stothers and Simon 1970). However, nonlinear pulsational
analysis shows the amplitude of oscillation remains finite,
resulting in a finite mass loss rate, which indicates that the
fractional mass lost over the hydrogen burning lifetime may be of
order unity (Appenzeller 1970; Papaloizou 1973; Talbot 1971;
Ziebarth 1970); hence at least some VMOs may survive this mass
loss period as VMOs which form massive helium cores. We (ABC1)
have some evidence from an implicit hydrodynamical code that the
exterior hydrogen envelope may be ejected in a 500 $M_\odot$ Pop I VMO,
leaving a bare helium core (of mass M_α) which would itself be
pulsationally unstable provided $M_\alpha > 13$ $M_\odot$ (Stothers and Simon
1970). It may then be possible to eject pure helium, which would
add to the synthesized helium ejected by winds in the earlier
hydrogen burning phase (Talbot and Arnett 1971) to enrich the
primordial helium supply. If the envelope remains attached to
the helium core, then the envelope can damp pulsations as occurs
in massive supergiant stars. In any case, the hydrogen envelope
will be of low density, and the evolution of the helium core
decouples from that of the envelope. Fricke (1973) finds that a
helium core will go GR unstable before it can be stabilized by
burning if $M_\alpha > 32000$ $M_\odot$. Helium burning under these high
entropy conditions produces primarily oxygen by the 4α process.
An oxygen core thus forms stably if $M_\alpha < 32000$ $M_\odot$. The oxygen
core mass (extrapolated from lower mass helium core results) is
$M_0 \sim 0.9$ M_α, and we take $M_\alpha \sim 0.5$ M, where M is the mass of the
initial hydrogen star. The relations $M(M_\alpha)$ and $M_\alpha(M_0)$ are very
uncertain due to large mass loss effects and the easy convective
mixing which occurs in a $\gamma \sim 4/3$ gas, especially if rotation can
drive it. We will assume that the final fate of a VMO is
determined entirely by its interior oxygen core.

3. OXYGEN CORE EVOLUTION

Fowler and Hoyle (1964) first proposed that a type of supernova can occur for massive stars that go pair unstable. Zeldovich and Novikov (1971) find that, if $M_0 > 8000 \, M_\odot$, oxygen cores will go GR unstable before they go pair unstable. A black hole will be the fate of these cores. Past numerical studies (Fraley 1968; Barkat, Rakavy and Sack 1967; Arnett 1973; Wheeler 1977), though not always agreeing with each other, indicate that massive oxygen cores collapse, less massive ones explode and cores with $M_0 \sim 30 \, M_\odot$, which are just marginally pair unstable, may undergo relaxation oscillations accompanied by a slow consumption of oxygen, with their final state uncertain.

Rakavy and Shaviv (1968) used the dimensionless entropy per baryon to characterize collapse. We subscribe to this entropic view of oxygen core evolution, and demonstrate its utility in what follows. The terrain through which the core passes in thermodynamic phase space is shown in Figure 1, along with adiabats for a plasma consisting of O^{16}, γ, e^+e^-. We assume that the oxygen core has been made spatially isentropic through convective burning and prior to oxygen burning is temporally isentropic (since $\nu\bar{\nu}$ losses are small on dynamical timescales). Though not true in detail (Arnett's (1973) $M_\alpha = 100 \, M_\odot$ collapse has a lower entropy in the core centre than in the outer regions), it is a reasonable approximation which simplifies our analysis.

We take the density structure of the core throughout the collapse to be that of an $n = 3$ polytrope. Until nuclear burning occurs this is a fairly good approximation due to the radiation dominance of the pressure. The pressure, p, and energy per baryon, ε, are functions of the density, entropy, and abundances. The polytropic mass computed using the assumption of local hydrostatic equilibrium, $M_3(\rho) \sim (p/\rho^{4/3})^{3/2}$, changes for different densities in the core due to equation of state changes, and hence cannot provide a criterion for choosing the core mass as a function of the entropy. We define the core mass, M_0, given the initial entropy of the core, s_i, by requiring global virial equilibrium holds when the core is just marginally unstable ($\langle \Gamma_1 \rangle \sim 4.3$). Over the range of primary interest to us, $8 < s_i < 15$, $M_0(s_i) = 56(s_i/10)^{2.5} \, M_\odot$. We describe the global dynamics of the core by an effective potential,

$$V_e[\rho_c, s_i] = \int_0^{M_s} dm \left[\varepsilon - \frac{Gm}{r} - \int_{\rho_{co}}^{\rho_c} \frac{P_s(M_s, \rho_c)}{\rho(m, \rho_c)} \frac{d\rho_c}{\rho_c} \right],$$

which is the binding energy of that part of the core interior to mass M_s plus the PdV work term involving the surface pressure P_s arising from contraction of the core from the central density

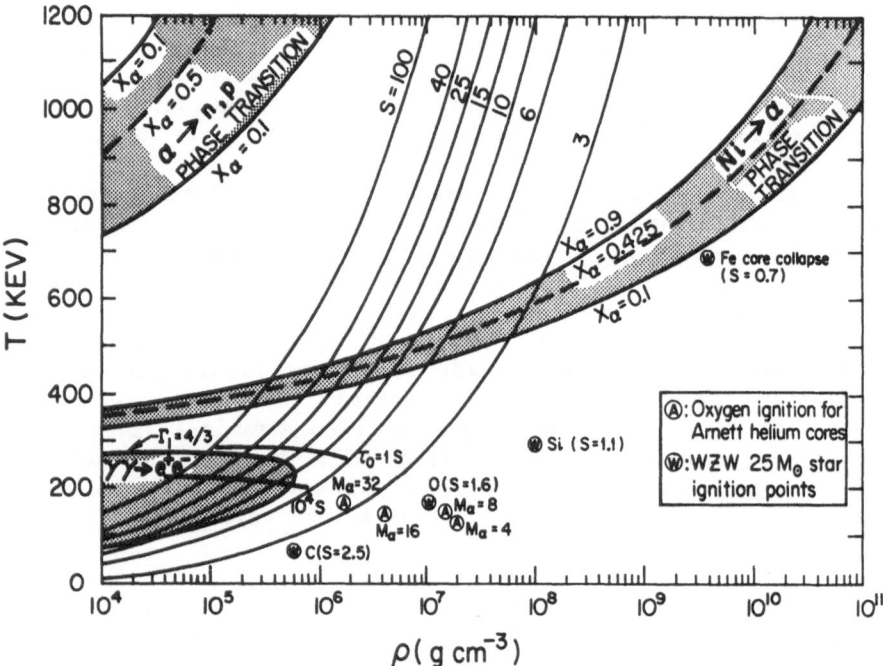

Figure 1.   Isentropic lines for a plasma of oxygen nuclei and
pairs are plotted in thermodynamic phase space.   Lines of
constant alpha abundance, X_α, through the nuclear phase
transitions, lines where the oxygen burning lifetime, τ_0, is 1s
and 10^4s, and points corresponding to Arnett's (1972) helium core
calculations and Weaver, Zimmerman, and Woosley's (1978) 25 $M_\odot$
star calculations are also shown.   The critical region for pair
instability supernova is s ~ 8–13.

at the onset of instability, ρ_{co}, to ρ_c. The energy, ε, includes
the nuclear potential energy and pair rest mass energy as well as
the thermal energies of photons, pairs, and ions. If we set the
zero of energy by $V_e[\rho_{co}] = 0$, then the kinetic energy is related
to the effective potential by $K = -V_e$ (assuming the luminosity in
$\nu\bar{\nu}$ pairs is negligible). This holds even through nuclear
burning.

The oxygen burning lifetime is about one second near the temperature $T \sim 280$ keV, and varies as T^{-25}. Dynamical collapse times are ~ 10 seconds. Thus, below the ignition temperature $T_{ign} = 280$ keV, burning is slow; above T_{ign}, it is very fast. Accordingly, we assume complete transmutation of oxygen into silicon and sulphur occurs at constant density once the ignition temperature is reached; ε is therefore conserved during transmutation, and since 472 keV of nuclear potential energy are released $(O \longrightarrow Si)$, the entropy jumps. We find $\Delta s(O \longrightarrow Si) = 1.19$; the entropy in ions goes down by ~ 0.5, since the final nuclear state is more ordered than the initial one, and the entropy in pairs and photons goes up by ~ 1.7, which is an expression of the extra pressure which burning generates. Silicon burning is complicated. For simplicity, we have assumed $Si \longrightarrow Ni$ also occurs instantaneously at $T_{ign} = 350$ keV, which results in an additional $\Delta s = 0.25$. Weaver and Woosley (1982) show that silicon burning under these conditions may be endoergic $(\Delta s < 0)$, and therefore it may be the first phase of alpha quenching, which drives collapse. We approximate alpha quenching by treating it as a $Ni \longrightarrow \alpha$ phase transition described by the Saha equation of nuclear statistical equilibrium. If alpha quenching is continuous through silicon burning, our critical mass will be slightly modified downward. This is not our biggest source of error.

The effective potential curves for various entropies as functions of the central density are plotted in Figure 2. We have divided V_e by the baryon number of the core to get our result expressed in keV/baryon; its magnitude should be compared with the core's gravitational potential (per baryon), $\Omega_G = -1467(M_0/100\ M_\odot)^{2/3}\ (\rho_c/10^6\ \mathrm{g\ cm^{-3}})^{1/3}$ keV. The behaviour of V_e can be understood by relating Figure 2 to the trajectories of stellar zones in Figure 1. V_e falls as the central regions enter the $\Gamma_1 < 4/3$ pair instability region. The centre passes through this region and crosses the oxygen ignition line, but the middle of the core is still pair unstable. Once enough of the central zones ignite oxygen, their higher entropy internal energy overcomes the middle zones' negative effect, and V_e rises. Depending upon the initial entropy and thus upon core mass, V_e may or may not hit zero before the devastating effects of alpha quenching in the collapsing central regions are experienced. With our approximations, we find that if $s_i > s_c \sim 12.4$, or $M_0 > M_{0c} \sim 95\ M_\odot$, V_e never reaches zero again: a black hole is the inferred fate for these cores. If $s_i < s_c$, complete thermonuclear disruption results. We also find that V_e is very small for $s_i \sim 8$ $(M_0 \sim 32\ M_\odot)$, since we just pass near the boundary of the instability region; a very small amount of burned oxygen makes V_e reach zero, which shows the possibility of relaxation oscillations in this mass region.

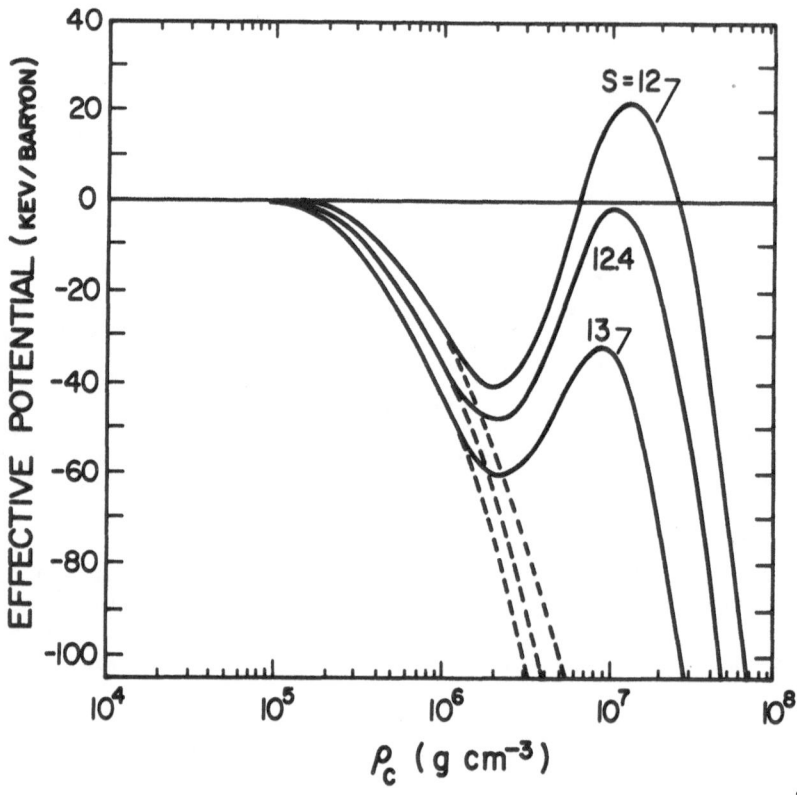

Figure 2. The effective potential (per baryon) is plotted for a
sequence of n = 3 polytropic density structures labelled by the
central density ρ_c. The three regimes of pair instability,
oxygen burning, and alpha quenching are apparent. The s = 12
core explodes; the s = 13 core collapses.

4. DISCUSSION

    The major source of error in this calculation is the
assumption that an entropy and pressure discontinuity can be
maintained between burned and unburned regions. The high entropy
interior will expand relative to our assumed polytropic density
structure; we thus overestimate the central density and
temperature, and thereby enter into silicon burning and then into
alpha quenching earlier than a realistic model would predict. As
a result of the latter, our estimate of M_{0c} is undoubtedly too
low. As a consequence of the former, we expect that our
predictions of the fraction of silicon burning products released
in those VMOs which explode will be too high. Our estimate of
the fraction of elements released that have undergone oxygen
burning is not subject to this uncertainty. In Figure 3, we plot

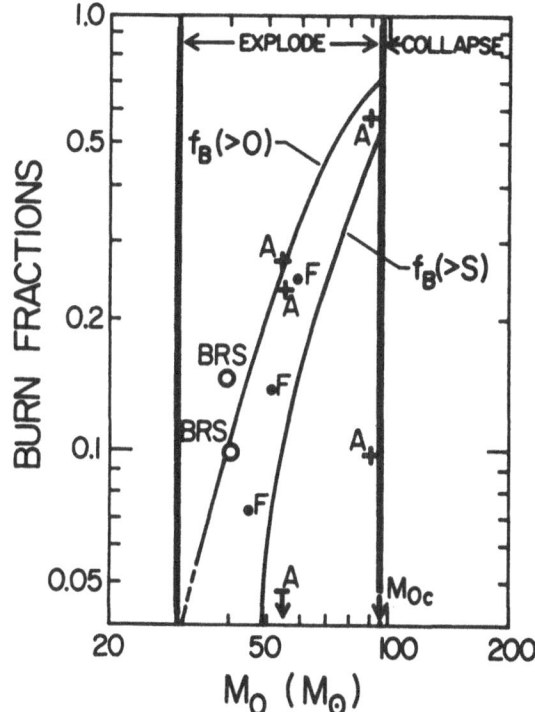

Figure 3. The predicted ejected mass fractions of the oxygen core in elements heavier than oxygen and heavier than sulphur. Points refer to the numerical calculations of Barkat, Rakavy, and Sack (1967), Fraley (1968), and Arnett (1973).

the burn fractions f_B (> 0) of elements heavier than oxygen and f_B ($> S$) of elements heavier than silicon and sulphur that we have obtained as a function of mass. As expected, f_B (> 0) compares well with the results available in the literature and f_B ($> S$) does not.

Rotation aids in stabilizing against collapse. For small angular momenta, J, we find $M_{0c}(J) \approx M_{0c}(0)(1 + a_M(J/J_0)^2)$, where $a_M = 3.5$ and $J_0/M_{0c} = 10^{19}$ cm^2/s: this compares with the specific angular momentum of an 05 star of ~ 10^{18} cm^2/s, which suggests rotation has a small effect on raising the mass. Once V_e drops due to alpha quenching, collapse is effectively ensured, since $\nu\bar{\nu}$ losses due to the pair process and later the Urca process further lower the effective potential, and the $\alpha \longrightarrow n,p$ phase transition transfers entropy from the pairs and photons to the nucleons to facilitate alpha breakup, thereby robbing pressure.

Collapse approaches freefall ($V_e \longrightarrow \Omega_G$). The last signature of collapse is emission from a neutrino fireball (Bond 1981), which forms within the last second before an event horizon forms. Since rotation becomes progressively more important as collapse proceeds, it is conceivable that the rotational energy can become of order the gravitational energy before a black hole forms. Interesting possibilities then arise for gravitational radiation emission, and, potentially, for ejection of some metals though $M_0 > M_{0c}(J)$.

Ober et al. (1982) and Weaver and Woosley (1982) report in this volume new numerical calculations of VMO evolution. Only by such detailed studies can the nucleosynthesis from VMOs and the exact value of M_{0c} and thereby M_c be obtained. We have sketched here a simple theory which explains the main features of their work and of previous work. We expect that complete collapse to a black hole will occur for oxygen core masses somewhat in excess of one hundred solar masses, and of initial star masses $> M_c \sim$ 220 $M_\odot$; M_c is quite uncertain due to the unknown amount of pulsationally driven mass loss. A signature for the formation of VMOs with $M_0 < M_{0c}$ in the past is the relative fraction of oxygen burning products to oxygen, which is fairly accurately predicted by our theory. We refer the reader to ABC1 for more details on this work.

The research was supported by grants NSF AST-79-23243 at Berkeley and NSF AST-80-22876 at Chicago. BJC gratefully acknowledges the support of the SRC. We thank Martin Rees for his hospitality at IOA.

REFERENCES

Appenzeller, I. 1970. Astr.Astrophys., 9, 216.
Arnett, W.D. 1972. Ap.J., 169, 681.
Arnett, W.D. 1973. In Explosive Nucleosynthesis, ed. W.D.
    Arnett and D.N. Schramm. (Austin: University of Texas Press)
Arnett, W.D., Bond, J.R. and Carr, B.J. 1981a. Preprint.
Arnett, W.D., Bond, J.R. and Carr, B.J. 1981b. Preprint.
Barkat, Z., Rakavy, G. and Sack, N. 1967. Phys.Rev.Lett., 18,
    379.
Bond, J.R. 1981. In preparation.
Carr, B.J., Arnett, W.D. and Bond, J.R. 1982. This volume.
Fowler, W.A. and Hoyle, F. 1964. Ap.J.Suppl., 9, 201.
Fraley, G.S. 1968. Ap.Space Sci., 2, 96.
Fricke, K.J. 1973. Ap.J., 183, 941.
Ober, W.W., El Eid, M.F. and Fricke, K.J. 1982. This volume.
Papaloizou, J.C.B. 1973. Mon.Not.R.astr.Soc., 162, 169.
Rakavy, G. and Shaviv, G. 1968. Ap.Space Sci., 1, 429.
Silk, J. 1977. Ap.J., 211, 638.
Stothers, R. and Simon, N.R. 1970. Ap.J., 160, 1019.

Talbot, R.J. 1971. Ap.J., <u>165</u>, 121.
Talbot, R.J. and Arnett, W.D. 1971. Nature <u>229</u>, 150.
Tohline, J.E. 1980. Ap.J., <u>239</u>, 417.
Weaver, T.A., Zimmerman, G.B. and Woosley, S.E. 1978. Ap.J.,
    <u>225</u>, 1021.
Wheeler, J.C. 1977. Ap.Space Sci., <u>50</u>, 125.
Woosley, S.E. and Weaver, T.A. 1982. This volume.
Zeldovich, Ya.B. and Novikov, I.D. 1971. <u>Relativistic
    Astrophysics</u>, (Chicago: University of Chicago Press).
Ziebarth, K. 1970. Ap.J., <u>162</u>, 947.

PREGALACTIC VERY MASSIVE OBJECTS AND THEIR COSMOLOGICAL CONSEQUENCES

B.J. Carr[1], W.D. Arnett[2,3] and J.R. Bond[4,5]

A 'Very Massive Object', henceforth referred to as a VMO, is defined to be a star that goes pair unstable during its oxygen burning phase. This means that it must have an initial mass in the range 10^2–10^5 $M_\odot$. The evolution of such a star, in both its hydrogen and helium burning phase (when it is subject to rapid mass loss) and its unstable oxygen burning phase has been summarized in an accompanying talk (Bond, Arnett and Carr 1982). The important qualitative conclusion is that oxygen cores with mass larger than about 10^2 $M_\odot$ collapse entirely to black holes, whereas oxygen cores smaller than this disrupt. The critical dividing mass M_c associated with the precursor hydrogen star would lie in the range 200–500 $M_\odot$, the large uncertainty reflecting our ignorance of how much mass loss occurs during the hydrogen and helium burning phase.

In this paper we will present arguments for believing that VMOs may have formed in the pregalactic era with important cosmological consequences. The arguments are described in more detail in Arnett, Bond and Carr (1981), henceforth referred to as ABC.

(1) The reason one might expect pregalactic stars to form is that the existence of galaxies implies that the early Universe must have contained density fluctuations. Providing the

1. Institute of Astronomy, Cambridge University.
2. Astronomy and Astrophysics Center, University of Chicago.
3. Max Planck Institute for Astrophysics, Munich.
4. Department of Astronomy, University of California, Berkeley.
5. Present address: Department of Physics, Stanford University.

M. J. Rees and R. J. Stoneham (eds.), Supernovae: A Survey of Current Research, 313–318.
Copyright © 1982 by D. Reidel Publishing Company.

fluctuations were isothermal, the form of the fluctuations required on a galactic scale and above, if extrapolated to smaller scales, would be of order unity on a scale 10^6-10^8 $M_\odot$ (Peebles 1974; Fall 1979). Thus bound regions could form well before galaxies and one would expect these regions to fragment into stars. Calculating the characteristic mass of the fragments is complicated but both the initial absence of metals and the influence of the background radiation would tend to make the stars considerably more massive than those which form in the present epoch (Silk 1980). In particular, calculations of Kashlinsky and Rees (1981) suggest masses of order 100 $M_\odot$. There are also observational reasons for believing that the mass spectrum falls off less steeply with decreasing metallicity and, for $Z < 10^{-3}Z_\odot$, it may be shallow enough for most of the mass to be in the largest stars (Melnick, Terlevich and Eggleton 1981). These considerations suggest that, if pregalactic stars do form, they could well consist predominantly of stars in the VMO range.

(2)   The black hole remnants of pregalactic stars would be natural candidates for explaining the missing mass in galactic halos and clusters (White and Rees 1978). However, the considerations above indicate that only VMOs larger than $M_c \sim 200$-500 $M_\odot$ could be expected to undergo complete gravitational collapse. Smaller stars might also leave black hole remnants but only a small fraction of their mass would be involved, so it would be difficult to explain how most of the Universe could get into these holes. The most important observational constraint on the sort of holes which could provide the missing mass in the galactic halo stems from considering the tidal disruption of loose star clusters by the holes as they pass through the disc. These considerations suggest that the holes which dominate the mass of the halo need to be smaller than 10^5 $M_\odot$ (Carr 1978). This means that the missing mass in the halo can comprise black holes only if the holes derive from VMOs rather than the sort of supermassive objects which go unstable to general relativistic effects.

(3)   An important consequence of pregalactic stars would be their nucleosynthetic effects. Calculations of Weaver and Woosley (1980) suggest that the mass fraction of a star in the mass range $15 < M/M_\odot < 100$ returned as metals should be $Z_{ej} = 0.5-(M/6.3M_\odot)^{-1}$, i.e. Z_{ej} should lie between 0.2 and 0.5. Since there exist population II stars with metallicity as low as 10^{-5}, we infer that either the fraction of the Universe going into pregalactic stars was tiny (less than $\sim 10^{-5}$) or that most of the stars were more massive than M_c so that they collapsed to black holes together with their nucleosynthetic products. This argument therefore supports the suggestion that the first pregalactic stars were VMOs. Of course, if all the stars were larger than M_c, one would not expect any pregalactic

enrichment, whereas from some points of view a burst of pregalactic enrichment would be desirable (e.g. to explain the G-dwarf problem, see Truran and Cameron 1971). Also any model which depends on grains to thermalize the 3K background, either in total (Rees 1978) or in part (Rowan-Robinson et al. 1979: Puget and Heyvaerts 1980), requires a pregalactic enrichment of at least 10^{-5}. However, it is not difficult to conceive of ways in which one can produce a small amount of enrichment even though most of the stars are in the collapsing mass range. For example, one might have a mass spectrum with $M_{min} < M_c < M_{max}$ which is weighted to the more massive end, so that only a small fraction of the Universe goes into stars smaller than M_c. Alternatively, one might envisage a small number of stars having sufficiently large rotation that the value of M_c is increased above M_{min} even though most stars have M_c below M_{min}. Both these situations are rather contrived. A somewhat more attractive scenario, in which an enrichment of 10^{-5} arises fairly naturally, is as follows: as the fraction of the Universe in stars increases, the amount of radiation generated by them will increase and eventually the Universe will be completely reionized. Simple energetic arguments (Hartquist and Cameron 1977) suggest that this happens when the fraction of the Universe in stars is of order 10^{-5}; more detailed arguments (ABC), accounting for the effects of the 3K background, give a somewhat larger fraction. Once the Universe is ionized, the formation of further bound regions and stars may be suppressed by the Compton drag of the 3K background until a redshift of order 140 (Hogan 1979). When the next generation of stars do form, they may be larger than the first generation because the Jeans mass has been boosted as a result of the ionization. This could explain why the second generation of stars were processed into black hole remnants without producing any further enrichment. Alternatively, if the 3K background is initially absent (so that Compton drag is inoperative), most of the Universe may go into black holes first (since no feedback process then operates), with the small fraction of metal producing stars forming later.

(4) One might also invoke pregalactic VMOs to explain various abundance features. For example, the well-known oxygen anomaly (Sneden et al. 1979), that metal poor stars have an (O/Fe) ratio which is about 3 times solar, suggests that O and Fe must have had different nucleosynthetic histories. Twarog (1980) has inferred that, whereas 90% of the Fe in the disc must have been produced during the disc's lifetime, only two-thirds of the O was so produced. Since VMOs explode during their oxygen burning phase, they would naturally produce a large amount of oxygen and the (O/Fe) ratio generated could plausibly be as large as 3 times solar. Another property of VMOs is that, unlike ordinary stars, they might produce nitrogen as a primary element due to convection effects. It is usually assumed that N is a

secondary element, so that $(N/H) \sim (Fe/H)^2$. However, recent observations of very metal poor stars (Edmunds and Pagel 1978: Barbuy 1981) suggest that $(N/H) \sim (Fe/H)$, in which case one must seek a way of generating nitrogen directly. Finally, of course, it should be pointed out that VMOs could be prolific generators of helium. A VMO might return 20-50% of its mass to the background medium as helium during its pre-oxygen-core phase (Talbot and Arnett 1971) and, in the mass range $M > M_c$, there would be no danger of overproducing heavy elements at the same time. Massive VMOs would therefore be natural candidates for generating the observed 20-25% helium abundance, even though this is usually assumed to be of cosmological origin. This is significant because, if one also wants the VMOs or their remnants to generate the 3K background, the early Universe would have been cold, in which case there may have been no cosmological helium production at all (Kaufman 1970).

(5)   Another important consequence of pregalactic stars is that they would have generated a lot of radiation during their nuclear burning phase. Background light limits therefore place interesting constraints on the pregalactic star mass spectrum and formation redshift. Roughly speaking, one percent of each star's rest mass energy may be converted into radiation, so if the stars burn at a redshift z_*, the present radiation density generated by them would be $\Omega_R \sim 10^{-2} \Omega_*(1 + z_*)^{-1}$, where Ω_R and Ω_* specify the radiation and stellar densities in units of the critical density. Since the radiation density over all wavebands (with the possible exception of the IR band) cannot exceed $\Omega_R \sim 10^{-4}$, we infer a limit $z_* > 10^2\Omega_*$. Thus, if the black hole remnants of these stars are to provide the missing mass in halos (which requires $\Omega_* > 0.1$), their precursors certainly need to form before $z = 10$. Furthermore, in order to burn their nuclear fuel before $z = 10$, they must have a mass of at least 10 $M_\odot$. This sort of limit has been discussed in more detail by ABC, Thorstensen and Partridge (1975) and Eichler and Solinger (1973).

(6)   More positive evidence for the effects of pregalactic starlight may come from certain distortions observed in the spectrum of the 3K background (Woody and Richards 1979): there appears to be an excess of energy, together with a distinctive dip, shortward of the peak. Rowan-Robinson et al. (1979) have suggested that these features can be explained if 25% of the 3K background density is radiation generated by pregalactic stars and thermalized by grains, the grains also being generated by the stars. This model requires that the 25% component of the 3K background be generated at a redshift of order 100 and that the stars have a density Ω_* of at least 0.1. The stars can burn their nuclear fuel by $z_* \sim 100$ only if they have a mass exceeding about 30 $M_\odot$. Puget and Heyvaerts (1980) have suggested another way in which pregalactic starlight could explain the Woody-

Richards distortion: their scheme requires that the stars generate about 30% of the 3K background and that they do so at a somewhat later epoch. Of course, if pregalactic stars generate such a large fraction of the 3K background, it is not implausible that they may have generated all of it. Pregalactic stars would be expected to span a range of masses and Carr (1981) has suggested that the initial component of the background may have been generated by the most massive ones. These would have completed their evolution earliest, before the Rowan-Robinson et al. or Puget and Heyvaerts stars, and perhaps early enough for the radiation to be thermalized by free-free processes rather than grains. In this scenario, energetic and thermalization criteria require that the first stars produce their light at $z \sim 10^3$. This corresponds to a time of order 10^6yr which, remarkably, is just the characteristic lifetime of stars more massive than $10^2 M_\odot$. Thus pregalactic stars could generate the entire 3K background providing they are in the VMO mass range. These same stars could give rise to black hole remnants which might today constitute the missing mass in halos and clusters.

(7)   It has been suggested that the formation of pregalactic objects could explain the origin of galaxies themselves. For example, Ostriker and Cowie (1981), Ikeuchi (1981) and Ostriker (1982) have suggested that the explosions of supermassive stars or clusters of stars could produce shock waves which would trigger the collapse of objects much larger than the stars or clusters themselves. They envisage 'seed' objects with mass of order $10^9 M_\odot$ generating galaxies at a redshift of order 5, but in principle the same process could occur at larger redshits and with a smaller seed mass. ABC show that bound objects as small as $10^6 M_\odot$ at decoupling could initiate a bootstrap process resulting in galaxy formation providing the bound objects contain stars which can produce explosive energy with sufficient efficiency. Again, this may require that the stars be VMOs. While one still needs initial density fluctuations to produce the pregalactic objects themselves, it is usually easier to produce small-scale fluctuations than large-scale ones. Indeed Carr and Silk (1981) have argued that even statistical effects associated with the Universe becoming 'grainy' at early times could generate the necessary pregalactic objects.

In conclusion, pregalactic VMOs may play a crucial role in explaining the origin of the 'missing mass', the element abundances, the 3K background, and galaxies, perhaps the four most important problems in cosmology. An understanding of the evolution of VMOs and of their final states, such as has been presented in accompanying papers at this conference (Bond et al. 1982; Ober et al. 1982; Weaver and Woosley 1982), is therefore vital. Even if pregalactic VMOs do not play the cosmological roles alluded to above, detailed analysis of the sorts of effects

discussed here allow one to place interesting constraints on the
mass spectrum of any pregalactic stars.

REFERENCES

Arnett, W.D., Bond, J.R. and Carr, B.J. 1981. Preprint.

Barbuy, B. 1981. Preprint.

Bond, J.R., Arnett, W.D. and Carr, B.J. 1982. This volume.

Carr, B.J. 1978. Comm.Astrophys., 7, 161.

Carr, B.J. 1981. Mon.Not.R.astr.Soc., 195, 669.

Carr, B.J. and Silk, J. 1981. Preprint.

Edmunds, M.G. and Pagel, B.E.J. 1978. Mon.Not.R.astr.Soc.,
     185, 77P.

Eichler, D. and Solinger, A. 1973. Ap.J., 203, 1.

Fall, S.M. 1979. Rev.Mod.Phys., 51, 21.

Hartquist, T.W. and Cameron, A.G.W. 1977. Astrophys.Sp.Sci.,
     48, 145.

Hogan, C. 1979. Mon.Not.R.astr.Soc., 188, 781.

Ikeuchi, S. 1981. Publ.Astr.Soc.Japan, 33, 211.

Kashlinsky, A. and Rees, M.J. 1981. Preprint.

Kaufman , M. 1970. Ap.J., 160, 459.

Melnick, J., Terlevich, R.J. and Eggleton, P.P. 1981. Preprint.

Ober, W.W., El Eid, M.F. and Fricke, K.J. 1982. This volume.

Ostriker, J.P. 1982. This volume.

Ostriker, J.P. and Cowie, L.L. 1981. Ap.J. (Letters), 243,
     L127.

Peebles, P.J.E. 1974. Ap.J. (Letters), 189, L51.

Puget, J.L. and Heyvaerts, J. 1980. Astr.Astrophys., 83,
     L10.

Rees, M.J. 1978. Nature, 275, 35.

Rowan-Robinson,M., Negroponte, J. and Silk, J. 1979. Nature,
     281, 635.

Silk, J. 1980. In Star Formation, 10th Saas Fee Lecture
     Course (Geneva, ed. L. Martinet and A. Maeder).

Sneden, C., Lambert, D.L. and Whitaker, R.W. 1979. Ap.J.,
     234, 964.

Talbot, R.J. and Arnett, W.D. 1971. Nature Phys.Sci., 229, 150.

Thorstensen, J.R. and Partridge, R.B. 1975. Ap.J., 200, 527.

Truran, J.W. and Cameron, A.G.W. 1971. Astrophys.Sp.Sci., 14,
     179.

Twarog, B.A. 1980. Thesis, Yale University.

Weaver, T.A. and Woosley, S.E. 1980. Ann.N.Y.Acad.Sci., 336,
     335.

White, S.D.M. and Rees, M.J. 1978. Mon.Not.R.astr.Soc., 183,
     341.

Woody, D. and Richards, P.L. 1979. Phys.Rev.Lett., 42, 925.

Woosley, S.E. and Weaver, T.A. 1982. This volume.

AUTOMATED SUPERNOVA SEARCH

Stirling A. Colgate

New Mexico Institute of Mining and Technology, Socorro,
New Mexico 87801
and
Los Alamos National Laboratory, Univ. of California,
Los Alamos, New Mexico 87545

ABSTRACT

An automated supernova search is under development at the New Mexico Institute of Mining and Technology to search several thousand galaxies per night to pick up SN at 15th and 16th mag. The digitally controlled telescope and imaging system are remote from the on-campus (mini) computer, a Prime 300. The most innovative aspect of the current project is the use of a multi-user virtual memory operating system modified to do multi-tasking (search tasks) in a real-time environment, while allowing multi-user program development and operation.

INTRODUCTION

A program for the development of an automated supernova search based upon complete remote computer control of a telescope and vidicon digital imaging system has been under development at the New Mexico Institute of Mining and Technology, Socorro, New Mexico, since 1968. This very long time has seen the multiple development of many systems as well as some failures. The primary problem has been the complexity of the software system (Moore et al. 1975). In the early days several different digital image tubes were developed, among which were Orthicons and Isocons, before the present system of an intensified Vidicon silicon target tube was put into place (Colgate et al. 1975a). The software problem is a more subtle problem than the single components of hardware (Colgate et al. 1975b). However, before going into software in depth let us consider the scientific objectives.

M. J. Rees and R. J. Stoneham (eds.), Supernovae: A Survey of Current Research, 319–323.

One would like to detect supernovae at an early stage of their light curve in sufficient numbers to establish some of the following criteria:

1.   One would like to understand if the early light curve is derived from a shock wave in an extended envelope or by an internal energy source, for example, the radioactive decay of Ni^{56}.

2.   One would like to understand if the luminosity at peak is consistent with an internal energy source that may be radio-activity or a hydrodynamic shock wave or early pulsar emission, and compare this to theories of the origin of supernovae involving thermonuclear energy release or dynamic collapse to a neutron star state.

3.   One would like to obtain sufficient statistics of the peak and width of all types of supernovae light curves to understand the consistency of the initial state and the consistency of the surrounding environment of the supernova. For instance, type I supernova occurring in E-galaxies with negligible gas and dust may exhibit sufficiently repetitive light curves to serve as a secondary standard in the measurement of the size and curvature (H_0, q_0) of the universe (Wagoner 1977: Colgate 1979).

4.   One would like to obtain sufficient statistics of supernovae concerning their location within galaxies as well as the galaxy type, and ultimately extend that study to large enough z to determine a possible evolutionary factor in the occurrence of supernova as a measure of stellar evolution in galaxies as a function of time.

These questions are best served by a system that images galaxies sequentially and attempts to find a difference in the galaxy image in the past and at present indicative of a point source of light increasing in luminosity with a time constant of days to weeks. This requires (1) storing a previous image and comparing it to a present (2) maintaining a catalogue of galaxies as well as part images (3) a catalogue system that keeps track of false alarms and contains sufficient logic to deal with these in a hierarchy of exclusion tests (4) a system which can continue unattended for long periods of time because of the intense boredom of monitoring a system that is performing a highly repetitive, although complex, function. It was therefore the objective from the beginning to develop a stand-alone system of a telescope in a dark, good seeing site controlled through a microwave link from a computer on a campus where students could take turns in the relatively routine function of computer operation.

This system almost reached operational status but was terminated in 1974 for several reasons.

1. No further external funding was available.

2. The computer, an IBM 360, Model 44 was completing its natural life after 10 years at the institution and the maintenance requirements were becoming so excessive that negligible night time or off hours use was feasible.

3. Probably most important, the software structure had become so complicated that debugging and modifying it became impractical because of the immense difficulty of tracing an inoperative interaction throughout the whole system, roughly 100,000 FORTRAN statements in size.

The question naturally arose whether the entire system should be dismantled. On the other hand, the hardware on the mountain, although constructed to a significant extent of surplus components, was still significantly beyond the state of the art at the time. The hardware included a telescope mount driven by digital stepping motors in a feedback mode that could slew the telescope a radian in several seconds of time with an accuracy of better than 10 s of arc. It used a special light-weight design with a self-supported central mirror and allowed the change of focal ratio from F-30 to F-12 to take place in less than 5 s. It contained a fully automated cross-dispersed eschell spectrograph and an intensified Vidicon digital image tube system that operated with attention for many months at a time. In addition, a microwave link had been developed with a drop-out rate of less than 1 in 10^8 bits which was adequate for the purpose in question. Finally, there was the relatively complicated decoding and query logic in place for all the many functions needed to operate such a system provided the appropriate 16-bit words were transmitted to the mountain. This seemed such a significant development in itself that the consideration of a computer with its software to direct this system seemed to be a feasible addition that could take place at a relatively modest level of investment over an extended period of time.

A possibility to do this arose with the funding to purchase a Prime 300 minicomputer for joint use with the atmospheric sciences with the specific capabilities of a multiuser and virtual memory operating system. The reason for this choice is the recognition that the software effort will be many times larger than the hardware effort. It therefore stands to reason that if you have one task 10 times greater than the others it is wise to have more than one person work on the problem simultaneously. This is why the operating system must be a

multiuser virtual memory time shared system (Colgate and Thompson
1980). Multiuser includes not only multiprogram access, but also
multitasking where a separate task (like a user) is assigned to
each of several telescope functions: e.g. telescope slewing, the
digital cameras (CCD or SIT) star field pattern recognition,
spectra, guiding, frame 'taking', reduction and calibration,
catalogue access and maintenance of stars, galaxies, quasars, and
nebulae. A multitasking operating system allows all of these
functions to be operating concurrently <u>provided</u> a real time
interrupt of the operating system can take place. Most multiuser
operating systems assign a sequential time slice to each user. A
telescope must be able to command attention with a priority above
other tasks.

   In addition there must be:

1. User observation program testing, development and scheduling.

2. User data formating and presentation.

3. User intervention.

4. Interlocks, telescope safety, and fault diagnosis.

   A multiuser environment allows any task to talk to another
task by passing parameters much like the function 'mail' where
any user passes messages to other users.

   Many may ask why not use a separate microcomputer for each
task? One then has the problem of networking all these computers
which can be done but is presently much more difficult, again
because the software effort is much greater than the hardware
effort. It is much more efficient to time-share one CPU that
'knows' where everything (everyone) is located and where the
operating system is a development used for thousands of
installations.

   We have been modifying a Prime 300 (a 16-bit minicomputer
with 196K core and 80 Mbyte disc) multiuser time sharing system
to handle priority interrupts so that we can perform multi-
tasking in a real time environment.

   Currently the Prime 300 operating system has been modified to
perform priority interrupt real time transfer of data to and from
the mountain via microwave link. All programs are imbedded in a
multi-user environment. In addition, message passing between
tasks has been implemented so that parameters are easily
available to all programs. Many other system changes have been
implemented to accomplish this and roughly 3/4 of the operating
routines have been rewritten and are operational.

A telephone modem interface with call-back using the state Watts telephone system allows users to log in with a remote terminal.

ACKNOWLEDGEMENTS

Bill Thompson of Lawrence Livermore National Laboratory and Bryan Edwards of Prolink, Boulder did the initial design of the Prime system. It has been developed and implemented Pearce and Kevin Meier at New Mexico Tech. It is supported by New Mexico Institute of Mining and Technology, Los Alamos National Laboratory and private support.

REFERENCES

Colgate, S.A. 1979. Astrophys.J., 232, 404.
Colgate, S.A. and Thompson III, W.C. 1980. Ground Based
    Automated Telescope. Paper presented at The Optical and
    Infrared Program, Tucson, Arizona, January 7-12, 1980 on
    'Telescopes for the 1980s'.
Colgate, S.A., Moore, E.P. and Colburn, J. 1975a. Applied
    Optics, 14, 1429.
Colgate, S.A., Moore, E.P. and Carlson, R. 1975b. Publ.Astron.
    Soc.Pacific, 87, 565.
Moore, E., Merillat, P., Colgate, S. and Carlson, R. 1975.
    Software for a Digitally Controlled Telescope. Proc. MIT.
    Conf. on Telescope Automation, pp.91-107, ed. M.K. Huguenin
    and T.B. McCord.
Wagoner, R.V. 1977. Astrophys.J., 214, L5.

THE BERKELEY AUTOMATED SUPERNOVA SEARCH

Jordin T. Kare, Carlton R. Pennypacker, Richard A. Muller,
Terry S. Mast, Frank S. Crawford, and M. Shane Burns

Lawrence Berkeley Laboratory
University of California, Berkeley 94720

The Berkeley automated supernova search employs a computer controlled 36-inch telescope and charge coupled device (CCD) detector to image 2500 galaxies per night. A dedicated minicomputer compares each galaxy image with stored reference data to identify supernovae in real time. The threshold for detection is m_v=18.8. We plan to monitor roughly 500 galaxies in Virgo and closer every night, and an additional 6000 galaxies out to 70 Mpc on a three night cycle. This should yield very early detection of several supernovae per year for detailed study, and reliable pre-maximum detection of roughly 100 supernovae per year for statistical studies. The search should be operational in mid-1982.

1. INTRODUCTION

Our observational knowledge of supernovae is still quite limited. Of the several hundred supernovae identified, less than 40% have even been classified as to type. Fewer still have good light curves or spectra, and virtually none have been observed well before maximum light. Yet the number of supernovae found per year is declining, as photographic surveys are terminated. The reason is that such surveys, while valuable, cannot gather the data needed to produce a qualitative increase in our knowledge of supernova explosions.

New searches are needed which can i) catch a few supernovae very early, for detailed study, ii) find a large number of nearby supernovae in a known sample of galaxies, and iii) find a very large number of distant supernovae for statistical and cosmological purposes.

M. J. Rees and R. J. Stoneham (eds.), Supernovae: A Survey of Current Research, 325–339.
Copyright © 1982 by D. Reidel Publishing Company.

The Berkeley automated supernova search is uniquely qualified to satisfy the first two of these goals, and is complementary to several other automated searches in achieving the third. Based on ideas proposed by Stirling Colgate, our system uses a computer controlled telescope and solid state imaging detector to collect images of up to 2500 galaxies per night. Images are processed in real time, to allow prompt detection and confirmation of supernovae and elimination of most false alarms. When fully operational, we should detect typically 100 supernovae per year well before maximum light, and several nearby supernovae per year at a fraction of a percent of maximum. Broadband and possibly standard filter (UBV) photometry will be performed directly, and cooperating observatories, both ground and space based, will be able to obtain spectral information at all frequencies from very early times.

2. A BRIEF REVIEW OF PAST AND PRESENT SEARCHES

The original search for extragalactic supernovae was implemented by Zwicky (1) on the 18-inch Palomar Schmidt in 1937, and extended using the 48-inch Schmidt in 1959. The search on the 48-inch has stopped, but photographic searches are continuing on the 18-inch and at other observatories (2,3). Schmidt plates of selected regions of the sky are exposed at regular intervals and compared with older plates using a blink comparator or film overlay technique. A large area of sky is covered by each plate, but the low quantum efficiency of photographic plates implies fairly long exposure times. The comparison process is tedious, and its detection efficiency is unknown — some supernovae are likely to be missed. The limited dynamic range of photographic plates leads to the loss of supernovae occurring against the bright cores of galaxies. However, the major limitation of photographic searches has been the long interval between successive plates, plus the long time required for analysis. In many cases, supernovae are not found until after they have faded into invisibility, precluding any further observations.

Hynek (4) employed an image tube detector and video comparison system to improve both sensitivity and speed of detection, and succeeded in finding several supernovae. However, the actual identification of supernovae was still done by eye. This search has been discontinued.

In recent years, several groups have turned to computers to analyse images. By digitally subtracting images, the sudden increase in brightness representing a supernova can be made conspicuous, even to a computer. Kibblewhite (5) has retained the photographic plate, but uses automated plate analysis, and has recently begun finding supernovae. This system should find large numbers of distant supernovae near maximum light, but currently has turnaround times (including plate shipment) of several days. Another system, under development by Angel and Sargent, employs a pair of charge coupled device (CCD) detectors on a fixed transit telescope to provide very deep scans of a narrow strip of sky every night (6). This system should be ideal for finding supernovae at high red shifts for cosmological purposes, but covers too small an area of sky to find nearby supernovae.

As early as 1971, Stirling Colgate began work on an automated telescope which could swing rapidly among a large number of foreground galaxies (7). An image tube supplied images to a remote computer for immediate processing. This system was quite ambitious, and proved to strain the limits of the available technology and manpower; work on it is still proceeding but it has not yet gone into operation.

Our own project is a direct descendant of the Colgate search, stimulated by the recent spectacular progress in solid state electronics. High performance CCD detectors, minicomputers cheap enough to dedicate to the search, and microcomputers for telescope control have all become available in the past few years. With the aid of limited funding from a large number of sources, we have assembled a group including one full time and several part time physicists, two graduate students, and a full time senior programmer. We have designed and built most of the special hardware required, and in collaboration with the Monterey Institute for Research in Astronomy (MIRA) have upgraded the MIRA 36 inch telescope drives for computer control. We are currently testing our hardware and writing software for the search, and we expect to be operational in mid 1982.

3. OVERVIEW OF THE BERKELEY SYSTEM

Figure 1 is a diagram of the major components of our system. Here we outline the properties of the various items.

A telescope in the 24 to 36 inch range is needed to provide reasonable sensitivity for the search. We have been fortunate in arranging to use the MIRA 36 inch telescope, an equatorially mounted f/10 Cassegrain. The telescope Guidance/Acquisition Package (GAP) contains a moveable diagonal which allows rapid switching among several detectors — in particular, between our CCD and a Reticon-based spectrophotometer for "instant" spectra of bright objects.

The telescope employs a roller drive in right ascension, and has a maximum slew rate of 11 degrees per second. In collaboration with MIRA, we are installing a digitally controlled servo system which will slew the telescope smoothly over 1-2 degrees in under 3 seconds. The main slew motors drive the telescope to within roughly 1 arc minute of the desired position; then pneumatic clutches engage the stepping motor drives to bring the telescope within a few arc seconds. The eventual aim is to have the telescope controller (a HEX microcomputer) calculate all refraction and flexure corrections and provide a "virtual telescope" capable of arc second pointing accuracy anywhere in the sky. Since our software calculates the telescope position on every field, errors do not accumulate, and a much more modest pointing accuracy of about 15 arc

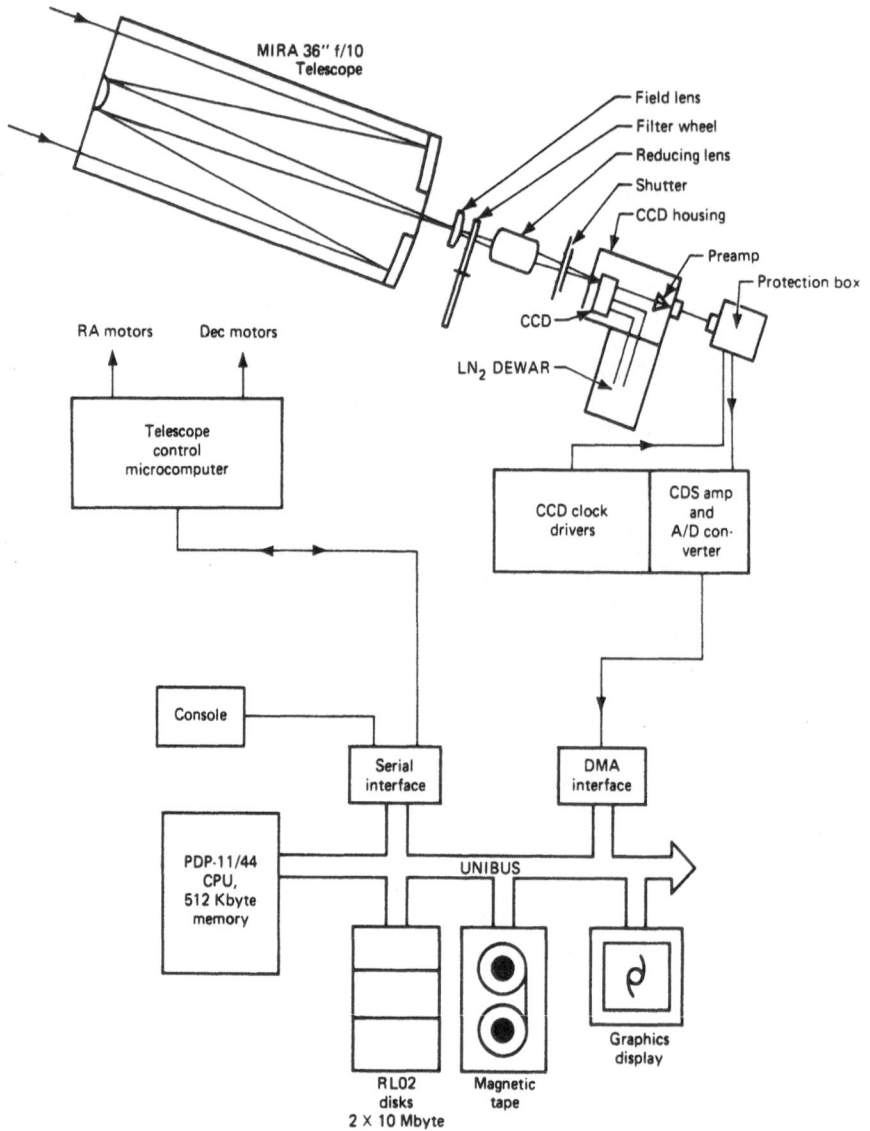

Figure 1 -- The Berkeley Automated Supernova Search System

seconds over a 2 degree move is sufficient. The telescope control system is diagrammed in figure 2. In consultation with Frank Melsheimer, the telescope designer, we have determined that telescope wear will not be a problem, being concentrated mainly in easily replaced components such as teflon o-rings.

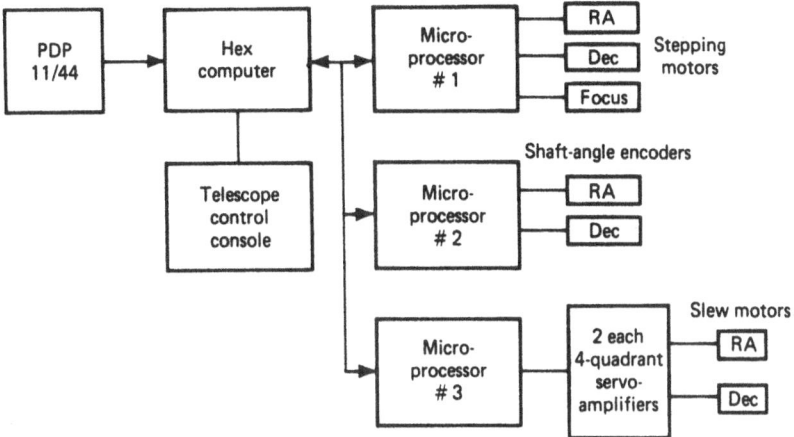

Figure 2 — MIRA Telescope Control System

The heart of the system is the CCD detector, RCA type number SID 53612. It is 1.5 cm by 1 cm, and is organized as an array of 512 by 320 photosensitive elements (pixels). Each pixel is 30 microns square. Photons striking the CCD generate photoelectrons which accumulate in each pixel until read out. Our CCD is a buried channel, thinned device which provides extremely high quantum efficiency and good blue response; the spectral response is shown in figure 3. At the end of each exposure, the collected electrons are shifted out one pixel at a time in a raster scan pattern, and detected by an on-chip FET amplifier. The CCD is cooled to -112°C by liquid nitrogen, with a feedback control system to maintain a

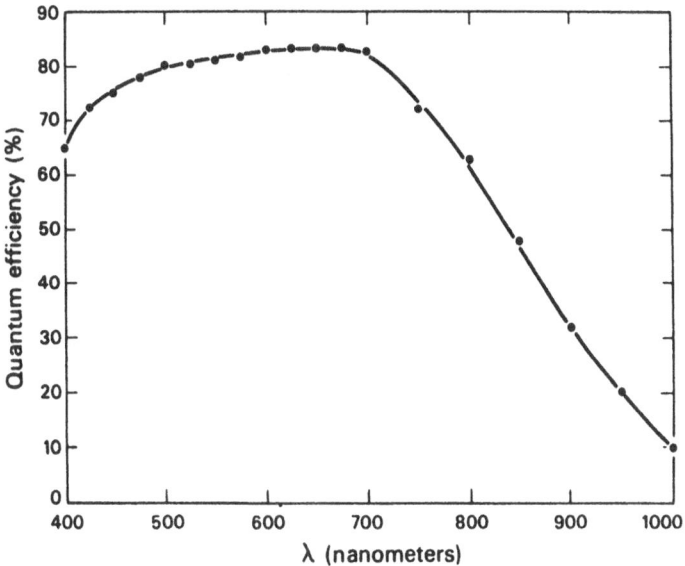

Figure 3 — Quantum Efficiency of RCA CCD

constant temperature. An evacuated housing provides insulation and protection from condensation. At such temperatures there is no detectable dark current, even over several hour exposures. Noise is introduced primarily by the output amplifier. The particular CCD we plan to use has a typical noise level equivalent to 40 electrons RMS per pixel, with substantial variation from chip to chip. Lower noise amplifiers are being developed; for now, we use the figure of 40 electrons. The pixel well depth, or charge capacity before overflow, is roughly 200,000 electrons, yielding a dynamic range approaching 5,000 to 1.

The telescope image is focused onto the CCD using a field lens and a Nikon reducing lens. The final effective f/number is f/2.25, which corresponds to a "CCD scale" of 3 arc seconds per pixel, or a total field of 16 by 25 arc minutes. This scale can be reduced as needed; see discussion below. A filter wheel can carry several filters for photometric purposes. Normally, a short-pass filter removes wavelengths longer than 750 nm to eliminate atmospheric OH lines which form interference fringes in the thin CCD chip. A long pass filter can also be installed to reduce the transmission of scattered moonlight on nights of full moon. A shutter, controlled by the computer, covers the CCD between exposures.

The output of the CCD is fed to a preamplifier inside the vacuum housing, and then to an electronics package (card cage) which houses a correlated double sampling amplifier. The correlated double sampling process subtracts noise associated with the CCD amplifier reset pulse from the signal. An analog to digital converter translates the amplified signal from each pixel to a 14 bit serial binary word. The card cage also contains the circuits which generate the precise clock pulses needed to drive the CCD. Signals from the card cage are routed through a "protection box" containing fuses and Zener diode clipping circuits, to reduce the risk of CCD damage due to circuit failure or transients. We have copied the circuitry used by John Geary of the Harvard-Smithsonian Observatory (8) to minimize the problems and risks of getting the CCD operating. Geary's CCD has been in use at the Mt. Hopkins observatory for over a year, so the design is well-tested.

The digital data from the card cage are fed into our computer via a custom-built buffer board and a standard Direct Memory Access (DMA) interface. We selected the PDP-11/44 computer from Digital Equipment Corporation as providing a large main memory space in a small physical space. The computer proper occupies a single foot-high chassis; the complete system of processor, disks, and power supplies fits easily into two 4 foot racks. We have 512 kilobytes of main memory, sufficient to store an entire image (160,000 pixels or 320 kilobytes) along with analysis software; this greatly reduces the time spent accessing disk memory. Two 10 megabyte disks provide mass storage for program development and for a limited number of reference images (see software discussion below); a magnetic tape drive for additional storage and data logging will be acquired soon. A large Grinnell graphics display with gray scale (512 x 512 pixels, 8 bits/pixel) allows monitoring of images for a variety of purposes.

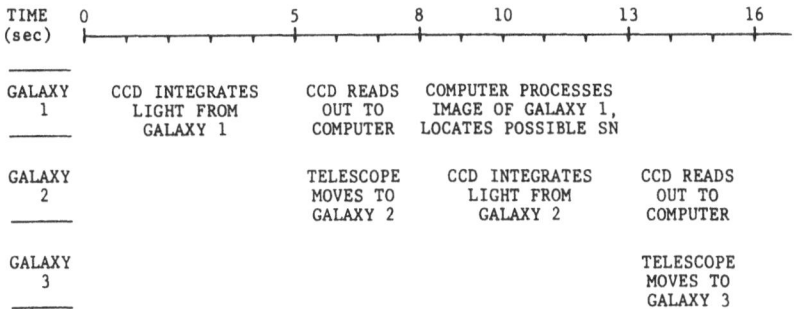

Figure 4 — Automated Search Sequencing

4. IMAGE ANALYSIS

Figure 4 shows the sequence of events during automated searching. The telescope is pointed at the first galaxy, and the CCD is exposed for a few seconds. As the telescope moves to the next galaxy, the CCD is read out and the image stored in the computer. During the new exposure, the stored image is processed to identify possible supernovae. If any are found, the computer can instruct the telescope to return to the galaxy in question and take a second look; otherwise the image is discarded and a new image read in, while the telescope moves to yet another galaxy.

The actual processing of the image is outlined in figure 5. An initial scan through the image locates bright pixels and collects them into a list of bright objects, mostly foreground stars brighter than 14th magnitude, of which there are typically 10 in each field. These stars are then matched against a reference list of stars to determine the telescope "offset" from the exact position desired. (This list, and all other reference data, are generated from CCD images acquired in the early stages of the search, rather than, for example, from catalogs). This offset is fed back to the telescope control software, so that pointing errors do not accumulate. However, the offset is used primarily to align the current image with a stored reference image for subtraction of constant features. To this end, simple fitting routines locate the position of each bright star to typically 0.2 pixels (0.6 arc sec). By averaging over 10 to 20 stars, the offset can be found to better than ±0.1 pixels.

Once the offset is known, the location of the galaxy of interest on the CCD is known also. Since the angular size of even nearby galaxies is typically less than the field of view of the CCD, substantial computing time is saved by processing only a rectangular section of the image containing the galaxy (with some margin to catch supernovae beyond the visual "edge" of the galaxy). Typically 10% of the full CCD field would be processed, although as the software is optimized such restrictions should become unnecessary. This processing involves:
   -subtraction of amplifier bias
   -subtraction of sky background, measured on a "dark" patch of sky in
    each field
   -removal of faulty pixels and overflow from very bright stars

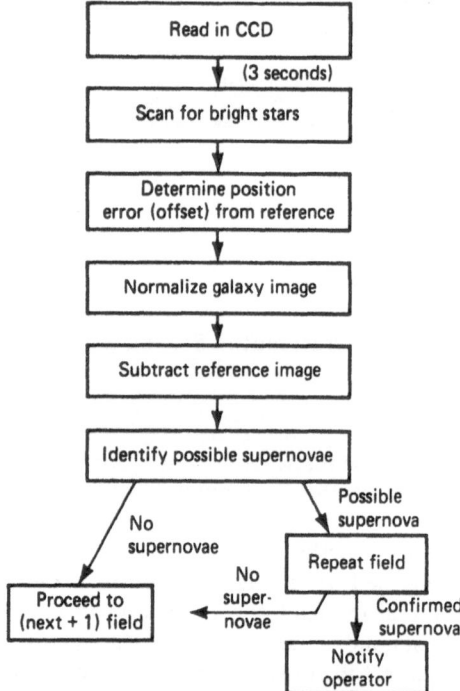

Figure 5 — Automatic Search Program

-scaling for pixel quantum efficiency variations (flat fielding)
-scaling for sky transmission variations, calculated from foreground
 star brightness
-alignment to the nearest whole pixel with the reference field.

Once this processing is complete, the system can search for supernovae in
the active field.   A variety of techniques can be used to store and
compare images. We originally intended to simply store an image of the
entire galaxy for reference, but this consumes a large amount of storage
and is highly sensitive to slight misalignments.   Since only the central
portions of galaxies, and foreground stars, are detectable above the sky
background with our system, we currently plan to store a list of
detectable point objects, of which there may be a few hundred, plus very
small maps of tens or hundreds of pixels around extended features such as
galaxy cores or jets. The software will thus process three types of pixels:
    -"ordinary" pixels, with a fixed threshold set roughly 5 standard
     deviations above mean sky background,
    -"stars", which will be either disregarded as possible supernovae
     (thus eliminating variable stars) or given individual thresholds.
    -"blobs", including galaxy cores, from which reference maps will be
     subtracted.
Based on simulations using 1-dimensional scans across galaxies obtained
from J. Geary, we estimate that even very simple interpolation between
pixels allows good subtraction with less than a factor of 2 increase in

threshold, except in the exact cores of galaxies, which must be handled separately. A number of more sophisticated techniques may be introduced once the search is operational. For example, the telescope optics can be easily shifted to provide a "high resolution" reference picture with a pixel size of 1 or 1.5 arc seconds. Such a picture could be processed using Fourier transform techniques to match both fractional offsets and seeing variations, and would allow very accurate subtraction.

5. SYSTEM PERFORMANCE

5.1. Sensitivity and parameter selection

The noise level of our detection system is set by two factors: the CCD readout noise, and the statistical $n^{1/2}$ fluctuations in the number of photons received. The CCD noise statistics are assumed to be Gaussian, with a standard deviation (RMS value) of n_{rn} electrons. For our current CCD, we take $n_{rn}=40$. Away from the cores of galaxies, most of the background photons come from diffuse sky light, which at the location of the MIRA telescope is equivalent to one 21st magnitude star per square arc second. By comparison with a known photometric system (8), we calculate that we collect approximately 30 photoelectrons (pe) per second from a star of $m_v=20$, so we expect 12 pe/sec/arc sec^2, or $N_{sky} = 108$ pe/sec/pixel. These noise sources are independent, so our total noise for a single pixel is

$$n_{tot} = (n_{rn}^2 + N_{sky} \, t)^{1/2} \qquad (1)$$

We choose an integration time t of 5 seconds. Shorter times clearly make inefficient use of telescope time, since the telescope requires 3 seconds to move between galaxies. Longer times allow fainter supernovae to be detected, but reduce the number of galaxies observable each night. The gain in sensitivity is decreased once the sky noise dominates the CCD noise. For our current system, $(n_{rn})^2$ is 1600, so the CCD noise dominates for exposures up to 15 seconds, but we hope eventually to obtain quieter CCD's. Meanwhile, 5 second exposures provide quite good sensitivity.

The signal from a supernova (or other object) of magnitude m (disregarding color variations) would be

$$N_{supernova} = 30 \text{ pe/sec} \times 10^{(20-m)/2.5} \times t \qquad (2)$$

If a fraction f of this light falls into a single pixel, the "signal to noise" ratio for a single pixel would be

$$S/N = N_{supernova} \times f \, / \, n_{tot} \qquad (3)$$

In order to make the fraction f large, it is desirable to make the "pixel scale" as large as possible, at least until the pixel size is equal to the expected seeing disk. Beyond that point, raising the pixel size increases the sky background noise while increasing the mean value of f (i.e. decreasing the chance that the supernova blur disk will fall on a pixel boundary) only slowly. However, as long as the CCD noise dominates the total noise, larger pixels are desirable. Large pixels also minimize computing time and increase our field of view. Our chosen value of 3 arc seconds/pixel is limited by vignetting in the MIRA telescope optics and by the f/number obtainable with standard optics. The expected seeing at the current telescope site is 2-3 arc seconds, and when the telescope moves to its permanent site on Chew's Ridge, the seeing should typically approach 1 arc second. The mean value of f is therefore close to 1. For the following, we take a conservative value of 0.5, so that half the light falls in one pixel.

Because of the large number of pixels involved, relatively rare statistical fluctuations can cause unacceptable numbers of false alarms. Setting a detection threshold of 5 standard deviations would, assuming ideal Poisson statistics, yield 1 false alarm per 1.7×10^6 pixels, or, at 10^4 pixels per galaxy, one per 170 galaxies. Since such false alarms are caught on the second look, this is a reasonable rate. In practice, we will vary the threshold to obtain a suitable false alarm rate. However, using S/N = 5, we can find the limiting magnitude for the search:

$$(20 - m_{limit})/2.5 = \log_{10} (S/N \times n_{tot})/(f \times t \times 30 \text{ pe/sec}) \qquad (4a)$$

$$m_{limit} = 18.8 \qquad (4b)$$

5.2 False alarms

In addition to false alarms caused by random noise, there will be a number of distinct events which can "look like" a supernova. Many of these can be eliminated by a prompt second look at the field, something possible only with a system of this type.

Cosmic rays will generate traces on our CCD an average of 1.5 to 3 times per minute, but only 1/10 of these will fall in the area of the galaxy. Thus the cosmic ray false alarm rate will be 1 per 25 to 50 fields. This figure may be reduced by adding simple software tests to rule out extended or improbably bright "supernovae."

Variable stars will be listed as stars if they spend much time above our threshold; we will probably monitor them as interesting objects in their own right. Cataclysmic variables which brighten drastically will cause false alarms on their first eruptions, but should be very rare. In the ecliptic, asteroids occur with a surface density of roughly 6 per square degree to 19th magnitude, enough to trigger a false alarm once in 40 fields. Unfortunately, Virgo is in the ecliptic; outside Virgo 10% to 20% of possible fields are affected. The proper velocity of asteroids is not enough to distinguish them promptly, but after 10 minutes they will move one or more pixels, and can be rejected by a third, delayed check. It may also be possible to identify them by their reflected solar spectrum.

At full moon, scattered moonlight raises the sky background to about 900 photoelectrons/pixel/second, increasing the system noise level by a factor of close to 2 for a 5 second exposure. The light level drops sharply away from full moon, however, so we lose only 2 to 3 days per month. A 500 nm long pass filter can be used to block as much as 90% of the scattered light with a loss of less than half of the supernova signal; this may allow low thresholds even during full moons.

5.3 Search strategies

With t = 5 seconds, and a telescope move time of 3 seconds, we can examine 450 fields per hour, or 2700 per six hour night. Allowing for 200 repeated fields (a 6% false alarm rate) we can check 2500 fields per night. There are three options available, corresponding to the three areas where data are needed:

-search 2500 closest galaxies every night to catch close supernovae very early,
-search 2500 x n nearby galaxies every n^{th} night (n < 5) to catch a moderate number of supernovae well before maximum, or
-search distance-limited sample of rich clusters, with typically 10 galaxies/CCD field, to obtain large numbers of distant supernovae near maximum light.

We plan initially to employ a mix of the first two options. We will examine galaxies in the Virgo cluster and closer (5-700 fields) every night, and an additional sample of 6000 galaxies at an average distance of 50 Mpc (H_o = 100 in all calculations) on a three night cycle. The third option depends critically on the system sensitivity, and requires additional software. In addition, other searches are competitive with ours in this area. We therefore ignore the "cluster strategy" for the moment.

We assume that the peak luminosity of a Type I supernova is -18.5, and that extinction due to dust in our galaxy and the supernova parent totals 0.2 magnitude. The observed magnitude at peak in Virgo (20 Mpc) is then m = 13.3. We can thus detect supernovae in Virgo 5.5 magnitudes below maximum, or at 0.6% of maximum light — lower if we choose to extend exposure times to 10 seconds. Type II supernovae are roughly 1.5 magnitudes fainter, and could be detected at 2-3% of maximum.

Since we check Virgo galaxies every night, an average of 12 hours elapses between the time a supernova crosses our threshold and the time we see it. The shape of the light curve is completely unknown at these early times, so the resulting increase in brightness at detection is uncertain; a factor of 2 in 24 hours is a reasonable guess, so the average brightness at detection may be a factor of 1.5 (0.4 magnitude) brighter than the minimum. Supernovae occurring in or near galaxy cores will not be detected until somewhat later, depending on the quality of our subtraction, but they will be detected well before maximum light.

The supernova rate in various types of galaxies is still uncertain. For convenience we assume 1 per galaxy per 50 years, a reasonable estimate. If we search 500 galaxies in Virgo, we would expect to find 10 supernovae per year. This number is sufficient to provide work for spectroscopists, satellite observers, and others interested in detailed observations, but too small to add appreciably to the statistical data base. We therefore also observe 6000 nearby galaxies selected from the Nilson and other catalogs. Thus we expect to see over 100 supernovae per year located in a known set of galaxies. This fixed sample should greatly reduce the statistical problems of brightness limited samples.

In these more distant galaxies, our detection thresholds are higher: at 50 Mpc we detect Type I supernovae only 3.5 magnitudes before maximum, and in the worst case a type II supernova at 70 Mpc would be detected only 0.5 magnitudes before maximum. The longer interval between scans of a given galaxy would also raise the average brightness at detection. Still, all supernovae not heavily obscured should be detected, and virtually all (especially type I's) detected several days before maximum.

Because the entire search, including the selection of galaxies, is governed by easily modified software, it will be straightforward to modify the search strategy as we learn more, or to meet specific requests. For example, if galaxies of a particular type or orientation prove interesting, a larger portion of the search can be devoted to them.

5.4 Additional observations

The primary function of our system in connection with the closest supernovae will be to serve as an alarm system, allowing other observatories to collect optical spectra and information outside the visible range at very early times. The very low level at which we detect supernovae may lead to entirely new observations; for example, detection of underluminous supernovae, up to 100 times fainter and 10 times rarer than type I's, which have been proposed to explain Cas A.

We will make nightly photometric measurements on all previously discovered supernovae, and on other objects of interest, such as variable stars. We can achieve reasonable accuracy through the use of foreground stars in the same field as standards. A limited number of fields could be examined repeatedly, using standard filters to obtain UBV data, or even using a grating to obtain rough spectra. Such photometry will be of particular interest in following the very early part of the light curve — for example, in catching the "glitch" in the Type I light curve which might occur when the shockwave reaches the stellar surface.

Finally, on more distant supernovae of primarily statistical importance, we expect to be able to provide light curves sufficient for classification by type, and approximate position within the parent galaxy. Position measurements accurate to about 1 arc second should be achievable by averaging over several observations. This should suffice for most applications, but a photographic backup will be desirable for maximum precision.

6. CURRENT STATUS

As of 1 August 1981, our hardware has been largely completed, and is being tested. The CCD driver electronics have been completely built and tested with an electronic CCD simulator. The computer has been operating for several months, and the CCD interface has been debugged. Images are being read from the CCD simulator and displayed on our graphics display.

The CCD detector head and dewar have been tested for vacuum tightness and thermal control, and we anticipate plugging in our current "loaner" CCD chip very soon. The telescope drive servo amplifiers have been designed and breadboarded, and final versions are being built. The microprocessor controllers are currently being programmed by MIRA.

The magnitude of the required software effort is indicated by figure 6; the real time analysis software already discussed must be surrounded with support programs. A substantial amount of software, between one and two man years' worth, has been written. However, we estimate that an even larger amount remains to be written. When hardware testing is complete, we will have three to four people programming essentially full time.

We have a commitment from MIRA for one night of telescope time per week for initial testing. Once our hardware is operating, we are guaranteed 3 nights per week for six months, sufficient to determine the efficiency of our search and find a significant number of supernovae.

Our timetable is as follows:

1 September 1981
    Hardware complete; CCD testing in lab
1 January 1982
    Telescope drives ready; CCD moves to telescope. Image data transferred to Berkeley via videotape for software development.
1 April 1982
    Computer moves to telescope. Minimal software complete. Automatic scanning tests begin.
4 July 1982
    SN 1982

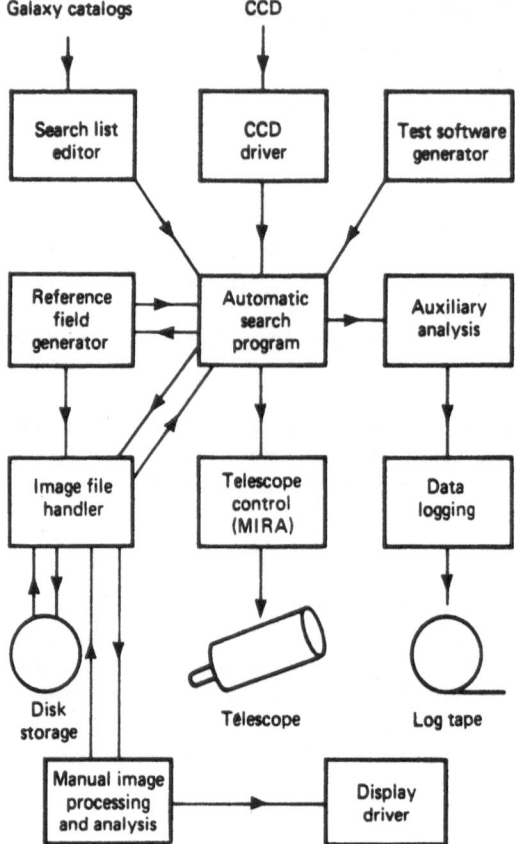

Figure 6 — Supernova Search System Software

Acknowledgements

We would like to thank Stirling Colgate for advice and strong encouragement. Roger Williams, our senior programmer, has been invaluable in developing our software, and Robert G. Smits and George Gibson of the LBL engineering staff are responsible for much of our progress in hardware. Undergraduates John Culver, David Smith, and Geordie Zapalac have contributed substantially to the project. John Geary at Harvard has been very helpful in getting our CCD operational. Luis Alvarez stimulated our original interest in this problem, and has been an enthusiastic supporter of our work. Finally, we would like to acknowledge the creative participation of Bruce Weaver and the other astronomers of MIRA in this project.

This work was supported in part by contracts and grants from the Department of Energy (contract W7405ENG48), the California Space Institute, the DuPont Corporation, the Research Corporation, the National Aeronautics and Space Administration, and the Alan T. Waterman Award of the National Science Foundation.

References

(1) W.L.W. Sargent, L. Searle, and C.T. Cowal, in Supernovae and
     Supernova Remnants, ed. C.B. Cosmovici, p. 33   (Reidel,
     Dordrecht 1974).

(2) L. Rosino, in Supernovae, ed. D. Schramm, p. 1 (Reidel,
     Dordrecht, 1977).

(3) J. Maza, in Proc. Texas Workshop Type I Supernovae,   ed.
     J.C. Wheeler, p. 7 (U. Texas, Austin 1980).

(4) J.R. Dunlap, J.A. Hynek, and W.T. Powers, in Advances in
     Electronics and Electron Physics, eds. J.B.McGee, D.McMullan,
     and E. Kahan, 33B, 789 (1972).

(5) M.G.M. Cawson and E.J. Kibblewhite. This volume.

(6) J.T. McGraw, J.R.P. Angel, and T.A. Sargent, in Applications
     of Digital Image Processing to Astronomy, ed. D.A. Elliott,
     p. 20 (SPIE, Washington, 1980).

(7) Information on this project is available from Stirling Colgate
     at the Los Alamos Scientific Laboratory. The automated tele-
     scope is described in: S.A. Colgate, E.P. Moore,   and
     R. Carlson, Publ. Astron. Pacific, 87, 565 (1975).

(8) H. Gursky, J. Geary, R. Schild, T. Stephenson, and T. Weeks,
     in Applications of Digital Image Processing to Astronomy, ed.
     D.A. Elliott, p. 14 (SPIE, Washington, 1980).

AUTOMATED SUPERNOVA SEARCH FROM PHOTOGRAPHIC PLATES

M.G.M. Cawson and E.J. Kibblewhite

Institute of Astronomy, University of Cambridge.

ABSTRACT

A new automated technique for detecting supernovae on pairs of IIIaJ Schmidt plates is described and the initial results presented. Analysis of 2144 galaxies on 4 plates produced 3 supernovae indicting a supernova rate per galaxy of one every 160 ± 50 years. Selection effects are discussed briefly and will be dealt with in more detail elsewhere. The technique will be used to determine an objective estimate for the supernova rate and to detect faint supernovae in real-time in order to measure the Hubble constant. A turn-around time of under 12 hours is possible after receipt of plates.

INTRODUCTION

Most supernovae have been discovered by blinking pairs of photographs taken with Schmidt telescopes. Such surveys require dedicated effort on the part of the observer and, even with a good alert observer, a variety of selection effects due to seeing differences or differences in the sky background or the characteristic curves of the emulsions cause the marked drop-off in supernovae detected in faint galaxies. The very marked fall in discoveries in spiral galaxies that are not face-on, reported by Tammann (1977), may also reflect the amount of time the observer spends looking at 'good' galaxies (i.e. face-on spirals). For these reasons, to obtain an objective estimate of the true rate of supernova occurrence in galaxies, new automated techniques are required.

Attempts to pick up supernovae using changes in the intensity of a galaxy clearly give rise to serious selection effects since

341

M. J. Rees and R. J. Stoneham (eds.), Supernovae: A Survey of Current Research, 341–353.

the supernovae have to increase the overall brightness of the
galaxy by 0.5 magnitude to be detected by this method. This
means that the brightest galaxy in which a Type I can be detected
in -19, and -17 for Type II, whereas Figure 1 shows that most
supernovae occur in galaxies that are between -20.5 and -22 in
absolute luminosity. The most promising method for automatically
detecting supernovae is therefore not to measure changes in the
integrated magnitude of the whole galaxy, but to subtract
two-dimensional pictures of galaxies taken at different epochs.
In this way we can detect changes in brightness over areas of a
seeing-disk rather than over the whole galaxy.

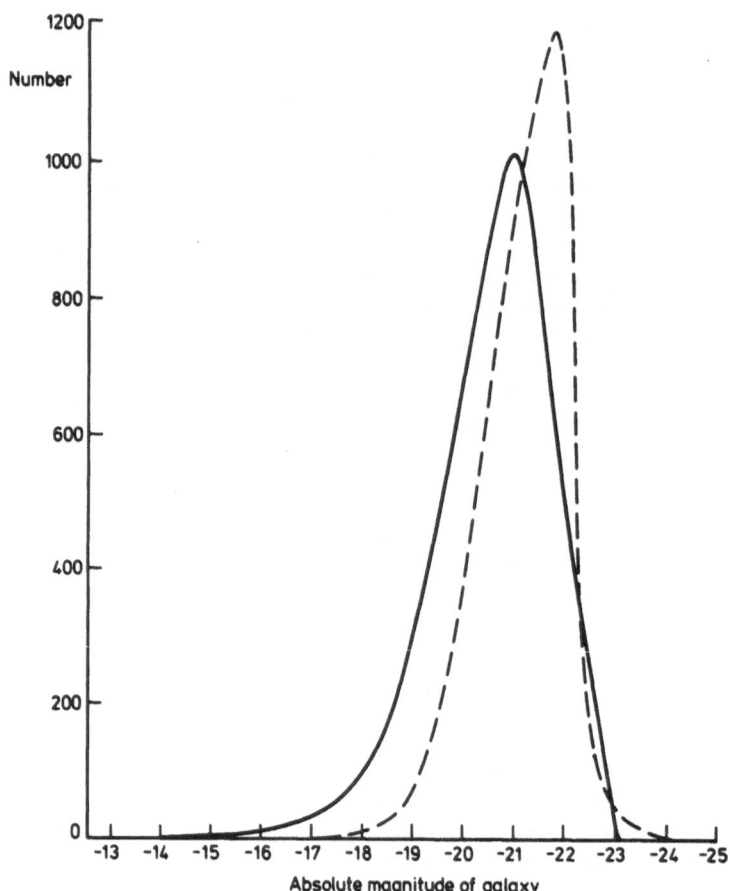

Figure 1. Solid curve shows the distribution of absolute
magnitudes of galaxies in an apparent-magnitude limited sample.
Broken curve shows the number of supernovae detectable per 100
years as a function of absolute magnitude of parent galaxy.

CHOICE OF DETECTOR

S.A. Colgate pointed out many years ago that if the search for supernovae were restricted to nearby objects, direct-imaging television techniques were superior to photographic plates. For these nearby searches a small area of sky can be sampled around known galaxies, and this is most efficiently achieved by steering a small sensitive detector to each of the galaxies in turn (Kare et al. 1982, Colgate 1982).

For fainter, more distant searches the large field of view of the photographic plate more than compensates for its low quantum efficiency as seen below.

The signal-to-noise ratio of a detector system is given by:

$$\frac{S}{N} = \frac{N.q.T}{\sqrt{n.p^2.q.T}}$$

where N is the number of photons from the supernova entering the detector system per second,

q is the quantum efficiency of the detector system,

T is the exposure time,

n is the number of photons per square arc-second entering the detector system per second,

and p^2 is the effective pixel area of the detector system. p is the convolution of the seeing-disk and the pixel size of the detector (i.e. the resolution for photographic plates).

Thus to achieve a required signal-to-noise ratio, the exposure time must be:

$$T = (\frac{p^2}{q})\ (\frac{n}{N^2})\ (\frac{S}{N})^2$$

For faint samples, the number of galaxies recorded is proportonal to the area of the detector:

$$N_{gal} = d.A, \quad (N_{gal} \gg 1)$$

where d is the number of galaxies per unit area of sky, and A is the area of sky covered by the detector.

The exposure time per galaxy in this case is:

$$t = (\frac{p^2}{qA})\ (\frac{n}{dN^2})\ (\frac{S}{N})^2$$

The first term in this expression depends entirely on the detector system and A/p^2 is the total number of independent pixels in the detector.

For a IIIaJ Schmidt plate having a resolution much finer than the seeing-disk, with a field of view of 36 square degrees, and a quantum efficiency of 1%, the detector term is:

1.7×10^{-7} for 1 arc-second seeing
and
6.7×10^{-7} for 2 arc-second seeing.

This can be compared with a large CCD array of over 160,000 pixels and a quantum efficiency of 70% where the detector term is:

3.5×10^{-5}.

This assumes that the scale of the image on the CCD array is adjusted so that the seeing-disk exactly matches the pixel size (which is difficult to achieve in practice), and that four pixels are required for detection since the light from the supernova may fall across a number of pixels. The extra bandwidth of a CCD is of limited use since the supernovae are always very blue and the sky and galaxy background are bright in the red. For deep samples therefore photography is between 50 and 200 times faster than using a single CCD array.

To determine the true supernova rate and for using supernovae to determine the Hubble constant (Wagoner 1979: Branch 1977), where distant supernovae are required, we have developed an automated technique for discovering supernovae on a set of blue Schmidt plates which are measured using the APM facility at Cambridge (Kibblewhite et al. 1975).

SELECTION EFFECTS

In addition to the problems of objectivity in any visual search for supernovae, it is extremely difficult to assess exactly how many galaxies are surveyed each time a pair of plates is compared by eye. However, such an estimate is essential in order to calculate the supernova rate per galaxy.

In an automated search the number of galaxies surveyed is well known, but if the galaxies are in a magnitude-limited sample it is possible that some of them are highly luminous and too distant for their supernovae to be detected. This is because for any given detection technique there will be a limiting apparent magnitude for the supernovae, m_{det}, fainter than which the technique will fail to detect them. The technique is thus able to detect supernovae to a given distance modulus:

$$D_I = m_{det} - M_I$$

$$\text{and } D_{II} = m_{det} - M_{II}$$

where M is the absolute magnitude of the supernova event.

Thus a deep magnitude-limited sample, to m_{lim}, will include some distant galaxies even though there is no possibility of detecting supernovae in them as they would be too faint for the detection technique being used. In this way the supernova rate per galaxy is always under-estimated.

The problem can be eliminated either by using a distance-limited sample of galaxies, which is difficult at these magnitudes, or else by using a magnitude-limited sample with a relatively bright cut-off so that even the most luminous galaxies are close enough to detect their supernovae. However, this latter approach eliminates many intrinsically fainter galaxies which are close enough to detect their supernovae, so fewer supernovae are found, and the errors become large due to 'small-number' statistics.

For a given sensitivity of the detection technique there will be an optimum magnitude limit for the sample of galaxies which can be calculated from a knowledge of the luminosity function of galaxies (see Cawson 1981 for a discussion of this).

From simulations using actual stellar images superposed onto galaxies at known positions, we have determined the limiting apparent magnitude for the detection of supernovae by our method to be about 21.5 ± 0.2 magnitude. If supernovae of that magnitude are to be detected in galaxies of absolute magnitude -22, then the magnitude limit of the galaxies should be about +16.5. However, in order to increase the number of galaxies sample and the number of supernovae detected we have used a magnitude limit of 18.5 and we calculate the expected incompleteness factor to be 1.5 (see Figure 2).

TECHNIQUE

The Automatic Plate Measurement facility at Cambridge is first used to define a magnitude-limited sample of galaxies on the Schmidt plate. The central region of the plate is scanned reducing each image detected to a set of parameters including its position, luminosity, eccentricity, radius of gyration and profile. These parameters are used to pick out the galaxies above some limiting magnitude using criteria such as surface brightness.

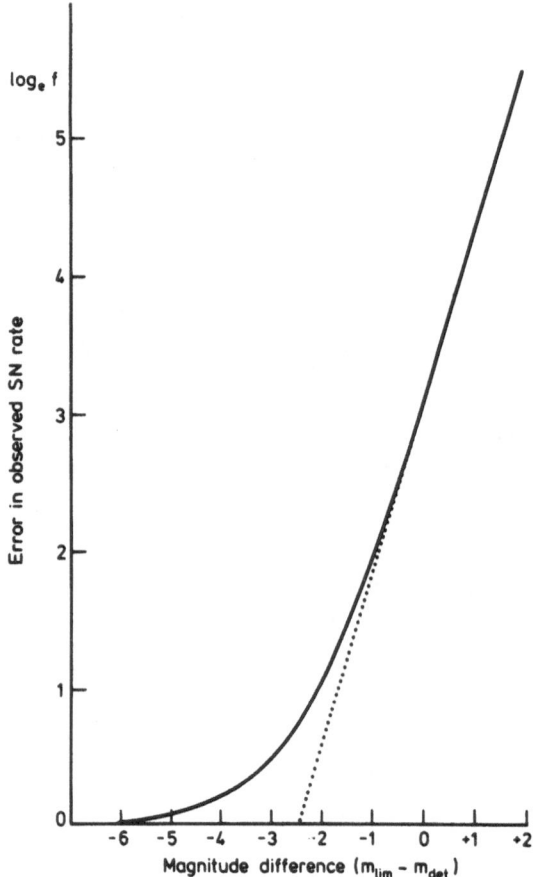

Figure 2.   Systematic errors.   The ratio of the observed
supernova rate to the 'true' rate, f, is a function of the
difference between the magnitude limit, m_{lim} of a sample of
galaxies and the limiting magnitude for detection of the
supernovae, m_{det}.

The galaxies are then sorted into regions of the plate (each
about 0.13 square degrees in area) and stored in luminosity order
within each region.   A small number of stars of each of several
different luminosities scattered over the whole plate are added
to the list in order to measure the seeing, to build up a set of
stellar images for simulations and for comparison with potential

supernovae, and to allow eventual photometry of candidate objects.

The machine then goes back to the positions of each object and produces a raster-scan of 2000 square arc-seconds around each one. This gives us 8100 pixels of density data per galaxy at four pixels per square arc-second, which is stored on disc. These original scans are used in effect as a template for comparison with similar scans of the same objects from a different plate taken a month or so later. The original scans are used to find the positions and sizes of foreground stars (see below), and are continually updated to the mean of all scans made to date so that the noise in them gets progressively less and less, thus making them suitable for doing deep photometry of a large sample of galaxies.

The immediate problem, and the most important one, is to ensure exact alignment of the two raster-scans. This we achieve using a bi-linear cross-correlation on the marginal sums of the scans. This seems to be an ideal compromise between speed and accuracy. The shifts calculated in this are constrained by the modal shifts for all galaxies in the same region of the plate.

The effect of misalignment depends on the slope of the density of the image being subtracted, the degree of misalign- ment, and the effective noise of the image. The degree of misalignment allowable is equal to the noise in the raster-data divided by the slope of the image being sampled. For extended low surface-brightness galaxies an arc-second or so is satisfactory, but for stars, the centres of most galaxies and for 'knotty' features, even alignment to the best pixel (i.e. 1/4 arc-second) can lead to significant difference images. Depending on the seeing we have between 4 and 16 pixels per seeing-disc. For larger pixel sizes more significant difference images are inevitable even with optimal alignment.

The second problem is calibration. Using our technique on raster-scans to detect differences, and particularly to detect supernovae, it is not necessary to calibrate all scans in terms of intensity before comparison so long as they are corrected for differences in the characteristic curves of the two plates.

To achieve this we have developed a statistical technique of Calibration Mapping whereby we build up a scattergram of pixel values on one plate against corresponding values on the other plate. A smooth curve is then fitted through the most dense region of the scattergram to produce a look-up table to map pixel values in the new scan into those of the original scan (see Figure 3). In effect second and subsequent scans do not have to be individually calibrated in terms of intensity, but merely in

terms of the original scans. It does ensure, however, that
plates can be co-added in the same procedure to build up a more
and more noise-free template for comparing new scans, and to get
better and better 2-D information on the light distribution in
the parent galaxies. A very deep photometric survey of a large
magnitude-limited sample of galaxies will be an important
by-product of the supernova search.

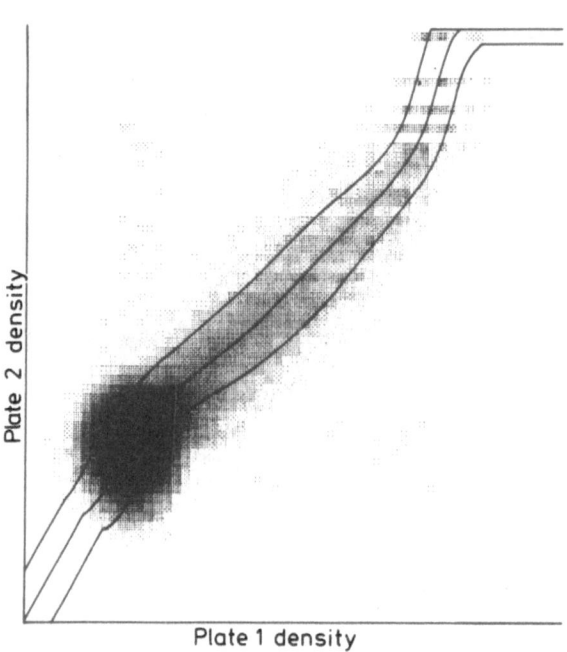

Figure 3.   Calibration Mapping.

    Measurement of the width of the densest part of the scatter-
gram gives the noise at different densities and enables
difference pictures to be exactly measured in terms of
significance. This automatic adjustment to the measured noise
allows detection of difference images right to the centres of
galaxies, and, indeed, the sensitivity of the technique is almost
flat over the entire range of density from the sky background to
the centres of the galaxies.

After alignment and Calibration Mapping, the two scans are subtracted and difference images are detected by looking for connected groups of pixels with a significant difference between the two scans. Parameters such as the total significance, size, and covariance matrix of these connected images are calculated and compared with the results of stimulated differences using the real stars scanned at the same time as the galaxies. A score is calculated depending on the similarity of the difference image to the simulations and if this is high enough a cross-correlation with one of the simulated difference images is calculated. Because foreground stars can cause very significant difference images even with optimal alignment, a check has to be made to eliminate difference images occurring over the surface of such stars. The positions of difference images are compared with the list of star positions and sizes made from the first set of raster-scans. Images passing all of these criteria are output as potential supernovae for visual assessment on the original photographic plates.

PLATE MATERIAL

At present we have 4 deep IIIaJ plates of the ESO–SRC sky-survey area number 406 (22h48,-35), the first taken at the UK Schmidt telescope in Australia on 19th August 1976 and the others taken 28, 334 and 388 days later. The plates were not taken for this project and in fact were considered to have too many defects for astronomical use. However they have been used successfully to check the technique and to determine the optimal exposure for plates and whether pairs of plates taken of the same field of the same night will be necessary to eliminate all spurious differences due to dirt and plate-flaws etc. Because the number of noise and spurious images decreases rapidly with size (as an error-function), we are currently using a larger size-filter than would be necessary if each plate had a pair to check for false-alarms.

On these plates 137 stars were scanned in 7 luminosity bins and 2144 galaxies in 121 regions of the plate. The scanning time is presently about 5 hours and analysis is currently done off-line on the Starlink network taking a further 1 hour of CPU-time per plate-pair. In the near future the analysis will be integrated into the APM machine and done as it moves from object to object cutting the turn-around time from 12 hours to about 6 hours.

For studies of the supernova rate, plates taken every two weeks are adequate since the supernovae will appear on at least two plates. For unambiguous detection in a single night in order to study the supernovae in real-time, two plates should probably be taken of each area on the same night.

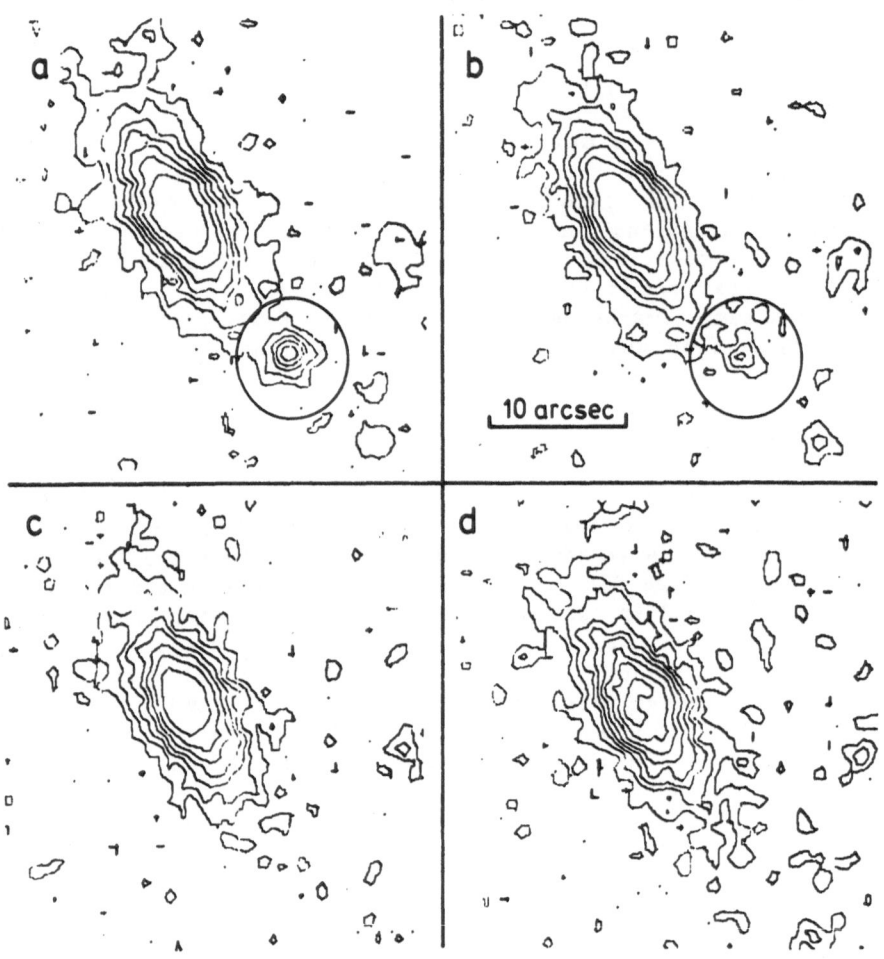

Figure 4.

RESULTS

At present we are ignoring all difference images smaller than
2.5 square arc—seconds (10 pixels) which produces about 2300
difference images per plate containing 2144 galaxies. Approximately half of these are negative images (i.e. brighter on the
first plate) and could no longer be studied in real—time. Of the

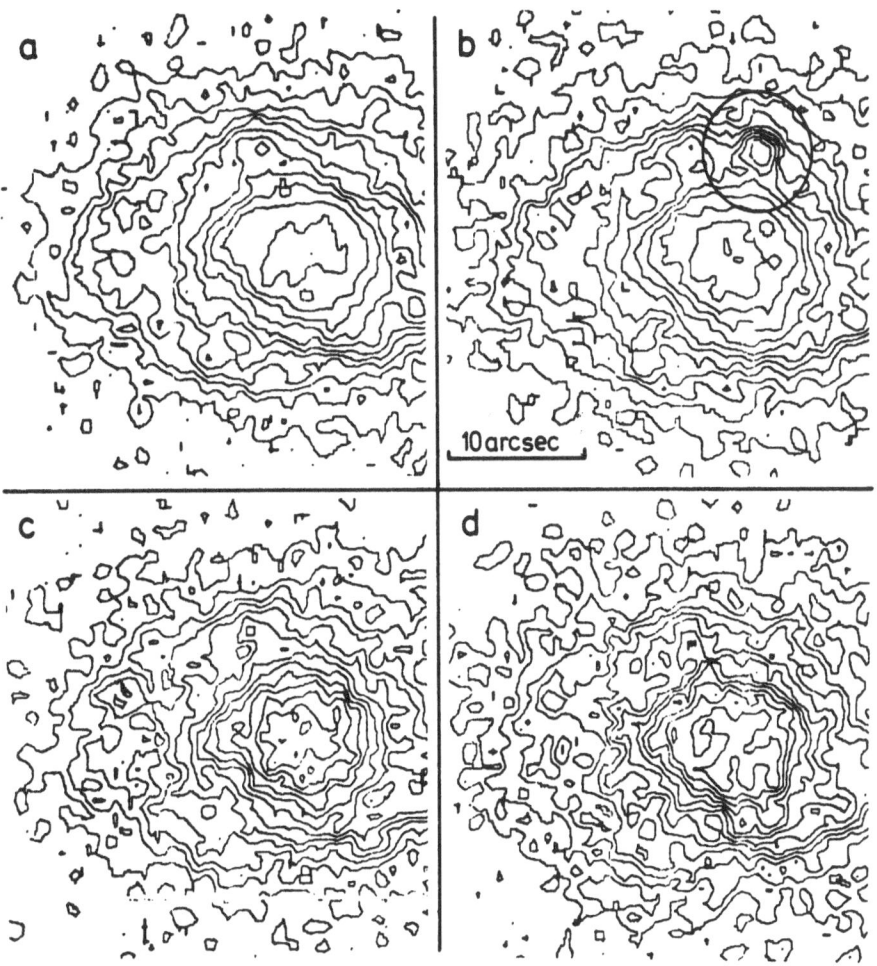

Figure 5.

remainder about 25% are associated with foreground stars and are eliminated automatically. 43% do not pass the parameter tests and a further 30% fail the cross-correlation tests. This leaves between 30 and 60 prospective supernovae per plate. At least 75% of these can be eliminated by visual inspection of the difference pictures on a colour monitor (dirt or slight misalignment) leaving a handful of possible astronomical objects which are looked at on the original plates using a microscope. The few

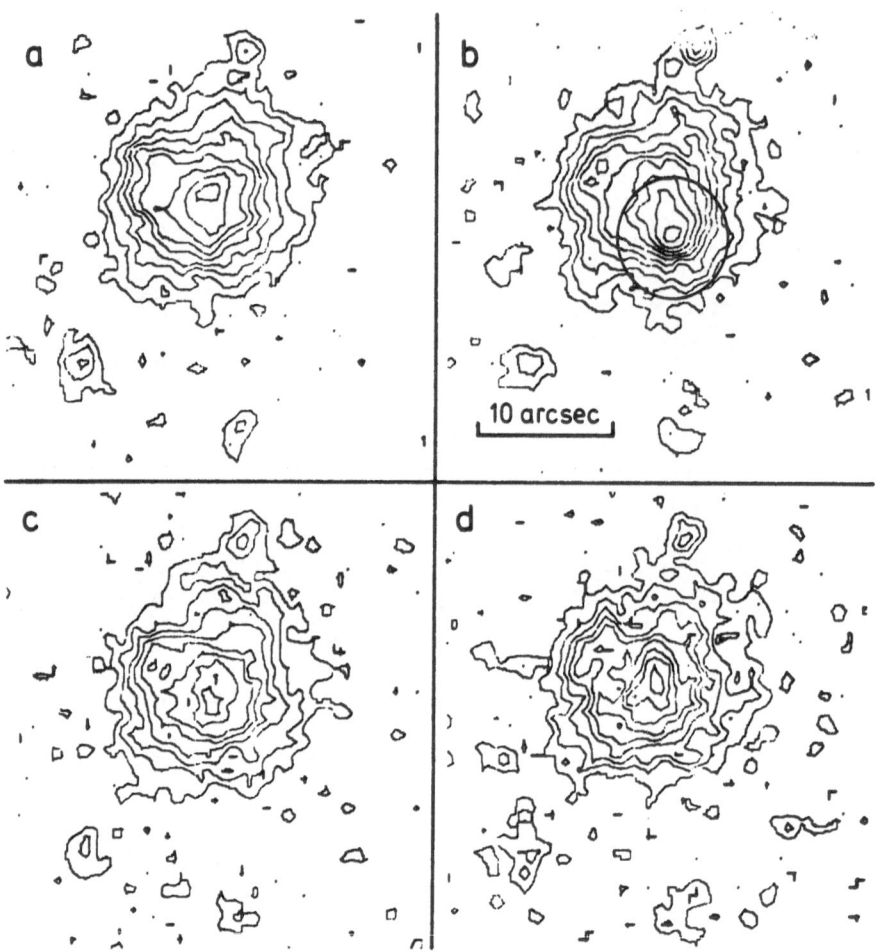

Figure 6.

objects remaining are then either supernovae or flare or variable stars, the possibility of the former increasing rapidly with proximity to the centre of the parent galaxy. Asteroids produce an extended image during the hour-long exposure and are therefore eliminated automatically.

After analysis of the 2144 galaxies on each of our 4 plates we have discovered 3 supernovae and a few other variable stars.

Assuming that the supernovae are detectable by this technique for about one month, this would suggest a supernova rate of one per galaxy every 160 ± 50 years. We expect to be detecting an average of 1.5 supernovae per plate in the near future.

One of the supernovae discovered appeared on the earliest epoch plate and it was still just detectable on the second plate taken 28 days later (see Figure 4). This supernova was well towards the outskirts of the parent galaxy but another has been detected where the light from the parent galaxy contributes about half of the total light per pixel (see Figure 5), and another where the supernova is within an arc-second or two of the centre of the galaxy (see Figure 6), indicating that the method is applicable even in the centres of the galaxies where detection by eye is most difficult.

CONCLUSIONS

Photographic searches coupled with automated measuring techniques are the best way of finding distant supernovae. A 48 Schmidt telescope can survey about 300 square degrees of sky per night and pick up supernovae out to $z = 0.15$. A 150 telescope could detect supernovae out to $z = 0.4$. Observations of these supernovae should give us estimates of H_0 from groundbased telescopes and q_0 from the space telescope. The most serious problem is the logistics. For measurements of supernova rates and light curves, immediate detection is not essential. Nevertheless, spectroscopic observations of supernovae in their early stages of expansion are most important. At present the plates have to be flown to Cambridge for analysis and this implies a delay of a few days between taking the plate and finding the supernovae. If the facility were moved to or duplicated at the telescope, supernovae could be observed the following night.

REFERENCES

Branch, D. 1977. Supernovae, ed. D.N. Schramm (Reidel: Dordrecht), pp. 21-28.
Cawson, M.G.M. 1981. Pub.Astr.Soc. Pac., in press.
Colgate, S.A. 1982. This volume.
Kare, J.T., Pennypacker, C.R., Muller, R.A., Mast, T.S., Crawford, F.S. and Burns, M.S. 1982. This volume.
Kibblewhite, E.J., Bridgeland, M.T., Hooley, T.A. and Horne, D. 1975. Image Processing Techniques in Astronomy. eds. C. de Jager/H. Nieuwenhuijzen. (Reidel: Dordrecht), pp.245-246.
Tammann, G.A. 1977. Supernovae, ed. D.N. Schramm. (Reidel: Dordrecht), pp. 95-116.

THE HISTORICAL SUPERNOVAE

David H. Clark[1] and F. Richard Stephenson[2]

1. SOURCES OF HISTORICAL ASTRONOMICAL RECORDS

Only a few civilisations throughout the world have contributed significantly to the <u>written</u> records of astronomical phenomena from the historical past. Pre-eminent among these must be Europe, China (together with Korea and Japan), Babylon and the Arab World. A wealth of pre-telescopic data from these sources is readily accessible, and it is in this that searches for historical supernova sightings have been concentrated.

1.1 <u>Ancient Europe (before AD 500)</u>

Astronomical observations from Europe in this period are essentially confined to two principal sources: (i) the Greek and Roman Classics and (ii) Ptolemy's <u>Almagest</u>. In both sources only a small fraction of the original observations has survived.

There are no certain records of 'new stars' (novae or supernovae) in the Classics. Pinder makes tantalising reference to a bright star in his First Olympian Ode –

>   'Look no more for another bright star
>   by day in the empty sky
>   more warming than the Sun...'

1.   Space and Astrophysics Division, Rutherford and Appleton
      Laboratories, Chilton, Didcot, Oxfordshire
2.   Department of Geophysics, The University, Liverpool

M. J. Rees and R. J. Stoneham (eds.), Supernovae: A Survey of Current Research, 355–370.
Copyright © 1982 by D. Reidel Publishing Company.

Even if the passage was inspired by a new star (or comet?) bright enough to be seen in daylight, poetic licence has undoubtedly destroyed any astronomical value. The first known star catalogue, compiled by Hipparchus in about 130 BC and giving the positions of more than a thousand stars, was supposedly inspired by the appearance of a new star. Pliny, however, though writing two centuries after the event, alludes to the motion of the 'star', so there is a distinct likelihood that it may have been a comet. The new star that Hadrian claimed to have seen (AD 130) after the death of his favourite Antinous may have been only a figment of the emperor's wishful imagination. The Roman poet Claudian supposedly witnessed a temporary star that was plainly visible in the daytime and which Claudian regarded as a portent of Honorius being made a co-emperor in AD 393. In this same year Chinese astronomers also recorded a spectacular new star, but unlike Claudian they left an account of its position and duration.

Two short passages in ancient Roman literature refer to celestial portents connected with the reign of the emperor Commodus (AD 180-192). While the Chinese did record a spectacular new star in the year AD 185, this was at a position in the sky where it would not have been visible from Italy or Gaul, but might have been seen from Antioch, Carthage or Alexandria. The records merely state:'There were certain portents which coincided with these events; some stars shone continuously by day, others became elongated and seemed to hang in the middle of the sky'; and 'A comet appeared. Footprints of the gods were seen in the Forum departing from it. Before the war of the deserters the heavens were ablaze'. It is far from obvious that either record was inspired by the AD 185 new star, and even if this was the case neither contains anything of astronomical value. Indeed, it remains doubtful whether there are any astronomically useful references at all to new stars in the Classics.

1.2 Mediaeval Europe (AD 500-1450)

Perhaps rather surprisingly, although there was little real interest in astronomy during the Middle Ages in Europe, numerous observations of certain kinds of phenomenon are recorded. This was very much the era of the chronicler. The early period is rather sparsely covered. However, after about AD 1000 numerous monasteries, as well as some towns, began to keep chronicles. These annals are by far the most prolific source of European astronomical records before the Renaissance. Most of the records come from England, France, Germany, and Italy, where the concentration of monasteries was highest. We are indeed fortunate that extensive compilations of mediaeval European chronicles exist.

Sightings of new stars, and sunspots, are rare in the chronicles. It would appear that scant notice was taken of these phenomena because their existence conflicted with contemporary cosmology, which, following Aristotle, supposed that outside of the immediate vicinity of the Earth the heavens were perfect and changeless. Comets escaped this censorship because they were believed to be atmospheric in origin. Nevertheless, records of a brilliant supernova in AD 1006 exist in the chronicles.

1.3 Renaissance Europe (AD 1450-1609)

By this time, some of the best scientific minds in Europe were breaking free of the Aristotelian concept of a changeless celestial vault, and in many respects the careful observations of the supernovae which appeared in AD 1572 and 1604 represent some of the greatest triumphs of the age.

1.4 Babylon

Babylonian astronomical records come from two distinct periods – a short interval covering the reign of a single king during the first half of the second millenium BC, and an incomplete span of about 600 years between 650 and 50 BC.

The first group is essentially only a single document and exists only in late copies. The second group is of considerable interest since the observations are so varied. Most of the observational texts in this later group are in the form of astronomical diaries, mainly referring to planetary motions. At this time Babylonian astrology, like its present-day counterpart, was dominated by prognostications on the basis of the positions and motions of the planets. The tables also record lunar and solar eclipses, occultations and a very few comets. So far, however, none of the texts studied makes reference to new stars. A vague allusion to a great star in Vela on an early tablet, but supposedly originating from observations several millennia earlier, has been obliquely associated with the Vela supernova remnant, but this is more by wishful thinking than by detailed scholarship. The lack of any mention of new stars is possibly a result of the fragmentary condition of the tablets. The tablets represent a far from complete astronomical record for the period of interest, and in any case bright novae as well as supernovae are rather rare events. It is to be hoped that future work on the tablets might reveal new star sightings.

1.5 The Arab World

The peak period for astronomical observations in the Arab World seems to have been from about shortly after the time of

introduction of Islam until about AD 1100. Observations were, in
the main, modelled on the observations made by the Greeks and
recorded by Ptolemy, but in general they were much more accurate.
The Arabs were restricted in their outlook by the influence of
Aristotle, but it was by no means a stranglehold as in Europe. A
number of accounts exist for the AD 1006 new star, but, perhaps
rather surprisingly, there is only a single record, hidden away
in the journal of a physician, for another event just 48 years
later, the new star of AD 1054.

1.6 The Far East

The situation in the ancient orient was very different from
that in the ancient occident. From very early times each Chinese
ruler was believed to be the 'Son of Heaven', appointed on a
mandate from 'Heaven' and able to rule only so long as he
fulfilled Heaven's wishes. To the ancient Chinese, Heaven was
more than merely the void of space, and was believed to have a
controlling influence on the destinies of men. Any departure by
the ruler from the 'path of virtue' would be expected to be
signalled by signs in the Heavens – such as the appearance of
comets, meteors, new stars and so on. A rigid system of
'political astrology' was thus established in China, and was
later copied after the tenth century in Korea. Professional
astronomers/astrologers were appointed to maintain a constant
watch of the sky, and to report and interpret any unusual events
which might happen. Many such events were recorded in the
official histories of the various dynasties. As a consequence,
remarkably detailed astronomical records exist from China going
back to about 200 BC.

Very few records remain from earlier times because of the
widespread 'burning of the books' in 213 BC. This was instigated
at the command of Ch'in Shih-huang, who had unified China in 221
BC and become its first emperor. The purpose of this literary
holocaust was to eradicate all memory of the former warring
states which had vied with the now-ruling state of Ch'in. To
ensure that the histories were not rewritten, the Emperor also
ordered that 400 of the country's top academics be buried alive.
It is hardly surprising that such an oppressive regime was
short-lived, lasting just seven more years, but even its
overthrow, with the systematic burning and sacking of the Ch'in
capital, further destroyed ancient records of possible
astronomical worth. Despite these tragedies, some astronomical
records from the earlier Shang (about 1500 to 1100 BC) and Chou
(about 1100 to 481 BC) dynasties have survived.

The earliest written records are divination texts from the
Shang dynasty. The people of Shang practised divination on a
large scale, using inscriptions on tortoise shells and the bones

of various animals. A question was written on the bones, and then heat applied to a chiselled-out cavity. The form of the pattern of cracks resulting was then interpreted as the answer. After use, the bones were buried to prevent defilement. Towards the end of the last century, vast quantities of these oracle bones bearing a very primitive form of ideographic writing were discovered near An-yang at the site of one of the main Shang capitals. They were originally sold to apothecaries as 'dragon bones' to be ground up as remedies for various ailments. Fortunately their historical significance was soon realised. It has been suggested that the inscription on one of the oracle bones may refer to a new star that appeared near Antares; if so, this would be the earliest oriental record of a new star. Some scholars, however, suggest that the text may be nothing more than a question as to whether a sacrifice should be made to the 'Great Star' (Venus or Sirius?) and the 'Fire Star' (Antares). The precise interpretation depends on a single character which could mean either 'new' or 'sacrifice'. Much later the Spring and Autumn Annals, possible edited by Confucius (551-479 BC), contain some astronomical records, including more than 30 observations of solar eclipses.

From the Han dynasty (202BC-AD220) onwards, we find astronomical records of all kinds: solar and lunar eclipses, comets, meteors, planetary conjunctions, occultations of stars and planets by the Moon, sun-spots, aurorae, sightings of Venus in daylight, certain atmospheric phenomena – and, of course, new star records in abundance. Later, such records are often duplicated from Korea and Japan. All in all, there are far more records relating to early astronomy extant from the Far Eastern civilization than from any others.

2. THE ORIENTAL OBSERVATIONS

2.1 Background

The great richness of astronomical data from the ancient orient results in part from the official character of Chinese astronomy. By the time of the Han Dynasty, an astronomical office had been established as a special subdepartment within the Ministry of State Sacrifices. Throughout subsequent Chinese history, as well as later in Korea and Japan, and even until modern times, the Astronomical Bureau existed as an important government office.

The Astronomical Bureau had two main functions. The first was to maintain an accurate calendar, an important consideration for a people so dependent on a stable agricultural system. The second function of the Bureau was to observe and interpret celestial portents – hence the large number of records of phenomena, such as new stars, of no calendrical importance. As already mentioned, the belief in astral influence on state events seems to have germinated very early in Chinese history, so that by the Han dynasty an elaborate system of political astrology had developed. An imperial observatory was built at each of the various capitals of China. When the capital was moved, as after the fall of a dynasty, a new observatory was built. For later dynasties, suspicion that subordinates could manipulate astrological data to their own purpose led the emperors to establish independent observatories within the confines of their palaces.

Another important department of the Chinese civil service was the Bureau of Historiography. It became standard practice from Han times for the Bureau of Historiography to compile an official dynastic history for the preceding dynasty. The compilers would have been given free access to all official records, including those of the Astronomical Bureau. So strong was the ancient Chinese belief in the importance of history that the dynastic histories are believed to represent reliable accounts of the main events from the deposed dynasties, although of course the written accounts merely reflect what the official historians saw fit to preserve. In particular, the astronomical records, which were usually contained in separate astronomical treatises, are just summaries of what presumably were very detailed first-hand accounts. Despite these reservations, the dynastic histories represent an almost continuous record covering nearly 2000 years, without equal from any other civilization.

2.2 Reliability and Continuity

Since astronomy was under political control and had such a strong astrological orientation, one is bound to ask whether the records were ever deliberately fabricated. The simplest check for reliability during a particular period is to check the accuracy of planetary observations - the precision and reliability of such obervations can be rarely faulted, and we have found very few examples of fabrication or falsification.

The continuity of the observations is of greater concern. the essential continuity of the record can be checked by records and predictions of eclipses required for calendrical purposes. However the influence of political and social pressures clearly influenced the recording of phenomena of astrological interest, and this must include new stars. An abundance of records of astrological significance with their accompanying admonitions and prognostications are often found to precede the overthrow of a dynasty. At the start of the dynasty, when 'warnings from heaven' seemed not to be required, few were recorded. With the knowledge of political and social influences on astronomical records in mind, one must consider with care how complete the historical new star data record may be - additional concerns are deliberate suppression of evidence, or loss of historical data. One is left with the hope that nothing as spectacular as a bright new star would have escaped the attention of either the astronomers or historians of ancient China.

3. THE SEARCH OF THE HISTORICAL RECORDS

3.1 Selection Criteria

The records show that three kinds of new stars were recognised in the Far East. The first were known as k'o-hsing - the 'guest stars' or 'visiting stars'. The well-known supernovae of AD 1054, 1572 and 1604 were all described as k'o-hsing, so that we might expect the term to be synonymous with novae and supernovae. Occasionally, however, the term is used where there is also allusion to motion, indicating that the object in question was in fact a comet. While therefore one expects to find novae and supernovae described as k'o-hsing, there is need for caution. The standard term for a comet with discernible tail (the second category of new star) was hui-hsing - the 'broom stars' or 'sweeping stars'. In the official history of the Chin Dynasty (AD 265-420) the following description of a hui-hsing is given: 'Its body is a sort of comet, while the tail resembles a broom. Small comets measure several inches in length, but the larger ones extend across the entire heavens.' The final kind of new star categorized in the oriental annals were the po-hsing - the 'rayed stars'or 'bushy stars'. The term seems to have been used to describe an apparently tail-less comet. Again in the history of the Chin Dynasty, we read: 'By definition a comet pointing towards one particular direction is a hui comet, and one that sends its rays evenly in all directions is a po comet.' Other miscellaneous terms were sometimes used - the AD 1006 supernova was called a chou-po star (an 'Earl of Chou' star), a special classification apparently to emphasize its exceptional brilliance and auspicious nature. The term ch'ang-hsing ('long star') was sometimes used as an alternative to hui-hsing.

By careful searching in oriental dynastic histories, encyclopedias, diaries and astronomical works, investigators have produced a large number of new star records. Almost all records of hui-hsing and po-hsing make definite mention of motion and can therefore immediately be rejected as comets. This reduces the list of potential novae and supernovae from a full two thousand years of records to just seventy-five. A remarkable although perhaps not totally unexpected fact is that for these seventy-five cases the records indicate that the new stars were seen for either less than twenty-five or more than about fifty

days. Events of intermediate duration are completely absent.
Because a conspicuous supernova of any type fades fairly slowly,
with a naked-eye visibility of several months or more, we can
ignore the short duration stars, which were probably novae, and
concentrate on the few which lasted for more than fifty days; the
list of supernova candidates is then shortened to twenty. This
list is given as Table 1.

3.2 Candidate Events

Nova surveys indicate an average distance to those bright
enough to be detected with the naked eye as about 500 pc; the
average distance from the galactic plane is 300pc. We would
therefore expect novae visible to the naked eye, and recorded
historically, to be distributed more or less uniformly over the
celestial sphere. By contrast, extragalactic supernova surveys
show those in spirals to be concentrated to flattened disc
populations. Because of the large average distance from us of
galactic supernovae which may be detected with the naked eye
(several kpc) they would be expected to lie close to the plane
(say $|b| < 25°$). It is therefore tempting to preclude the high
galactic latitude events in Table 1 as likely novae.

In this type of selection process there is the obvious danger
that a genuine historical supernova record may have been
rejected. What about a supernova so close to the Earth that even
if it lay near to the galactic plane its angular displacement was
still large? In such a case extreme brightness would be expected
to have been mentioned, yet this is not the case for a single one
of the records that were rejected because they lay more than
twenty-five degrees from the galactic plane. What about new
stars which were genuine supernovae but which just happened to be
short-lived and therefore did not satisfy the selection criterion
of a duration of more than about fifty days? Even for such short
duration events, some remnant would be expected - but none are
obvious for any of the short duration new stars for which
positional estimates are possible. Some doubts remain, but in
the absence of further evidence one must necessarily accept that
genuine supernova records may have been inadvertently rejected.

Of the historical new stars from Table 1 at high galactic
latitude ($|b| > 25°$), the three objects of AD 61, 64 and 402 may
well have been comets, since there is a possible allusion to
motion despite the fact that they were specifically referred to
as 'K'o-hsing'. The objects of 5 BC and AD 247 were described as
'hui-hsing', so may have been comets, although there is no actual
mention of motion. In all five cases either a comet or a nova
may have been referred to; it is impossible to determine which
with any certainty. However, it is extremely unlikely that any
could have been supernovae since, as argued earlier, a supernova

Table 1

New Stars of Medium and Long Duration

(1) Date	(2) Where sighted	(3) Duration	(4) Approximate coordinates 1	b	(5) Remarks
Mar/Apr 5 BC	China	70 + days	30°	−25°	comet/nova?
Sep 27 AD 61	China	70 days	60	+70	comet/nova?
May 3 AD 64	China	75 days	290	+55	comet/nova?
Dec/Jan AD 70	China	48 days	215	+45	Nova?
Dec 7 AD 185	China	20 months	315	− 2	Supernova
Jan 16 AD 247	China	156 days	295	+40	comet/nova?
Mar/Apr AD 369	China	5 months	−	−	Position estimate impossible
Apr/May AD 386	China	3 months	10	0	Possible supernova
Feb/Mar AD 393	China	8 months	345	0	Supernova
Jul/Aug AD 396	China	50 days	175	−25	Nova?
Nov/Dec AD 402	China	2 months	240	+60	comet/nova?
May 3 AD 837	China	75 days	280	+65	Nova?
May 1 AD 1006	Arab lands, China, Europe, Japan	several years	330	+15	Supernova
Jul 4 AD 1054	Arab lands, China, Japan	22 months	185	− 6	Supernova
Aug 6 AD 1181	China, Japan	185 days	130	+ 3	Supernova
Nov 8 AD 1572	China, Europe, Korea	16 months	120	+ 1	Supernova − Type I
Nov 28 AD 1592A	Korea	15 months	150	−70	Mira Ceti?
Nov 30 AD 1592B	Korea	3 months	125	0	?
Dec 4 AD 1592C	Korea	4 months	115	0	?
Oct 8 AD 1604	China, Europe, Korea	12 months	5	+ 7	Supernova − Type I

at such high galactic latitude would be expected to be nearby, extremely brilliant, and long-lived.

Because of comparatively 'short' duration, high galactic latitude and no mention of extreme brilliance, it is tempting to exclude the 'guest stars' (k'o-hsing) of AD 70, AD 396, and 837 as being likely novae. The final record of a high galactic latitude new star, that of AD 1592A, may refer to Mira Ceti. In looking for possible supernovae, we are thus left with the new stars of AD 195, 369, 386, 393, 1006, 1054, 1572, 1592B, 1592C, and 1604. With the possible exception of the object of AD 369, where the vagueness of the historical record makes it impossible to make a definite positional determination, all are low galactic latitude ($|b| < 15°$) objects. Some doubt might be expressed as to the possible authenticity of the remaining two recorded new stars of AD 1592. It is somewhat surprising that 3 supposedly independent new star discoveries should have been made within the space of only 6 days - and only in Korea, at a time when the emergence of European Renaissance astronomy and the continued diligence of Chinese astronomers might have led one to expect an independent discovery had they been supernovae.

It is from the final shortlist of eight remaining historical new stars (those of AD 185, 386, 393, 1006, 1054, 1181, 1572 and 1604) that we might hope to establish associations with known galactic supernova remnants.

AD 185: The single Chinese record of this new star has been subjected to very careful interpretation. In particular, the period of visibility can be established to have been 20 months, making it an almost certain supernova, and its position can be determined to have been 'between' the stars α and β Cen. A planetary nebula with high velocities, suggestive of an expansion initiated less than 2 millenia ago, lies exactly between α and β Cen, but it seems inconceivable in view of the long duration that the event was a nova. Our preferred identification is the supernova remnant RCW 86 (G315.4-2.3) lying approximately one degree south of the line joining α and β Cen. (epoch AD 185), but with properties commensurate with it being the probable remnant of a supernova event some 2000 years ago.

AD 386: With a duration of 3 months, the AD 386 new star is the shortest-lived of the candidate historical supernovae. The single Chinese record of the event places it 'in the vicinity of' the star group 'Nan-tou' (southern ladle) comprising μ, φ, σ, and τ Sgr. If AD 386 was indeed a supernova, then any 'young' supernova remnant in the vicinity of Nan-tou must be considered a possible candidate remnant. One such object does exist, the radio remnant G11.2-0.3; however, the short duration makes the

AD 386 new star a rather weak supernova candidate, and the poor positioning makes G11.2-0.3, at best, its possible remnant.

AD 393: The only extant Chinese record of an 8 month duration new star is strongly suggestive of it being a supernova. It appeared 'within' the feature known as the 'tail' of the occidental constellation Scorpio. Unfortunately this covers a full 50 square degrees of sky, and includes 7 radio supernova remnants (none of which have been identified optically). Of these, only three, G350.0-1.8, G348.5+0.1, and G348.7+0.3, are close enough (within ~ 6 kpc) to have made an historical detection of the supernova likely — and only the last two have properties commensurate with an age of ~ 1600 year. Either G348.5+0.1 or G348.7+0.3 could be the possible remnant of the supernova of AD 393, but neither have been detected optically so that they could reveal properties suggestive of the remnant of the historical supernova.

AD 1006: The new star of AD 1006 was the most extensively recorded from the whole period of written history prior to the Renaissance. Numerous records remain from the Far East, the Arab lands, and from Europe. These records describe the star's extreme brilliance and duration in terms such that one can be left in little doubt that it must indeed have been the most spectacular stellar outburst seen by man for which written records still exist. The duration of the event from first discovery down to the naked-eye limit was reported from China to be 'several' (i.e. more than two) years. At maximum it was said to be a little brighter than the quarter-Moon (apparent magnitude ~ -9.5); one could supposedly see things clearly by its light (again requiring an apparent magnitude ~ -9.5). All the historical descriptions are consistent with the position of the radio supernova remnant G327.6+14.5 (PKS 1459-41). PKS 1459-41 must be considered the certain remnant of the supernova of AD 1006.

AD 1054: The AD 1054 supernova gave birth to the Crab Nebula. A positional discrepancy in one record placing the event 'south-east' of ζ Tau although the Crab is in exactly the opposite direction is we believe of no consequence. Other examples of exact translation of direction exist in the planetary data, indicating that such confusion can occur. If we do not accept this interpretation, we have to explain why the supernova which did produce the Crab Nebula is not recorded, and also what became of the remnant of the AD 1054 star.

The ease of the new star's detection in day-light (it was 2 hours ahead of the sun at discovery) in China, Japan, and Constantinople, plus comparisons with Venus, suggest a maximum apparent magnitude of -4 to -5. Chinese astrologers reported that it was seen in daylight for 23 days, with a total duration of order 22 months. Attempts to construct a light-curve from the scant historical data are fraught with difficulties, and its Type must therefore remain uncertain.

AD 1181: The new star of AD 1181 was recorded in China and Japan. A six month period of visibility, and proximity to the plane, make it a strong supernova candidate. The period of visibility, and historical comparisons with Saturn, suggest a maximum apparent magnitude of order 0, or slightly brighter. There are uncertainties in the historical positional descriptions, three different star groups separated by a full ten degrees being referred to - the reason for this may have been astrological. The most specific positional description refers to the new star 'invading' (fan) the star group Ch'uan-shê (the 'guest houses'). An interpretation of lunar and planetary data suggests thàt the term 'fan' was never used unless a positional proximity of less than 1 degree was being referred to. The only known radio supernova remnant within 1 degree of the 'guest houses' is G130.7+3.1, detected in the optical and X-rays; this object must be considered the <u>probable</u> remnant of the supernova of AD 1181.

AD 1572: The careful records of the AD 1572 new star made by Tycho Brahe, Thomas Digges, Michael Maestlin, and others prove beyond any reasonable doubt that it was a supernova of Type I, and allow us to establish with certainty that its remnant is G120.1+1.4.

AD 1604: The observations of the AD 1604 new star made by Johannes Kepler, David Fabricius, and others compare in detail and accuracy with those for the spectacular new star just thirty years earlier. Like its predecessor, the new star was undoubedly a supernova of Type I, which can be conclusively identified with the remnant G4.5+6.8.

The above conclusions are summarised in Table 2.

The historical data for the AD 1572 and 1604 supernovae allow their light curves to be determined. This is not possible for any of the other historical events. It is interesting to note that for the case of Kepler's supernova, we would have excellent photometric and positional information from the Far East alone, even without the European observations.

It is possible that one or more of the high galactic latitude historical new stars of unexceptional brightness, rejected earlier, might have been a supernova of lower than typical absolute magnitude at maximum. However, if so, no remnants of such outbursts are known - the highest latitude catalogued supernova remnant is G327.6+14.5, the remnant of AD 1006. Undoubtedly the historical new star records are incomplete. Some

Table 2

The Historical Supernovae and their Remnants

Supernova	Estimated magnitude at max	Radio remnant	Remarks on association	Approx. distances from Sun	from Plane	Detections Opt.	X-ray
AD 185	-8	G315.4-2.3 (RCW 86)	Probable	<2kpc	<80pc	Yes	Yes
AD 386(?)	+1	G11.2-0.3	Possible	>5kpc	>25pc	No	No
AD 393	-1	G348.5+0.1 or G348.7+0.3 (CTB37 A and B)	Possible	>6kpc	>20pc	No	No
AD 1006	-8 to -10	G327.6+14.5 (PKS 1459-41)	Certain	~1 kpc	~250pc	Yes	Yes
AD 1054	-4 to -5	G184.6-5.8 (Crab)	Certain	~2kpc	~200pc	Yes	Yes
AD 1181	0	G130.7+3.1 (3C58)	Probable	~8kpc	~400pc	Yes	Yes
AD 1572 Type I	-4	G120.1+1.4 (Tycho's SN)	Certain	~3kpc	~80pc	Yes	Yes
AD 1604 Type I	-2.5	G4.5+6.8 (Kepler's SN)	Certain	~5kpc	~600pc	Yes	Yes

records must have been lost. Certain new stars would have been too far south to be seen by northern hemisphere civilizations (particularly in the orient) with their written histories. Some might have been daytime objects which were not bright enough to be seen by day and which would have faded sufficiently so as not to be recognised as new stars when next they were night-time objects. Perhaps certain supernovae are not intrinsically bright in the optical, or lie in regions of very high obscuration. The question then arises as to whether any known supernova remnants are definitely young enough (less than 2000 years old) and close enough (closer than, say, 6 kpc) for a historical detection to have been expected. The answer is, only one - Cas A. The objects

RCW 103 and MSH 11-54 are almost certainly 'young' (less than 1000 years old) supernova remnants; however, both lie so far south as to have made it impossible that either could have been seen from China. Any one of the other reasons given above could be the explanation of why the birth of Cas A was not recorded.

One other historical new star might be mentioned — that of AD 1408. It did not make our 'shortlist' of supernova candidates, since in the dynastic history there is no mention of period of visibility. Recently other references to the event have been found in Chinese regional records, again leading to speculation on its true nature. None of the newly discovered records give duration, or improved positions. The outburst was in a region of the sky containing Cygnus X-1 and the peculiar supernova remnant CTB 80. There are persuasive reasons for rejecting either object as a young supernova remnant. The case of AD 1408 remains unresolved.

4.   THE FREQUENCY OF GALACTIC SUPERNOVAE FROM HISTORICAL RECORDS

Seven or eight galactic supernovae detected from the last two millenia suggests an average interval of about 250-300 years between supernovae within a distance usually taken to be ~ 6kpc, such that they might still be detected at Earth despite the effects of obscuration. (Approximately 6 kpc is the maximum distance for which an absolute magnitude -19 supernovae might be expected to be discovered, assuming a mean visual interstellar absorption of 1.5 mag per kpc.) Thus only those supernovae within a mere one-seventh to one-eight portion of our Galaxy might be expected to reveal themselves, implying that if supernovae had been detectable to the outer bands of the Galaxy a rate of about one every 40 years would have been obtained. (This assumes that the surface density of supernovae projected onto the galactic plane is constant.) Further allowance, however, must be made for the incompleteness of the historical records. It is difficult to quantify such effects accurately, but collectively they might account for an incompleteness of a factor of order two in the historical records. An incompleteness factor of order two suggests a 'true' Galactic supernova rate of about one every 20 years on average (with large uncertainty). But if supernovae occur as often as one every 20 years, why has none been detected since the advent of the telescope? And why are the historical detections 'bunched' (3 in the 11th-12th centuries, then a break until the 2 Renaissance supernovae, and nothing detectable subsequently)?

In an attempt to investigate these problems, we set up a simple computer simulation. A model thin-disk Galaxy, 16 kpc radius (comparable with the radii for the Galactic HI distribution, and Population I objects) was envisaged in which supernovae exploded regularly at 20 year intervals. The surface density of supernovae, occurring randomly in position over the complete galactic disk, was assumed constant (1 supernova per 1.6 x 10^{10} years per pc^2), as is suggested by statistical studies of external galaxies, and the position of the supernova was chosen at each time step by generating two random numbers and relating them to the polar coordinates of the supernova. But only the light pulses from supernovae lying within a distance of 6 kpc were recorded as reaching the Earth. The light travel time was then the dominant factor in estimating the frequency of detected supernovae. This point is emphasised in the case of the two eleventh century historical supernovae; the AD 1054 supernova, at a distance of 2.2 kpc, actually exploded some 3000 years before the AD 1006 supernova, at a distance of 1.3 kpc. The arrival times of light pulses from supernovae were thus highly sporadic, reflecting the effect of pulse travel time plus the fact that only supernovae in a 1/7th portion of the model Galaxy were recorded. The effect of making the supernovae occur at regular intervals instead of randomly with a mean rate of 1 every 20 years is small as the difference in light travel times arising from the random space distribution already gives rise to a Poissonian distribution of arrival times.

An 180,000 year simulation demonstrated the 'bunching' of detected supernovae that arises naturally in such a model – occasionally two or more supernovae were seen within a few decades in the simulation, followed by centuries with no detection. The number of detected supernovae per millenium in the simulation showed a peak in the range 6 to 8, although the actual number ranged from 2 to 15 over the 180 millenia considered. The seven historical supernova detections from a full two millenia thus appear to be compatible with a Galactic supernova rate of one every 20 years and an incompleteness factor of 2, although higher rates with lower incompleteness or lower rates with a more modest allowance for incompleteness are obviously possible.

The main factor that increases the length of time between observed supernovae is the reduction in the surveyed volume of the Galaxy caused by extinction. In our simple model, this increases the expected average time between detections to 142 years (20 x $16^2/6^2$). The Poissonian distribution of arrival times, caused by the random distribution of supernova events in space and time, is then easily able to account for the observed wide range of intervals between detected supernovae. The chance of no supernova being detected in an interval T is just $P_0(T)$ = e^{-RT}, where R is the supernova rate (= 1/142). Then $P_0(T)$ = 0.495 for T = 100 years, 0.245 for T = 200 years and 0.121 for T = 300 years. An interval exceeding the 377 years since the last observed Galactic supernova is expected in about 7 per cent of cases. Allowing for an incompleteness factor of 2 (R = 1/284) raises this figure to 27 per cent, and so the length of the interval since the last supernova is not in the least surprising. It is rather the close proximity of Kepler's and Tycho's supernovae that is surprising; the chance of 2 or more supernovae being detected within 32 years is only 0.6 per cent (allowing for incompleteness).

Thus a simple model shows that the long time-interval since the last observed Galactic supernova neither negates inferences from extragalactic and historical studies of an average interval between Galactic supernovae as low as 20 years nor implies that we are seriously overdue for a Galactic supernova.

The study in Section 4 was a collaborative exercise with Peter Andrews (RGO) and Robert Smith (University of Sussex). The reader is referred for detailed references to:- The Historical Supernovae, David H. Clark and F. Richard Stephenson, Pergamon, Oxford. 1977.

SUPERNOVA STATISTICS AND RELATED PROBLEMS.

G. A. Tammann

Astronomisches Institut der Universität Basel
European Southern Observatory, Garching

In the following the frequencies of SNe in external galaxies
(Section 1) and in our Galaxy (Section 2) are discussed. The
galactic frequencies are compared with SNRs, pulsars, and
white dwarfs. In Section 3 the absolute maximum magnitudes of
SNeI and SNeII are discussed, as well as the evidence of SNeI
being good standard candles. Speculations on the progenitors of
SNe in Section 4 are consistent with the assumption that SNeII
come from stars with masses $> 8 \, \mathfrak{M}_\odot$, while SNeI could come
from accreting white dwarfs in close binary systems.

Heuristically SNe are defined as variables which reach at
maximum at least $M_B \approx -15^m$. It is assumed that only two types
of SNe exist, SNeI and SNeII (Oke and Searle, 1974). A blind eye
is turned toward stars which remain very bright over extended
periods like η Car or SN 1961 v in NGC 1058 (classified by F.
Zwicky as Type V); they are anyhow statistically insignificant
- A Hubble constant of 50 km s^{-1} Mpc^{-1} is adopted throughout.
The absolute magnitude of the Sun is adopted as $M_B = 5^m_\cdot 48$.

1. THE FREQUENCY OF SNe IN EXTERNAL GALAXIES.

The frequency of SNe has been determined for external
galaxies from two large samples, i. e.

A. 400 Shapley-Ames galaxies with $v_0 <1200$ km s^{-1} (including
the Virgo cluster members irrespective of v_0) and $\delta > -36^0$

M. J. Rees and R. J. Stoneham (eds.), Supernovae: A Survey of Current Research, 371–403.
Copyright © 1982 by D. Reidel Publishing Company.

with 77 known SNe (up to March, 1976), - of which 31 are classified as Type I, and 24 as Type II. This <u>distance-limited</u> sample has led to <u>absolute</u> SN frequencies in <u>"SN units"</u> (1 SNu = 1 SN per $\overline{10^{10}}$ $L_{B\odot}$ per 100 yr) under the assumption that all SNe in the sample galaxies were caught during the 13-year interval from 1960 to 1972 (cf. Tammann, 1974). If this assumption is incorrect, the derived frequencies must be considered as lower limits. (Tammann, 1977).

B. 2955 galaxies from the RC2 (de Vaucouleurs, de Vaucouleurs, and Corwin, 1976) with 173 SNe (known by 1975), - of which 44 and 28 are classified as SNeI and SNeII, respectively. [Note that 75 % of all classified SNe are common to sample A and B]. This <u>open</u> sample was used to derive <u>relative</u> SN frequencies per unit luminosity for the different morphological types of the parent galaxies. (Oemler and Tinsley, 1979).

To compare the results from samples A and B the numbers must be homogenized. To this end the binning of morphological types in sample B is rearranged; the necessary interpolations introduce only negligible errors. In sample A the luminosities of the sample galaxies were calculated after a correction for the total internal absorption (following Holmberg, 1958). In sample B the magnitudes B_T^0 from the RC2 were used; these magnitudes are corrected to face-on orientation in the case of spiral (S) and Magellanic-type irregular galaxies (Im), - however, they still contain the neutral, inclination-independent contribution to the internal absorption. This affects also the relative SN frequencies, because the neutral absorption varies with the morphological type. The additional corrections for neutral absorption (Holmberg, 1958; cf. also Sandage and Tammann, 1981), which must be applied to the luminosities of the sample B galaxies are shown in Table 1, col. 2. The correction is particularly severe for I0 galaxies, now classified as "amorphous" (Sandage and Brucato, 1979), for which the neutral absorption is very large and poorly known.

Another effect, which influences the observed SN numbers, and which has not been taken into account for sample B so far, is the inclination of the parent galaxy. In nearly face-on Sc galaxies the SN frequencies appear 6.4 times higher than in their inclined counterparts (Tammann, 1974; the size of the effect has been questioned by Maza and van den Bergh, 1976; see however Shaw, 1979). This can only be explained as a selection effect discriminating against the discovery of SNe in inclined spirals.

Table 1

Relative SN Frequencies from Sample B

Type	Neutral absorption	inclination factor	relative frequencies					
			B			A		
			all	I	II	all	I	II
(1)	(2)	(3)	(4)	(5)	(6)	(7)	(8)	(9)
E	0.ᵐ0	1	1.82	1.82	0	1.4	1.4	0
S0	0.0	1	0.58	0.58	0	1.3	1.3	0
S0/a, Sa	0.22	1.5	3.20	3.20	0	0.7	0.7	0
Sab, Sb	0.43	2.8	8.30	4.84	3.46	3.6	1.8	1.8
Sbc, Sc, Scd, Sd	0.28	2.8	10	6.25	3.75	10	5	5
Sdm, Sm, Im	0.15	1.5	2.91	2.39	0.52	9.6	9.6	<3
I 0	large	?	undetermined			-		

The effect must be dependent on the morphological type of the parent galaxy, because it must (nearly) vanish for elliptical galaxies. Unfortunately present SN members do not suffice to determine the inclination effect beyond Sc galaxies. The correction factors shown in Table 1, col. 3, for different morphological types must therefore be taken only as educated guesses. But neglecting the inclination effect altogether would certainly result in a disproportion of the SN frequencies in Sc and E galaxies.

After an application of the corrections shown in Table 1, col. 2 and 3, the relative SN frequencies transform to the relative frequencies (normalized arbitrarily to 10 for Sc) of all SNe (col. 4), of SNe I (col. 5), and of SNeII (col. 6). The corresponding values from sample A are shown in col. 7-9.

The advantage of sample B is that it is based on a somewhat larger sample of SNe. Its disadvantage is that it is vulnerable to all distance-dependent selection effects. It is for instance suspicious that the ratio of SNeI : SNeII is higher in sample B than in sample A, because the brighter SNeI are expected to dominate in an unbound sample. Sample B gives also very high relative SN frequencies for Sab/Sb galaxies compared to Sbc-Sd galaxies, which is unacceptable from the nearby field and Virgo cluster galaxies in sample A. Balancing the merits of sample A (better definition) and sample B (larger sample), one concludes that the best relative frequencies are probably obtained by the average frequencies of the two samples.

The average SN frequencies from sample A and B are calibrated in absolute frequencies (in SNu) by requiring that Sbc-Sd galaxies have a true overall SN frequency of 1.38 SNu (Tammann, 1977). This value carries a statistical error of only 20 %, because it is based on 30 SNe. The resulting absolute frequencies are shown in Table 2.

Table 2

Adopted SN Frequencies in SNu (per 10^{10} $L_{B\odot}$[1] per 100 yr) for Different Types of Galaxies

	All SNe	SNe I	SNe II	n_{SN}[2]
E	0.22	0.22	0	13
S0	0.12	0.12	0	6
S0a, Sa	0.28	0.28	0	9
Sab, Sb	0.69	0.37	0.32	38
Sbc, Sc, Scd, Sd	1.38	0.77	0.61	93
Sdm, Sm, Im	1.02	0.83	0.19	11
I0	undetermined			7

1) The B-luminosities are in the B_r-system of the RC2, they are corrected for galactic and internal absorption as outlined in Sandage and Tammann (1981).

2) n_{SN} is the number of SNe in sample B; these numbers give an estimator of the statistical error of the listed frequencies.

The main conclusions from Table 2 are

1) The overall SN rate increases from elliptical (E) galaxies to late-type galaxies (Sdm-Im). The relatively low values of S0's (as compared to E's) and of Sdm-Im's (as compared to Sbc-Sd's) can be due to statistical fluctuations.

2) As is well known SNe I occur in all types of galaxies. They do therefore belong to the old stellar population (unless one makes the artificial assumption, that E galaxies have an important fraction of young population), but they are not typical for that population, because their frequency increases for later-type galaxies.

3) SNe II occur only in Sab and later galaxies. In Sab-Sd galaxies the ratio SNe I:SNe II = 100:82. SNe II are about 3 times rarer in Sdm-Im galaxies than in Sbc-Sd galaxies. The effect, which

is hardly due to small-sample statistics, is surprising because late-type galaxies are expected to have a relatively
high death rate of massive stars.

4) There is no basis for the assumption that the SN rate is high
   in I0 or amorphous galaxies. It remains only the fact that NGC
   5253, an amorphous galaxy (Sandage and Brucato, 1979), has
   produced two SNeI, on which, however, one cannot base any
   statistics.

Predicted mean SN intervals τ are calculated in Table 4 for
the four brightest external members of the Local Group. The
values are not in contradiction with the historical evidence. The
fact that 1 (instead of 5) SNe were observed in M 31 during the
last century (SN 1885) could be a statistical fluctuation and there
is no guarantee that all SNe were found in this galaxy (at $m_v <$
6^a, - allowing for internal absorption even fainter). A SN event
in LMC 200-300 yr ago would not have been recorded (δ = -70°!).

Table 3

Predicted SN Frequencies in the Local Group

Galaxy	Type[1]	M_B[1]	τ(yr)
M 31	Sb	-21.61	21
M 33	Sc	-19.07	110
LMC	SBm	-18.43	268
SMC	Im	-16.99	1008

[1]From Sandage and Tammann (1981).

The systematic errors of the frequencies in Table 2 are determined by the following effects:

a) Optimistic assumptions as to the control times of the sample
   galaxies have been made. The true frequencies must be higher.
   The size of this effect is limited by the observation that the
   galactic SN frequencies, determined from external and historical evidence (Section 2), agree within a factor of 1.5.

b) No allowance has been made for SNe which might have been
   lost in the inner parts of the sample galaxies due to overexpo-
   sure. This does not seem to be an important effect because
   the radial SN distribution conforms with the distribution of
   visible mass in nearby parent galaxies (Tammann, 1974; the

radial SN distribution in distant galaxies is biased by selection, cf. Shaw, 1979).

c) The true SN frequencies may be higher due to absorption losses. The effect is probably not important (cf. Section 2. 3).

d) The adopted correction factors for inclination (Table 1, col. 3) could be off by a factor ~ 1.5.

Summarizing these systematic error sources leads to the conclusion that the SN frequencies are probably correct within factors of 2, and that the true frequencies are - if anything - higher.

The random errors of the SN frequencies vary with morphological type: they are negligible for types with many SNe (i. e. Sc), and they are in the order of 30 % for early-type and very-late-type galaxies. Of course, the separate frequencies for SNe I and SNe II carry even larger errors.

A substantial improvement of the SN frequencies requires at least a doubling of the present SN numbers in nearby galaxies. This implies a very extensive and systematic SN search over many years. If truly complete search techniques become available down to fainter limits, as suggested at this Conference, well defined, more distant galaxy samples and clusters of galaxies shall also become useful. Less ambitious search programs may help to improve the ratio SNe I:SNe II in different galaxy types and to improve the frequencies particularly in S0-Sa and Sdm-Im galaxies.

An instructive, limited test could be provided by the Virgo cluster: for its inner region with radius 6° the present frequencies imply 2. 4 SNe yr^{-1} (cf. Tammann, 1976). Over the last 20 yr the discovery rate has been 3 times lower, which seems reasonable in view of a) incomplete surveillance, b) observability of the cluster during only 9 month, and c) the inclination effect of spiral galaxies, which alone accounts for a factor ~ 2.4 for the overall sample. A complete surveillance of the Virgo cluster over a decade should result in a statistically useful sample between 10 and 25 SNe, - depending only on selection effects and the correctness of the present SN frequencies.

2. THE SN FREQUENCY IN OUR GALAXY.

2.1. From External Evidence.

In order to apply the SN frequencies from the previous section to our Galaxy, its morphological type and its luminosity
must be known. The crucial parameter for the (sub-)classification
of spirals is the disc-to-bulge ratio. From this criterion our
Galaxy is between an Sb and Sbc spiral (Schmidt-Kaler and
Schlosser, 1973). Assuming that the total light of the Galaxy is
composed of an exponential disc component and a bulge component, - the latter contributing ~ 35 % of the light (Freeman, 1970;
de Vaucouleurs, 1979), - one can determine the luminosity of the
disc in two different ways:

a) The luminosity surface density at the position of the Sun ($R_\odot =$
8.7 kpc; Oort and Plaut, 1975; Graham, 1979) is $l = 28$ $L_{B\odot}$
pc^{-2} (Lequeux, 1979; and from data in Miller and Scalo, 1979).
The galactic exponential scale length of the disc from HI data
is $\Lambda = 3.4$ kpc (Gordon and Burton, 1976) or 3.2 kpc (Knapp,
Tremaine, and Gunn, 1978). The disc luminosity L_{disc} follows
then from:

$$L_{disc} = 2 \, L_o \, e^{1/\eta} \eta^2, \qquad (1)$$

where $\qquad \eta \quad = \Lambda/R_\odot$

and $\qquad L_o \quad = \pi \, l \, R_\odot^2.$

Inserting the above values (using a mean of $\Lambda = 3.3$ kpc) yields

$$L_{disc} = 2.7 \cdot 10^{10} \, L_\odot.$$

b) Freeman (1970) has found that many spirals have a central
surface brightness of $m_c^i = 21.65$ mag arcsec^{-2} ($l_c = 145$ $L_\odot$
pc^{-2}). Although this value has many exceptions (Kormendy,
1977), we assume for the moment that it holds also for the
Galaxy. The local surface brightness of $l = 28$ $L_\odot pc^{-2}$ or $m' =$
23.43 mag arcsec^{-2} is dimmed for an external observer by
the absorption along a column in the z-direction through the
Sun. This absorption is estimated to be $0^m.35$ (i.e. the mean
internal absorption of face-on Sbc and Sb galaxies; Holmberg,
1958). The central surface brightness and the absorption-
dimmed surface brightness of $m' = 23.78$ mag arcsec^{-2} at $R_\odot =$
8.7 kpc yield a scale length of $\Lambda = 4.4$ kpc. The total disc luminosity, as it appears to an external observer, is given by:

$$L_{disc} = 2 \pi l_c \Lambda^2 = 1.8 \cdot 10^{10} L_\odot. \qquad\qquad (2)$$

This leads with an internal absorption correction of $0\overset{m}{.}35$ to a true total disc luminosity of $L_{disc} = 2.4 \cdot 10^{10} L_\odot$.

Adopting $L_{disc} = 2.55 \cdot 10^{10} L_\odot$ as the best value and increasing it by the contribution of the bulge component, we find for the total luminosity of the Galaxy $L_T = 3.9 \cdot 10^{10} L_\odot$ ($M_B = -21\overset{m}{.}00$).

The appropriate SN rate of the Galaxy is 1.04 SNu (the mean for Sbc-Sd and Sab-Sb galaxies). This and the above value of L_T gives one SN per 25 years or a frequency of $\nu = 0.041$ SNe yr^{-1}.

The combination of 1.04 SNu and the local luminosity surface density of 28 $L_\odot$ pc^{-2} gives a local rate of $n_0 = 2.9 \cdot 10^{-11}$ SNe pc^{-2} yr^{-1}. If, however, the disc component is more prolific in SNe than the bulge component, the true value could be up to $\sim$ 30-40 % higher, because the external SN rates are based on total luminosities, whereas the local surface brightness is dominated by the disc component.

2.2. From Internal Data.

Within the wider solar neighbourhood seven SNe have occurred during the last millennium (Table 4).

The first five SNe of Table 4 are optically well documented; their correct identification, even in the case of SN 1181, can hardly be questioned. The SN of CasA has not been observed; this may hint to the possibility that the SN was underluminous (or simply a faint Type II event). The object poses the additional problem, that it is by far the most powerful radio source among the galactic SNRs. In spite of that, CasA is accepted as a statistically representative object. The historical records fail for the most southern part of the sky ($\delta < - 50^\circ$); one should therefore increase the sample by 20 % or one object. The SNR MSH 11-54, which apparently has indeed an age of less than 10^3 yr (Clark, 1981), was adopted here as this one object. The specific choice of this additional SNR has no significant effect on the following conclusions. The identification of a recorded, optical event of AD 1408 with the suspected SNR CTB 80 is still sufficiently uncertain (van den Bergh, 1980; Angerhofer, 1981; and references therein) to be neglected here. Even if eventually confirmed, the SN would only marginally increase the galactic frequency and decrease (?) the mean $|z|$-distance of historical SNe.

The individual distances r in Table 4 are uncertain, and in

Table 4

Galactic SNe during the last 10^3 yr.

SN	Name	m_v(max)	r(kpc)	$(m-M)^o$	A_V	M_V(max)	l	b	R (kpc)	Z (pc)
1006		-8[1][2]	1.5[3]	10.90	0.6[2]	[-19.5]	327.6	+14.5	7.5	+376
1054	Crab	-4.8[2]	2.0[4]	11.50	1.55[5]	-17.85	184.6	-5.8	10.7	-202
1181	3C58	0[6]	8.2[6]	14.57	3.1[6]	-17.67	130.7	+3.1	15.4	+443
1572	Tycho	-4.4[2]	~4[7]	13.0	2.1[8]	-19.51	120.1	+1.4	11.3	+ 98
1604	Kepler	-3.5[2]	~4[9]	13.0	3.47[10]	-19.98	4.5	+6.8	4.7	+474
(1667)	CasA	>0	2.8[11]	12.24	4.3[12]	>-16.5	111.7	-2.1	10.1	-103
-	MSH 11-54	-	3.7[13]	-	-	-	292.0	+1.8	8.1	+116

1) Clark and Stephenson, 1977 (p. 135).
2) Pskovskii, 1978.
3) Assuming an absolute magnitude of M_V(max) = $-19\overset{m}{.}5$.
4) Trimble, 1968.
5) Reina and Tarenghi, 1973; Miller, 1973.
6) Panagia and Weiler, 1980.
7) The determinations range from 2.3 kpc (Chevalier, Kirshner, and Raymond, 1980) to 7.2 kpc (Clark and Caswell, 1976), cf. also Goss, Schwarz, and Wesselius, 1973; Williams, 1973; Goss, W.M.: 1979, cited in Chevalier et al., 1980; and Milne, 1979.
8) Chevalier et al. (1980); Pskovskii, 1978.
9) Milne, 1979; Caswell and Lerche, 1979.
10) Danziger and Goss, 1980.
11) van den Bergh, 1971.
12) Searle, 1971.
13) Clark, 1981.

some cases mere guesses. It is, however, unlikely that the
distances, which are based on a number of different methods,
are systematically off by as much as a factor of 1.5. The calculated distances R from the galactic centre and the distances z
from the galactic plane should therefore be representative to
within this factor.

The seven SNe of Table 4 are shown projected on the galactic
plane in Fig. 1. They fall within a sector of opening angle 50°.
Six of the SNe lie within a solar distance of 5 kpc. This raises the
question whether SNe which occurred outside these boundaries
would indeed not have been observed. If the historical detection
limit is m_v(max) $\approx +1^m$ and the galactic absorption 1 mag kpc^{-1},
a SN with M_V(max) = -18^m would be detectable at a distance of 5
kpc for roughly 2 weeks. From this and the additional fact that
some SNe II are probably fainter than M_V(max) = -18^m, it appears
plausible why only one historical SN lies outside r = 5 kpc (SN 1181).

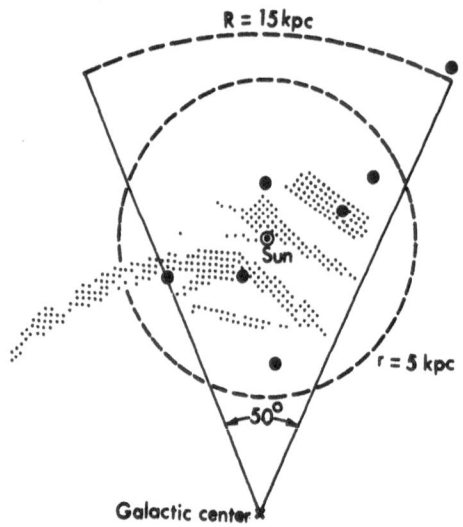

Fig. 1. The seven galactic SNe of the last millennium
projected onto the galactic plane. The local spiral
structure is schematically shown.

The galactocentric distribution of the historical SNe is shown
in Fig. 2 together with the corresponding radial distribution of SNe
in external, face-on Sc spirals (Tammann, 1974). The compari-

son shows that the historical SNe lie in fact at radial distances
where most of the SNe are expected, however the external evidence suggests that a fraction of 20 % of inner- and outermost
SNe have historically been missed. The 50°-sector should therefore be assigned 8. 8 SNe.

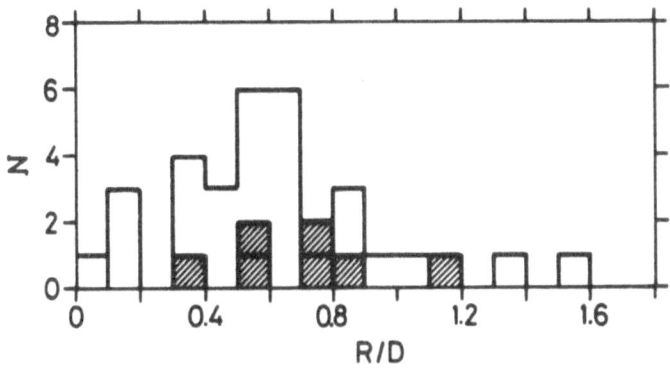

Fig. 2. The radial distribution of historical SNe in the
Galaxy (hatched) and of SNe in external, face-on Sc
galaxies (open squares). The radial distance R is expressed in units of the optical radius D. (For the
Galaxy adopted D = 14 kpc).

From this follows that the whole galactic disk has produced
8. 8 · $(360^{\circ}/50^{\circ})$ = 63. 4 SNe in 10^3 years, corresponding to a
mean interval between SNe of τ = 16 years, or ν = 0. 063 SNe yr^{-1}.
If one excludes CasA from this extrapolation because of its high
radio power, one finds ν = 0. 054 SNe yr^{-1}. We shall adopt from
the historical evidence ν = 0. 058 SNe yr^{-1}. The error of this
determination from n = 7 objects is probably dominated by the
statistical $\sqrt{n}$-error, amounting to $\pm$ 40 %.

The historical SNe are particularly suited to determine the
local SN rate. Seven SNe in 10^3 yr within a 50°-sector with radius
~ 16 kpc indicate n_0 = 6. 3 · 10^{-11} SNe pc^{-2} yr^{-1}. An even pre-
ferable basis is provided by the six SNe within 5 kpc from the
Sun; they lead to n_0 = 6. 4 · 10^{-11} SNe pc^{-2} yr^{-1}. The uncertainty
of this value is probably governed by the error of the distance
limit. With an estimated error of this limit of a factor of 1. 3,
the uncertainty of the local rate becomes a factor of 2.

2.3. Comparisons.

The galactic SN rate determined under 2.1. and 2.2. from completely independent data agree very satisfactorily. The best estimate is probably the mean rate, as listed in Table 5. The table shows also the corresponding rates of SNeI and SNeII (assuming a ratio of 1:0.82).

Table 5

Adopted Galactic SN frequencies

	$\nu\,(\mathrm{yr}^{-1})$	$\tau\,(\mathrm{yr})$	$n_o(\mathrm{pc}^{-2}\mathrm{yr}^{-1})$
all SNe	0.050	20	$5.2 \cdot 10^{-11}$
SNeI	0.027	36	$2.9 \cdot 10^{-11}$
SNeII	0.023	44	$2.3 \cdot 10^{-11}$

There is a factor of 2.2 discrepancy between the local SN rates derived in sections 2.1. and 2.2. The former value, however, is a lower limit and should be given the weight $1/2$. The resulting mean local rate is also listed in Table 5.

One important caveat must be noted to the values in Table 5: they could still be lower limits. If SNe occurred preferentially in very-high-absorption regions they would have been missed historically as well as in external galaxies. This would imply, however, - because at least some SNe are only moderately obscured, - that the SNe in any given external spiral would appear with a very wide scatter in m(max). There are a few spirals, which are sufficiently close that even some of their heavily obscured SNe would be detectable, and which have known m(max) for more than one SN. They are listed in Table 6 (from Cadonau and Tammann, 1981).

The maximum range of m(max) for a given galaxy is $2^{m}_{.}8$. This range can be accounted for by the intrinsic luminosity scatter of SNe (see below) and speaks against the occurrence of very high absorption values. There is an independent argument which points into the same direction: some of the highly obscured SNe would be at the detection limit for even the nearest galaxies, and their counterparts would be lost in more distant galaxies. This would result in a strong distance dependence of the SN rates. However,

Table 6

Spirals with more than two SNe with known m(max)

Galaxy	SN	m_B(max)	Type
NGC 3184	1921c	11.22	I
	1937f	14.0	II
NGC 3913	1963j	13.49	I
	1979	13.00	I
NGC 3938	1961u	14.30	II
	1964b	13.52	I
NGC 4303	1926a	14.50	II
	1961i	13.41	II
NGC 5236	1923a	12.3	II
	1968 l	11.71	II
NGC 5457	1909a	12.5	II
	1970g	11.58	II
NGC 6946	1939c	13.56	II
	1948b	14.40	II
	1980	11.80	II

between a distance limit of r = 10 pc and r = 22 pc the observed SN rate decreases only by a factor of 1.3 ± 0.6 (Tammann, 1974). A stronger distance dependence was found by Oemler and Tinsley (1979), but their sample includes many intrinsically fainter galaxies, whose surveillance for SNe is probably quite uneven and distance-dependent. The conclusion then is, that the absorption scatter is not very large and that therefore very highly absorbed SNe are quite rare.

In the following we exclude the possibility that absorption dominates the SN statistics. In that case not only the SN frequencies derived above are realistic, but also the z-distribution perpendicular to the galactic plane of the historical SNe in Table 4 is representative. Their z -distribution is shown in Fig. 3.

The high z -values of the historical SNe are quite surprising. They give $\sqrt{<z^2>}$ = 302 pc. If one assumes that they follow an exponential distribution of the form

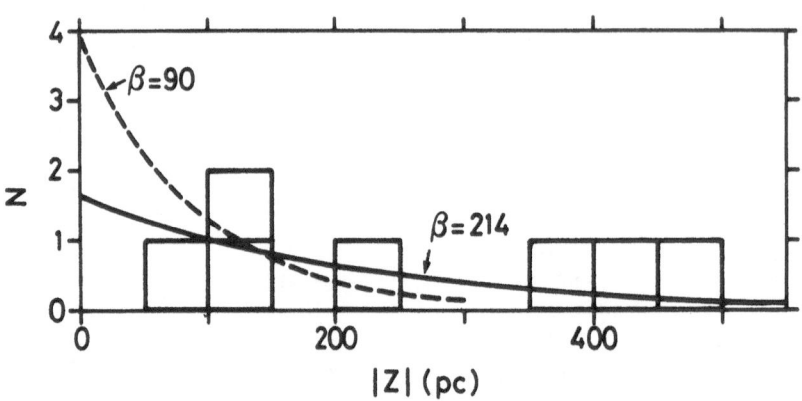

Fig. 3. The distances $|z|$ perpendicular to the galactic
plane of the seven galactic SNe during the last millennium.

$$N(z) = N_0 \exp(-(|z|/\beta)),\hspace{4em}(3)$$

then one finds for the scale height $\beta = 2^{-\frac{1}{2}} \cdot \sqrt{<z^2>} = 214$ pc. A χ^2-
test shows that the assumption of an exponential distribution is
acceptable indeed. However, the same test strongly rejects the
possibility that the scale height could be as low as $\beta = 90$ pc, - a
value which would be expected for massive stars (Miller and Scalo,
1979). This sets stringent limits on the possible progenitors of
SNe, as it will be discussed in section 4.

The large scale heights of SNe are seemingly in contradiction
with the flat distribution of SNRs ($\beta = 60$; Clark and Caswell, 1976).
However, it is now clear that their z-distribution is dictated by
the small scale height of the interstellar medium; high-$|z|$ SNe do
not form long-lived SNRs. The surface density of SNRs (with diameters < 36 pc) at the position of the Sun is about $\rho' = 1.1 \cdot 10^{-6}$
pc^{-2} (Clark and Caswell, 1976), and therefore their space density
at z = 0 is $\rho = 0.5 \, \rho' \beta_{SNR}^{-1} = 9.2 \cdot 10^{-9}$ pc^{-3}. From the local SN
density n_0 in Table 5 one finds the SN space density at z = 0, i.e.
$N_0 = 0.5 \, n_0 \beta_{SN}^{-1} = 1.2 \cdot 10^{-13}$ SNe pc^{-3} yr^{-1}. A comparison of
these numbers gives a characteristic lifetime of low-$|z|$ SNRs of
$8 \cdot 10^4$ yr.

The best estimates of the galactic pulsar birth rate, with careful allowance for all selection effects, give τ (Pulsars) = 30 yr

(Gunn and Ostriker, 1970) and 20-50 yr (Lyne, 1982; cf. also
Guseinov and Kasumov, 1981). These values are very similar to
the best estimate of the total SN rate of τ(SNe) = 20 yr. Also the
pulsar surface rate of n_0(Pulsars) = 5 $\cdot$ 10^{-11} pc^{-2} yr^{-1} (Gunn
and Ostriker, 1970) is fortuitously close to the adopted value for
SNe (cf. Table 5). But considering the uncertainties of the different
estimates, one can conclude with confidence only that τ(Pulsars)
= $a \tau$(SNe), where 0. 3 < a < 2. In spite of this wide margin, it is
tempting to speculate that only a fraction of all SNe, - most likely
the SNe II, - form pulsars. Two of the historical SNe (SN 1054,
SN 1181) are exceptional in as far as they have filled radio remnants ("plerions"; Panagia and Weiler, 1980), as they have been
relatively faint at maximum (Table 4), and as their remnants
contain a pulsar and a point X-ray source, respectively (Helfand,
1981). Although this circumstantial evidence cannot replace a definite classification from spectroscopy, it is consistent to assume,
that they derived from massive stars and that they were Type II
events. On the other hand SN 1572 and SN 1604 are conventionally
assumed to be of Type I; they were probably quite bright at maximum (Table 4), and their light curves conform with this assumption (Tammann, 1977). Down to very faint limits they do not contain a point X-ray source (Helfand, 1981), and therefore they have
probably not formed a pulsar.

    An independent piece of evidence makes it unlikely that all SNe
form pulsars. The available data on space velocities of pulsars
indicate that they are formed close to the galactic plane (Lyne, 1982;
- the Crab pulsar being the one clear exception). However, the
seven historical SNe have typically high $|z|$-values. This almost
necessitates the conclusion that preferrentially that fraction of
SNe, which occur near to the galactic plane, - i. e. probably those
with massive progenitors, - do form pulsars.

    Finally there is some evidence that pulsars are concentrated
toward spiral arms (Harding, 1981). The same holds for SNe II
(Maza and van den Bergh, 1976), making a relationship between
pulsars and SNe II quite possible. At the same time this argument
casts doubt on a relationship of pulsars and SNe I, because the latter
do not follow a spiral pattern (Maza and van den Bergh, 1976).

    The present birth rate of white dwarfs is N_0 = 2 $\cdot$ 10^{-12} WD
pc^{-3} yr^{-1} (at z $\approx$ 0) to within a factor of 2 (Weidemann, 1978).
Their probable scale height is $\beta \approx$ 300 pc. With this value one finds
a local surface rate of n_0 = 2 βN_0 = 1. 2 $\cdot$ 10^{-9} WD pc^{-2} yr^{-1}.
A comparison with the corresponding value for SNe, n_0 = 5. 2 $\cdot$

10^{-11} SNe pc^{-2} yr^{-1}, shows that ~ 4 % of all dying stars in the solar neighbourhood go through the SN stage.

3. THE ABSOLUTE MAGNITUDES M_B(max) OF SNe.

On the assumption of an ideal Hubble flow and H_o = 50 the absolute magnitudes M_B(max) can be calculated for SNe with observed or extrapolated apparent maximum brightness m_B(max). Values of m_B(max) are taken for 50 SNe I and 29 SNe II outside the Local Group from a compilation by Cadonau and Tammann (1981). These magnitudes are corrected for galactic absorption following Sandage (1973; cf. Tammann, 1980). With the exception of SNe I in E and S0 galaxies, SNe are additionally dimmed by internal absorption in the parent galaxies. The internal absorption can be estimated in cases where the maximum colour (B-V)$_{max}$ is observed. Then, assuming (B-V)$^o_{max}$ = - 0.15 for all SNe (Barbon, Ciatti, and Rosino, 1973; 1979), one obtains E_{B-V} = (B-V)$_{max}$ + 0.15 and $A_B \approx 4 E_{B-V}$. Note that for the unusual spectra of SNe the ratio between A_B and E_{B-V}, R = 4, is only a rough approximation. It should also be noted that because of the high value of R the available colour observations are too poor in most cases to derive a useful absorption value.

3.1. The Absolute Magnitude of SNe I.

The nearly dust-free E and S0 galaxies offer the best possibility to determine M_B(max) independent of internal absorption. For 17 SNe I in such galaxies one finds (cf. also Fig. 4)

$$< M_B(max) > = - 19.69 \pm 0.14, \ \sigma (M) = 0.58.$$

Using only the 9 best observed SNe I gives

$$< M_B(max) > = - 19.73 \pm 0.14, \ \sigma (M) = 0.43.$$

The small luminosity scatter of 0.43 can be fully explained by observational errors, such as
a) poor photometry,
b) extrapolation of m_B(max) along the standard SN I light curve (Barbon, Ciatti, and Rosino, 1973),
c) deviations from a pure Hubble flow (random motions and Virgo-centric streaming).

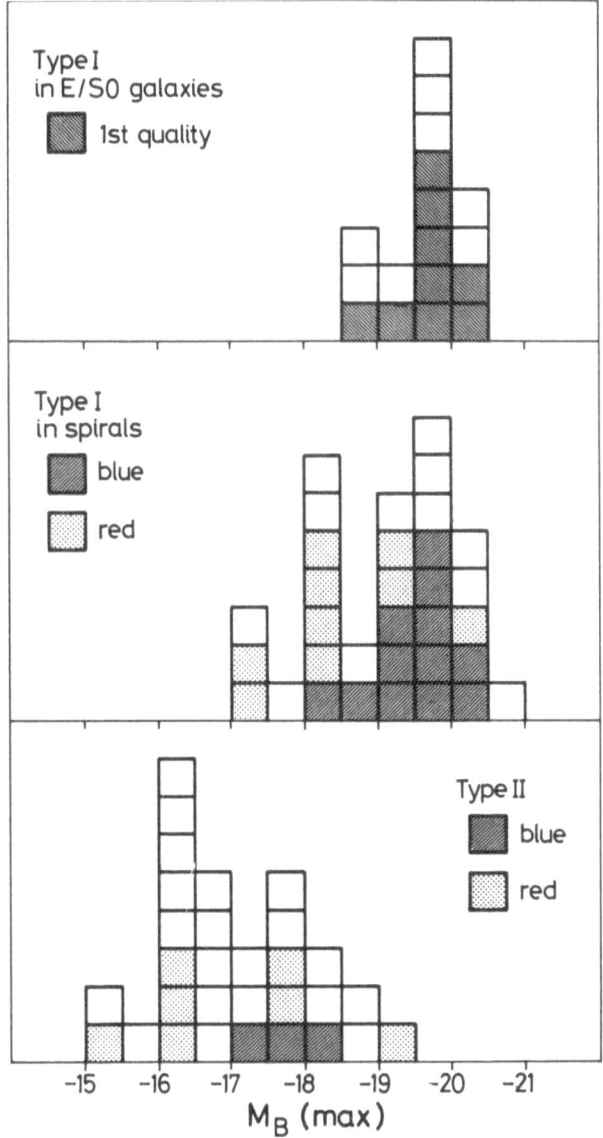

Fig. 4. The distribution of M_B(max) for SNeI in E/S0 galaxies (top) and in S/Im galaxies (middle) and of SNeII (bottom). Blue SNe are on the average brighter than red ones as an effect of interstellar absorption in the parent galaxy.

The small scatter is also supported, independently of any assumptions on distances, by SNeI in E/S0 cluster galaxies, i.e.

Virgo cluster (n = 6): $< m_B(max) > = 12\overset{m}{.}02$, $\sigma = 0\overset{m}{.}43$
Coma cluster (n = 5): $< m_B(max) > = 16\overset{m}{.}24$, $\sigma = 0\overset{m}{.}62$

(The somewhat larger scatter in the Coma cluster is typical for the poorer photometry at fainter levels). Also the case of NGC 5253, an apparently dust-poor amorphous galaxy, is worth mentioning, which has produced two well observed SNeI:

Table 7

The SNeI of NGC 5253

	$m_{pg}(max)$	$m_B(max)$	$v_0(km\ s^{-1})$	$M_B(max)$
SN 1895 b	$8\overset{m}{.}0$[1]	$8\overset{m}{.}24$[2]	232 ± 21[3]	-20$\overset{m}{.}$07
SN 1972 e		8.5 [1]	"	-19.83

1) Sargent, Searle, and Kowal (1974)
2) From $m_{pg}(max) = B - 0\overset{m}{.}24$ for SNeI (Branch and Bettis, 1978; Cadonau and Tammann, 1981).
3) Mean velocity of the NGC 5128 group (Tammann and Kraan, 1978); NGC 5253 proper has $v_0 = 147\ km\ s^{-1}$ (Sandage and Tammann, 1981).

Their magnitudes are the same within $\Delta m_B(max) = 0\overset{m}{.}26$. It could be argued that their absolute magnitudes were too bright to conform with the value of SNeI in E/S0 galaxies. However, a reduction of only 25 $km\ s^{-1}$ of the group velocity (i.e. one standard deviation) is sufficient to bring them exactly to the mean value of $M_B(max) = -19\overset{m}{.}7$.

The important conclusion is that un-absorbed SNeI have little luminosity scatter at maximum, - certainly less than $\sigma(M) = 0\overset{m}{.}43$, - and that they are possibly the best standard candles in the universe.

There is an independent check on the nature of the luminosity scatter of SNeI via their Hubble diagram (cf. Fig. 5): if the observed scatter is due to random observational errors, standard candles still follow the <u>mean</u> relation

$$m = 5 \log v_0 + const\ (v_0 \ll c). \tag{4}$$

If, however, the scatter is intrinsic (due to true scatter in lumi-

nosity or variable absorption) only exceptionally bright objects
are observed at large distances, i.e. with large v_o. This results
in a slope of less than 5 in eq.4, provided v_o is treated as the
independent variable (cf. Tammann, 1978). The fact that the SNe I
in E/S0 galaxies define very well the expected slope of 5 is additional evidence for their being standard candles (Fig.6).

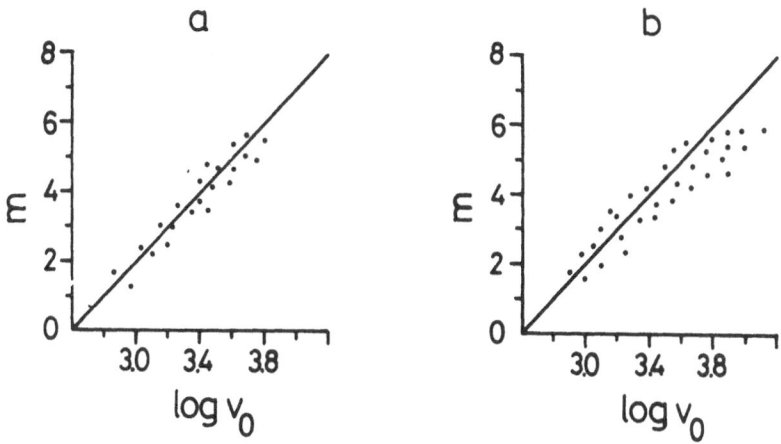

Fig. 5. Schematic illustration of the Hubble diagram for
standard candles with a) random scatter in m due to observational errors, and b) intrinsic scatter. In the latter
case the points define a curve with slope < 5.

The SNe I in spiral and Im galaxies are generally fainter and
show much larger scatter in the Hubble diagram than their counterparts in E/S0 galaxies. They define a slope of less than 5, indicating that they do not appear as standard candles to an external
observer. If they are faint because of internal absorption, their
absolute magnitude must be correlated with the maximum colour
$(B-V)_{max}$ of the SN. Indeed the blue $[(B-V)_{max} < 0^m\!.2]$ SNe I in
S/Im galaxies are considerable brighter than the red $[(B-V)_{max}$
$> 0^m\!.2]$ ones, as the following numbers show:

$(B-V)_{max} < 0^m\!.2$ (n = 12): $< M_B(max) > = -19^m\!.56$, $\sigma (M) = 0^m\!.65$
$(B-V)_{max} > 0^m\!.2$ (n = 9): $= -18.53$, $\sigma (M) = 1.18$
all SNe I in S/Im (n = 33): $= -19.11$, $\sigma (M) = 1.00$

The blue SNe I in S/Im's are only slightly fainter than the
SNe I in E/S0's and they have still relatively small luminosity
scatter. Their mean absorption is only $< A_B > \approx 0^m\!.15$. The red

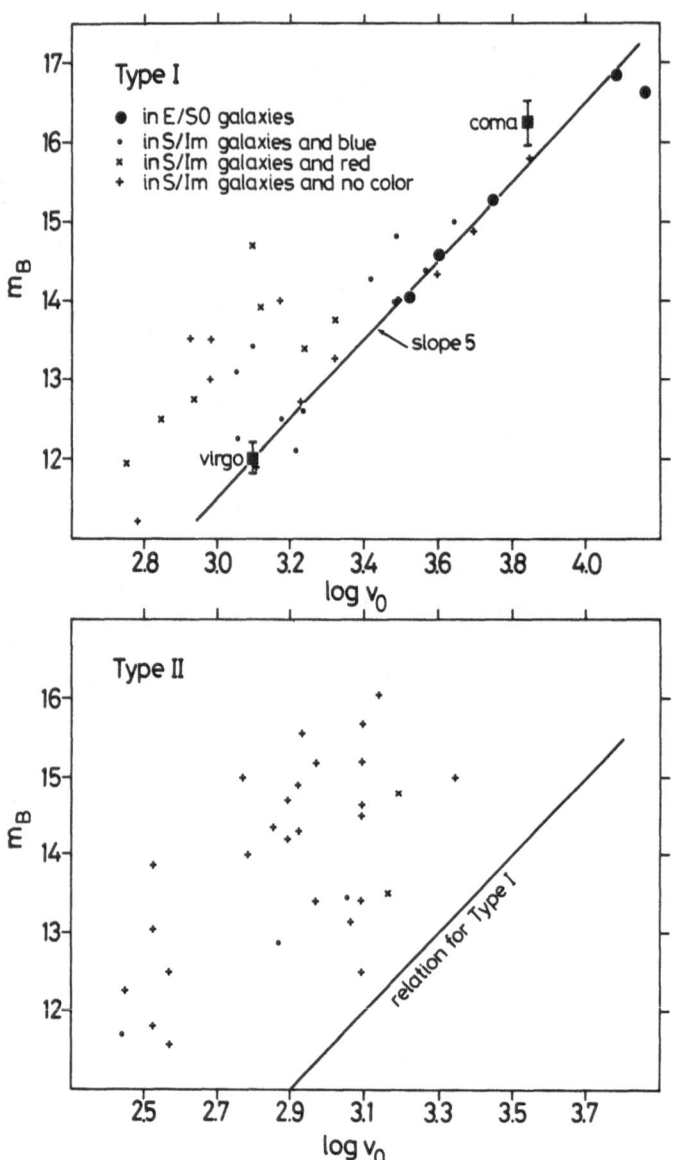

Fig. 6. The Hubble diagram of SNeI (top) and SNeII (bottom). The absorption-free SNeI in E/S0 galaxies are shown with large symbols (those in Virgo and Coma cluster members, respectively, are combined). Blue and red SNe and SNe without colour information are shown with different symbols.

SNe I are on the average stronly absorbed, $< A_B > = 1^m.2$, and they have correspondingly very large luminosity scatter.

It has been suggested that there are two distinct classes of SNe I, which differ in their post-maximum decline rate (Barbon et al., 1973; Barbon, 1980). There is, however, little reason to devide the SNe I into two distinct classes; rather one could think of a continuous intrinsic spread of decline rates. However, the reality of this spread is still quite doubtful in view of the vulnerability of the decline rates against observational errors. Incorrect allowance for background effects in photoelectric observations and even small magnitude errors of the comparison stars in photographic photometry lead necessarily to an almost arbitrary deformation of the lightcurves. Little confidence therefore can be put into the suggestion (Pskovskii, 1977; Branch, 1982) that the decline rate correlated with the maximum luminosity. Moreover, the evidence for SNe I being good standard candles (after absorption corrections in S/Im galaxies) leaves little room for any such correlation.

3.2. The Absolute Magnitude of SNe II.

The distribution of M_B(max) of SNe II is illustrated in Fig. 4, and their Hubble diagram is shown in Fig. 6. It is obvious that SNe II are fainter on the average than SNe I, and that they have large scatter $\sigma(M)$. The slope of the Hubble diagram defined by SNe II is less than 5, indicating that the luminosity scatter is intrinsic (either true luminosity scatter and/or variable internal absorption).

In the presence of intrinsic luminosity scatter care must be taken to avoid selection effects, if the mean luminosities are calculated. An apparent-magnitude-limited sample would contain some excessively bright, distant objects, and would lead to an overestimate of the average luminosity. A distance-limited, unbiased sample is tentatively defined by all SNe II with $v_0 < 1200$ km s^{-1} (including the Virgo cluster). From this sample one finds:

$(B-V)_{max} < 0^m.2$ (n = 3): $< M_B(max) > = -17^m.92$, $\sigma(M) = 0.41$:
$(B-V)_{max} > 0.2$ (n = 7): $= -16.98$, $\sigma(M) = 1.33$
all SNe II (n = 26): $= -16.96$, $\sigma(M) = 1.06$

In order to decide whether the relative faintness of SNe II and their large luminosity scatter is in fact an intrinsic property, or whether these properties are imposed by interstellar absorption, it is helpful to consider those SNe II for which individual determinations of E_{B-V} are available. They are listed in Table 8.

Table 8

SNe II with individual E_{B-V} determinations

SN	Galaxy	$M_B(max)$	A_B	$M_B^0(max)$	Source
1959 d	NGC 7331	$-18.^m04$	$2.^m0$	-20.04	1
1969 l	NGC 1058	-17.74	0.56	-18.30	2
1970 g	NGC 5457	-17.82	0.60	-18.42	2, 3
1972 q	NGC 4254	-16.03	4.8	-20.83	2
1973 r	NGC 3627	-15.36	0.68	-16.14	3
1979 c	NGC 4321	-20.10	0.36	-20.46	4, 5
1980 k	NGC 6946	-17.3	1.24	-18.54	4
				-18.96, $\sigma(M) = 1.^m62$	

1) Arp (1961).
2) Schurmann, Arnett, and Falk (1979).
3) Kirshner and Kwan (1974).
4) Kirshner (1982); Benvenuti (1981).
5) de Vaucouleurs et al. (1981).

The mean error of the values A_B in Table 8 are estimated to be $< 1^m$. If this is realistic, then there can be no doubt that the mean absolute magnitude of SNe II is intrinsically fainter by ~ $0.^m7$ than that of SNe I and that their true scatter $\sigma\overline{(M)}$ is considerable. This conclusion is not new, but the data here show that it still depends on (reasonable) assumptions concerning absorption. (The smaller luminosity scatter of $\sigma(M) = 0.^m78$ found by Barbon, Ciatti, and Rosino (1979) is not significant, because they excluded the fainter SNe II).

3.3. The Absolute Calibration of SN Luminosities.

The foregoing calibration of the absolute magnitudes of SNe I and SNe II depends on the choice $H_O = 50$. For any other value of H_o the magnitudes scale like:

SNe I: $M_B(max) = -19.^m7 + 5 \log H_O/50$ (5)
SNe II: $= -19.0 + 5 \log H_O/50$ (6)

It would be of great interest to turn the argument around, viz. to determine the distance of some SNe (and $M_B(max)$) independently and to derive from this the value of H_O.

The discussion under 3.1 and 3.2 makes it clear that SNe I are much more suitable for the determination of H_0 than SNe II. Knowing $M_B(max)$ of any one SN I gives the absolute magnitude of all SNe I to within better than $0\overset{m}{.}43$ because of the small luminosity scatter of unobscured SNe I. In other words the well defined Hubble line of absorption-free SNe I (cf. Fig. 4) allows one to carry the luminosity calibration out to $v_0 \approx 10\ 000$ km s^{-1}, i.e. to distances where all possible deviations from an ideal Hubble expansion are negligible. The corresponding possibility does not exist for SNe II because their Hubble diagram (Fig. 4) is strongly affected by large scatter and corresponding luminosity selection. Any one SN II can provide only a local value of H_0, unless it lies itself outside of all local deviations from a cosmic expansion law, say at $v_0 > 4000$ km s^{-1}.

The distances to historical SNe I in our Galaxy have become, if anything, less reliable in recent years. Taking the values in Table 4 at face value one finds $M_B(max) = -19\overset{m}{.}74$ for the mean of SN 1572 and SN 1604, and through eq. 5 $H_0 = 49$. But this value carries an uncertainty factor of 1.5.

The distance of SN 1885a in M 31 is well known, $(m-M)_{AV} = 24\overset{m}{.}60$ (Baade and Swope, 1963). Also the value $m_V(max) = 5\overset{m}{.}7$ (Jones, 1976) is relatively well determined. From this follows $M_B = -19\overset{m}{.}05 - A_B$, where A_B is the unknown amount of internal absorption. Eq. 5 leads then to an upper limit only of $H_0 < 70$. The assumption of SN 1885 being a SN I is marginally supported by the visual descriptions of its spectrum, as compiled by Jones (1976). The time scale of its light decline (Jones, 1976; de Vaucouleurs and Buta, 1981), however, is about three times faster than for a typical SN I (Tammann, 1977). (On the other hand the decline is roughly three times slower than that of a smooth-curve Nova (Dürbeck, 1981), if ever one wanted to identify the object with a galactic nova in spite of its nearness to the nucleus of M 31).

An increasingly exciting possibility to calibrate SN luminosities comes from expansion parallaxes. The state of the art leaves still much to be desired (cf. Wagoner, 1982; Branch, 1982), but it is worthwhile to compile here the present results (Table 9 and 10).

For reasons stated above the magnitudes of SNe II listed in Table 10 cannot be used to derive a value of H_0 from eq. 6. But the magnitudes cover about the same range as those in Table 8, which are calculated with $H_0 = 50$. An interesting route is offered by SN 1979 c, whose absolute magnitude corresponds to a distance

Table 9

The Luminosity of SNe I from Expansion Parallaxes

SN	Galaxy	M_B(max)	Source
several	–	-20.25	Branch (1977)
1975 a	NGC 2207	-19.12	Arnett (1982)
1981	NGC 4536	-20.5	Branch (1982)

The unweighted mean of the magnitudes in Table 9 gives M_B (max) = -19.96 and through eq. 5 $H_o \approx 44$.

Table 10

The Luminosity of SNe II from Expansion Parallaxes

SN	Galaxy	M_B(max)	Source
1969 b	NGC 1058	-17.80	Kirshner and Kwan (1974)
		-18.13	Schurmann et al. (1979)
1970 g	NGC 5457	-18.07	Kirshner and Kwan (1974)
		-18.32	Schurmann et al. (1979)
1979 c	NGC 4321	-20.61	Branch et al. (1981)

of 23 Mpc (Branch et al., 1981). The SN appeared in NGC 4321, i.e. in a member of the Virgo cluster, and gives therefore the distance of that cluster. This distance, corresponding to a modulus of $31^m.81$, can then be used to calibrate the six SNe I in the Virgo cluster (see above). From this one obtains for SNe I M_B(max) = $-19^m.79$, and hence from eq. 5 $H_o = 48$.

It is noteworthy that, in spite of the incomplete under-standing of SN atmospheres, the expansion parallaxes of SNe I and SNe II give distance scales which agree better than one could have expected: that is $H_o = 44$ and $H_o = 48$ respectively.

If SNe I are indeed good standard candles, all efforts should be made to find more SNe I particularly in E galaxies and to obtain good m(max) values for them. They would, in addition to setting better limits on $\sigma(M)$, extend the Hubble diagram and improve the definition of the Hubble line. Because SNe I at redshifts of ~ 0.5 can effectively be searched for and photometered with the Space Telescope, this route will eventually lead to an elegant determination of q_o (Tammann, 1979a, b). If ever this route should be

hampered by luminosity evolution of SNe I over cosmic timescales, any single high-redshift SN could still lead to a value of
q_0 via its expansion parallax (Wagoner, 1980). Moreover, larger
samples of SNe I in nearby E galaxies ($v_0 < 4000$ km s^{-1}) would be
of prime importance to set up relative, velocity-independent
distances between clusters, and to map deviations from a pure
local expansion field.

In contrast to SNe I the rôle of SNe II in cosmology is judged
pessimistically. They can provide useful expansion parallaxes in
a limited number of cases, but their Hubble diagram is hope-
lessly biased by luminosity segregation, and at high redshifts
their search is extremely time-consuming because they avoid
(together with their parent spiral galaxies) the rich clusters.

4. SN PROGENITORS.

The possible progenitors of SNe II and SNe I are discussed here
from a purely statistical point of view.

4.1. SNe II from Massive Stars.

SNe II occur only in galaxies with young population (Table 1),
they concentrate along spiral arms (Maza and van den Bergh,
1976), their frequency increases with the blueness of the parent
galaxies (Tammann, 1977; Oemler and Tinsley, 1979), and their
radial distribution agrees roughly with that of hydrogen (Tammann,
1977 b).

If, hence, SNe II come from young stars they most probably
are the death mode of massive stars. This is even more probable
because at least some stars with initial masses $\mathfrak{M}_i = 6 - 7 \, \mathfrak{M}_\odot$
are known to die as white dwarfs (Romanishin and Angel, 1980;
Wegner, 1981; Köster and Reimers, 1981). The lower mass
limit of SNe II can be determined on the simplistic assumption,
that $\mathfrak{M}_i$ alone determines the final fate of a star (cf. Weidemann,
1981) by comparing the SN II frequency with the initial mass
function in the solar neighbourhood (IMF). Unfortunately the latter,
as determined by different authors, leaves factor-of-3 discrepancies
for the relevant mass range (Fig. 7).

The steady-state situation for short-lived, massive stars
makes the IMF equal to the present death rate. To account for a
SNe II frequency of $n_0 = 2.3 \cdot 10^{-11}$ pc^{-2} yr^{-1} it is then necessary
to invoke all stars with $\mathfrak{M}_i > 8.4 \, \mathfrak{M}_\odot$. This value holds for the

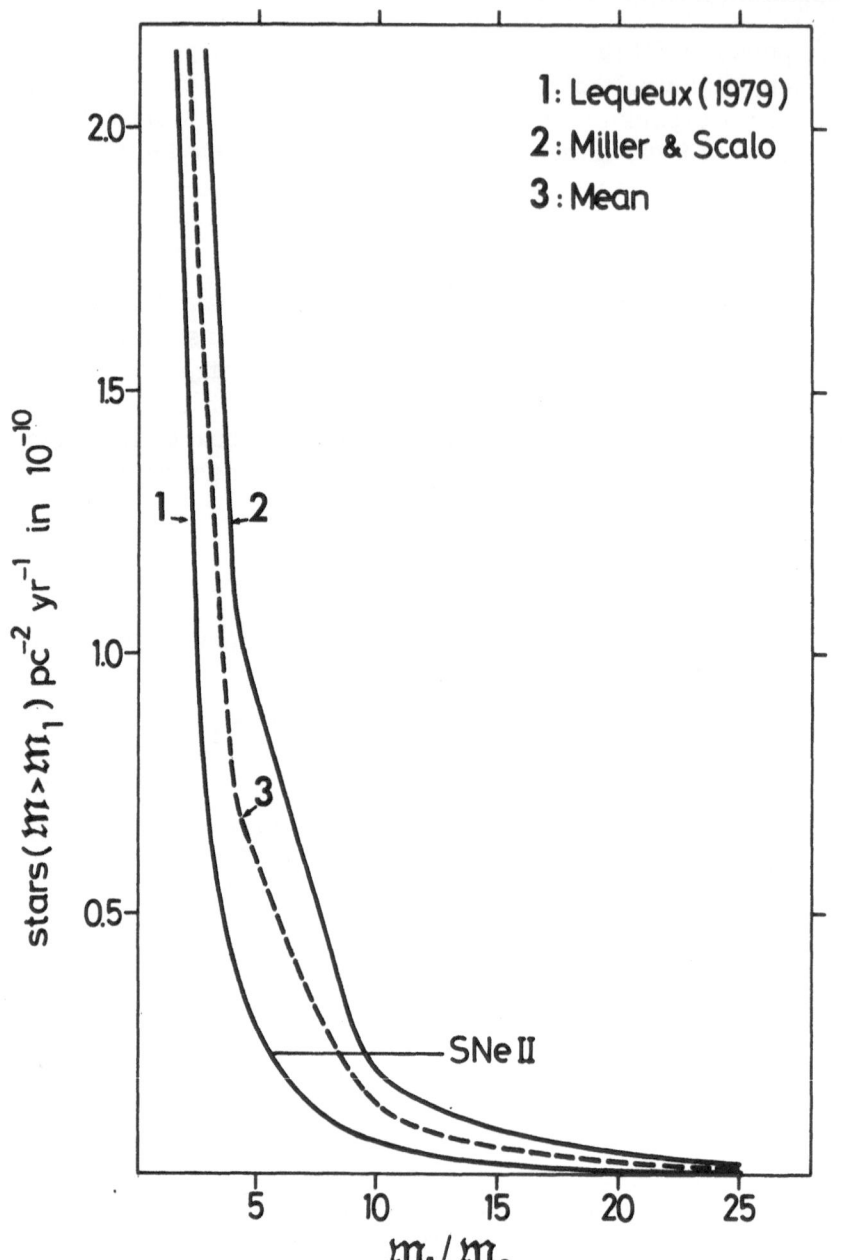

Fig. 7. The IMF of the solar neighbourhood according to Lequeux (1979; and similarly Ostriker, Richstone, and Thuan, 1974) and to Miller and Scalo (1979). The integral number is shown of stars with mass $> \mathfrak{M}_1$ (in solar units) born per pc^2 per yr.

mean IMF of Lequeux (1979) and Miller and Scalo (1979). Using the separate IMFs of these authors one finds $5.6 < \mathfrak{M}_i < 9.6\, \mathfrak{M}_\odot$. If in addition an error in n_0 of a factor of 2 is allowed for, the mass range becomes $3.7 < \mathfrak{M}_i < 13\, \mathfrak{M}_\odot$. Even if one accepts the extreme, improbably high limit $\mathfrak{M}_i > 13\, \mathfrak{M}_\odot$ it is clear that only a very small fraction of SNe II can have progenitors with $\mathfrak{M}_i > 20\, \mathfrak{M}_\odot$. This limits the significance of SN II models involving very high masses.

Stars with $\mathfrak{M}_i > 3\, \mathfrak{M}_\odot$ have scale heights in the z-direction of $\beta = 90$ pc. This calls for an explanation why the two suspected galactic SNe II, viz. SN 1054 and SN 1181, have $|z| = 202$ and 443 pc, respectively. As a solution the following picture emerges.

Low-$|z|$ and low-space velocity ($v < 25$ km s^{-1}) O stars with $\mathfrak{M}_i > 20\, \mathfrak{M}_\odot$ develop a He core with $> 3\, \mathfrak{M}_\odot$, become Wolf-Rayet stars (WC, WN), which also have $< |z| > = \beta = 90$ pc (Moffat and Seggewiss, 1979), and explode as SNe II (e.g. Maeder, 1981). These stars are normally in double or multiple systems; only $< 12\%$ are single (Stone, 1981). The O stars in close binary systems transfer mass during their evolution, which typically results in a massive (O-type) secondary at the time of the primary's explosion. In most cases the system is not disrupted during explosion, but kicked off (e.g. van den Heuvel, 1976).

About 50 % of the most massive O stars ($20 < \mathfrak{M}_i < 100\, \mathfrak{M}_\odot$) have high space velocities ($v = 25$-100 km s^{-1}) and high $|z|$ ($\beta \approx 150$ pc) (Stone, 1981). The majority of them have no observable companion (Stone, 1981) and they have no red supergiant counterparts (Lequeux, 1979). They are the secondaries of the low-$|z|$ O stars, after mass accretion and kick-off. Their former primaries, presumably now collapsars, are not expected to be directly observable. The secondaries evolve on into a second Wolf-Rayet stage (a good candidate is the single-line binary HD 197 406 at z = 1000 pc; Moffat and Seggewiss, 1979), into massive X-ray binaries, 50 % of which have $|z| > 95$ pc (van den Heuvel, 1980), and SS 433 objects (z > 140 pc; van den Heuvel, 1980), and finally erupt as high-$|z|$ SNe II (like SN 1054 and SN 1181), leaving a second pulsar (like the pulsar 0527 and possibly the X-ray point source in 3C 58; Helfand 1981). In this scenario the possible pulsar pair 0527 and 0531 (Gott, Gunn, and Ostriker, 1970; Trimble and Rees, 1971) is only atypical in so far as the system was disrupted during the first SN explosion.

We have seen above that stars with $> 20\, \mathfrak{M}_\odot$ can account for only a fraction of all SNe II (even if every O star led to two SN II events). The majority of SNe II must therefore come from B stars

(roughly B0 to B3 with $\sim$ 20-8 $\mathfrak{M}_\odot$). Their incidence of multiple
systems is significantly lower and they rarely form high-velocity
stars (Stone, 1981). The bulk of the radio SNRs ($\beta \approx$ 60 pc !) and
of pulsars, which are born near to the galactic plane (Section 2.3),
must therefore come from B stars. Here as well as in any other
scenario it remains to be explained why pulsars obtain high space
velocities at birth.

4.2. SNe I as Old Population Objects.

SNe I do occur in E galaxies. This is clear evidence that their
progenitors are old stars and cannot be massive. It has been tried
to bypass this evidence by postulating some star formation in E
galaxies, but to explain the SN frequencies (Table 2) the star formation rate per unit luminosity in E galaxies could be down by
only a factor of 6 as compared to Sc galaxies. To test this hypothesis a special hydrogen search program in ellipticals was
started, which has resulted in the first 3 σ-detection of HI in E
galaxies at hydrogen mass-to-luminosity levels of $\mathfrak{M}_{HI}/L_B < 10^{-3}$
(Huchtmeier, Tammann, and Wendker, 1975), which is at least
300 times lower than in Sc galaxies (Tammann, 1980b), arguing
strongly against significant amounts of star formation. A concentration of the little HI in E galaxies into the centre would not
help, because SNe have a surface distribution similar to that of
the total light of ellipticals (Maza and van den Bergh, 1976). If
only a fraction of all E galaxies, e.g. the bluest ones, should
contain massive SN progenitors, they would have to be unreasonably
prolific to account for the overall SN frequency in ellipticals.

An additional argument against massive SN I progenitors has
so far not been given sufficient weight: the z-distribution of historical SNe in the Galaxy (β = 214 pc) makes it highly improbable
that they could be related with any stars with $\mathfrak{M}_i > 3 \mathfrak{M}_\odot$ (β = 90 pc).
For the two presumable Type II SNe in the Galaxy (SN 1054 and
SN 1181) one could invoke in the previous Section the high-velocity
O stars. But not all seven galactic SNe of the last millennium can
come from these rare stars! Then they must come from stars with
$\beta \approx$ 200 pc, i.e. from stars with $\mathfrak{M}_i < 1.5 \mathfrak{M}_\odot$ (cf. Miller and Scalo,
1979).

Further support for the assumption that SNe I do not have massive progenitors comes from the lack of correlation between SNe I
frequencies and galaxy colour (Tammann, 1977; see, however,

Oemler and Tinsley, 1979; Caldwell and Oemler, 1981), from
their not being concentrated toward spiral arms (Maza and van
den Bergh, 1976), and from their radial distribution, which anti-
correlates with the radial hydrogen distribution (Tammann, 1977 b).
SNe I, on the other hand, are not halo objects, because they concentrate towards the disks of spiral galaxies (Tammann, 1977),
and for six SNe I in edge-on spirals one finds an upper limit of
$< |z| > <$ 700 pc (Tammann, 1977 b). In the intracluster medium
their frequency per unit luminosity is more than 5 times lower
than in E/S0 galaxies (Crane, Tammann, and Woltjer, 1977).

SNe I come therefore from objects common to the intermediate-
age or old disk population and to the population of E galaxies. It is
evident that in view of the vast numbers of stars in these populations,
only a tiny fraction can end up as SNe I, i.e. those which fulfill a
very specific condition. An attractive possibility are white dwarfs
which are driven over the Chandrasekar limit by accretion from
an evolving companion star. Since the first proposal of this model
by Schatzman (1963) it has gained increasing consideration by
theorists, - although it is not clear how an accreting white dwarf
can secularly gain mass, without loosing this mass intermittently
in nova or dwarf-nova explosions, and without forming an extended
red-giant-like atmosphere (e.g. Sugimoto and Nomoto, 1980;
Wheeler, 1981; Ritter, 1981).

The known statistical properties of SNe I are all in perfect
agreement with the white-dwarf hypothesis. It could be argued
that in this case the SN I rate in E galaxies should be higher than
in any other type of galaxy - contrary to observation - because the
fraction of white dwarfs must be highest in E galaxies. This ob-
jection, however, is invalid. Not the number of white dwarfs is
important, but the (unknown) number of white dwarfs in evolving
close double systems as a function of galaxian type. It is well
possible that such binaries become relatively rare in highly
evolved galaxies.

The main conclusion of this Section is that all known statistical
properties of SNe can be explained if one assumes that SNe II come
from stars with $\geqslant 8 \, \mathfrak{M}_\odot$, and that SNe I are the result of accreting
white dwarfs.

The author thanks all his colleagues, who were willing to discuss the topics of this talk with him. Dr. J. P. Ostriker was one of
them. He expresses his gratitude to the organizers of this conference and particularly to Prof. Martin Rees for hospitality in Cambridge.
His work was supported by the Swiss National Science Foundation.

REFERENCES.

Angerhofer, P.E.: 1981, Archaeoastronomy 4, 23.

Arnett, W.D.: 1981, preprint.

Arp, H.C.: 1961, Ap. J.133, 874.

Baade, W., and Swope , H.H.: 1963, A. J.68, 435.

Barbon, R., Ciatti, F., and Rosino, L.: 1973, Astron. Astro-
        phys. 25, 241.

Barbon, R., Ciatti, F., and Rosino, L.: 1979, Astron. Astro-
        phys. 72, 287.

Barbon, R.: 1980, "Type I Supernovae", ed. J.C.Wheeler, Austin:
        University of Texas, p. 16.

Benvenuti, P.: 1981. Paper presented at this Conference.

Bergh, S. van den: 1971, Ap. J.165, 457.

Bergh, S. van den: 1980, Publ. Astron. Soc. Pacific 92, 768.

Branch, D.: 1977, "Supernovae", ed. D. N. Schramm, Dordrecht:
        Reidel, p. 21.

Branch, E.: 1982, this volume.

Branch, D., and Bettis, C.: 1978, A. J.83, 224.

Branch, D., Falk, S.W., McCall, M.L., Rybski, P., Uomoto,
        A.K., and Wills, B. J.: 1981, Ap. J.244, 780.

Cadonau, R., and Tammann, C.A.: 1981, to be published.

Caldwell, C.N., and Oemler, A.: 1981, preprint.

Caswell, J. L., and Lerche, I.: 1979, M.N.187, 201.

Chevalier, R.A., Kirshner, R.P., and Raymond, J.C.: 1980,
        Ap. J.235, 186.

Clark, D.H.: 1981, private communication.

Clark, D.H., and Caswell, J.L.: 1976, M.N.174, 267.

Clark, D. H., and Stephenson, F. R.: 1977, "The Historical
        Supernovae", Pergamon Press, Oxford, 10 + 233 pp.

Crane, P., Tammann, G.A., and Woltjer, L.: 1977, Nature 265,
        124.

Danziger, I. J., and Goss, W. M.: 1980, M.N.190, 47P.

Freeman, K.C.: 1970, Ap. J.160, 811.

Gordon, M.A., and Burton, W.B.: 1976, Ap. J.208, 346.

Goss, W. M., Schwarz, U.J., and Wesselius, P.R.: 1973, Astron.
        Astrophys. 28, 305.

Gott, J.R., Gunn, J.E., and Ostriker, J.P.: 1970, Ap. J. Letters
        160, L91.

Graham, J.A.: 1979, I.A.U.Symp. 84, 195.

Gunn, J.E., and Ostriker, J.P.: 1970, Ap. J.160, 979.

Guseinov, O.H., and Kasumov, F.K.: 1981, I.A.U.Symp. 95, 437.

Harding, A.K.: 1981, I.A.U.Symp. 95, 439.

Helfand, D. J.: 1981. Paper presented at this conference.

Heuvel, E. P. J. van den: 1976, I. A. U. Symp. 73, 35.

Heuvel, E. P. J. van den: 1980, preprint.

Holmberg, E.: 1958, Medd. Lund Obs. Ser. II, No. 136.

Huchtmeier, W. K., Tammann, G. A., and Wendker, H. J.: 1975, Astron. Astrophys. 42, 205.

Jones, K. G.: 1976, J. Hist. Astron. 7, 27.

Kirshner, R. P.: 1982. This volume.

Kirshner, R. P., and Kwan, J.: 1974, Ap. J. 193, 27.

Knapp, G. R., Tremaine, S. D., and Gunn, J. E.: 1978, A. J. 83, 1585.

Kormendy, J.: 1977, Ap. J. 217, 406.

Köster, D., and Reimers, D.: 1981, in press.

Lequeux, J.: 1979, Astron. Astrophys. 80, 35.

Lyne, A. G.: 1982. This volume.

Maeder, A.: 1981, Astron. Astrophys., in press.

Maza, J., and Bergh, S. van den: 1976, Ap. J. 204, 519.

Miller, J. S.: 1973, Ap. J. Lett. 180, L83.

Miller, G. E., and Scalo, J. M.: 1979, Ap. J. Suppl. 41, 513.

Milne, D. K.: 1979, Austr. J. Phys. 32, 83.

Moffat, A. F. J., and Seggewiss, W.: 1979, I. A. U. Symp. 83, 447.

Oemler, A., and Tinsley, B. M.: 1979, A. J. 84, 985.

Oke, J. B., and Searle, L.: 1974, Ann. Rev. Astron. Astrophys. 12, 315.

Oort, J. H., and Plaut, L.: 1975, Astron. Astrophys. 41, 71.

Ostriker, J. P., Richstone, D. O. and Thuan, T. X.: 1974, Ap. J. Lett. 188, L87.

Panagia, N., and Weiler, K. W.: 1980, Astron. Astrophys. 82, 389.

Pskovskii, Yu. P.: 1977, Astron. Zh. 54, 1188.

Pskovskii, Yu. P.: 1978, Astron. Zh. 55, 737.

Reina, C., and Tarenghi, M.: 1973, Astron. Astrophys. 26, 257.

Ritter, H.: 1981, private communication.

Romanishin, W., and Angel, J. R. P.: 1980, Ap. J. 235, 992.

Sandage, A.: 1973, Ap. J. 183, 711.

Sandage, A., and Brucato, R.: 1979, A. J. 84, 472.

Sandage, A., and Tammann, G. A.: 1981, "A Revised Shapley-Ames Catalog of Bright Galaxies", Washington: Carnegie Institution.

Sargent, W. L. W., Searle, L., and Kowal, C. T.: 1974, "Super-novae and Supernova Remnants", ed. C. B. Cosmovici, Dordrecht: Reidel, p. 33.

Schatzman, E.: 1963, "Star Evolution", ed. L. Gratton, New York: Academic Press, p. 389.

Schmidt-Kaler, Th., and Schlosser, W.: 1973, Astron. Astrophys. 29, 409.

Schurmann, S. R., Arnett, W. D., and Falk, S. W.: 1979, Ap. J. 230, 11.

Searle, L.: 1971, Ap. J. 168, 41.

Shaw, R. L.: 1979, Astron. Astrophys. 76, 188.

Stone, R. C.: 1981, A. J. 86, 544.

Sugimoto, D., and Nomoto, K.: 1980, Space Sci. Rev. 25, 155.

Tammann, G. A.: 1974, "Supernovae and Supernova Remnants", ed. C. B. Cosmovici, Dordrecht: Reidel, p. 155.

Tammann, G. A.: 1976, "Proc. 1976 DUMAND Summer Workshop", ed. A. Roberts, Batavia: Fermi National Accelerator Laboratory, p. 137.

Tammann, G. A.: 1977, "Supernovae", ed. D. N. Schramm, Dordrecht: Reidel, p. 95.

Tammann, G. A.: 1977b, Ann. New York Acad. Sci. 302, 61.

Tammann, G. A.: 1978, Mem. Soc. Astron. Ital. 49, 315.

Tammann, G. A.: 1979a, "Astronomical Uses of the Space Telescope", ed. F. Macchetto, R. Pacini, and M. Tarenghi, Geneva: ESA/ESO, p. 329.

Tammann, G. A.: 1979b, "Scientific Research with the Space Telescope", I. A. U. Coll. No. 54, ed. M. S. Longair and J. W. Warner, Washington: U. S. Government Printing Office, p. 263.

Tammann, G. A.: 1980, "Physical Cosmology," ed. R. Balian, J. Audouze, and D. N. Schramm, Amsterdam: North-Holland, p. 127.

Tammann, G. A.: 1980b, "Dwarf Galaxies", ed. M. Tarenghi, and K. Kjär, Geneva: ESO/ESA, p. 73.

Tammann, G. A., and Kraan, R.: 1978, I. A. U. Symp. 79, 71.

Trimble, V.: 1968, A. J. 73, 535.

Trimble, V., and Rees, M. J.: 1971, Ap. J. Letters 166, L85.

Vaucouleurs, G. de: 1979, I. A. U. Symp. 84, 203.

Vaucouleurs, G. de, and Buta, R.: 1981, Publ. Astron. Soc. Pacific 93, 294.

Vaucouleurs, G. de, Vaucouleurs, A. de, Buta, R., Ables, H. D., and Hewitt, A. V.: 1981, Publ. Astron. Soc. Pacific 93, 36.

Vaucouleurs, G. de, Vaucouleurs, A. de, and Corwin, H. G.: 1976, "Second Reference Catalogue of Bright Galaxies", Austin: Univ. of Texas Press.

Wagoner, R. V.: 1980, "Physical Cosmology", ed. R. Balian, J. Audouze, and D. N. Schramm, Amsterdam: North-Holland, p. 179.

Wagoner, R. V.: 1982. This volume.

Wegner, G.: 1981, A. J. 86, 264.

Weidemann, V.: 1978, I. A. U. Symp. 76, 353.
Weidemann, V.: 1981, I. A. U. Coll. No. 59, 339.
Wheeler, J. C.: 1981, Rep. Progress Phys., in press.
Williams, D. R. W.: 1973, Astron. Astrophys. 28, 309.

THE GALACTIC DISTRIBUTION OF PULSARS

A.G. Lyne

Nuffield Radio Astronomy Laboratories
Jodrell Bank

ABSTRACT

The galactic distribution of pulsars follows the general
form of many population I objects in galactocentric radius, but
has a wide distribution above and below the plane due to high
space velocities imparted to the pulsars at birth.

Statistical studies of the properties of large numbers of
pulsars and proper motion measurements demonstrate that the
effective magnetic dipole moments decay on a timescale of about
8 million years. This work provides a better knowledge of pulsar
evolution and ages and shows that a birthrate of one pulsar
every 20 to 50 years is required to sustain the observed galactic
population of 300,000. This rate is comparable with most recent
estimates of the galactic supernova rate, but requires nearly
all supernovae to produce active pulsars.

1. INTRODUCTION

The identification of radio pulsars as the collapsed cores
remaining after supernova explosions is generally accepted because
supernova events provide a theoretically acceptable formation
mechanism. The direct experimental evidence however is rather
weak: only two pulsars are firmly identified with supernova
remnants by both position and age, PSR 0531+21 and PSR 0833-45
with the Crab and Vela supernova remnants respectively.

This paper discusses what is known about the galactic
distribution, evolution and origin of pulsars and what we can

M. J. Rees and R. J. Stoneham (eds.), Supernovae: A Survey of Current Research, 405–417.
Copyright © 1982 by D. Reidel Publishing Company.

learn of the formation event. Some of this work is the preliminary result of an analysis with Dick Manchester and Joe Taylor
(Lyne, Manchester and Taylor 1981), while other parts come from
a study with Bryan Anderson and Mike Salter at Jodrell Bank concerning the proper motions (Lyne, Anderson and Salter 1981).

There are now 330 known pulsars, most of which have been
discovered in the five main surveys carried out at Arecibo, Jodrell
Bank, Molonglo and Greenbank, all at frequencies of about 400 MHz.
(Hulse and Taylor 1974, 1975; Davies, Lyne and Seiradakis 1972,
1973; Large and Vaughan 1971; Manchester et al. 1978 and
Damashek et al. 1978). Figure 1 shows a plot of the known pulsars
in galactic coordinates and in this the concentration of pulsars
along the galactic plane is clear. However, although there is
now fairly good coverage of the whole sky, the depth of coverage
is not completely uniform. For instance the high density of
pulsars around longitude 50⁰ is due to the very deep Arecibo survey
which covered only a very limited area of sky close to the galactic
plane. The low density of pulsars towards the centre of the
Galaxy is due, at least in part, to the reduced receiver sensi-
tivities caused by the very high galactic background noise in
this direction.

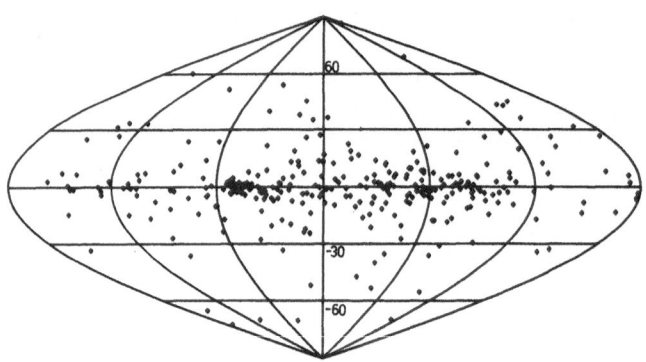

Figure 1: The distribution of pulsars in galactic coordinates.

2. THE DISTANCE SCALE

In order to study the distribution of pulsars through the
Galaxy, it is necessary to know their distances and fortunately
the dispersion measure provides an approximate indicator when
combined with some model of the galactic distribution of electrons.
Also, independent distance estimates can be obtained for a number

of pulsars. The main independent distance estimates are obtained
from the absorption of the 21 cm radiation from the pulsar by
interstellar HI when combined with a model for the differential
rotation of the Galaxy. Such observations have been carried out
for about 30 objects. See for example Graham et al. (1974),
Ables and Manchester (1976) and Booth and Lyne (1976).

These observations have been used, together with a certain
amount of other information, to develop and to calibrate a model
for the distribution of the electron density, n_e, through the
Galaxy. Hall (1980) gives a useful survey of our knowledge of
the electron distribution in the Galaxy. In the model most
recently adopted, n_e has the form

$$n_e = 0.025 + 0.015 \exp\left(-|z|/70\text{pc}\right) \frac{2}{(1 + R/R_0)} \quad \text{cm}^{-3}$$

where z is the distance above the galactic plane, R is the
galactocentric radius and R_0 is taken to be 10 kpc (Lyne,
Manchester and Taylor 1981).

In order to determine the distance to a pulsar using this
model we move out from the Sun along the line of sight to the
pulsar until we reach the distance at which the integrated column
density is equal to the dispersion measure of the pulsar. The
distances so obtained agree satisfactorily with other measurements,
although there are a number of discrepancies, usually associated
with the line of sight passing through large HII regions. For
this reason, where the lines of sight pass through the Gum Nebula,
the most notable contributor to these discrepancies, esimates of
the nebular contribution are used to modify the calculated
distances.

3. THE GALACTIC DISTRIBUTION

From the distance of a pulsar, d, determined either from
HI absorption measurements or from the dispersion measure as
described above, it is possible to calculate 3 astrophysically
useful parameters.

     1) The galactic z-distance, z
     2) The galactocentric radius, R
and   3) A luminosity parameter, $L = S_{400}d^2$
where S_{400} is the flux density at 400 MHz.

Figure 2 shows the positions of the 330 known pulsars
projected onto the galactic plane, where the Sun is assumed to
be at a distance of 10 kpc from the Galactic Centre. We note the

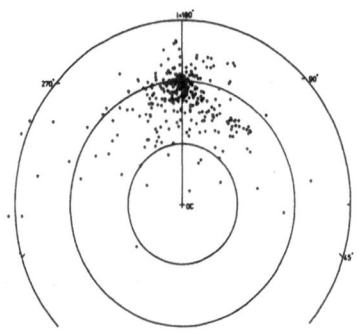

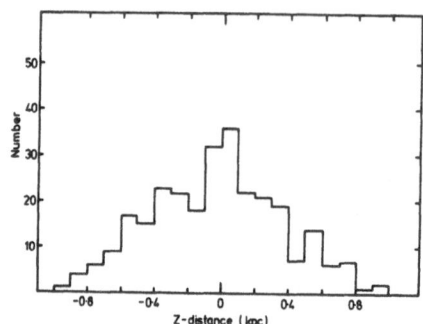

Figure 2:   The distribution of
pulsars projected onto the
galactic plane

Figure 3:   The observed z-
distribution of 321 pulsars

strong selection effect caused by the inverse square law and it
is also obvious that there are more pulsars within the solar
circle than outside it.

The observed distribution in z-distance is shown in figure
3 which is affected only slightly by selection effects, the most
important being a reduction in sensitivity to low z-distance
pulsars because of the high galactic background radiation at low
latitudes. We see an approximately exponential distribution
having a scale height of about 400 pc. The observed distributions
in the galactocentric radius R and luminosity L are shown in
figures 4 and 5, both of which are of course severely affected
by selection effects. Again the inverse square law is the main

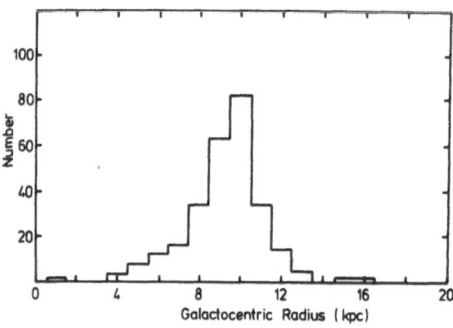

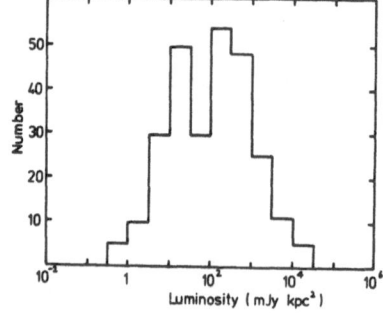

Figure 4:   The Observed galactocentric radius distribution

Figure 5:   The observed
luminosity distribution

culprit causing the peak at R ≈ 10 kpc and the small number of observed low luminosity pulsars because of the small volume of Galaxy searched in which such weak pulsars can be detected from the Earth.

As we have seen, the main selection effects at play in the distributions in z-distance z, galactocentric radius R and luminosity L (Figs. 3,4 and 5) are due to the inverse square law, the areas of sky surveyed and the changing receiver sensitivity due to the galactic background noise variations across the Galaxy. Various methods have been used to correct for these observational selections (Lyne 1981). In effect they consist of calculating the volume of space searched for pulsars having luminosity L and lying between z and z+dz and R and R+dR in any surveys (Large 1971, Taylor and Manchester 1977 and Davies, Lyne and Seiradakis 1977). From the number of pulsars observed and the volume searched it is possible to calculate a space density at a particular z, R and L. The results of such an analysis carried out for the five main surveys mentioned in section 1 are summarised for the distributions in R and L in figures 6 and 7. The true distribution in z has a scale height of about 350 pc and is not much different from the observed distribution (Figure 3) and is not shown.

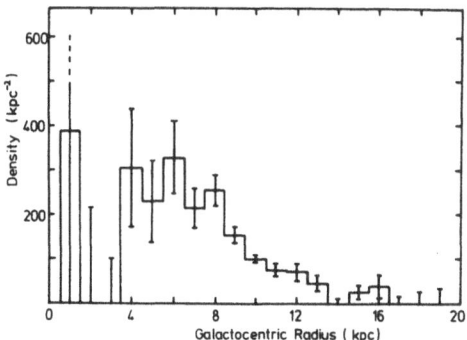

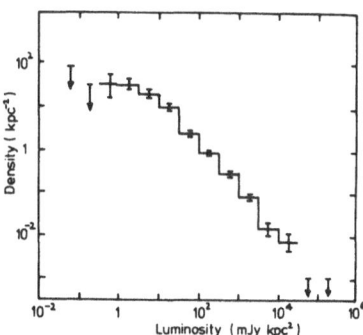

Figure 6: The derived galacto-centric radius distribution

Figure 7: The derived luminosity distribution

The luminosity distribution in Figure 7 shows the density
of pulsars at the position of the Sun as the number of pulsars
in each semi-decade range of luminosity per square kiloparsec
of the plane. This plot increases steadily with decreasing
luminosity and although it shows signs of flattening at low
luminosities there is no well-defined low luminosity cut-off,
so that it is possible only to derive a lower limit on the total
density of pulsars. However, assuming that there are no pulsars
with luminosity less than 0.3 mJy kpc^2 then integration under
this curve gives a density of 98 observable pulsars kpc^{-2} at
the position of the Sun (R $\simeq$ 10 kpc).

It can be shown that errors in the assumed distances
cause a systematic overcount of low luminosity pulsars. A rough
estimate of the errors in the distance determinations causes
this figure to be adjusted downwards to 65 kpc^{-2}.

The galactocentric radial distribution in Figure 6 is a
strong function of radius, R, and shows a marked resemblence
to that of supernova remnants and HII regions and probably also
of massive stars. It is reassuring that the distribution varies
reasonably continuously through the position of the sun (R =
10 kpc) showing that the powerful selection effects responsible
for the peak in Figure 4 have been substantially removed by
the analysis.

The scale height of about 400 pc which we see in z is
substantially greater than that of most massive stars which
are likely to be the progenitors of pulsars, and also than
that of supernova remnants, both of which have scale heights
of about 70 pc. Gunn and Ostriker (1970) provided the explanation for this as pulsars being runaway stars which have
moved large distances since birth.   This can now be seen
quite explicitly from recent proper motion measurements on
26 pulsars using interferometry at Jodrell Bank (Lyne,
Anderson and Salter 1981).

In Figure 8 the significant proper motions have been
converted to velocities using the pulsar distance estimates
and are shown as a function of galactic longitude and galactic
z-distance. It is clear that with one exception all the pulsars
outside a "progenitor" layer of width of about 70 pc above and
below the galactic plane are migrating away from the plane.
The one exception is PSR 1237+25 which lies very close to the
galactic pole so that the observed transverse motion is
dominated by the velocity component parallel to the galactic
plane. The mean transverse velocity is about 145 km s^{-1} for
the whole of the Jodrell sample (assuming those with only upper
limits to have zero proper motion) and corresponds to a Max-
wellian distribution having a velocity dispersion of 120 km s^{-1}.
This compares with estimates from previous, smaller samples of
pulsars of 130 km s^{-1} (Taylor and Manchester 1977), 150 km s^{-1}
(Helfand and Tademaru 1977) and 70 km s^{-1} by Hanson (1979).
These smaller samples were certainly influenced by some extent
by a selection tendency towards large proper motions, and the
work of Hanson (1979) attempted to take into account the
effects of the selection which now seem to have been rather
over-estimated.

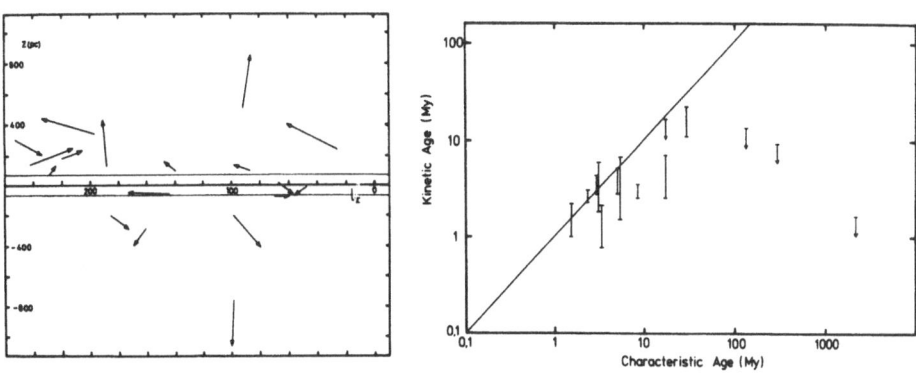

Figure 8: The velocity   vectors
of 18 pulsars plotted as a function
of galactic longitude and z-
distance. The pulsar's position is
at the back of the arrow. The
length of the arrow represents
the velocity, the largest being
400 km s^{-1}.

Figure 9: The kinetic age
plotted against the
characteristic age for 14
pulsars

The mean component of velocity normal to the galactic plane is close to 100 km s^{-1} and the deduced mean space velocity of the pulsars in the sample is 180 km s^{-1}. We thus have the picture of pulsars being born close to the plane from progenitors having a population I form of distribution and being given a velocity "kick" at birth of a few hundred km s^{-1}; an order of magnitude greater than the random motions within such a population.

The total number of observable pulsars is obtained by integrating the distribution in R through the galactic disc and gives a value of 54 x 10^3. Note that the rather poorly determined density near the galactic centre does not affect the accuracy of this value much because of the relatively small volume of space involved. Because of geometrical effects, we can expect to observe only a small fraction, f, of all the active pulsars in the Galaxy, i.e. those whose radiation beams cross the Earth during a rotation of the pulsar. If the radiation beams have the same width in latitude as in longitude (and for most polar cap models this is the case) then f ∿ 0.2. For most relativistic beaming models it seems that a similar value of f is also appropriate. Certainly, it is difficult to see how a value of f much less than this can come about. Adopting a value for f of 0.2, then the total number of active pulsars in the Galaxy is N_G = (270 ± 160) x 10^3 for L > 0.3 mJy kpc^2. The main contributions to the error in this figure are.

a)   Statistical   30%
b)   The local distance scale   40%
c)   The beaming factor   30%

It should be noted that although only about 4% of the observed pulsars have luminosities between 0.3 and 3 mJy kpc^2, they represent between 80 and 90 percent of the galactic population. This highlights the fact that the pulsars observed from the Earth are mostly a small minority of very high luminosity objects which have a mean luminosity of about 100 times the mean luminosity of the galactic population (see Figures 5 and 7).

We see therefore that the estimates of the galactic population rely heavily upon the statistical measurements of only a small number of low luminosity pulsars which lie within a few hundred parsecs of the Sun. They are too close for HI absorption measurements to determine their distance and unfortunately we have had to rely upon the dispersion measure as described in section 2. However, both Lyne (1974) and Hall (1980) have concluded that the local mean electron density is similar to that elsewhere in the Galaxy and provided that this is the case the population estimates will not be greatly in error. In fact if Hall (1980) is correct and the electrons are more closely confined to the plane than previous models have suggested and the population

figures will have to be revised upwards. The recent measurement
of the trigonometric parallax on PSR 1929+10 (Salter, Lyne and
Anderson 1979), one of the low luminosity pulsars, indicates a
mean electron density of 0.069 cm^{-3} and lends some support to
Hall's model.

4. THE LIFETIMES AND BIRTHRATE OF PULSARS

In order to determine the birthrate of pulsars which is
required to sustain the galactic population, it is necessary to
investigate the evolution and "lifetimes" of pulsars.

Let us consider again the proper motion observations and the
migration from the plane. Clearly it is possible to extrapolate
a pulsar path back onto the galactic plane and thus estimate an
"age" since the initial ejection from the progenitor population.
There will be errors in such estimates due to the uncertainty of
z at birth, the unknown component of velocity along the line of
sight (and hence the precise form of the track on the sky) and
the errors in the proper motion measurements themselves. In
figure 9, these "kinetic" ages are compared with the "characteristic" ages $t_c = P/2\dot{P}$, obtained from an assumption of magnetic
braking by a constant magnetic field. The form of this plot
confirms the doubts expressed by Lyne, Ritchings and Smith (1975)
and Taylor and Manchester (1977) that the characteristic age is
not a good estimate for ages of greater than a few million years.
Figure 9 is consistent with the effective magnetic dipole
moment decaying on a timescale, t_d, of between 4 and 10 million
years. This was first proposed by Gunn and Ostriker (1970) but
the experimental evidence in its favour has up to now been weak.
In their model, the surface magnetic field decays exponentially
with time and it is possible to calculate the true age, t, from
the characteristic age, t_c. Thus

$$B = B_0 \exp (-t/t_d)$$

$$t = \tfrac{1}{2}t_d \ln (1 + 2t_c/t_d)$$

Gunn and Ostriker also suggested that the luminosity
depended upon the square of the magnetic field. Their simple
model fits the data remarkably well. For instance in Figure 10
we see the distribution of characteristic ages and the expected
distribution for a magnetic field characteristic decay time of
8 million years. Figure 11, showing the luminosity as a function
of true age, lends some further support to the model where the
data is reasonably consistent with a luminosity decay time of 4
million years, i.e. a magnetic field decay time of 8 million years.

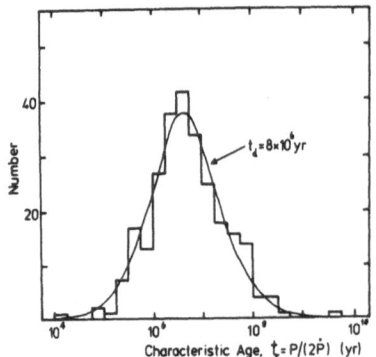

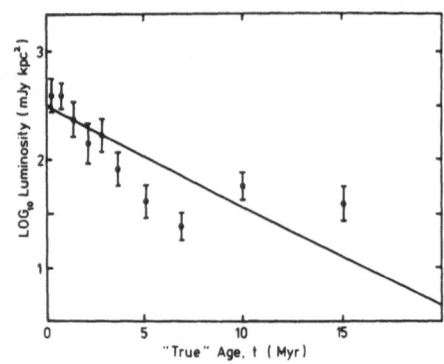

Figure 10: The characteristic age distribution of pulsars and the fit of the model of Gunn and Ostriker (1970).

Figure 11: The mean luminosity of pulsars as a function of the true (or reduced) age, t.

In this model therefore pulsars move from one luminosity bin to the next (figure 7) in 4.6 My. Then the number required to be born in 4.6 million years in order to sustain the observed density at the Sun is

$$\sum_{L=L_{min}}^{\infty} \left[N(L) - N(L+1) \right] = N(L_{min}) \, kpc^{-2} = 38 \, kpc^{-2}$$

leading to a birthrate of $8 \pm 3 \, kpc^{-2} \, My^{-1}$. Integrating through the Galaxy and applying a 0.2 beaming factor gives a required birthrate of 1 every 28 years. Taking into account the various uncertainties we believe the birthrate must lie in the range of 1 every 20 to 1 every 50 years.

Many other estimates of the birthrate (e.g. Davies, Lyne and Seiradakis 1977; Taylor and Manchester 1977) have used the concept of a pulsar "lifetime", the pulsar dying suddenly after a certain age, the death probably being associated with the phenomenon of pulse nulling (Ritchings, 1976). These lifetimes have then been estimated from the ages of the <u>observed</u> pulsars, which as we have seen are the superluminous minority and which we have now seen to be relatively young. This has caused the underestimation of the age of the pulsar population as a whole and an overestimation of the birthrate in the past.

5. DISCUSSION AND CONCLUSIONS

The main facts pertinent to the origin and evolution of pulsars are as follows:-

1) The galactic radial distribution is the same as that of massive stars and supernova remnants.

2) The scale height of pulsars in z is about 350 pc, and most are moving away from the galactic plane.

3) The scale height of the progenitors is <100 pc, implying a population I origin.

4) The number of active pulsars in the Galaxy is about 300,000.

5) The magnetic fields decay on a timescale of about 8 My.

6) The radio luminosity is roughly proportional to B^2 and decays on a timescale of 4 My.

7) Ejection velocities from the birth event are about 200 km s^{-1}.

8) Only 3 pulsars out of 330 known are in binary systems.

9) A birthrate of 1 pulsar every 20-50 years is required to sustain the observed population.

This birthrate is just about compatible with recent estimates of the rates of supernova from historical and extragalactic evidence, but rather higher than those derived from supernova remnants (Tammann 1977, Caswell and Lerche 1979, Clark and Stephenson 1977).

Although it is now generally accepted that pulsars are formed in the supernovae of massive stars, the direct experimental evidence for this is still not very convincing. The main evidence is the almost incontrovertible association of the pulsars PSR 0531+21 and PSR 0833-45 with the Crab Nebula and Vela supernova remnants respectively because of the positional agreements and similarity of ages. The supernova events also provide a theoretically acceptable formation mechanism as well as an explanation for the rather high space velocities of pulsars and the small number in binary systems. The similarity in the galactic radial distributions of pulsars and supernova remnants also provides some weak circumstantial evidence, although there are many other populations of galactic objects which have the same distributions. It seems that if supernovae are solely responsible for pulsar production, every supernova must produce an active radio pulsar.

From the evidence as it stands', we must still be prepared
to question whether all pulsars are born in supernovae and to
consider instead whether they may be born in rather less
violent and less obvious events which are nevertheless violent
enough to provide high space velocities and disrupt most binary
systems.

REFERENCES

Ables, J.G. and Manchester, R.N.: 1976, Astron.Astrophys.
    50, p.177.
Booth, R.S. and Lyne, A.G.: 1976, Mon.Not.R.astr.Soc. 174,p.53P.
Caswell, J.L. and Lerche, I.: 1979, Mon.Not.R.astr.Soc. 187,p.201.
Clark, D.H. and Stephenson, F.R.: 1977, Mon.Not.R.astr.Soc.
    179, p.87.
Damashek, M., Taylor, J.H. and Hulse, R.A.: 1978, Astrophys.J.
    Letters 225, p.L31.
Davies, J.G., Lyne, A.G. and Seiradakis, J.H.: 1972, Nature
    240, p.229.
Davies, J.G., Lyne, A.G. and Seiradakis, J.H.: 1973, Nature
    Phys.Sci. 244, p.84.
Davies, J.G., Lyne, A.G. and Seiradakis, J.H.: 1977, Mon.Not.
    R.astr.Soc. 179, p.635.
Graham, D.A., Mebold, U., Hesse, K.H., Hills, D.L. and
    Wielebinski, R.: 1974, Astron.Astrophys. 37, p.405.
Gunn, J.E. and Ostriker, J.P.: 1970, Astrophys.J. 160, p.979.
Hall, A.N.: 1980, Mon.Not.R.astr.Soc. 191, p.751.
Hanson, R.B.: 1979, Mon.Not.R.astr.Soc. 186, p.357.
Helfand, D.J. and Tademaru, E.: 1977, Astrophys.J. 216, p.842.
Hulse, R.A. and Taylor, J.H.: 1974, Astrophys.J.Letters 191, p.L59.
Hulse, R.A. and Taylor, J.H.: 1975, Astrophys.J.Letters 201, p.L55.
Large, M.I.: 1971, The Crab Nebula, I.A.U. Symposium No. 46,
    Reidel, p.165.
Large, M.I. and Vaughan, A.E.: 1971, Mon.Not.R.astr.Soc. 151,
    p.277.
Lyne, A.G.: 1974, Galactic Radio Astronomy, I.A.U. Symposium
    No. 60, Reidel, p.87.
Lyne, A.G.: 1981, Pulsars, IAU Symposium No. 95, Reidel, p.423.
Lyne, A.G., Anderson, B. and Salter, M.J.: 1981, in preparation.
Lyne, A.G., Manchester, R.N. and Taylor, J.H.: 1981, in
    preparation.
Lyne, A.G., Ritchings, R.T. and Smith, F.G.: 1975, Mon.Not.R.
    astr.Soc. 171, p.579.
Manchester, R.N., Lyne, A.G., Taylor, J.H., Durdin, J.M.,
    Large, M.I. and Little, A.G.: 1978, Mon.Not.R.astr.Soc.185,p.409.
Ritchings, R.T.: 1976. Mon.Not.R.astr.Soc. 179, 249.
Salter, M.J., Lyne, A.G. and Anderson, B.: 1979, Nature 280,
    p.477.

Tammann, G.A.: 1977, in Supernovae, ed. D.N. Schramm, Reidel, p.95.
Taylor, J.H. and Manchester, R.N.: 1977, Astrophys.J. 215, p.885.

YOUNG SUPERNOVA REMNANTS

Roger A. Chevalier

University of Virginia

INTRODUCTION

There are two basic ingredients which lead to an under-standing of young supernova remnants. The first is the nature of the expanding supernova material which was originally a part of the progenitor star. The density and composition structure of this expanding matter are important parameters, while, in many cases, the velocity structure is given by free expansion. The second ingredient is the nature of the environment into which the supernova is expanding.

The interaction of the supernova with its environment leads to a rich set of phenomena, which can be observed at radio, optical, ultraviolet, and X-ray wavelengths. The purpose of this paper is to review these phenomena, with particular emphasis on the interpretation of recent observations of young supernova remnants.

SUPERNOVA ENVIRONMENTS

There are three aspects to the problem of supernova environments: the general nature of the interstellar medium, the question of whether the supernovae occur in a particular component of the interstellar medium, and the circumstellar medium generated by mass loss from the progenitor star.

The interstellar medium contains a number of components (e.g. Spitzer 1978): molecular clouds with densities of about 10^3 cm^{-3} and denser cores, diffuse clouds with diameters of about 5 pc and densities of 20 cm^{-3}, a warm, neutral medium with a

M. J. Rees and R. J. Stoneham (eds.), Supernovae: A Survey of Current Research, 419–432.
Copyright © 1982 by D. Reidel Publishing Company.

temperature of about 8000 K and a density of 0.2 cm^{-3}, and hot, low density gas with a temperature greater than a few times 10^5 K. The question of whether most of the volume of the interstellar medium contains warm, neutral gas or hot, ionized gas is controversial.

The progenitors of type I supernovae are long-lived and these explosions can be expected to occur anywhere in the interstellar medium. However, the progenitors of type II supernovae are probably massive OB stars which have relatively short lives and are born within dense, molecular clouds. The B stars have lifetimes of about 10^7 years or greater and they are probably not within the clouds at the end of their lives either because they outlive their parent clouds or because they leave the clouds due to their random motions. The O and early B stars are copious producers of ionizing radiation, which is capable of substantially changing their cloud environment. It is possible to show that the HII region around an O star in a medium of density 10^3 cm^{-3} is about 7 pc in radius at the end of the star's life. The size of a dense cloud is typically several pc, although they can occur in complexes which are considerably larger (Blitz 1979). In any case, it is likely that the HII region reaches the edge of an individual cloud. Even if the star is within the core of the cloud where the density is about 10^5 cm^{-3}, the final HII region size should be larger than the core size (about 0.5 pc).

Once the ionization front reaches the cloud boundary, it rapidly moves into the intercloud medium and a pressure drop is established at the cloud boundary. The outer edge is expected to move out from the cloud at a speed of 30 km s^{-1}: detailed numerical calculations show how the ionized cloud material fans out from the boundary (Bodenheimer, Tenorio-Tagle and Yorke 1979). A rarefaction wave moves back through the HII region at the local sound speed. It is expected to cross the HII region in a time of 10^5 R_{pc} years where R_{pc} is the size of the region in pc. Once the rarefaction wave has crossed the region, a continuous flow is set up in which material is ionized from the edge of the HII region and expands away from the cloud. The cloud material must receive an impulse away from the flow in order to conserve momentum. If a few percent of the mass of a dense cloud goes into stars, it is possible that ionizing radiation from the massive stars results in the disruption of the cloud (Whitworth 1979). These considerations indicate that when an O star ends its life, the density of the surrounding interstellar medium is considerably smaller than that of the dense cloud in which it is born.

Another important factor for the environments of type II supernovae is mass loss from the progenitor star. Massive stars spend most of their lives as hot, luminous stars which are capable of driving mass loss by radiation pressure. Observationally derived rates of mass loss show that the mass loss rate is quite sensitive to the stellar luminosity and that large rates of mass loss ($M \gtrsim 10^{-6} M_\odot yr^{-1}$) only occur for stars more massive than about 40 $M_\odot$ (e.g. Conti and Garmany 1980). A typical terminal velocity for the stellar wind is 2000 km s^{-1}. The interaction of the stellar wind with the surrounding medium is expected to create a number of distinct regions (Pikelner 1968: Weaver et al. 1977): 1) the stellar wind in which the density falls as r^{-2}: 2) a hot region of shocked stellar wind material: 3) a narrow region of shocked, swept-up ambient material: and 4) the ambient medium. A star with a 2000 km s^{-1} wind losing $10^{-6} M_\odot yr^{-1}$ in a medium of density 1 cm^{-3} is expected to create a cavity about 25 pc in radius after 10^6 years.

While shells are expected around most O stars, they are not commonly observed: the reason for this is not clear. Shells are more commonly observed around Wolf-Rayet stars. While these stars do have strong stellar winds (mass loss in excess of 10^{-5} $M_\odot yr^{-1}$), it is uncertain whether the shells are swept up surrounding mass driven by the wind or whether they are the ejected envelopes of the massive stars. The estimated masses of the expanding shells are comparable to what might be expected for the ejected envelopes of massive stars. Also Parker (1978) has found an overabundance of nitrogen in the shell NGC 6888 by a factor of 3, which is compatible with the expected N enhancement in the envelope of a massive star. An overabundance of He is also indicated. It is possible that an instability results in the ejection of the extended envelope of the star, leaving behind the hot interior (Conti 1978).

The Wolf-Rayet stars, and other stars that lose their envelopes, are expected to end their lives as hot stars. However, most massive stars probably end their lives as red supergiants. The presence of an extended envelope appears to be required to give the characteristics of a type II supernova. Red supergiants are observed to lose mass in winds with velocities of about 10 km s^{-1} at a rate of about $10^{-6} M_\odot yr^{-1}$. The maximum extent of the mass loss region, 1 pc, is obtained by multiplying the velocity by the timescale for the red supergiant phase, about 10^5 years.

Mass loss is well documented for massive stars, which are the progenitors of type II supernovae. On the other hand, the progenitors of type I supernovae are not well known. In one evolutionary picture, the progenitors are massive white dwarfs in low mass binary systems (e.g. Whelan and Iben 1973). Mass must be lost from the binary system during its evolution, but the form that the mass loss takes is not known.

THE INITIAL INTERACTION WITH THE CIRCUMSTELLAR MEDIUM

The initial interaction of a type II supernova envelope with the surrounding medium is with the slow red supergiant wind. The expanding supernova envelopes drives a fast shock wave into the wind gas, which is heated to a temperature of about 10^9 K. Branch et al. (1981) and Chevalier (1981b) estimated that the bremsstrahlung emission from this hot gas has a luminosity of about 10^{37} erg s^{-1} when the supernova is near maximum light. This is well below the upper limits for the X-ray emission from the type II supernova SN 1979c in M100 (Palumbo et al. 1981). The X-ray emission may be even stronger than the above estimate if a reverse shock wave develops on the inside of the hot region.

The radiation from the hot shell dominates the supernova photospheric radiation beginning at far ultraviolet wavelengths and it is capable of making Si IV, C IV and N V ions comprise a substantial fraction of the Si, C and N ions in the preshock region (Chevalier 1981b). These ions have transitions which can absorb ultraviolet photospheric radiation in a narrow preshock region. Although the absorbed radiation is immediately reemitted, the initial absorption accelerates the gas. The radiatively accelerated shock precursor can reach velocities of several thousand km s^{-1}. This theory predicts that P Cygni-type lines of Si IV, C IV, and N V should appear in the ultraviolet spectrum of the supernova near maximum light. These lines may have been observed in the spectrum of SN 1979c (Panagia et al. 1980), giving indirect evidence for the presence of the hot shell.

Models of type II supernova explosions have shown that the shock wave accelerates as it passes through the steep density gradient in the outer layers of the progenitor star (Chevalier 1976: Falk 1978). This results in low density, high velocity tail to the main supernova envelope. As emphasized by Jones, Smith and Straka (1981), the early part of the interaction of the supernova with its surroundings involves the deceleration of this low density material. In his models for the early evolution of supernova remnants, Gull (1973a) included several initial density distributions for the expanding supernova matter. In the model with high-velocity, low density tail, the deceleration of the low density material led to the onset of the Rayleigh-Taylor instability quite early (before 30 years into the evolution). When the presence of circumstellar material is taken into account, the onset of the instability is expected to occur even earlier. In a model for the Cas A remnant, Gull (1973c) noted that if the unstable regions efficiently put energy into magnetic fields and relativistic particles, the radio luminosity of Cas A can be reproduced. Chevalier (in preparation) found that if similar efficiencies apply to the early stages when the low density gas is interacting with circumstellar matter, the radio luminosities of recent extra-galactic supernovae (Weiler et al. 1981: Sramek, van der Hulst and Weiler 1980) can be reproduced. If this model is correct, the radio observations of supernovae give information on physical processes which are very similar to those occurring in the young galactic remnants.

YOUNG GALACTIC REMNANTS

a) Crab Nebula

One of the fundamental problems with the Crab nebula is to associate its characteristics (and those of SN 1054) with a particular kind of supernova explosion. One possibility is that it was a rare type of explosion involving a helium star. Another possibility is that it was a normal type II supernova and the fast envelope of the exploded star has not yet been detected.

Attempts to detect a fast shell around the Crab nebula at radio and X-ray wavelengths have failed. The most recent attempt at X-ray wavelengths is that of Schattenburg et al. (1980), who obtained upper limits on the line emission from the nebula and its surroundings. These limits led to an upper limit on the X-ray emitting mass of about 1 to 2 $M_\odot$ for a 7' diameter shell, depending on the temperature of the emitting gas. If SN 1054 was a normal type II supernova, the fast envelope should have swept up the circumstellar matter from the supergiant wind, but may not have interacted with much other matter, particularly if the strong wind from the hot star created a cavity around the progenitor object. The amount of mass in the supergiant wind is expected to be approximately 10^{-6} $M_\odot$ yr^{-1} times 10^5 yrs, or 0.1 $M_\odot$, which is not ruled out by the present observations. The helium star model is also not ruled out.

The X-ray observations show that the supernova has not interacted with much surrounding matter, and if a fast undetected envelope does exist, most of the mass is in the free expansion phase. This cool gas can be photoionized by ultraviolet radiation from the Crab nebula and it should emit broad optical emission lines. It does not appear that there have been observational attempts to detect this radiation.

b) Cassiopeia A

In recent years, Cas A has been the best studied of the supernova remnants and there has been substantial progress in both theory and observations. The observations suggest that the remnant is in a stage where the deceleration of the ejecta and a reverse shock wave are important, as originally discussed by

Gull (1973a,b: 1975) and McKee (1974). The best evidence for this is the X-ray structure of the remnant (Fabian et al. 1980). The Einstein HRI picture shows two shell-type structures. The inner edge of the inner shell is at radius, r_1, of about 100" and the outer shell is at $r_2 = 140"$. These two radii presumably give the limits of the high pressure region in Cas A. Inside of r_1, there is expected to be cool, freely expanding gas that was part of the progenitor star. The X-ray picture is consistent with no emission inside of r_1. Taking the distance and age of Cas A from studies of the optical knots (Kamper and van den Bergh 1976), the velocity of the freely expanding gas at r_1 is about 4000 km s^{-1}. The fast optical knots have proper motions approximately proportional to their distance to the centre of expansion, indicating that they have not been decelerated. There are no fast knots observed with space velocities less than 4000 km s^{-1} (van den Bergh 1971), which is consistent with the knot material becoming visible only when it enters the high pressure region. The optical emission is then from shock waves which are driven into the knot gas. Knots are observed with space velocities up to 8500 km s^{-1}, indicating that some knots have moved out ahead of the high pressure shell and are interacting directly with circumstellar matter.

An analysis of the X-ray structure, spectrum, and flux shows that the X-ray emitting mass is at least 15 M$_\odot$ (Fabian et al. 1980): a good model for the emission gives 20 M$_\odot$ as the mass estimate. If the ejecta are interacting with circumstellar matter that was lost from the progenitor star, as is indicated by optical spectra of slow clouds, then 15 M$_\odot$ is a lower limit to the initial progenitor mass. Additional mass could be ahead of the main shock wave and within r_1. Comparison of the observed value r_2/r_1 with the models of Gull (1973a) suggests that the mass ratio (ratio of swept up mass to ejected mass) is about unity. Gull (1973b) obtained a similar value on the basis of the radio structure. In the models, this implies that most of the ejected mass has been shocked in the reverse shock wave. However, Gull used fairly simple density distributions in his models. If the density rises sharply for velocities less than 4000 km s^{-1}, there could be considerable hidden mass in Cas A.

Bell (1977) has set an upper limit of 1700km s^{-1} to the mean radial expansion velocity of the radio peaks on the basis of two maps made 5 years apart. The radio peaks are probably regions of enhanced magnetic field strength and their motions presumably reflect the motions of the thermal gas. While this radial velocity is much less than that of the fast optical knots, it is not unusual for a region that has recently been shocked by the reverse shock. In Gull's (1973a) models, the gas velocity at the reverse shock drops by a factor of 0.55 when the mass ratio is 1.9. For Cas A, this corresponds to a postshock gas velocity of

2200 km s^{-1}. The velocity difference increases with time until eventually the postshock gas moves inward.

The Einstein solid state spectrometer shows lines of Mg, Al, Si, S, Ar, Ca, and Fe in the spectrum of Cas A (Becker et al. 1979). Assuming equilibrium ionization and that the continuum radiation is due to hydrogen, the Si group elements (Si through Ca) are overabundant by factors of 2 to 7 relative to solar abundances, while the other elements are approximately solar. If the continuum radiation is not due to hydrogen but to helium or heavier elements, the overabundances of the observed heavy elements become larger, factors of 10 to 20 (Fabian et al. 1980). While the abundances relative to hydrogen are not well determined, the abundances of the observed heavy elements relative to each other are probably reliably estimated and can be compared to abundance estimates obtained from optical spectra of the fast moving knots (Chevalier and Kirshner 1979). Among the Si group elements S, Ar, and Ca, both the X-ray and the optically emitting gas show abundance distributions that are similar to solar. However, the S/Fe ratio (by mass) is found to be 1 in the X-ray emitting material while it is estimated to be greater than 30 in the optical filament no. 1. Lines of Fe ions have not been observed in the optical spectra. A similar discrepancy occurs for Mg, although the optical abundance determination is quite uncertain in this case. While Fe could be in grains in the optical knots, the presence of Ca argues against this explanation.

The difference in the abundances between the X-ray and optical emitting gas applies to elements of gas which have similar space velocities. One model that is suggested by these observations is that knots of heavy element gas have become embedded in more uniform matter which is mainly composed of light elements (oxygen or lighter than oxygen). As the reverse shock wave moves into the uniform gas, it heats it to X-ray emitting temperatures. Slower shocks are driven into the knots because of their higher densities. The gas cools behind the shocks giving the fast optical knots. The knots are undecelerated unlike the uniform shell because of their high densities. The knots may have formed as a result of a Rayleigh–Taylor instability during the initial expansion of the star. This can occur when an inner fast layer is decelerated by an outer low density layer of the star (Chevalier 1979: Jones, Smith and Straka 1981). This instability may also explain the fact that the knot abundances are uncorrelated with their space velocities (Chevalier and Kirshner 1979).

Observations of the slow moving 'flocculi' in Cas A show that they are N and possible He rich. This suggests that much of the H envelope was lost from the progenitor star and that the fast knots moving at about 4000 km s^{-1} are embedded in either a He, C, or O layer. If this shocked gas was the only X-ray emitting gas, substantial overabundances of all the elements observed in the SSS spectra would be required, which would cause problems (especially for Fe). However, this can be avoided by attributing the X-ray continuum to H emission from the shocked circumstellar gas. The Fe abundance is then approximately solar, and the enhancement of Si group elements is due to some admixture of knot gas. This description of the emission is speculative and requires quantitative calculations.

The fast knots in Cas A appear to be composed entirely of heavy elements and provide a test for theories of nucleosynthesis in the inner layers of a massive star which has exploded. Johnston and Joss (1980) have addressed this problem, using parameterized forms for the adiabatic expansion of a gas element after it is shocked in the supernova. They found that the observed relative abundances of O, S, Ar, and Ca could be reproduced starting from a wide range of initial conditions. In particular, the abundance inhomogeneities could be reproduced by the same initial composition undergoing nucleosynthesis at different peak temperatures and densities. However, the results are not so promising when they are placed in the context of stellar models. Weaver and Woosley (1980) have calculated the complete evolution of a 25 $M_\odot$ star, which may be close to the mass of the Cas A progenitor star. The final composition structure of the expanding star shows fairly distinct regions containing O and Si group elements. In Cas A, there are many knots which have substantial amounts of both O and Si group elements. While the elements present are those expected, the mixture is not.

There are a number of possible reasons for this discrepancy. First, the Cas A progenitor may have been more massive than 25 $M_\odot$. More massive stars have greater mass loss, as appears to have occurred in Cas A. The mass limit at which pair instability supernovae occur is not well known. This type of supernova has the advantages that there is no remnant neutron star and almost no Fe is ejected. Murray et al. (1979) have noted the absence of thermal X-rays from a neutron star in Cas A. Second, the initial explosion may have been highly asymmetric, which could change the character of the explosive nucleo-synthesis. In fact, Markert, Canizares and Winkler (1981) have found evidence for a Doppler shift in the X-ray lines, moving across the remnant in one dimension. They interpret these observations as evidence for expansion of the heavy elements in a ring. However, there is no clear evidence for this asymmetry in the velocities of the fast optical knots (e.g. van den Bergh 1971), although a detailed study of this problem may now be warranted. The spatial resolution of the X-ray observations is

poor and the observed asymmetry could be due to a small number of
X-ray emitting regions rather than a systematic expansion effect.
Third, there may be problems with the physics used to calculate
the stellar model. Examples of uncertain areas are nuclear
reaction rates and the theory of convection. Also, the model did
not include mass loss, which appears to have been important for
Cas A: but the nuclear reactions near the centre of the star are
probably not much influenced by the presence or absence of a
hydrogen envelope. Fourth, the interpretation of the optical line
spectra may be incorrect. Plausible assumptions were made in the
derivation of abundances, but self-consistent shock models have
not yet been calculated. A start on this problem has been made by
Itoh (1981) who has calculated the emission spectrum from a shock
wave moving into a pure oxygen gas. He was able to reproduce the
relative line intensities observed in the Cas A oxygen knot. It
is important to add additional heavy elements to these models.

The nature of the slow moving material in Cas A can be
compared with the circumstellar material observed around massive
stars. The slow moving gas is at a radial distance of about 1.7
pc from the centre and has abundance anomalies which suggest that
it was initially part of the stellar envelope. These properties
cannot be reconciled with an origin in a stellar wind, either a
fast wind from a hot star or a slow wind from a supergiant, and
an origin from an envelope instability seems more likely. The
observed properties are fairly close to those found in the nebula
NGC 6888 around a Wolf-Rayet star, as noted by Parker (1978),
although NGC 6888 has a somewhat larger radius, about 3 pc. If
this hypothesis is correct, the ejecta in Cas A are currently
interacting with a shell instead of with a uniform medium. Gull's
models are for interaction with a uniform medium: interaction
with a shell may change some of the model results, e.g. the
expansion rates of the radio and X-ray shells. The study of Cas A
has evolved to the stage where detailed models particular to the
object in question are required.

c) Remnant of Tycho's Supernova

The remnant of SN 1572 is the best studied of the remnants
which are thought to be the result of type I supernovae. It is
thus of interest to compare its properties with those of observed
extragalactic type I supernovae and with models of those events.
Current models for type I supernovae generally use the idea
proposed Colgate and McKee (1969) that the radioactive decay of
Ni^{56} to Co^{56} to Fe^{56} gives rise to the energy input for the
observed luminosity. There is direct observational evidence for
Fe in the supernova spectra both at late times (Kirshner and Oke
1975) and at earlier times. Yet the solid state spectrometer
spectra of Tycho's remnant showed very strong Si group lines
relative to those of Fe (Becker et al. 1980a): assuming

collisional equilibrium and a hydrogen continuunm, the Si group
elements are overabundance relative to solar by factors of 10 to
200, while Fe is slightly underabundant.

The proper motions of the optical filaments in Tycho's
remnant show that the radius of the shock wave is increasing at
t^n where n = 0.38 $\pm$ 0.01 (Kamper and van den Bergh 1978). This
rate of increase is very close to that of an adiabatic blast
wave, where n = 0.4. This suggests that the remnant is in the
blast wave stage of evolution with most of the X-rays being
emitted in the immediate postshock region. With this in mind,
Hamilton, Sarazin and Chevalier (in preparation) calculated
nonequilibrium ionization models for blast waves moving into a
medium with normal abundances. None of the models were able to
reproduce the strong Si group lines observed in Tycho's remnant,
indicating that there is a real abundance effect.

Another possibility is that the X-ray emission is from a
reverse shock wave which has moved into the Si group zone but has
not yet penetrated the Fe zone (Arnett 1980). At first sight, it
is not clear whether the reverse shock model is consistent with
the outer shock radius increasing as $t^{0.4}$. Yet Gull's (1973a)
models show that the n = 0.4 behaviour begins when the mass ratio
is about 2 and the supernova ejecta extend to about 80% of the
outer shock radius. The reason for this is that the mass in the
blast wave is strongly concentrated at the shock wave so that the
very outer parts can begin to behave like a blast wave when the
inner regions are still far from blast wave condition. While the
reverse shock wave model appears to be promising, it is important
to check it by analysis of the X-ray structure. Most of the
X-ray line emission should be from the shocked ejecta gas, which
may be at about 80% of the shock radius. If the emission were
from the blast wave, it should be closer to the shock wave.
Another prediction of the reverse shock model is that there
should be no X-ray emission from inside the reverse shock, which
may be at a radius of 60-80% of the shock radius.

Arnett (1980) applied the reverse shock model to a specific
type I supernova model involving the explosion of the core of a
star which initially was in the mass range 8-12 $M_\odot$. The reverse
shock model should also apply to other type I models, such as the
carbon deflagration of a white dwarf star (Chevalier 1981a). The
exploding white dwarf model was hypothesized to have outer layers
containing Si group elements because Si and Ca lines have been
observed in the spectra of type I supernovae. These elements
appear to have velocities of about 9000 km s^{-1} and greater. In
either the white dwarf or the massive star core model, the Fe
rich material has velocities extending to 7000 or 8000 km s^{-1}.
The lack of Fe in the X-ray emitting material implies that the
reverse shock has not yet reached material moving at this

velocity. Considering Gull's model, the velocity of the outer
shock wave must be at least 4000 or 5000 km s^{-1}. These limits
correspond to limits on the distance to Tycho's remnant because
the proper motion of the shock front is known. The lower limit is
of order 4 or 5 pc. The distance to the remnant is controversial.
On the basis of a theory for the optical shock wave emission,
Chevalier, Kirshner and Raymond (1980) derived a distance of 2.3
kpc with a large uncertainty. On the other hand, Schwarz et al.
(1980) obtained a lower limit of 4 kpc on the basis of the HI
absorption line observed toward Tycho's remnant.

d) Remnant of SN 1006

The X-ray spectrum of the SN 1006 remnant is surprisingly
different from that of Tycho's remnant: no lines appear to be
present (Becker et al. 1980b). This suggests that the emission is
nonthermal in nature. Pye et al. (1981) have analyzed the X-ray
structure of the remnant and while the probable nonthermal
spectrum casts doubt on certain of their conclusions, others
still stand. In particular, they found that the emission could
be divided into two spatial components: a shell and a uniform
interior. Whatever the source of the interior emission, it
indicates that the pressure throughout the interior is high. In
the absence of a central energy source, the reverse wave must
have made its way to the centre of the remnant.

The SN 1006 remnant has an optical filament with a pure
hydrogen spectrum similar to that observed in Tycho's remnant
(Schweizer and Lasker 1978).   Chevalier, Kirshner and Raymond
(1980) developed a theory for the emission from a fast shock wave
which reproduced the essential features of the Tycho spectrum and
predicted that the Hα line profile in the SN 1006 remnant should
have narrow and broad components of approximately equal stength.
Lasker (1981) obtained an accurate line profile for Hα and set an
upper limit of 0.25 on the strength of the broad component
relative to the narrow component.   Lasker suggested that the
discrepancy with the theory could be explained by having the two
remnants interacting with clouds of systematically different
density. While this proposal seems implausible, I have no better
suggestions. Further observations of both the Tycho and SN 1006
optical filaments would be useful. In the SN 1006, it should be
possible to measure the proper motion of the filament, which will
place important constraints on the shock velocity. In Tycho's
remnant, it is important to obtain spectra in more than one
position so that uniformity of the ratio of broad component to
narrow component can be checked.

DISCUSSION AND FUTURE PROSPECTS

From our knowledge of the evolution of massive stars, we expect type II supernovae to evolve through a number of phases. Initially, the outer regions of the expanding envelope interact with the dense supergiant wind. This interaction can lead to substantial radio and X-ray fluxes. The emission drops as the shock wave moves into regions of decreasing density. A low point in the emission is attained when the shock has passed through the slow wind and it begins to interact with either a cavity created by a fast wind or a low density interstellar medium. The Crab nebula may be in this stage of evolution. Eventually the ejected envelope interacts with its own mass of circumstellar or interstellar matter and the remnant again becomes a luminous object. The Cas A event was probably different from a type II supernova in that there was extensive presupernova mass loss giving rise to a shell of matter about 2 pc from the exploding star. We are now observing the interaction of the ejecta with this shell.

This review suggests several theoretical problems that need further investigation: the interaction of the type II supernova envelope with the supergiant wind, the emission from a shock wave moving into a gas composed of heavy elements, the nonequilibrium ionization of heavy elements in young remnants, and the calculation of further stellar models for comparison with the Cas A abundances. There are also observational projects to be carried out. With current optical techniques, it should be possible to investigate emission from a faint shell around the Crab nebula. Future techniques, such as those involving the Space Telescope, may be necessary to make progress on spectra of Cas A and the remnants of Tycho and SN 1006. In Cas A, access to the ultraviolet will provide estimates of Si and Mg abundances, which are important for understanding the nucleosynthesis leading to the knot abundances. In Tycho and SN 1006, measuring the Hα line profile in various positions should give information on the shock wave structure. X-ray studies hold great promise, especially when it becomes possible to obtain spectra with good spatial resolution. The models discussed here for Cas A and Tycho predict considerable inhomogeneity in the X-ray spectra. It is clear that the investigation of young supernova remnants will be a rich subject for future study.

REFERENCES

Arnett, W.D. 1980. Astrophys.J., 240, 105.
Becker, R.H., Holt, S.S., Smith, B.W., White, N.E., Boldt, E.A.,
    Mushotzky, R.F. and Serlemitsos, P.J. 1979. Astrophys.J.
    (Letters), 234, L73.

Becker, R.H., Holt, S.S., Smith, B.W., White, N.E., Boldt, E.A., Mushotzky, R.F. and Serlemitsos, P.J. 1980a. Astrophys.J. (Letters), 235, L5.

Becker, R.H., Szymkowiak, A.E., Boldt, E.A., Holt, S.S. and Serlemitsos, P.J. 1980b. Astrophys.J.(Letters), 240, L33.

Bell, A.R. 1977. Mon.Not.R.astr.Soc., 179, 573.

Blitz, L. 1979. in Giant Molecular Clouds in the Galaxy, ed. P.M. Solomon and Edmunds (Oxford: Pergamon Press).

Bodenheimer, P., Tenorio-Tagle, G. and Yorke, H.W. 1979. Astrophys.J., 233, 85.

Branch, D., Falk, S.W., McCall, M.L., Rybski, P., Uomoto, A.K. and Wills, B.J. 1981. Astrophys.J., 244, 780.

Chevalier, R.A. 1976. Astrophys.J., 207, 872.

Chevalier, R.A. 1979. Mem.della Soc.Astr.Ital., 50, 65.

Chevalier, R.A. 1981a. Astrophys.J., 246, 267.

Chevalier, R.A. 1981b. Astrophys.J., in press.

Chevalier, R.A. and Kirshner, R.P. 1979. Astrophys.J., 233, 154.

Chevalier, R.A., Kirshner, R.P. and Raymond, J.C. 1980. Astrophys.J., 235, 186.

Colgate, S.A. and McKee, C. 1969. Astrophys.J., 157, 623.

Conti, P.S. 1978. Ann.Rev.Astron.Astrophys. 16, 371.

Conti, P.S. and Garmany, C.d. 1980. Astrophys.J., 238, 190.

Fabian, A.C., Willingale, R., Pye, J.P., Murray, S.S. and Fabbiano, G. 1980. Mon.Not.R.astr.Soc., 193, 175.

Falk, S.W. 1978. Astrophys.J.(Letters) 225, L133.

Gull, S.F. 1973a. Mon.Not.R.astr.Soc., 161, 47.

Gull, S.F. 1973b. Mon.Not.R.astr.Soc., 162, 135.

Gull, S.F. 1975. Mon.Not.R.astr.Soc., 171, 263.

Itoh, H. 1981. P.A.S.Japan, 33, 1.

Johnston, M.D. and Joss, P.C. 1980. Astrophys.J., 242, 1124.

Jones, E.M., Smith, B.W. and Straka, W.C. 1981. Astrophys.J. 249, 185.

Kamper, K and van den Bergh, S. 1976. Astrophys.J.Suppl., 32, 351.

Kamper, K.W. and van den Bergh, S. 1978. Astrophys.J., 224, 851.

Kirshner, R.P. and Oke, J.B. 1975. Astrophys.J., 200, 574.

Lasker, B.M. 1981. Astrophys.J., 244, 517.

Markert, T.H., Canizares, C.R. and Winkler, P.F. 1981. Talk at meeting of N.Y. Astr.Soc.

McKee, C.F. 1974. Astrophys.J., 188, 335.

Murray, S.S., Fabbiano, G., Fabian, A.C., Epstein, A. and Giacconi, R. 1979. Astrophys.J.(Letters), 234, L69.

Palumbo, G.G.C., Maccacaro, T., Panagia, N., Vettolani, G. and Zamorani, G. 1981. Astrophys.J., in press.

Panagia, N. et al 1980. Mon.Not.R.astr.Soc., 192, 861.

Parker, R.A.R. 1978. Astrophys.J., 224, 873.

Pikelner, S.B. 1968. Astrophys.Letters, 2, 97.

Pye, J.P., Pounds, K.A., Rolf, D.P., Seward, F.D., Smith, A. and
    Willingale, R. 1981. Mon.Not.R.astr.Soc., 194, 569.
Schattenburg, M.L., Canizares, C.r., Berg, C.J., Clark, G.W.,
    Markert, T.H. and Winkler, P.F. 1980. Astrophys.J.
    (Letters), 241, L151.
Schwarz, U.J., Arnal, E.M. and Goss, W.M. 1980. Mon.Not.R.astr.
    Soc., 192, 67P.
Schweizer, F. and Lasker, B.M. 1978. Astrophys.J., 220, 167.
Spitzer, L. 1978. Physical Processes in the Interstellar
    Medium, (New York: Wiley).
Sramek, R., van der Hulst, J.M. and Weiler, K.W. 1980. IAU
    Circular No. 3557.
van den Bergh, S. 1971. Astrophys.J., 165, 457.
Weaver, R., McCray, R., Castor, J., Shapiro, P. and Moore, R.
    1977. Astrophys.J., 218, 377.
Weaver, T.A. and Woosley, S.E. 1980. Ann.N.Y.Acad.Sci., 336,
    335.
Weiler, K.W., van der Hulst, J.M., Sramek, R.A. and Panagia, N.
    1981. Astrophys.J.(Letters), 243, L151.
Whelan, J. and Iben, I.,Jr. 1973. Astrophys.J., 186, 1007.
Whitworth, A. 1979. Mon.Not.R.astr.Soc., 186, 59.

THE EVOLUTION OF SUPERNOVA REMNANTS AND THE STRUCTURE OF THE
INTERSTELLAR MEDIUM

Christopher F. McKee

University of California, Berkeley

1. INTRODUCTION

Supernova remnants (SNRs) play a key role in astrophysics.
Young SNRs inject the nucleosynthesis products of supernova
explosions into the interstellar medium (ISM). SNRs are thought
to produce most of the cosmic rays and the nonthermal radio
emission in the Galaxy (Blandford 1982). They are also one of
the major energy sources for the medium and thereby affect its
structure and ultimately the star formation process which leads
to supernovae.

Theoretical studies of SNRs have generally assumed a
homogeneous ambient medium, but in looking at pictures of SNRs –
whether in the radio, optical, or X-ray region of the spectrum –
one is struck by the importance of inhomogeneities. Indeed, Cox
and Smith (1974) have argued that SNRs lead to the creation of a
hot, low density component of the ISM, and McKee and Ostriker
(1977: hereafter, MO) developed a three phase model of the ISM
incorporating this idea. The evolution of SNRs is significantly
altered by the presence of this hot phase. Hence the evolution
of SNRs and the structure of the ISM are a single self-consistent
problem.

After discussing the homogeneous case from a new perspective
(section 2) I shall summarize the results of analytic and
numerical studies of SNR evolution in a three phase ISM (section
3,4), describe the observational implications (section 5), and
discuss the effects of SNRs on the ISM (section 6).

433

M. J. Rees and R. J. Stoneham (eds.), Supernovae: A Survey of Current Research, 433–
Copyright © 1982 by D. Reidel Publishing Company.

2. SNR EXPANSION IN A HOMOGENEOUS MEDIUM

The expansion of an SNR in a homogeneous medium goes through several well-defined stages (Woltjer 1972): (1) Initially, the ejecta from the supernova act like a spherical piston which drives a shock into the ambient medium at almost constant velocity (see Chevalier 1982). (2) When the mass swept up by the expanding shock exceeds that ejected by the SN, the expansion approaches the blast wave discussed by Sedov (1959) and Taylor (1950) in which the dynamics are controlled only by the energy of the explosion E and the density of the ambient medium ρ_0. (3) Eventually radiative losses become important at the outer edge of the SNR, where the cooling time is shortest, and a dense shell forms. The subsequent expansion is dominated by the momentum of the expanding shell. (4) The expansion of the SNR ceases when it comes into pressure equilibrium with the ambient medium.

a) Virial Theorem

Several approximate analytic techniques have been developed to describe SNR expansion. Here we follow the approach of Ostriker, McKee, and Cowie (1981) and use the virial theorem. In the absence of gravity, and for negligible ambient pressure, the virial theorem for a $\gamma = 5/3$ gas is simply

$$\frac{1}{2} \frac{d^2 I}{dt^2} = 2(E_k + E_{th}) = 2E, \qquad (1)$$

where

$$I = \int r^2 \, dm \qquad (2)$$

is the moment of inertia, and E_k and E_{th} are the total kinetic and thermal energy, respectively. Since the virial theorem refers to a fixed mass, the integral in equation (2) must extend from the origin to just beyond the radius R of the remnant. Ostriker et al. demonstrate that the virial theorem retains this form even if the interior mass increases due to evaporation from stationary clouds. Define the structure parameter $\kappa_{//}$ as

$$\kappa_{//} = \frac{1}{M R v_{ps}} \int r v \, dm, \qquad (3)$$

where M is the mass of the SNR and v_{ps} is the post-shock fluid velocity. In terms of the shock velocity v_s, we have $v_{ps}/v_s = 3/4$ for a nonradiative shock and $v_{ps}/v_s = 1$ for a radiative shock. We now assume that the expansion is self-similar, so that dimensionless integral properties of the expansion such as $\kappa_{//}$

are independent of t. The virial theorem for SNR expansion then becomes

$$\frac{d}{dt} MRv_s = \frac{2E}{K_{//}} \left(\frac{v_s}{v_{ps}}\right),$$ (4)

where $v_s = dR/dt$.

This equation is exact under the assumptions specified above. Two techniques for estimating the constant $K_{//}$ for nonevaporative remnants are described by Ostriker et al. One is accurate to within 1/2%. The other is based on approximating the fluid velocity v as a linear function of r: then $I = K_{//}R^2M + R^2\Delta M$, where ΔM is the mass of the infinitesimal shell just beyond R which must be included in evaluating dI/dt so that the total mass remains constant. To allow the possibility of spatial variation in the ambient density we write

$$\bar{\rho}(R) = \bar{\rho}(1)R^{-k}\rho,$$ (5)

where $\bar{\rho}(R)$ is the mean density inside the SNR, and $\bar{\rho}(1)$ and k_ρ are constants. In the absence of evaporation or condensation (see section 3 below), the mean density is related to the ambient density $\rho_0(R)$ by $\bar{\rho} = \rho_0/(3-k_\rho)$: hence we require $k_\rho < 3$ in order that $\bar{\rho}$ be finite. Equating the expression for dI/dt based on the approximation above to the exact expression $2K_{//}MRv$, obtained from equations (2) and (3), we find

$$K_{//} = \frac{6-2k_\rho}{7-2k_\rho}.$$ (6)

For a uniform medium ($k_\rho = 0$), this gives $K_{//} = 0.86$, as compared to the exact value inferred from Sedov (1959) of 0.79.

An alternate approach to describing SNR evolution is based on the observation that most of the mass in an SNR is concentrated in a thin shell at the outer edge of the remnant (Chernyi 1957: cf. Zel'dovich and Raizer 1966). The equation of motion for such a thin shell is

$$\frac{d}{dt} Mv_s = (\gamma + 1) 2\pi R^2 \bar{p}(R),$$ (7)

where $\bar{p}(R)$ is the mean pressure in the SNR and where we neglected the ambient pressure. Ostriker et al. demonstrate that the virial theorem (equation 4) with approximation (equation 6) for $K_{//}$ is equivalent to this equation for most self-similar expansions of interest.

b) Self-similar solutions

It is straightforward to integrate the virial equation (4) to obtain explicit solutions in both the nonradiative and radiative phases of SNR evolution. In the nonradiative, Sedov-Taylor phase the radius is given by

$$R = \left[\frac{5 - k_\rho}{\pi K_{//}} \frac{E}{\bar{\rho}(1)} \right]^{-\frac{1}{5-k_\rho}} t^{\frac{2}{5-k_\rho}} \tag{8}$$

In order to determine the mean temperature in the SNR, it is necessary to know what fraction of the energy is thermal. Making the linear velocity approximation ($v \propto r$) implies the kinetic energy is $1/2\ K_{//}\ m\ [(3/4)v_s]^2$, or $(3/2)E/(5 - k_\rho)$ by virtue of equation (8). The ratio of remaining, thermal energy to the total is then

$$\frac{E_{th}}{E} = \frac{7 - 2k_\rho}{10 - 2k_\rho}. \tag{9}$$

For $k_\rho = 0$, this is quite close to the exact Sedov result of 0.71; for $k_\rho = 5/3$, this gives 0.55, whereas the exact result is 0.56 (Chieze and Lazareff 1981). In this approximation one can show that the mean pressure in the SNR is $\bar{p} = 1/2\ p_s$, the post shock value (Cavaliere and Messina 1976) and the mean temperature is $\bar{T} = 2(3 - k_\rho)\ T_s/3$.

In order to determine how a SNR evolves in the radiative phase, we set $v_{ps} = v_s$ and $K_{//} = 1$, which is appropriate for a thin shell. If the radiative losses from the SNR are dominated by the emission from the radiative shock at the edge of the remnant, then we have

$$\frac{dE}{dt} = -\ 4\pi R^2\ \frac{(\rho_0\ v_s^2)}{2}. \tag{10}$$

Equations (4) and (10) have two power-law solutions

$$R \propto t^{2/(7-k_\rho)}, \tag{11a}$$

$$R \propto t^{1/(4-k_\rho)}. \tag{11b}$$

The first solution is the pressure-driven snowplow (MO): the energy is predominantly thermal and decreases adiabatically as R^{-2}. The second solution is the classical Oort solution for a shell moving with constant momentum and no internal energy. Numerical calculations by Chevalier (1974) for the case of constant ambient density ($k_\rho = 0$) give $R \propto t^{0.31}$ in this phase, quite close to the $t^{2/7}$ result above. There was no indication of significant radiative cooling of the hot interior in his calculations: lower ambient pressures would be needed to allow such cooling to become manifest. The transition from the Sedov-Taylor phase to the radiative phase occurs at $R = 10^{1.29} E_{51}^{0.29} n_0^{-0.41}$ pc, where E_{51} is the SNR energy in units of 10^{51} erg and n_0 is the ambient density of hydrogen nuclei, which has been assumed constant.

c) Entropy of Hot Gas in SNRs

The evolution of the hot gas in the interior of an SNR is most conveniently described in terms of its entropy (Kahn 1976). After being shocked to a high pressure by the SNR, the entropy of a parcel of gas is conserved during the subsequent expansion, except that if the entropy is sufficiently high it will be reduced by thermal conduction, whereas if it is sufficiently low, radiative losses become important.

We define the 'reduced entropy' $s*$ as

$$s* = T^{3/2}/n ;\tag{12}$$

in terms of $s*$, the entropy per unit mass is $(k/\mu) \ln s*$, where μ is the mean mass per particle. Just behind the SNR shock, the Rankine-Hugoniot conditions (e.g. McKee and Hollenbach 1980) give

$$s* = \frac{1}{4n_0} \left(\frac{3}{16} \frac{\mu v_s^2}{k}\right)^{3/2} .\tag{13}$$

In young SNRs such as Cas A and Tycho, temperatures in excess of 10^8K are observed (Pravdo and Smith 1979), so that values of $s_{10}* \equiv s*/(10^{10} K^{3/2} cm^3)$ up to $25/n_0$ can be expected in SNRs. If the SNR occurs in the hot intercloud medium (the HIM: MO, Cox and Smith 1974), then n_0 is of order $3 \times 10^{-3} cm^{-3}$ and $s_{10}*$ could reach 10^4. The mass of gas in an adiabatic blast wave with a reduced entropy of at least $s*$ is readily shown to be (Kahn 1976)

$$M(s*) = \frac{80 \ E_{51}}{(n_0 s_{10}*)^{2/3}} M_\odot\tag{14}$$

for $k_0 = 0$. The mean reduced entropy of this gas, defined as $\bar{s}* \equiv \bar{T}^{3/2}/n_0$, is $2^{7/2} s*$ since $T = 2T_s$ and $n_0 = 1/4 \ n_s$ (see the discussion after equation 9). At any time during the subsequent adiabatic expansion of the SNR, the temperature is $T = 2.9 \times 10^5 (s_{10}* \ \tilde{P}_4)^{2/5}$ K, where $\tilde{P}_4 = k^{-1}p/(10^4 \ K \ cm^{-3})$ measures the pressure.

The effect of radiative losses on the entropy is governed by

$$p \frac{d\ln s*}{dt} = - n^2 \Lambda ,\tag{15}$$

where Λ is the cooling function in erg cm^{-3} s^{-1}. For a gas of cosmic abundances in collisional ionization equilibrium, the cooling function can be approximated as Λ = 1.6 x 10^{19} $T^{-1/2}$erg cm^{-3} s^{-1} over the temperature range 10^5K $\lesssim$ T $\lesssim$ $10^{7.5}$K (cf. Raymond, Cox and Smith 1976). The time t_c for the gas to cool from a temperature T ($10^{7.5}$ K $>$ T $\gg$ 10^5 K) down to 10^5 K is then directly proportional to s* (Kahn 1976): integrating equation (15) gives

$$t_c = 6.3 \times 10^5 \ s_{10}* \ \beta^{-1} \ yr, \tag{16}$$

where β is the ratio of the actual cooling rate, including the effects of non-equilibrium ionization, to the approximate Raymond et al. value. We have assumed a fully ionized gas with 10% He by number. Note that this result is quite general: in particular, it is valid for both isochoric and isobaric cooling. In the three-phase model of the ISM (MO), the mean interval between SNRs is about 4 x 10^5 yr. Hence gas with $s_{10}* \lesssim \beta$ will cool to low temperatures (T ~ 10^4 K), whereas gas with $s_{10}* \gtrsim \beta$ will be steadily heated. The effect of this high entropy gas on the ISM will be discussed in section 6 below.

Thermal conduction can play an important role in redistributing the entropy of the hot interior gas. If conduction is unimpeded in fairly young remnants, then the interior becomes nearly isothermal and nearly isentropic: the limiting case of an isothermal blast wave has been analyzed by Solinger, Rappaport and Buff (1975). An estimate of the maximum entropy which survives the effects of conduction can be obtained by equating the characteristic conduction time $t_\kappa = uR^2/\kappa T$, where u is the energy density and κ the thermal conductivity, to the age of the remnant: we obtain

$$s_{10}* \lesssim 10 \ E_{51}^{3/14} \ n_0^{-4/7} \ \phi_c^{-9/14} \ . \tag{17}$$

Here ϕ_c $\langle$ 1 is the ratio of the actual conductivity to the Spitzer (1962) value, which was calculated assuming no magnetic fields and no anomalous resistivity. For low ambient densities n_0 this is not a very stringent limit. However, as pointed out by Lerche and Vasyliunas (1976), the electrons and ions are not necessarily in equipartition at the high temperatures required for conduction to be important. The electron-ion equipartition time (Spitzer 1962) is directly proportional to s*:

$$t_{eq} = 2600 \ s_e*_{10} \ yr, \tag{18}$$

where $s_e* = T_e^{3/2}/n_e$ and where we set the Coulomb logarithm equal to 30. Equating this to the age of the remnant, we find that equipartition holds only for $s_{10}* \lesssim 3 \ E_{51}^{3/14} \ n_0^{-4/7}$. Since s is comparable to our estimate (equation 17) at which conduction becomes unimportant, it is clear that the electron and ion temperatures must be treated separately in evaluating the effects of conduction (Cowie 1977: see also Itoh 1978). In most SNRs, however, the ambient medium is not homogeneous, and conduction between the hot intercloud gas and the embedded clouds can have a far more important influence on the evolution of the SNR than conduction in the hot gas itself. We turn to a study of this topic now.

3. SNR EXPANSION IN A CLOUDY MEDIUM

Observational evidence and theoretical arguments suggest that most of the mass in the ISM is confined to a small fraction of the volume (e.g. Field, Goldsmith and Habing 1969, MO). In such a situation, the dynamics of an SNR is governed primarily by the low density intercloud medium (Cox and Smith 1974: McKee and Cowie 1975: Bychov and Pikel'ner 1975). Since the pressure inside a SNR is approximately constant and since the pressure behind a shock is approximately $\rho_0 v_s^2$, if follows that $v_s \propto \rho_0^{-1/2}$: the blast wave advances most rapidly into the low density gas, and slower shocks subsequently compress the embedded clouds.

An accurate theory of SNR evolution must then include an account of the mass, momentum, and energy exchange between the intercloud medium and the clouds. Energy exchange is relatively straightforward: it is predominantly cloud crushing, and the energy stored in the compressed clouds is radiated away if the shock velocity in the clouds is less than about 200 km s^{-1}. Momentum exchange is cloud acceleration (McKee, Cowie and Ostriker 1978) and its converse, cloud drag on the intercloud medium. The coupling between the clouds and the ambient gas depends to some extent on the strength of the interstellar magnetic field.

Mass exchange from the clouds to the intercloud medium is effected by evaporation and hydrodynamic ablation: the reverse processes are condensation onto clouds and cloud formation via thermal instability (Field 1965). Since thermal conduction is significant only along the magnetic field, the rate of evaporation depends on the topology of the magnetic field and is therefore uncertain. If the intercloud medium is moving relative to the cloud, then viscosity and hydrodynamic instabilities (both Kelvin-Helmholtz and Rayleigh-Taylor) can strip the outer layers of the cloud. Such turbulent viscous stripping has been analyzed by Nulsen (1981) for the case of stripping of gas from galaxies in clusters, and is somewhat less effective than evaporation in the absence of magnetic inhibition. Note that evaporation and stripping do not remove energy from the SNR, but rather share the energy with more mass, thereby reducing the entropy. These processes are typically important only when the intercloud density is very low, so that the transfer of a small amount of mass from the clouds to the intercloud medium can have a significant effect on the latter.

Here we shall discuss the effects of cloud evaporation under the assumption that magnetic fields and plasma instabilities impede conduction by at most a factor of a few. Stripping contributes somewhat to the mass exchange, but it will not be considered explicitly. Cox (1979) has analyzed the opposite case, in which evaporation is negligible: more recently, however, he has argued that mass exchange must be efficient (Cox 1981). Cloud evaporation can have a major effect on SNR evolution and on the ISM as a whole, and it is crucial to resolve the issue of its importance observationally.

a) Evaporation of Clouds

The theory of the steady evaporation of cool clouds in a hot ambient gas has been developed over the past several years (Cowie and McKee 1977: McKee and Cowie 1977: Cowie and Songaila 1977: Balbus and McKee 1981: for application to laser fusion, see Max et al. 1980). When the blast wave of a SNR crosses a cloud, a shock is driven into the cloud and a sharp temperature gradient is produced at the contact discontinuity between the hot intercloud gas and the shocked cloud. This gradient relaxes due to thermal conduction: a steady state is set up in which the outer layers of the cloud heat up, flow out, and merge with the intercloud medium. The rate of this evaporation depends on the ratio λ/a, where λ is the electron mean free path in the hot ambient gas and a is the cloud radius. For $\lambda/a \lesssim 0.015$ the hot gas cools and condenses onto the cloud: for $0.015 \lesssim \lambda/a \lesssim 0.5$, the classical Spitzer (1962) conductivity applies and the cloud undergoes classical evaporation: for $1/2 \lesssim \lambda/a \lesssim 50$, the heat flux saturates at a value which can be estimated only roughly, and the cloud undergoes saturated evaporation: and finally, for $\lambda/a \gtrsim 10^3$ the ambient particles are sufficiently energetic that they penetrate into the cloud like low energy cosmic rays, and the cloud undergoes suprathermal evaporation.

For SNRs in the late stages of evolution, classical evaporation is most important. The inward heat flux $\sim KT/a$ balances the outward enthalpy flux $(5/2)pv$, giving an evaporation rate $\dot{m} = 4\pi\, a^2\rho v \sim 8\pi\, a\, K\, \mu/5k$. The exact calculation for spherical clouds gives a result which is 2.5 of this:

$$\dot{m} = 2.75 \times 10^4\ T^{5/2}\ a_{pc}\ \phi_c\quad g\ s^{-1}. \tag{19}$$

As remarked above, even a very weak magnetic field can inhibit evaporation since the conductivity perpendicular to the field is effectively zero. Hence there will be no evaporation along closed field lines. For open field lines, however, the evaporation is significantly inhibited only if the field lines are nearly tangential to the surface of the cloud, since the first order effect of having field lines at an angle to the temperature gradient is to steepen the gradient without materially altering $\dot{m}$. We adopt $\phi_c = 1/3$ as the uncertain allowance for magnetic inhibition. Note that turbulent stripping is unaffected by weak tangential fields, so it is unlikely that the effective value of ϕ_c in expression (19) for $\dot{m}$ is very much smaller than this.

b) Dynamics of Evaporative SNRs

The effects of cloud evaporation on SNR evolution can be determined by allowing for the variation in the mean density in the SNR (MO). In the limit of negligible ambient density, the internal density is entirely due to evaporated gas:

$$\frac{dM}{dt} = N_{cl} \, \dot{m} \, \frac{4\pi}{3} \, R^3, \qquad (20)$$

where N_{cl} is the number of clouds per unit volume. Assume that the SNR is nonradiative and that the SNR has expanded to the point that the evaporation is classical ($R \gtrsim 20$ pc for typical conditions). In the late stages of SNR evolution, clouds are not destroyed by evaporation so that N_{cl} is independent of time; we assume that it is independent of position as well. The mass in the SNR varies as

$$R^{3-k_\rho}$$

(cf. equation 5); since the thermal and kinetic energies are separately conserved in a self similar expansion, we find

$$T \propto v^2 \propto R^{-(3-k_\rho)}.$$

Equating powers of R in equation (20) then gives $k_\rho = 5/3$, or

$$\bar{\rho} \propto R^{-5/3}, \quad T \propto R^{-4/3} \qquad (21)$$

in an evaporative SNR. By contrast, the standard Sedov-Taylor blast wave in a uniform medium has $\bar{\rho} = $ const, $T \propto R^{-3}$. With approximation (6) for $\kappa_{//}$, equation (8) gives

$$R = \left[\frac{1.46 \, E}{\bar{\rho}(1)} \right]^{3/10} t^{3/5}, \qquad (22)$$

which is equivalent to the result obtained by MO. The parameter $\bar{\rho}(1)$ can be determined by solving equation (20). The $t^{3/5}$ behaviour for evaporative SNRs is to be compared with the $t^{2/5}$

behaviour for the Sedov-Taylor blast wave. Exact similarity
solutions have been obtained by Chieze and Lazareff (1981) which
reveal the internal structure of the remnant (see below) and
confirm these results.

The condition that the ambient density be low enough that the
SNR be evaporation dominated can be expressed in terms of Σ =
$\alpha\ a^{2}/3 f_c\ \phi_c$, which is inversely related to the rate of cloud
evaporation. Here $\alpha = v_s/c_h$ is the ratio of the shock velocity
to the mean isothermal sound speed of the hot interior (= 2.89
according to Chieze and Lazareff 1981) and $f_c = 4\pi N_{c1}\ a^3/3$ is the
cloud filling factor. The mass of evaporated gas exceeds the
swept up mass provided

$$n_o < 0.038\ \left(\frac{E_{51}^{2}}{\Sigma_{pc}}\right)^{1/3}\ \left(\frac{100\ pc}{R}\right)^{5/3}\ cm^{-3} \qquad (23)$$

where Σ_{pc} is in units of pc^{2}; MO inferred Σ_{pc} = 50 in the ISM.
Hence evaporation is important only if the ambient density is low
(recall that the assumption of classical evaporation requires R $\geqslant$
20pc). Correspondingly, since the mean density of the ISM is 0(1
cm^{-3}), evaporation has a relatively small effect on the clouds,
which contain most of the mass.

c) The Rim of an Evaporative SNR

Perhaps the most striking difference between evaporative and
non-evaporative SNRs occurs at the rim of the remnants. In the
standard Sedov-Taylor blast wave, the density and pressure peak
at the shock front. As pointed out by Cox (1979), however, the
density behind the shock in an evaporative SNR - four times the
ambient value - may be much less than the mean interior value.
The pressure at the outer edge of the remnant must also be
smaller than the mean interior value. These effects are clearly
shown in the similarity solutions of Chieze and Lazareff (1981):
in the limit of vanishing external density, the density and
pressure rise from zero just behind the shock and peak at 0.7 R
and 0.8 R, respectively; the temperature is a maximum at the
shock and falls monotonically toward the centre of the remnant.

The observational importance of this result has been
emphasized by Cowie and McKee (in preparation), who have
developed the following simple analytic picture for the structure
of the rim. Consider the region near the shock where $z \equiv$ R-r $\ll$ R
and the flow is nearly planar. In the shock frame, the equation
of continuity for steady flow is $\rho v = \rho_o v_s + \int N_{c1}\dot{m}dz$. Assuming
negligible cloud acceleration implies the evaporated gas is
injected at velocity v_s in the shock frame, so that the momentum
equation gives $p + \rho v^2 = \rho_o v_s^2 + v_s \int N_{c1}\dot{m}dz$, where we assumed the
shock to be strong. Combining these equations, we find

$$p + \rho v(v - v_s) = 0. \tag{24}$$

Conduction should iron out the temperature gradients, and even in the absence of conduction in the hot interior, Chieze and Lazareff (1981) find that the temperature is a slowly varying function of radius. Setting T = const. in equation (24), we conclude that v = const. as well; just behind the shock v = $1/4v_s$ so that

$$p = \frac{3}{4} \rho v v_s = \frac{3}{4} (\rho_0 v_s^2 + N_{cl} \dot{m} v_s z). \tag{25}$$

Hence, in this approximation the pressure increases linearly behind the shock. This pressure gradient is necessary to accelerate the gas evaporated from the clouds up to the expansion velocity of the remnant. If we consider the rim to be the region of the SNR beyond the pressure maximum, then Chieze and Lazareff (1981) find its thickness to be about 0.2 R. This would be reduced if the effects of cloud drag were included, and numerical calculations by Cowie, McKee and Ostriker (1981: hereafter, CMO) indicate an actual rim thickness of about 0.15 R. The appearance of the rim will be discussed in section 5 below.

4. NUMERICAL MODELS

Although analytic models are suitable for studying individual processes affecting SNR evolution in an inhomogeneous medium, numerical modelling is required to include all the important effects associated with mass, momentum, and energy exchange among the different components of the ISM. The first attempt at doing this has been made by CMO, who focussed on SNR evolution in the three phase ISM.

The principal assumptions in their work were: (1) Spherical symmetry. This ignores variations on scales of order 100pc due to the scale height of clouds in the galaxy and due to inter-actions with other SNRs. (2) Negligible magnetic field. In the late stages of evolution when the interior pressure is not much greater than the interstellar magnetic pressure, the inclusion of magnetic fields would increase the drag between the clouds and the intercloud medium. (3) Dynamical effects of cosmic rays are ignored. (4) A significant fraction of the volume of the ISM is filled with low density gas, typically n ~ $10^{-2.5}$ cm^{-3}. (5) Denser phases of the ISM are represented by two types of clouds: cold HI clouds with (n,T,a,f) = (40 cm^{-3},80K,0.4-1.6 pc, 0.02) and warm clouds with (n,T,a,f) = (0.3 cm^{-3},8000K,2 pc, 0.23). The warm clouds account for the warm HI seen in 21 cm studies as well as the HII seen in diffuse Hα emission and inferred from measurements of pulsar dispersion. (6) Effects of the pre-SN star on the ambient medium are neglected.

(7) Radiative cooling is calculated assuming ionization equilibrium (Raymond <u>et al</u>. 1976). (8) Electrons and ions assumed to have equal temperatures behind the shock. (9) Ambient temperature is held constant despite the fact that the cooling time is less than the age of the SNR at late times. However, within the confines of these assumptions, all important processes were considered, including cloud compression, radiation from the three phases, cloud acceleration and drag, and evaporation and condensation of clouds.

The evolution of a 3×10^{50} erg SNR in a medium of density 2.4×10^{-3} cm^{-3} is shown in Figure 1. These numerical calculations confirm the principal result of the analytic studies of SNR evolution in the three phase ISM (Cox and Smith 1974; MO): the SNRs expand to a very large radii (R $\gtrsim$ 150 pc) before cooling sets in. The $R^{-5/3}$ behaviour of the mean internal density found by MO was also confirmed.

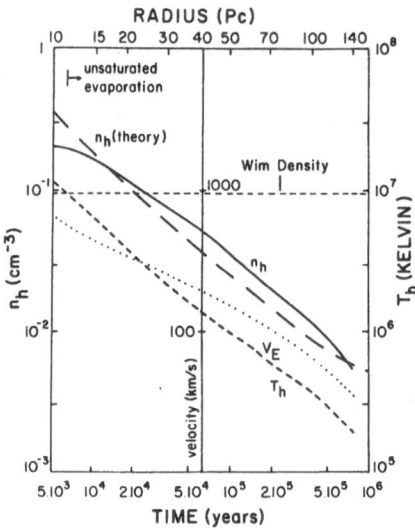

Figure 1. Evolution of a 3×10^{50} erg SNR in a three phase ISM, taken from CMO. The mean density of the hot intercloud medium (HIM) is shown by the solid line, the mass-averaged temperature of the HIM by the dashed line, and the peak velocity in the SNR by the dotted line; all are plotted as a function of the age and radius of the SNR. The value of the mean density predicted by MO is shown by the long dashed line. The ambient hot medium was assumed to have $n_0 = 2.4 \times 10^{-3}$ cm^{-3}, $T = 1.4 \times 10^5$K, and $f = 0.71$. Warm clouds of radius a = 2 pc filled 27% of the volume and had a mean density $n_w f_w = 0.1$ cm^{-3}. The cold clouds had a = 1.6 pc and f = 0.02. The evaporative efficiency ϕ_c was set equal to unity in this calculation.

Two important new results emerged from their study. First, the mean density of the hot interior gas n_h is limited to about twice the mean density of the warm clouds, ~ 0.2 cm^{-3}. This limitation is due both to the complete destruction of the readily evaporated warm clouds and to saturation of the evaporation at high temperatures. For young SNRs, this means that the transition from the ejecta-dominated stage to the Sedov-Taylor stage is delayed compared to the case of a homogeneous ambient medium, since only the evaporated gas can effectively absorb energy from the ejecta. Furthermore, the Sedov-Taylor stage now divides into two parts: in 'adolescent' SNRs (10^3 yr $\lesssim$ t $\lesssim$ 2 x 10^4 yr, 5 pc $\lesssim$ R $\lesssim$ 20 pc for the model in Fig. 1) such as the Cygnus Loop and IC443, the mean interior density remains approximately constant so that the evolution mimics that of an adiabatic blast wave in a two-phase ISM. On the other hand, in 'middle-aged' SNRs (2 x 10^4 yr $\lesssim$ t $\lesssim$ 3 x 10^5 yr, 20 pc $\lesssim$ R $\lesssim$ 90pc) such as the North Polar Spur and the Eridanus hot spot, the evaporation is classical and not limited by the amount of warm clouds, so $n_h \propto R^{-5/3}$ as found by MO.

The second new result is that radiative losses lead to the formation of a dense shell in the <u>interior</u> of the remnant and not at its outer edge. For an SNR in a homogeneous medium, the entropy and hence the cooling time (see section 2c) are a a the outer edge; as a result, cooling leads to a dense shell forming just behind the shock. In an evaporative SNR, however, the entropy has a maximum at the outer edge and has a minimum in the interior. The actual location of the dense shell in an evaporative remnant is also affected by variations in the density of clouds caused by cloud motions induced by the blast wave. By adopting an equilibrium cooling curve, CMO underestimated the actual cooling rate and hence overestimated the time of shell formation: the ionization state of evaporating gas lags its temperature and hence leads to greater cooling. MO assumed β = 10 in equation (16) whereas CMO estimated β = 1; the higher value is likely to be closer to the truth. More detailed calculations are required to determine the correct cooling rate.

5. COMPARISON WITH OBSERVATION

It is a somewhat daunting task to attempt to explain the complex and sometimes beautiful structure observed in SNRs with the simple models described above. The most basic idea - that a significant portion of an SNR's evolution can be described by a spherical, non-radiative blast wave - has been amply confirmed. As emphasized above, however, a more detailed understanding of the evolution of an SNR depends upon the structure of the ISM (the manner in which this structure is in turn determind by SNRs will be discussed in section 6). A fundamental aspect of the structure of the ISM is the number of distinct components, or phases, which comprise it, and we shall address this issue here. Another important aspect is the morphology of these components, which is crucial for determining the appearance of the SNRs; rather than considering this issue explicitly here, we shall adopt the simple model of spherical clouds embedded in an intercloud medium.

Three models for SNR evolution in an inhomogeneous ISM which have been considered in the literature are: (1) evolution in a two phase ISM of the type described by Field et al. (1969); (2) evolution in a three phase medium without evaporation (Cox and Smith 1974: Cox 1979); and (3) evolution in a three phase medium with evaporation (MO). Although these models are conceptually quite different, it has proven remarkably difficult to distinguish them observationally. Each has the same relation between pressure and radius, $p \propto E/R^3$; each predicts that X-ray emission from remnants such as the Cygnus Loop arise from gas with $n \sim 0.1 - 0.2$ cm^{-3}; and each assumes that the ISM in the vicinity of the SNR has not been significantly modified by the pre-SN star, which is not necessarily valid even for SNRs as large as the Cygnus Loop.

Unequivocal evidence that at least two phases must be considered in describing SNR evolution was provided by the discrepancy between the shock velocities inferred from optical and X-ray observations. For example, the bright optical filaments in the Cygnus Loop have velocities of about 100km s^{-1}, whereas the X-ray spectrum implies a velocity of about 400 km s^{-1}. Such a result occurs naturally if the X-ray emission arises in the shocked intercloud medium and the optical emission in shocked clouds (Woltjer 1972; McKee and Cowie 1975; Bychkov and Pikel'ner 1975; De Noyer 1975). This picture has been confirmed by the observation of faint, fast optical filaments moving at several hundred km s^{-1} (Kirshner and Taylor 1976) and [Fe XIV] emission (Woodgate et al. 1977) in the Cygnus Loop.

The evidence for the effect of a third phase - the hot, low density HIM - on the evolution of SNRs is ambiguous, and it will be a considerable challenge to observers to resolve the issue. Here we shall discuss four sets of observations which are most easily explained in terms of a three phase ISM.

Large SNRs. The primary prediction of the three phase model is that the non-radiative stage of SNR evolution extends to R $\gtrsim$ 150 pc, far greater than the 40 pc for the two phase model. Possible examples of such large remnants are the North Polar Spur, the Eridanus hot spot, and the local SNR surrounding the Sun which produces much of the soft X-ray background (Hayakawa et al. 1977; Tanaka and Bleeker 1977; MO). The North Polar Spur can be explained as a single energetic (E $\sim$ 3 x 10^{51} erg) SNR, or

as several 10^{51} erg SNRs, in the three phase ISM with R ~ 80 pc
and T ~ 2-3 x 10^6K. Adherents of the two phase model would have
to invoke in excess of 10 SNRs to achieve the same result. Such a
large number of SNRs might be expected if the Spur is in the Sco
Cen Association (H. Weaver, private communication), so this
evidence is inconclusive.

The Eridanus hot spot appears as an SNR in both X-rays
(Naranan et al. 1976) and HI (Heiles 1976). Interpreting the
velocity structure seen in 21 cm in terms of an SNR in a
homogeneous medium leads to an unsatisfactory result: with an
estimated radius of 50 pc, the remnant should be radiative and
have its mass concentrated in a thin shell. Heiles found that an
energy much greater than 10^{51} erg and a mean density of 4 cm^{-3}
was required; the absence of HI emission at the high positive and
negative velocity caps was not explained. On the other hand,
Cowie, McKee and Ostriker (in preparation) have shown that
Eridanus is consistent with a standard SNR in a three phase ISM.
The 21 cm observations are accounted for in terms of structure in
velocity space rather than physical space, with each HI cloud
receiving an outward kick as it is passed by the SNR shock. The
inferred HI column density is lower, explaining the absence of
the caps at high speeds: the caps should be accessible to higher
sensitivity observations, however. Unfortunately, this evidence
for a SNR in the three phase ISM is ambiguous as well, since
Reynolds and Ogden (1979) have suggested that Eridanus is
connected with the Orion OB1 association.

Absence of radiative shells. SNRs form a dense radiative
shell at a radius $R_c \approx 20 E_{51}^{0.29} n_0^{-0.41}$ pc uunless n_0 is so low
and R_c so large that they collide before cooling. Cowie and York
(1978) carried out a sensitive search for such shells by looking
for high velocity UV absorption lines. They found a conservative
lower limit to the mean free path to a 100 km s^{-1} absorption
feature in the galactic disk of 12 kpc; they assumed an SNR rate
of 1 per 50 yr in the Galaxy, and concluded that this requires an
ambient density of at last 0.1 cm^{-3} to keep the SNRs small enough
to satisfy this constraint. This result does not rule out the
three phase models, however, since the non-evaporative models
generally have negligible radiative losses from the hot gas and
hence do not form shells, whereas the evaporative model has the
shell in the interior of the SNR, where the velocity is well
below 100km s^{-1}.

In a two phase ISM, the radiative shells should also be
observable in emission. The effective emission measure of a
planar shock may be defined as EM = $(H_{rec} n_0 v_s)/(\alpha^{(2)} \times 3 \times 10^{18}$
cm pc$^{-1})$, where H_{rec} is the number of recombinations per incoming
H atom and $\alpha^{(2)}$ is the hydrogen recombination rate to excited
states, which we shall evaluate at 10^4 K. Averaging this

emission measure over the surface of a sphere increases it by 4; at $v_s = 100$ km s^{-1}, we have H$_{rec} = 1.5$ (Shull and McKee 1979) and $\overline{EM} = 75$ n_0. For $n_0 > 0.1$ cm^{-3}, this gives $\overline{EM} > 7.5$ cm^{-1} pc which is readily observable (cf. Reynolds, Roesler and Scherb 1974). There should be approximately $[t(v = 100)/30yr]/15^2 = 30(0.1/n_0)^{0.36}$ such shells within 1 kpc of the sun, where $t(v = 100)$ is the age of a radiative SNR with $v_s = 100$ km s^{-1}, n_0 is the intercloud density $(0.1$ cm$^{-3} \leqslant n_0 \leqslant 0.3$ cm$^{-3})$ and we have assumed 1 SN per 30 yr in the 15 kpc disk of the Galaxy. This is far greater than the number of reported radiative remnants near the Sun (S147 and the Monocerotis Loop have been suggested; they lie at distances of 1 and 0.7 kpc: Woltjer 1972). Until an explicit search for such low surface brightness features is carried out, however, this cannot be regarded as conclusive evidence against SNRs in a two-phase ISM.

Finally, we note that radiative shells have been oberved in absorption in the Orion and Carina associations, and have been interpreted as evidence for SNR expansion in the three phase ISM (Cowie, Hu and York 1981).

SNR rims. A unique prediction of evaporative SNR models is that once the SNR radius significantly exceeds the typical cloud radius, then the SNR has a rim of thickness $(0.1 - 0.2)$ R in which the pressure increases inward from a relatively low value just behind the shock (section 3c). Possible evidence for such a rim in the Cygnus Loop comes from a number of different observations of emission extending about 1 pc outside the bright optical filaments: X-ray emission (Tuohy et al. 1979); [Fe XIV] emission (Woodgate et al. 1977); and faint Balmer line filaments (Raymond et al. 1980). A quantitative interpretation of these observations in terms of the evaporative model depends on the morphology of the clouds near the Cygnus Loop and has not been completed.

There is another important difference between the outer regions of evaporative and non-evaporative SNRs in a three phase ISM: for SNRs such as the Cygnus Loop, the mean intercloud density is about 50 times lower in a non-evaporative SNR, and the blast wave velocity about 7 times greater. Such a large velocity (~ 3000 km s^{-1} for the Cygnus Loop) corresponds to 10^{17} cm in 10 yr and should be readily observable. No such motions have been seen in the Cygnus Loop.

Spectra of evaporating clouds. Dopita (1981) has observed a cloud in the SNR N49 in the LMC which appears both in [OIII] $\lambda 5007$ and [Fe XIV] $\lambda 5303$, and is a strong candidate for an evaporating cloud. Theoretically, the predicted spectrum of an evaporating cloud depends upon its ionization (McKee and Shull, in preparation). As the evaporating gas is heated by

conduction, its ionization will lag its temperature. For a
partially ionized cloud this implies a spectrum dominated by
collisionally excited Balmer lines, as in the case of excitation
in collisionless shock (Chevalier, Kirshner and Raymond 1980).
If the cloud is ionized, however, the ratio of collisionally
excited lines to Balmer recombination lines will be much greater
than in normal nebulae because of the high temperature.

It is to be hoped that future studies of large SNRs,
radiative shells, SNR rims, and evaporating clouds will resolve
the issues posed here. Studies of SNRs in the LMC (Long and
Helfand 1979) and Andromeda (Blair, Kirshner and Chevalier 1981)
will be particularly helpful for the first two items. Existing
statistical studies of radio SNRs (e.g. Clark and Caswell 1976)
have been restricted to small SNRs (R $\lesssim$ 16 pc) and hence are not
directly relevant to these problems. The discrepancy between the
observed number of radio SNRs and the number expected on the
basis of an SN rate of 1 per 10-40 yr (Tammann 1982) has been
interpreted in terms of a three phase ISM (e.g. Higdon and
Lingenfelter 1980), but a better understanding of radio emission
from old SNRs is necessary to make this conclusive.

6. THE EFFECT OF SNRs ON THE ISM

To this point we have discussed the evolution of SNRs in
the ISM as if SNRs were isolated events and the properties of the
ISM were fixed. In fact, SNRs are sufficiently energetic and
frequent that they probably determine the structure of the ISM.

a) SNR interiors almost fill the ISM.

A quantitative measure of the effects of SNRs on the ISM is
their volume filling factor f_{SNR}. Let Q(R) be the probability of
being inside a remnant with a radius less than R, and let S be
the supernova rate per unit volume (S = 2.4 x 10^{-13} pc^{-3} yr^{-1}
corresponds to one SN every 30 years in a disk 15 kpc in radius
and 200 pc thick). After passing through the adiabatic and
radiative stages of evolution, the SNRs reach pressure
equilibrium with their surroundings at a radius R_E. Neglecting
interactions among different remnants, we have f_{SNR} = 1
- exp[-Q(R_E)] and

$$Q(R_E) = \frac{4\pi}{3} R_E^3 \, S \, t_m, \tag{26}$$

where t_m is the lifetime of the remnant in the final stage. Cox
and Smith (1974) termed this value of Q the porosity of the
medium. For the lifetime, we adopt a sound crossing time, based
on the sound speed $(p_0/\rho_0)^{1/2}$ in the ambient medium: then (MO)

$$Q(R_E) = \frac{0.5\ E_{51}^{1.3}}{n_o^{0.14}\ \bar{p}_{04}^{1.3}} \quad (\frac{S}{10^{-13}\ pc^{-3}\ yr^{-1}}) \tag{27}$$

For observed values of the parameters (intercloud density $n_o \lesssim$
0.3 cm^{-3}, $\bar{p}_{04} \lesssim$ 1), this gives $Q(R_E) \gtrsim$ 1. Hence, as originally
demonstrated by Cox and Smith (1974), SNRs occupy a substantial
fraction of the volume of the ISM. If initially the ISM were in
a two-phase equilibrium (cold clouds in a warm interstellar
medium – Field et al. 1969), then supernovae would rapidly
destroy this equilibrium and create a three phase medium in which
the third phase – the hot intercloud medium (HIM) inside the SNRs
– has a significant filling factor. For $f_{SNR} <$ 1/2, the HIM may
be visualized as bubbles of hot gas which sometimes connect
together to form tunnels (Cox and Smith 1974): for $f_{SNR} >$ 1/2,
however, it is more convenient to consider the cooler phases, the
cold neutral medium at T ~ 10^2 K and the warm medium (both
neutral and ionized) at T ~ 10^4 K, as clouds embedded in hot gas
(MO).

The argument which led to equation (27) can be inverted: if
SNRs are the dominant energy source for the ISM, then f_{SNR} must
be of order unity in order that they can determine the pressure.
Hence the pressure in equation (27) must be low enough that
$Q >$ 1.

Observations of the soft X-ray background (Fried et al. 1981
and references therein) and of measurements of Lyman α absorption
(e.g. Bohlin 1975) have established that the filling factor of
hot gas within about 100 pc of the Sun is indeed of order unity.
However, contrary to the assumption usually made in astronomy,
the local ISM may not be typical: Tanaka and Bleeker (1977) have
argued that less than 5% of the galactic disk could have the same
X-ray emissivity as the local ISM, and Fried et al. (1981) have
inferred an unusually small number of cold clouds in the local
ISM based on the lack of X-ray shadows. The strongest observational argument that the HIM is widespread is based on the OVI
absorption lines seen in many stars, most of which are well
beyond 100 pc (Jenkins and Meloy 1974: York 1974). The interpretation that this absorption arises in the conductive
interfaces of clouds embedded in the HIM (McKee and Cowie 1977:
MO) is supported by correlations of the OVI line profiles with
the lower excitation SiIII and NII profiles (Cowie et al.
1979). The OVI data does not determine f_{SNR} very accurately,
however; Jenkins (1978) obtains an uncertain estimate f_{SNR} ~ 0.2.
It is of major importance to obtain more accurate measurements of
the filling factor of the HIM in other parts of the Galaxy, and
observations of SNRs (section 5) may provide the best method. 21
cm observations such as those of Heiles (1980) can determine the
filling factor of large (R > 100 pc) holes in the HI, but it is

difficult to measure the much smaller structure expected in the three phase ISM by this technique.

What are the possible holes in the theoretical arguments which led to a large value of f_{SNR}? If SNRs are not the dominant energy source for the ISM, then the pressure could be large enough to squeeze the porosity down to a small value. The Galaxy puts out about thirty times more energy in starlight than in supernovae, and if this energy could be harnessed for heating – for example, by photoelectric heating due to dust – then the pressure could in fact be raised. However, the warm medium cannot be in equilibrium at pressures $\bar{p} > 10^4$ cm^{-3} K (de Jong 1980), so this mechanism is of limited effectiveness. Another assumption made in equation (27) was that the SNRs are uncorrelated. This assumption is violated in young stellar associations, and supernovae there may lead to large bubbles (e.g. Bruhweiler et al. 1980) filling ~ 10% of the ISM. However, Type I SNe and those Type II's due to intermediate mass stars and runaway stars, which together comprise most SN, should be uncorrelated.

The most serious potential flaw in the argument is the assumption that the lifetime of an old SNR is determined by the sound crossing time. There are a large number of clouds embedded in the remnant, and if the interior of the SNR cools to the point that the pressure drops well below the typical interstellar pressure, then the clouds could expand to well above their typical volume, reducing the local value of f_{SNR} as a result. Thus negative fluctuations in the rate of supernovae could result in regions of low f_{SNR} which might require several million years at a normal supernova rate to recover a normal porosity.

b) Energy balance in the ISM

Time dependent studies of the effects of SNRs on the ISM show that a steady state is established in a time of $1-3 \times 10^7$ y (Habe, Ikeuchi and Tanaka 1981). In a steady state, conditions of energy balance and mass balance may be applied to determine the steady state conditions.

The energy injected by supernovae must be either radiated or convected away in a wind. Four types of models have been considered: (1) Radiation from shocked cool gas. In the two phase ISM, the hot gas inside a SNR loses much of its energy by driving a radiative shock into the ambient cool gas. In the three phase ISM, energy may be dissipated by radiative shocks driven into clouds (cloud-crushing). CMO found that cloud crushing dissipated no more than about half the energy for cloud filling factors of order 1/4: this fraction would have been less had non-equilibrium cooling been considered. (2) Radiation from

HIM in the disk. SNR energy is radiated away by the hot gas itself before it can escape into the halo (MO). (3) Galactic fountain: radiation from HIM in the halo (Shapiro and Field 1976; Chevalier and Oegerle 1979; Bregman 1980; Habe and Ikeuchi 1980; Cox 1981; Kahn 1981). The hot gas bubbles up out of the disk, cools in the halo, and falls back into the disk as clouds. (4) Galactic wind. Although a weak galactic wind may be driven by halo supernovae (MO), Chevalier and Oegerle (1979) showed that a galactic wind cannot be important for the energy balance of the disk (see below). Each of these four energy balance conditions gives a mean internal energy density in the disk of about $SE\tau$, where τ is the cooling time or the time for the gas to flow out of the disk.

c) Mass balance

In a steady state, the rates of creation and destruction of each phase must be in balance. First consider the rate at which the HIM is converted into the cooler phases. In order to satisfy the energy balance condition, an amount of gas

$$\Delta M = E/(\frac{5}{2} \frac{k\Delta T}{\mu}) = 1500 \frac{E_{51}}{\Delta T_6} M_\odot \tag{28}$$

per SNR must cool by an amount ΔT (this assumes the cooling is isobaric and is therefore a lower limit on ΔM). If this gas cools without reheating, as in fountain models and in the MO model, then $\Delta T \sim T \sim 10^{5.5-6}$K and a significant fraction of the mass cools to low temperatures. On the other hand, if SNRs overlap before cooling (Cox 1979), then the fact that the cooling above 10^5K is thermally unstable again ensures that a substantial fraction of 1500 $M_\odot$/SNR will cool to low temperatures. In both cases, the thermally unstable nature of the cooling implies that the gas which cools first will be compressed to substantially higher densities and form clouds (Schwarz, McCray and Stein 1972). Typically at least half the cooling mass goes into clouds, so that the rate at which HIM is converted to clouds is about 1500 E_{51} $M_\odot$/SNR, where we took $T = 10^{5.7}$ K.

Now we evaluate the rate of the converse process. If cloud evaporation is not important, then the rate at which clouds are converted to HIM is simply the rate at which they are shocked to an entropy high enough that their cooling time exceeds the mean time between SNRs ($\gtrsim 4 \times 10^5$ y) or the time it takes to rise into the halo ($\gtrsim 10^6$ y); this requires $s_{10}^* > 0.7$ (equation 16). The cloud mass shocked to this entropy is given by equation (14), reduced by the cloud filling factor f_c. Conservatively setting $f_c = 1/2$ and $n_0 = 0.2$ cm^{-3}, we find that at most 150 $M_\odot$/SNR of cloud material is converted to HIM. This is 10 times less than the converse rate, and we conclude that a steady state is impossible in non-evaporative models which rely on radiative losses from the HIM to achieve energy balance. A similar conclusion was reached by Habe et al. (1981) on the basis of their numerical calculations. In evaporative models, on the other hand, 540 $E_{51}^{6/5} \Sigma_{pc}^{-3/5} n_h^{-4/5}$ $M_{\odot}$/SNR is evaporated from clouds and incorporated into the HIM, and thus mass balance is readily achieved (MO). The results of CMO show that the cloud formation occurs in the interior of the SNRs rather than at the outer edge as assumed by MO, however.

The thermal instability of the hot gas provides another difficulty for non-evaporative models: in the disk, where the gas is exposed to SNR heating, the gas is driven away from the unstable equilibrium to high temperatures as well as to low temperatures. Just as Cox and Smith (1974) argued that SNRs form hot bubbles in a homogeneous ISM, so very hot bubbles will form in the HIM. Such bubbles would rise out of the plane at a velocity well below their sound speed due to the inhibiting effects of magnetic fields and drag, so a significant filling factor could result. Since conduction has been assumed to be negligible, the gas must cool radiatively, and it is quite possible that the resulting radiation would exceed the observed soft X-ray flux. Observations of soft X-ray emission in external galaxies (cf. Ikeuchi, Tomisaka and Ostriker 1981) would provide a valuable constraint on this effect.

Two further implications of equation (28) bear mentioning. Chevalier and Oegerle (1979) used this relation to demonstrate that the SN energy in the galaxy cannot be dissipated by a galactic wind: the mass loss rate would be catastrophic unless the temperature were so high ($> 10^7$ K) that it would violate observations. Second, the cloud formation rate ($\sim$ 50 $M_{\odot}$ y^{-1} for 1 SN per 30 y) is so large that it severely strains fountain models. Cox (1981) circumvented this difficulty by adopting a lower SN rate and SN energy so that only 7 $M_{\odot}$ y^{-1} was required. On the other hand, if most of the energy is dissipated in the disk, then a fountain consistent with observation can occur: MO found that SNRs expanded to a radius comparable to the disk scale height so that some mass and energy would be injected into the halo, and Kahn (1981) implicitly assumed that most of the SNR energy was radiated in the disk in his study of the dynamics of the fountain.

We conclude that the only models of the ISM that are consistent with both energy and mass balance are: (1) Modified two phase ISM in which the porosity of the HIM is low, the pressure of the ISM is not directly determined by SNRs, and the SNR energy is radiated by the dominant warm and cold phases after shock heating: or (2) Three phase ISM in which the porosity is high, evaporation significant, and energy balance is due to radiation from the HIM, as in the model of MO. The arguments give in Section 6a indicate a large porosity and hence favour the latter model, at least for that part of the Galaxy within a kiloparsec or so of the Sun. Since the supernova rate and cloud density vary with position in the Galaxy, it is quite possible that both models are correct, but are applicable to different parts of the Galaxy.

7. SUMMARY

Theoretical studies of SNRs have generally focussed on expansion in a homogeneous medium, and the virial theorem provides a good technique for obtaining accurate solutions in this case. However, SNRs produce large variations in the entropy of the ISM and thereby make it inhomogeneous. Thermal conduction acts to reduce the temperature gradients associated with these entropy gradients, but its effectiveness may be reduced by unfavourable magnetic field topology or by plasma instabilities. For SNRs expanding in a low density medium with embedded clouds, the evaporation of clouds due to thermal conduction significantly alters the evolution from the standard Sedov-Taylor blast wave and gives $R \propto t^{3/5}$, $\bar{n}_h \propto R^{-5/3}$, and $T \propto R^{-4/3}$. This applies only to remnants somewhat older than the Cygnus Loop: in younger remnants, limitations on the mass which can evaporate cause the evolution to mimic that in a two phase medium. SNRs expand to large sizes, $R \gtrsim 150$ pc for $n_0 \sim 10^{-2.5}$ cm^{-3}, before radiative losses in their interiors become important.

SNRs may well determine the energetics and structure of the ISM. If so, then evaporation is important in the mass balance of the resulting three phase ISM. On the other hand, if the filling factor of the hot gas is small, then the ISM is predominantly a two-phase medium and evaporation is relatively unimportant in SNR evolution. A variety of observations are required to explicate the coupled issues of SNR evolution and ISM structure, and to determine how these vary with external conditions in our galaxy and in other galaxies.

ACKNOWLEDGEMENTS

I am deeply grateful to Len Cowie and Jerry Ostriker for much of the inspiration and perspiration that went into the work described here. This work was completed during my stay at the Intitute of Astronomy, and I thank M. Rees for his hospitality. This research was supported in part by NSF Grant ASI 79 2324S.

REFERENCES

Balbus, S.A. and McKee, C.F. 1981. Astrophys.J., in press.
Blair, W.P., Kirshner, R.P. and Chevalier, R.A. 1981.
    Astrophys.J., 247,
Blandford, R.D. 1982. This volume.
Bohlin, R.C. 1975. Astrophys.J., 200, 402.
Bregman, J.N. 1980. Astrophys.J., 236, 577.
Bruhweiler, F.C., Gull, T.R., Kafatos, M. and Sofie, M. 1980.
    Astrophys.J.(Letters), 238, L27.
Bychkov, K.V. and Pikel'ner, S.B. 1975. Sov.Astr.Lett., 1, 14.
Cavaliere, A. and Messina, A. 1976. Astrophys.J., 209, 424.
Chernyi, G.G. 1957. Dokl.Akad.Nauk.SSSR, 112, 213.
Chevalier, R.A. 1974. Astrophys.J., 1881, 501.
Chevalier, R.A. 1982. This volume.
Chevalier, R.A., Kirshner, R.P. and Raymond, R.C. 1980.
    Astrophys.J., 235, 186.
Chevalier, R.A. and Oegerle, W.R. 1979. Astrophys.J., 227, 398.
Chieze, J.P. and Lazareff, B. 1981. Astr.Astrophys., 95, 194.
Clark, D.H. and Caswell, J.L. 1976. Mon.Not.R.astr.Soc., 174,
    267.
Cowie, L.L. 1977. Astrophys.J., 215, 226.
Cowie, L.L., Hu, E.M. and York, D.G. 1981. Astrophys.J.
    (Letters), in press.
Cowie, L.L., Jenkins, E.B., Songaila, A. and York, D.G. 1979.
    Astrophys.J., 232, 467.
Cowie, L.L. and McKee, C.F. 1977. Astrophys.J., 211, 135.
Cowie, L.L., McKee, C.F. and Ostriker, J.P. 1981. Astrophys.J.
    237, 908.
Cowie, L.L. and Songaila, A. 1977. Nature, 266, 501.
Cowie, L.L. and York, D.G. 1978. Astrophys.J., 223, 876.
Cox, D.P. 1979. Astrophys.J., 234, 863.
Cox, D.P. 1981. Astrophys.J., 245, 534.
Cox, D.P. and Smith, B.W. 1974. Astrophys.J. (Letters), 189,
    L105.
de Jong, T. 1980. In Highlights of Astronomy, Vol. 5., ed.
    P.A. Wayman (Reidel:Dordrecht), p. 301.
De Noyer, L.K. 1975. Astrophys.J., 196, 479.
Dopita, M. 1981. Private communication.
Field, G.B. 1965. Astrophys.J., 142, 531.
Field, G.B., Goldsmith, D.W. and Habing, H.J. 1969. Astrophys.
    J.(Letters), 155, L49.
Fried, P.M., Nousek, J.A., Sanders, W.T. and Kraushaar, W.L.
    1981. Astrophys.J., in press.
Habe, A. and Ikeuchi, S. 1980. Prog.Theor.Phys., 64, 1995.
Habe, A., Ikeuchi, S. and Tanaka, Y. 1981. Pub.Astr.Soc.Japan,
    33, 23.
Hayakawa, S., Kato, T., Nagase, F., Yamashita, K., Murakami, T.
    and Tanaka, Y. 1977. Astrophys.J.(Letters), 213, L109.
Heiles, C. 1976. Astrophys.J.(Letters), 208, L137.

Heiles, C. 1980. Astrophys.J., <u>235</u>, 833.

Higdon, J.C. and Lingenfelter, R.E. 1980. Astrophys.J., <u>239</u>, 867.

Ikeuchi, S., Tomisaka, K. and Ostriker, J.P. 1981. Preprint.

Itoh, H. 1978. Pub.Astr.Soc.Japan, <u>30</u>, 489.

Jenkins, E.B. 1978. Astrophys.J., <u>220</u>, 107.

Jenkins, E.B. and Meloy, D.A. 1974. Astrophys.J.(Letters), <u>193</u>, L121.

Kahn, F.D. 1976. Astr.Astrophys., <u>50</u>, 145.

Kahn, F.D. 1981. In <u>Investigating the Universe</u>, ed. F.D. Kahn (Reidel:Dordrecht).

Kirshner, R.P. and Taylor, K. 1976. Astrophys.J.(Letters),<u>208</u>, L83.

Lerche, I. and Vasyliunas, V.M. 1976. Astrophys.J., <u>210</u>, 85.

Long, K.S. and Helfand, D.J. 1979. Astrophys.J.(Letters), <u>234</u> L77.

Max, C.E., McKee, C.F. and Mead, W.C. 1980. Phys.Fluids., <u>23</u>, 1620.

McKee, C.F. and Cowie, L.L. 1975. Astrophys.J., <u>195</u>, 715.

McKee, C.F. and Cowie, L.L. 1977. Astrophys.J., <u>215</u>, 213.

McKee, C.F., Cowie, L.L. and Ostriker, J.P. 1978. Astrophys.J. (Letters), <u>219</u>, L23.

McKee, C.F. and Hollenbach, D.J. 1980. Ann.Rev.Astron. Astrophys., <u>18</u>, 219.

McKee, C.F. and Ostriker, J.P. 1977. Astrophys.J., <u>218</u>, 148.

Naranan, S., Shulman, S., Friedman, H. and Fritz, G. 1976. Astrophys.J., <u>208</u>, 718.

Nulsen, P.E.J. 1981. Mon.Not.R.astr.Soc., in press.

Ostriker, J.P., McKee, C.F. and Cowie, L.L. 1981. In preparation.

Pravdo, S.H. and Smith, B.W. 1979. Astrophys.J.(Letters), <u>234</u> L195.

Raymond, J.C., Cox, D.P. and Smith, B.W. 1976. Astrophys.J., <u>204</u>, 290.

Raymond, J.C., Davis, M., Gull, T. and Parker, R.A.R. 1980. Astrophys.J.(Letters), <u>238</u>, L21.

Reynolds, R.J. and Ogden, P.M. 1979. Astrophys.J., <u>229</u>, 942.

Reynolds, R.J., Roesler, F.L. and Scherb, F. 1974. Astrophys.J. (Letters), <u>192</u>, L53.

Schwarz, J., McCray, R. and Stein, R.F. 1972. Astrophys.J., <u>175</u>, 673.

Sedov, L. 1959. <u>Similarity and Dimensional Methods in Mechanics</u> (New York: Academic), p. 210.

Shapiro, P.R. and Field, G.B. 1976. Astrophys.J., <u>205</u>, 762.

Shull, J.M. and McKee, C.F. 1979. Astrophys.J., <u>227</u>, 131.

Solinger, A., Rappaport, S. and Buff, J. 1975. Astrophys.J., <u>201</u>, 381.

Spitzer, L. 1962. <u>Physics of Fully Ionized Gases</u>, 2nd ed. (New York: Wiley).

Tammann, G. 1982. This volume.

Tanaka, Y. and Bleeker, J.A. 1977. Space Sci.Rev., 20, 815.

Taylor, G.I. 1950. Proc.R.Soc.Lond.A, 201, 175.

Tuohy, I.R., Nousek, J.A. and Garmire, G.P. 1979. Astrophys.J. (Letters), 234, L101.

Woltjer, L. 1972. Ann.Rev.Astron.Astrophys., 10, 129.

Woodgate, B.E., Kirshner, R.P. and Balon, R.J. 1977. Astrophys. J.(Letters), 218, L129.

York, D.G. 1974. Astrophys.J.(Letters), 193, L127.

Zel'dovich, Ya.B. and Raizer, Yu.P. 1966. Physics of Shock Waves and High Temperature phenomena (New York: Academic) Vol.I, p. 97.

NON-THERMAL ASPECTS OF SUPERNOVA REMNANTS

R.D. Blandford

California Institute of Technology, Pasadena

1. INTRODUCTION

Supernova remnants have played a vital part in the development of modern astronomy. The discovery and identification
of the radio sources Taurus A and Cassiopeia A provided a
crucial impetus to the then infant science of radio astronomy.
The prediction of linear polarisation by Shklovskii and its
subsequent detection furnished the first evidence for the
widespread importance of non-thermal processes within
astronomical objects. However supernova remnants are now most
often viewed as lying at the interface between the physics of the
explosion and the astronomical description of the interstellar
medium, telling us about the energy of the former and the ambient
structure of the latter.

From the point of view of non-thermal processes, I believe
that their usefulness will ultimately be seen in a quite
different light. Supernova remnants provide a crucial
intermediate scale that enables us to extrapolate our understanding of astrophysical particle acceleration from the
relatively well-probed solar system to the distant quasars and
radio galaxies. More specifically, the particle density ($n \sim$
5 cm^{-3}), velocity ($V \sim 400$ km s^{-1}) and field strength ($B \sim 10^{-5}$)
characteristic of the solar wind are all within an order of
magnitude of the corresponding quantities within the well-studied
remnants. However, there is one important difference. The Larmor
radius, r_L, of a ~ 1 GeV particle is $< 3 \times 10^{11}$ cm which is not
especially small compared with, say, the size of planetary bow
shocks or even an astronomical unit. By contrast supernova
remnants have radii up to 10^{20} cm and so many of the finite

M. J. Rees and R. J. Stoneham (eds.), Supernovae: A Survey of Current Research, 459–474.
Copyright © 1982 by D. Reidel Publishing Company.

Larmor radius effects that complicate matters in the solar system should be suppressed within the remnants.

In this review, I will follow convention by considering separately young remnants (e.g. Cas A) dominated by material derived from the parent star either explosively or in the form of a wind, old remnants (e.g. Cygnus Loop) which are dominated by interstellar material and Plerions (e.g. Crab Nebula) which may be powered by a central pulsar. The relevant particle acceleration processes are probably quite different in all three cases.

Useful reviews that discuss non-thermal aspects of supernova remnants include those of Woltjer (1972) and Chevalier (1977b). Particle acceleration is discussed in the conference proceedings edited by Arons, Max and McKee (1979) and the book by Melrose (1980). Astrophysical shock waves are discussed in the review by McKee and Hollenbach (1980). More comprehensive lists of references can be found in these sources.

2. ENERGETICS

The polarised, non-thermal radio emission from supernova remnants is generally assumed to be produced by the synchrotron process. The minimum pressure associated with a homogeneous synchrotron source which obtains when the relativistic electron energy density $\propto B^{-3/2}$ is approximately equal to the magnetic energy density $\propto B^2$. In useful units the pressure is given approximately by

$$P_{min} \simeq 2 \times 10^{-10} \left(\frac{\Sigma_\nu}{10^{-16} \mathrm{erg\,cm}^{-2}\mathrm{s}^{-1}\mathrm{ster}^{-1}\mathrm{Hz}^{-1}} \right)^{4/7} \left(\frac{\nu}{1\mathrm{GHz}} \right)^{2/7} \left(\frac{\ell}{1\mathrm{pc}} \right)^{-4/7} \mathrm{dyne\ cm}^{-2} \qquad (1)$$

where Σ_ν is the surface brightness at frequency ν and ℓ is the path length through the source (e.g. Moffat 1975).

As an example consider Cas A. The X-ray observations analysed by Fabian et al. (1980) yield an estimate of the gas pressure $p_g \sim 1.5 \times 10^{-7}$ dyne cm^{-2}. However using the average radio surface brightness $\Sigma_{408} \sim 3 \times 10^{-14}$ erg cm^{-2}s^{-1}ster^{-1}Hz^{-1}, we compute a minimum synchrotron pressure only of $\sim 2 \times 10^{-9}$ dyne cm^{-2}.

Using these values, we find that Cas A is underluminous by a factor $\sim 10^{-4}$ compared with a maximally radiating homogeneous source of the same total energy ($\sim 10^{51}$ erg) and volume. This is important to bear in mind in studies of extragalactic sources in which it is commonly assumed the pressure is minimal. In fact Cas A is not homogeneous and the radio maps (Bell, Gull and

Kenderdine 1975: Bell 1977: Dickel and Greisen 1979) show that about a third of the emission comes from compact 'knots' in which the minimum synchrotron pressure is comparable with the gas pressure inferred from the X-rays. This means that the energy density in relativistic protons, turbulence, etc. within the knots should not greatly exceed the relativistic electron energy density. This is in common with the Crab Nebula but in contrast to what is observed within the general interstellar medium.

Supernova remnants are commonly supposed to be responsible for the maintenance of the galactic pool of cosmic rays. We can estimate the energy requirement fairly directly. The energy density of local cosmic rays is $e_{CR} \sim 10^{-12}$erg cm^{-3} residing in $\sim$ 1 GeV protons ($\sim$ 70 per cent), alpha particles ($\sim$ 1.7 per cent), heavier species ($\sim$ 5 per cent) and electrons ($\sim$ 2 per cent). From an analysis of the ratio of primary to secondary nuclei, it has been concluded that cosmic rays of this energy traverse roughly $\lambda_{CR} \sim 5$ g cm^{-2} of interstellar material before leaving the galaxy. Now the mean grammage through the galactic disk is $\lambda_d \sim$ 2 mg cm^{-2} and so an estimate of the local flux of cosmic ray energy leaving the galaxy is $F_{CR} \sim e_{CR}c(\lambda_d/\lambda_{CR}) \sim 10^{-5}$erg cm^{-2}s^{-1}. Averaging over the galactic disk gives a total cosmic ray power $\sim 3\times10^{40}$erg s^{-1}, independent of the properties of any halo and good to a factor $\sim$ 3. If one supernova remnant is created in the galaxy every 30 years then we require 3×10^{49} erg per supernova of cosmic ray energy. That is to say roughly 3 per cent of the energy normally associated with a supernova explosion ($\sim 10^{51}$ erg) must ultimately emerge as cosmic rays.

We can hope to learn about the present relativistic particle content of observed supernova remnants by detecting the gamma rays that they produce. Cosmic ray protons with kinetic energy in excess of 280 MeV will create π^0 s which decay into gamma rays. The gamma ray production rate above 100 MeV is $q_\gamma^p \sim 3\times10^{-13}$ p_p (H atom)$^{-1}$s^{-1} where p_p is the relativistic proton pressure. The corresponding relativistic electron bremsstrahlung production rate is $q_\gamma^e \sim 3\times10^{-12}p_e$(H atom)$^{-1}$s^{-1} (and rather larger for gamma rays of energy below 100 MeV). These estimates are good to no better than a factor three and somewhat sensitive to the nature of the particle spectrum and the composition. If we use Cas A again as an example and assume that the cosmic rays are well mixed with $\sim$ 20 M$_\odot$ of gas (Fabian et al. 1980) then we can use gamma ray observations to limit the cosmic ray content. Cowsik and Sarkar (1980) argue that the fact that Cas A is not a Cos B source limits the gamma flux at earth to 10^{-6}cm^{-2}s^{-1}. This implies that the proton pressure is less than 10^{-7}dyne cm^{-2} which is of order the estimated gas pressure. The electron pressure must be less than 10^{-8}dyne cm^{-2} which implies that the magnetic field within the remnant exceeds $\sim 8\times10^{-5}$G. The Gamma Ray

Observatory should be able to improve upon these limits and
perhaps provide a direct measurement of the field strength.

3. PARTICLE ACCELERATION

The problem of particle acceleration in supernova remnants is
to find a mechanism for efficiently converting the energy of
ordered motion into relatively few high energy electrons and
nucleons. An important quantity for assessing the importance of
various schemes is $\beta = 8\pi p_g/B^2$, the ratio of the gas pressure to
the magnetic pressure. In the ambient interstellar medium, β is
believed to lie in the range 1-5 (e.g. Spitzer 1978). If the
magnetic field in young supernova remnants is close to the
minimum pressure value, then $\beta \sim 10$-100 except in the knots
where it must be of order unity. By contrast, in the compressed
gas of old supernova remnants, β probably lies in the range of
0.1-1. We must therefore consider particle acceleration in both
magnetically-dominated and gas-dominated plasmas.

The particle acceleration mechanism must be fairly efficient
in accelerating a power-law distribution function of relativistic
electrons and protons. From the observed radio spectral indices,
we conclude that the electron energy distribution function within
the remnants satisfies $N_G \propto \gamma^{-q}$ for $100 \leqslant \gamma \leqslant 10^4$, where $2 \leqslant q \leqslant$
2.5. This value of q is similar to the slope directly observed
for cosmic rays. Observations of extragalactic sources provide
further evidence that this range of slopes is preferred. A
further requirement is that we have some means of injecting fresh
relativistic particles. If these particles are drawn from the
thermal pool then the acceleration rate must be fast enough to
overcome ionisation losses. (For a non-relativistic particle of
energy E, $(dE/dt)_{ion} \simeq E^{-1/2}$.) The electron-proton ratio and the
elemental abundances of galactic cosmic rays must also be
explained.

Most mechanisms of particle acceleration are, in a sense,
derived from the Fermi (1949) mechanism. In this process
particles are imagined to be scattered by much heavier clouds or
magnetic inhomogeneities moving with speed u. A relativistic
particle random walks in energy space doubling its energy after
$\sim (v/u)^2$ scatterings. More formally, the particle distribution
function $f(p,\underline{r},t) = d^6N/d^3pd^3r$ evolves according to

$$\frac{\partial f}{\partial t} = \frac{1}{p^2} \frac{\partial}{\partial p} p^2 D_{pp} \frac{\partial f}{\partial p} \tag{2}$$

plus terms describing injection, spatial transport and loss where
we have assumed isotropy. The diffusion coefficient in momentum
space is given by

$$D_{pp} = \frac{1}{3} \left(\frac{u}{v}\right)^2 p^2 \, \nu \qquad (3)$$

with ν being the scattering rate, e.g. Melrose (1980). (Note that the more general Fokker-Planck equation with two independent coefficients simplifies to equation (2) when the 'recoil' of the scatterer can be ignored.) Fermi acceleration has been invoked by several authors, including Scott and Chevalier 1975, Chevalier, Robertson and Scott 1976, Chevalier, Oergerle and Scott 1978 and Cowsik 1979, to account for particle acceleration in young supernova remnants and, by extension, the origin of galactic cosmic rays. Here the scattering centres are usually identified with the optical knots.

The Fermi acceleration must be powerful enough to overcome expansion losses due to divergence of the background flow velocity $\underline{u}$, expressible as a term

$$\frac{1}{3} \nabla \cdot u \; \frac{\partial f}{\partial \ln p}$$

added to the right hand side of equation (2). Power-law distribution functions can indeed by maintained by this mechanism but the slope of the power law is dictated by the balance between gain and loss (including escape). The near constancy of the observed slope within many different sources is then a very surprising coincidence. These models can however account for the decay and flattening of the radio spectrum in Cas A (Shklovskii 1960).

In order that a Fermi process occur it is necessary to invoke some level of pitch angle scattering to break the first adiabatic invariant. Resonant Alfven waves with $k \sim r_L^{-1}$ are usually invoked to do this (e.g. Skilling 1975). If the amplitude of resonant waves is δB, then every Larmor period the particle undergoes a stochastic change in pitch angle $\delta \phi \sim (\delta B/B)$. The scattering rate is then given by $\nu \sim (\delta \phi)^2 / \delta t \sim (\delta B/B)^2 \Omega_g$ where Ω_g is the Larmor frequency. More formally,

$$\nu = \frac{2\pi^2 e^2 v}{c^2 p} \; \varepsilon_k \qquad (4)$$

where ε_k is the wave energy density at $k = eB/pc$. There is a corresponding spatial diffusion coefficient along the magnetic field

$$D_{//} = \frac{1}{3} v^2/\nu \qquad (5)$$

and an associated Fermi acceleration rate

$$D_{pp} = \frac{1}{3} \ p^2 \ (a/v)^2 \ \nu$$

Note that Alfven waves are far more efficient at scattering relativistic particles than accelerating them. This is because the ratio of energy to momentum in the wave is the Alfven speed, a, which is smaller than the same ratio $\sim$ v for the particles.

All of these diffusion coefficients, as well as several that we shall shortly quote, are calculated using a form of perturbation theory known as quasi-linear theory. Essentially this involves both the small amplitude approximation and the random phase approximation and is known to have only mixed success in accounting for observations of laboratory and interplanetary plasmas. Its validity in the case of supernova remnants must be questionable. This theory does however predict that the waves necessary for scattering will be generated by the particles themselves if they have a mean streaming velocity in excess of the Alfven speed (e.g. Cesarsky 1980 and references therein). However, in a high β plasma, there is a difficulty in scattering a particle of momentum p through a pitch angle $\phi \sim 90° \pm M_p V_i/p$ because the necessary resonant waves are strongly damped by the thermal ions (moving with velocity V_i). For this reason, Holman, Ionson and Scott (1979) have argued that high energy particles stream freely through a high β plasma. (Note that in a high β plasma the Alfven modes are significantly modified by the thermal particles (Foote and Kulsrud 1979) and also by the relativistic particles (Zweibel 1979).) However, if longer wavelength modes have an amplitude $(\delta B/B) \gtrsim (m_i V_i/p)^2$ then mirroring (or resonance broadening) can occur and the problem is circumvented (e.g. Achterberg 1981). We conclude that sufficiently energetic particles should be scattered through $\phi \sim 90°$. In any case, the solar wind appears to solve this problem. (See Goldstein 1976 for a discussion.)

Alfven waves may also be damped. In a partially neutral gas of Hydrogen density n_H, the damping rate is $\sim n_H \ \sigma_{io} V_i \sim 10^{-8}(n_H/1 \ cm^{-3}) \ s^{-1}$, where σ_{io} is the ion-neutral collision cross section. Non-linear Landau damping, in which the beat frequency of two waves resonates with a thermal particle, occurs at a rate $\sim (V_i/c)(\delta B/B)^2 \Omega_i$ and can also cause saturation of the level of Alfven turbulence (Achterberg 1981: Stoneham 1981).

There is an alternative mechanism for scattering suprathermal particles in a high β plasma that has been analysed by Hall (1981). If the thermal particles are anisotropic so that the difference between the parallel and perpendicular components of gas pressure exceeds $B^2/4\pi$ then the plasma is unstable to MHD firehose or mirror instabilities. In an environment as violent as the interior of a supernova remnant, this high a pressure difference can probably be maintained and random magnetic inhomogeneities will grow which can be quite efficient at scattering suprathermal particles.

As should be apparent by now, we do not have a good and accepted theory of particle transport. The pragmatic approach is to rely upon the observation that galactic cosmic rays with energies up to 1000 GeV are observed to be as isotropic as we could reasonably expect (to 1 part in 10^4). Particle scattering, by whichever mechanism is dominant, should be far more effective within a remnant than in the relatively quiescent interstellar medium and its existence is a reasonable hypothesis which I shall henceforth adopt.

Granted the existence of this scattering, longer wavelength waves may also be able to effect a Fermi-like acceleration. The usual acceleration process invoked is magnetic pumping (e.g. Kulsrud and Ferrari 1971) and this occurs via an $\omega = k_\parallel v_\parallel$ resonance for particles with pitch angle $\phi \sim 90°$, when $\nu \ll \omega$. However, only the compressive magnetosonic waves can heat the particles and then only in a very low β plasma will these waves escape damping by the <u>thermal</u> electrons. This is therefore unlikely to be important within a young supernova remnant.

If the opposite limit, $\nu \gg \omega$ applies, then efficient wave acceleration is possible. In a high β plasma, the compressive modes are basically sound waves which will be damped in the usual manner by thermal conduction of heat from the compressed to the rarefied regions. Now if the relativistic particles have a sufficiently longer mean freepath than the thermal particles then they will absorb most of the wave energy. More formally, if the effective spatial diffusion coefficient is D(p) then D_{pp} in equation (2) is given by

$$D_{pp} = \frac{p^2 D(p) \int \varepsilon_k k^2 dk}{20 \varepsilon_0}$$

where ε_k is the wave energy density and ε_0 is the background thermal energy density. Only if the spatial diffusion coefficient is energy independent for the relativistic particles, will the acceleration be Fermi-like ($\dot{\gamma} \propto \gamma$). The sound waves will probably be emitted at long wavelengths by large scale mass motion and will cascade down to shorter wavelengths where they can be damped. Again there is no natural way to account for the 'universal' spectrum and indeed a single power law is not guaranteed with this mechanism.

The acceleration mechanism which at present seems to be most promising in young supernova remnants is first order Fermi acceleration at a shock front. This process has been recently reviewed by Axford (1981, 1982), Blandford (1979) and Toptygin 1980). The papers of Bell (1978a,b) are particularly clear and important. The basic idea is that particles be accelerated by scattering in the converging fluid flow on either side of a shock front. A particle of speed u will change its energy by a fractional amount ~ (u/v) every time it crosses a shock moving with speed v. The average particle must cross the shock ~ (v/u) times if it is to remain at rest on average with respect to the background medium. However, the process is statistical and an exponentially small fraction of the particles will receive an exponentially large gain in energy thus generating a power law distribution function. A straightforward calculation gives the important result that the slope of this power law is dictated by the shock compression r and is $d\ln f/d\ln p = -3r/(r-1)$. This is independent of the diffusion coefficient (save that it be positive), the particle speed v (save that it be sufficiently larger than the shock speed), the shock velocity u (save that other energy changing process in the vicinity of the shock front are unimportant) and the angle made by the magnetic field line with the shock front (as long as it is larger than u/v). A simple adiabatic shock in an ionized gas has a compression of r = 4 and so for relativistic particles $N_\gamma \propto p^2 f(p) \propto \gamma^{-2}$. For a lower compression of 3 in a slightly weaker shock the exponent is changed to 2.5 and we see that the approximately universal character of the spectrum is roughly reproduced. If we accept the results of quasi-linear calculations then scattering by self-excited Alfven waves can produce efficient scattering for proton energies ranging from several MeV to greater than 100 GeV, provided the unshocked gas is ionized (Bell 1978b: Morfill, Volk and Forman 1981).

We have argued that particle acceleration by supernova remnants should be fairly efficient. This implies that if shock Fermi acceleration is responsible, then the post-shock cosmic ray pressure may become comparable with the total momentum flux. The cosmic rays can no longer be regarded as test particles and the transmitted spectrum will be altered. The net effect is sensitive to the energy dependence of the spatial diffusion coefficient and is unclear. On the one hand the lower effective specific heat ratio will lead to a greater overall compression, whereas on the other, the cosmic ray pressure will decelerate the background medium ahead of its shock reducing the compression. Eichler (1979, 1981) has argued that high Mach number collisionless shocks are always strongly turbulent and mediated by the cosmic ray pressure with the post-shock cosmic ray pressure being of order half the momentum flux. A prominent solar flare occurring in August 1972 appears to have accelerated

protons with this high on efficiency. If this is generally true
within the solar system, then it will almost certainly be the
case in supernova remnants. Remnants like Cas A would then
presumably be full of relativistic protons and, as shown in
Section 2, their γ-ray flux should be detectable.

Supernova shock waves must not just replenish the energy of
the cosmic rays, they must also supply the particles themselves.
As emphasised by Bell (1978b) it is far harder to inject
electrons from the thermal pool than protons, because the Larmor
radius of a 'thermal' electron is $\sim (m_e/m_p)^{1/2}$ times the proton
Larmor radius which is the minimum thickness of a collisionless
shock front. Wave-particle interactions behind the shock front
undoubtedly control the injection, but as yet we have no way of
quantifying this. What is qualitatively clear is that the strong
underabundance of electrons and overabundance of medium and heavy
nuclei in galactic cosmic rays relative to the solar-mix has a
natural explanation in the shock wave model: the former being due
to the difficulty in injecting electrons and the latter being
caused by the higher rigidity (momentum per charge) of high Z
nuclei than that of protons of the same energy (Eichler 1979).
Indeed the strong similarities between the composition of solar
and galactic cosmic rays argues for their both being injected
from coronal gas of similar ionization structure.

4. YOUNG SUPERNOVA REMNANTS

In young supernova remnants, the dynamics of the exanding gas
is dominated by mass ejected from the star, either during the
advanced stages of stellar evolution or during the explosion. The
best studied remnant is Cas A which shows similar emission at
X-ray and radio frequencies. There is a broad outer ring of
emission which is presumed to lie within the expanding blast
ave. As we have argued, the magnetic field strength in the
remnant probably exceeds 8×10^{-5}G. Now a standard interstellar
field of 3 μG will only be amplified by a factor ~ 3 on passing
through a strong shock. The post-shock magnetic field lines must
then be stretched by a factor > 8 by turbulent motions perhaps
associated with the knots. (The high field strength might find
an alternative explanation if the gas ahead of the shock were
endowed with a magnetic field > 30 μG which could possibly have
been convected outwards by an MHD wind with velocity ~ 330 km s^{-1}
and discharge $\sim 5 \times 10^{-5}$ M$_\odot$ yr^{-1}.)

A better case for direct interaction with the interstellar
medium can be made for the Tycho remnant which appears to have
decelerated. However, here the surface brightness is ~ 300 times
smaller than in Cas A and the <u>equipartition</u> field is only
2×10^{-5} G. As argued by Bell (1978b), the remnant field behind the
shock could plausibly be compressed interstellar field and need

not violate the γ—ray constraints discussed in Section 2. It is
very striking that the radio map of Tycho shows such steep outer
contours indicating that a high brightness is achieved within ~
10^{17} cm of the shock front. Apart from the γ—ray flux an obvious
prediction of the idea that the radio emission occur in swept—up
interstellar field is that the polarisation vector be radial just
behind the shock within the Tycho remnant. Conversely, Chevalier
(1977a) and Reynolds and Chevalier (1981) have suggested that the
flow behind an adiabatic shock is turbulent and that a high
magnetic pressure is always transmitted. Although this is
certainly not a feature of gas dynamical shocks propagating
through a uniform medium, it could conceivably be caused by
inhomogeneity or magnetic stresses.

In addition to this outer ring of emission there is a
brighter inner ring again visible in both X—rays and radio. The
X—rays are probably produced from the hot gas behind a reverse
shock propagating into the decelerated stellar ejecta. The radio
emission from this region is probably enhanced as a consequence
of Rayleigh—Taylor instability of the contact discontinuity
separating the ejecta from shocked circumstellar material thus
corroborating the model put forward by Gull (1973a,b) and Shirkey
(1979). The non—linear development of this instability enhances
the magnetic field strength and produces the predominantly radial
magnetic field observed in young remnants. A further increase in
field strength by a factor ~ 5 in going from the outer to the
inner ring can account for the increase in brightness. The
alternative particle acceleration processes discussed in Section
3 may also operate within the Rayleigh—Taylor zone. In
particular it appears that in the brightest compact knots there
must be some selective electron acceleration. Note that the
particle energy content of the bright knots is only a small
fraction of the total for the nebula and so they are probably not
an important contributor to the overall cosmic ray production.

These considerations are probably also relevant to the radio
supernovae whose interpretation is discussed by R.A. Chevalier in
this volume.

5. OLD SUPERNOVA REMNANTS

The traditional explanation for old supernova remnants is due
to van der Laan (1962) who argued that the outer shock would
become radiative and lead to high post—shock compression. This in
turn would amplify the magnetic field and heat the relativistic
electrons leading to an enhanced emissivity. The direct
association of optical emission line filaments and radio features
in old remnants like the Cygnus Loop and IC443 (e.g. Duin and van
der Laan 1975) provides impressive support for this general
model.

However, it has long been known that many old supernova remnants are also sources of soft X-rays and coronal line emission indicating the presence of gas at temperatures in the range 10^6–10^7 K. This gas cannot cool on the expansion timescale. A natural modification of the van der Laan mechanism that is also compatible with $\geqslant$ 2 phase models of the interstellar medium is to localise all the gas that is dense enough to cool into clouds. When these clouds are passed by the primary blast waves, slower secondary shocks will propagate into them.

If the explosion energy is 10^{51} E_{51} ergs then the mean pressure when the remnant is R pc in radius is $\bar{p} \sim 5{\times}10^{-6}$ E_{51} R^{-3} dyne cm^{-2}. Now for normal composition a shock will become radiative on the expansion time scale if its speed V is less than $\sim$ 220 km s^{-1}. The shock will be pre-ionizing if V $\gtrsim$ 100 km s^{-1}. The hydrogen density in a marginally cooling cloud is

$$n_{Hcool} \sim 6000 \ E_{51}^{0.8} \ \bar{n}_{-1}^{-0.2} \ R^{-2.8} \ cm^{-3}$$

where 0.1 $\bar{n}_{-1}$ is the mean internal density outside the clouds (Blandford and Cowie 1981). X-ray observations indicate that $\bar{n}_{-1}$ $\sim$ 1–2 in the remnants under consideration. If the clouds have a hydrogen density n_H above n_{Hcool} then they will be crushed until the magnetic pressure is built up to the ambient pressure p. Interstellar cosmic ray electrons contained within the cloud will firstly be accelerated at the shock front and then have their energies increased proportional to the cube root of the density as the post-shock gas cools. Total cloud compression factors $\sim$ 50–500 can occur. The particle escape time from the crushed cloud probably exceeds the remnant expansion time and so the electrons will radiate continuously.

It is interesting that particle acceleration by the first order Fermi process can still occur at the shock front if the scattering ahead of the shock is inhibited on account of ion-neutral damping of Alfven waves and this furnishes some insight into the mechanism. To see that this is so, consider a volume V of non-scattering unshocked gas bounded by a contracting shock of compression r. (The post-shock gas is strongly scattering.) The momentum p of particles reflected off scattering centres in the post-shock gas will increase p $\propto$ $v^{-1/3}(1 - 1/r)$. To first order, the cosmic ray particle density n will be the same on either side of the shock fronts so that the total particle number N = nV will change according to dN = ndV/r. The transmitted distribution function is then

$$f(p) \propto p^{-2} \ dN/dp \propto p^{-3r/(r-1)}$$

as in Section 3.   Of course the result is more general than this
as the transmitted spectrum is independent of the pre-shock
scattering.   First order Fermi acceleration followed by
compression probably represents the minimum amount of particle
acceleration to be expected. Mechanisms like those described in
Section 3 may also be important for increasing the electron
energy within the compressed zone.

The radio powers of old supernova remnants are conventionally
displayed on a mean surface brightness-diameter diagram.   If we
make the conservative assumption that the mean external density
of gas in dense clouds is $\langle n \rangle \sim 1$ cm^{-3}, the conventional
interstellar value, and only the first order process is operative
then the computed surface brightness is approximately ten times
too faint to account for the brightest observed remnants. In
order to explain these, we must either set $\langle n \rangle \sim 10$ cm^{-3} or
invoke a degree of additional particle acceleration. It is not
necessary to bring the relativistic electron energy into equipartition however and indeed acceleration mechanisms that do so
will produce remnant brightnesses in excess of the observations.
It should be borne in mind that both the low observed radio
remnant birth rate within our galaxy and X-ray observations of
the Large Magellanic Cloud indicate that there is a substantial
population of radio-quiet supernova remnants. The prominent old
remnants may well be associated with the few supernovae occurring
in overdense regions of the interstellar medium.

It is possible to quantify this model further by relating the
radio brightness to the optical emission line intensity and also
to predict the expected gamma ray flux (see Chevalier 1977b). A
confirmation of the idea that the compression is controlled by
magnetic stresses is contained in the observation that old
remnants display an overall small radio polarisation which will
occur if, as expected, the dense gas is compressed most quickly
in the radial direction. A very striking feature of both optical
and radio remnants is the concentration of the emission into long
filaments rather than quasi-spherical clouds. This may reflect
the disposition of the gas in the ambient interstellar medium.
The particle acceleration and radio emission is of course not
sensitive to the topology of the cold phase.

There is an important corollary to this model of old
remnants. A significant fraction of the volume of the
interstellar medium is believed to be filled with warm ionised
medium of electron density ~ 0.3 cm^{-3} (e.g. McKee and Ostriker
1977). If most of the volume comprises hot coronal gas then
remnants can expand out to R $\gtrsim$ 100 pc before they become fully
radiative. However, the denser warm phase will be crossed by a
cooling shock wave when R $\sim$ 50 pc and can therefore also be
crushed after it has been enveloped by the primary blast-wave.

Radio loops, which are prominent arc-like features in the radio
background, may be identified with warm gas that has been crushed
by nearby remnants in their very late stages of expansion.
Furthermore, there has long been a discrepancy between the
directly observed cosmic ray electron spectrum and the spectrum
inferred from the intensity of the galactic radio background
calculated assuming a uniform standard interstellar field of B ~
3 μG (e.g. Webber, Simpson and Cane 1980). This discrepancy can
be removed if the radio background is in fact dominated by the
superposition of crushed warm gas along the line of sight. (See
Brown and Marscher 1977 for a discussion of the complementary
suggestion that the radio background be produced in dense
molecular clouds.)

6. PLERIONS

We next turn to the filled or plerionic remnants (Weiler and
Shaver 1978: Weiler and Panagia 1978), the prototype for which is
the Crab Nebula. Plerions are commonly assumed to be powered by
a rapidly spinning pulsar. In the case of the Crab Nebula, the
pulsar supplies the nebula with ~ 2×10^{38} erg s^{-1} of power through
a relativistic MHD wind (e.g. Piddington 1957: Kardashev 1965:
Rees and Gunn 1974: Kundt and Krotschek 1980). This wind
presumably passes through a strong shock at a radius ~ 10^{17} cm
when its momentum flux balances the ambient nebula pressure. It
also is believed to convect an ordered component of field which
is wound up and accounts for the nebular field. As the pulsar has
a 33 ms period it has rotated N ~ 10^{12} times in its ~ 1000 year
lifetime. The total flux in the nebula is ~ 3×10^{33} G cm^2 and so
the ordered flux emerging from the pulsar is ~ 3×10^{21} G cm^2 or
about 10 per cent of the total flux on the open field lines. The
remaining field lines form an oscillatory component of wavelength
10^9 cm in the wind. The fate of this wave field is still a matter
of controversy (e.g. Arons 1981) but it seems unlikely that it
can propagate far into the nebula beyond the shock. The
associated energy flux probably goes into particle acceleration
either through direct absorption of the low frequency waves or
again by Fermi acceleration at the shock. Note that it is far
harder to investigate this latter possibility because the shock
is relativistic (Peacock 1981).

We can infer the particle energy distribution from the shape
of the nebula spectrum. It appears that the injected spectrum is
a power law with index −1.5 extending from γ ~ 100 to γ ~ 3×10^6
where it appears to steepen. (The presently observed electron
spectrum in fact steepens above γ ~ 10^5 where the synchrotron
losses exceed the expansion losses.) If the wind is supplying
the required rate of ~ 10^{40} s^{-1} of freshly accelerated

relativistic particles (electrons and positron) then the wind Lorentz factor is probably $\sim 10^4$.

At least five other remnants show strong similarities to the Crab Nebula (filled, flat spectrum and ordered magnetic field) although only Vela has an associated pulsar. It is interesting to ask whether our understanding of the Crab Nebula can allow us to draw any conclusions about these objects. One relation which does follow is that if the same fraction of the open flux is wound up within the plerion as in the Crab pulsar, the plerion field should be given by

$$B \sim 10^{-5} \, L_{34}^{1/2} \, t_3 \, R_{pc}^{-2} \, G,$$

where $10^{34} \, L_{34}$ erg s^{-1} is the total pulsar luminosity, R_{pc} is the remnant radius, and the age is $10^3 \, t_3$ year. This expression is independent of the pulsar period and surface field strength. L may be related to the X-ray luminosity reported by D. Helfand at this conference. Dynamical models describing the evolution of plerions have been calculated by Pacini and Salvati (1973).

7. SUMMARY

It seems that plausible, although primitive, explanations can be offered for the main non-thermal properties of supernova remnants. In the young shell-like remnants, relativistic cosmic rays are injected and accelerated by the first order Fermi process at the outer shock front with roughly ten percent of the remnant energy going into relativistic protons and one percent into electrons. The interstellar magnetic field will be compressed by a factor ~ 3 behind the shock. Young remnants like Cas A that are just starting to be decelerated have Rayleigh-Taylor unstable zones within which the magnetic field is amplified and perhaps some auxiliary particle acceleration is occurring. This instability should be and appears to be less important in older remnants that have entered the Sedov phase. These remnants have a lower surface brightness and there is probably not much field amplification. When the external field is low, these remnants may be radio quiet. As the remnant expands further and the internal pressure falls, progressively more tenuous clouds of interstellar gas can be crushed to radiate optical emission lines with coincident high brightness radio emission by the trapped cosmic ray electrons. These remnants may only be visible when they expand into regions of high mean gas density. Very old remnants can crush clouds of ionised gas which may radiate the galactic radio and γ-ray backgrounds.

Several supernova explosions (possibly just the type II class) leave active central pulsars which power the plerionic remnants. Here the particles are probably injected by the pulsar and accelerated at the terminal shock of the relativistic hydro-magnetic wind emanating from the pulsar.

The galactic pool of cosmic rays can be maintained by acceleration at supernova shock waves. The transmitted distribution function is $f(p) \simeq p^{-4.2}$. Propagation within the galaxy will then produce the observed cosmic ray intensity. In particular the bounding shocks of old supernova remnants like the Cygnus loop still have a substantial Mach number, and should be efficient at accelerating cosmic rays.

ACKNOWLEDGEMENTS

I am indebted to R. Chevalier, L. Cowie, A. Fabian, C. McKee, E. Phinney and R. Tufts for informative discussions. Hospitality at Nordita and the Institute of Astronomy is gratefully acknowledged. I thank the National Science Foundation (AST 80-11752) and Alfred P. Sloan Foundation for support.

REFERENCES

Achterberg, A. 1981. Astr.Astrophys., 98, 161.

Arons, J. 1981. In IAU Symp. No. 94, The Origin of Cosmic Rays, ed. G. Setti, G. Spada and A.W. Wolfendale (Reidel: Dordrecht), p. 175.

Arons, J., Max, C. and McKee, C. 1979. (Eds.) Particle Acceleration Mechanisms in Astrophysics (American Institute of Physics: New York).

Axford, W.I. 1981. In IAU Symp. No. 94, The Origin of Cosmic Rays, ed. G. Setti, G. Spada and A.W. Wolfendale (Reidel: Dordrecht), p. 339.

Axford, W.I. 1982. Preprint.

Bell, A.R. 1977. Mon.Not.R.astr.Soc., 179, 573.

Bell, A.R. 1978a. Mon.Not.R.astr.Soc., 182, 147.

Bell, A.R. 1978b. Mon.Not.R.astr.Soc., 182, 443.

Bell, A.R., Gull, S.F. and Kenderdine, S.F. 1975. Nature, 257, 463.

Blandford, R.D. 1979. In Particle Acceleration Mechanisms in Astrophysics, ed. J. Arons, C. Max and C. McKee (American Institute of Physics: New York), p.333.

Blandford, R.D. and Cowie, L.L. 1981. In preparation.

Brown, R. and Marscher, A.P. 1977. Astrophys.J., 212, 659.

Cesarsky, C.J. 1980. Ann.Rev.Astr.Astrophys., 18, 289.

Chevalier, R. 1977a. Nature, 266, 701.

Chevalier, R. 1977b. Ann.Rev.Astr.Astrophys., 15, 175.

Chevalier, R.A., Oergerle, W.R. and Scott, J.S. 1978. Astrophys.J., 222, 527.

Chevalier, R.A., Robertson, J.W. and Scott, J.S. 1976. Astrophys.J., 297, 450.

Cowsik, R. 1979. Astrophys.J., 227, 856.

Cowsik, R. and Sarkar, S.   1980.   Mon.Not.R.astr.Soc., 191, 855.

Dickel, J.R. and Greisen, E.W.   1979.   Astr.Astrophys., 75, 43.

Duin, R.M. and van der Laan. 1975. Astr.Astrophys., 40, 111.

Eichler, D.   1979.   Astrophys.J., 229, 419.

Eichler, D.   1981.   Astrophys.J., 244, 711.

Fabian, A.C., Willingale, R., Pye, J.P., Murray, S.S. and
    Fabbiano, G.   1980.   Mon.Not.R.astr.Soc., 193, 175.

Fermi, E.   1949.   Phys.Rev., 75, 1169.

Foote, E.A. and Kulsrud, R.M.   1979.   Astrophys.J., 233, 302.

Goldstein, M.   1976.   Astrophys.J., 204, 400.

Gull, S.   1973a.   Mon.Not.R.astr.Soc., 161, 47.

Gull, S.   1973b.   Mon.Not.R.astr.Soc., 171, 263.

Hall, A.N.   1981.   Mon.Not.R.astr.Soc., 197, 977.

Holman, G.D., Ionson, J.A. and Scott, J.S.   1979.   Astrophys.J.,
    228, 576.

Kardashev, N.S.   1965.   Soviet.Astr.A.J., 8, 643.

Kulsrud, R.M. and Ferrari, A.   1971.   Astrophys.Sp.Sci., 12, 302.

Kundt, W. and Krotschek, E.   1980.   Astr.Astrophys., 83, 1.

McKee, C.F. and Hollenbach, D.J.   1980.   Ann.Rev.Astr.Astrophys.,
    18, 219.

McKee, C.F. and Ostriker, J.P. 1977.   Astrophys.J., 218, 148.

Melrose, D.B.   1980.   Plasma Astrophysics (Gordon and Breach:
    New York).

Moffat, A.T.   1975.   In Stars and Stellar Systems IX: Galaxies
    and The Universe, ed. A. Sandage, M. Sandage and J.
    Kristian (Univ. Chicago Press), p. 211.

Morfill, G., Volk, H. and Forman, M.   1981.   Preprint.

Pacini, F. and Salvati, M.   1973.   Astrophys.J., 186, 249.

Peacock, J.A.   1981.   Mon.Not.R.astr.Soc., 196, 135.

Piddington, J.H.   1957.   Aust.J.Phys., 10, 530.

Rees, M.J. and Gunn, J.E.   1974.   Mon.Not.R.astr.Soc., 167, 1.

Reynolds, S. and Chevalier, P.   1981.   Astrophys.J., 245, 912.

Shirkey, R.C.   1979.   Astrophys.J., 232, 826.

Shklovskii, I.S.   1960.   Sov.Astr.A.J., 4, 23.

Skilling, J.   1975.   Mon.Not.R.astr.Soc., 172, 557.

Spitzer, L.   1978.   Physical Processes in the Interstellar
    Medium (Wiley: New York).

Stoneham, R.J.   1981.   In IAU Symp. No.94, The Origin of Cosmic
    Rays, ed. G. Setti, G. Spada and A.W. Wolfendale (Reidel:
    Dordrecht), p. 255.

Toptygin, I.N.   1980.   Space Sci.Rev., 26, 157.

van der Laan, H.   1962.   Mon.Not.R.astr.Soc., 124, 175.

Webber, W.R., Simpson, G.A. and Cane, H.V.   1980.   Astrophys.J.,
    236, 448.

Weiler, K.W. and Panagia, N.   1978.   Astr.Astrophys., 70, 419.

Weiler, K.W. and Shaver, P.   1978.   Astr.Astrophys., 70, 389.

Woltjer, L.   1972.   Ann.Rev.Astr.Astrophys., 10, 129.

Zweibel, E.G.   1979.   In Particle Acceleration Mechanisms in
    Astrophysics, ed. J. Arons, C. Max and C. McKee (American
    Institute of Physics: New York), p. 319.

SUPERNOVA REMNANTS IN M31

William P. Blair

The University of Michigan

ABSTRACT

We have used Hα + [N II] and [S II] interference filter
photography to identify nebulosities which we believe to be supernova remnants in M31. Spectrophotometry has been used to confirm
this identification for 12 of the candidates. Since all the remnants are at the same distance, they can be directly compared.
Using the measured diameter and an estimate of the pressure in the
optical filaments, an initial energy for each remnant has been calculated. However, this energy appears to be correlated to the remnant's diameter, an effect which may be related to magnetic pressure in the filaments.

The spectra have also been used in conjunction with shock
model calculations to study abundance gradients in M31. Comparison of these results to abundance gradients obtained from H II
regions allows an appraisal of the accuracy of the two methods.
We find gradients in nitrogen and oxygen abundances that are comparable to those seen in other intermediate to late type spiral
galaxies.

1. MOTIVATION

Optical supernova remnants (SNRs) are most often identified
by the coincidence of a non-thermal radio source with a filamentary optical nebula whose spectrum shows emission lines of [S II]
λλ6717,6731 of comparable strength to Hα. These optical nebulae
are often very faint and only about one quarter of the 130 galactic radio SNRs have known optical counterparts. Hence, one might

M. J. Rees and R. J. Stoneham (eds.), Supernovae: A Survey of Current Research, 475–482.

ask why it is worthwhile to investigate a sample of extragalactic
SNRs since it will likely be very difficult to do so.

The reasons are several and have implications for both the
study of SNR evolution and galactic structure. Distances
to galactic SNRs are poorly determined in many cases. This makes
studies of relative remnant evolution difficult since even the
basic parameter of linear diameter is not well known. SNRs in
another galaxy can be directly compared to each other since they
are all at the same distance. Also recent advances in shock wave
modeling techniques are making it possible to interpret observed
line intensities as abundances. If the SNRs are found over a
range of galactocentric distance, they can be used to study
abundance gradients in the parent galaxy. H II regions have conventionally been used in this regard, but results from H II regions have recently been questioned because of a number of unknown factors in the analysis (e.g. effects of dust content,
clumpiness, and changes in the ionizing stars as a function of
position in the galaxy). If gradients from both SNRs and H II
regions can be obtained, they can be compared as a consistency
check. M31 is one of the only galaxies for which this comparison
can be made.

2.   BACKGROUND

Mathewson and Clarke (1972, 1973a and b) used Hα and [S II]
interference filter photography in conjunction with multifrequen-
cy radio surveys to detect 14 SNRs in the Magellanic Clouds. As
techniques have improved, additional candidates have been found
(Davies, Elliott and Meaburn 1976) and, as discussed by Helfand
and Long (1982), the HEAO-B X-ray survey of the Magellanic Clouds
in proving to be a fruitful way of detecting additional remnants.

Beyond the Magellanic Clouds, it becomes difficult to apply
radio or X-ray criteria for SNR identification because of resolution and/or sensitivity limitations. However, Hα and [S II] interference filter photographs of nearby galaxies by D'Odorico,
Dopita and Benvenuti (1980) have demonstrated that SNR candidates
can be found using the optical criterion alone. Generally,
spectral information is obtained to confirm that the nebulae
resemble galactic SNRs. This has been done for 12 of 19 M33 SNR
candidates by Dopita, D'Odorico and Benvenuti (1980).

3.   OBSERVATIONS AND ANALYSIS

We have investigated M31, which is a difficult case because
of its large angular extent and small angle between its galactic

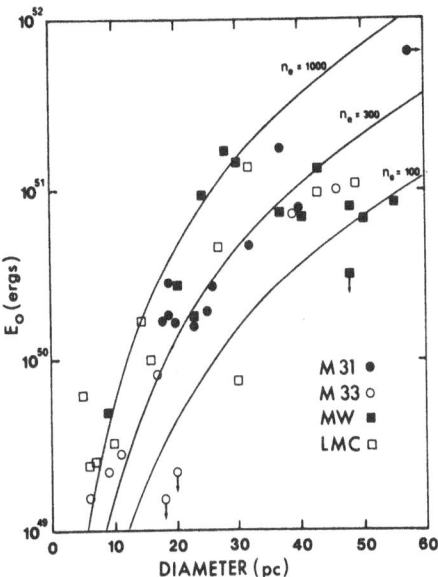

FIGURE 1: Initial supernova energy calculated from optical data
vs. diameter for SNRs in M31 (filled circles), M33 (open circles),
the Milky Way (filled squares), and the Large Magellanic Cloud
(open squares). The curves are lines of constant density and are
not intended to fit the data. The uppermost M31 point is plotted
at the correct energy, but has a diameter of 90 pc. (See text
for references.)

plane and the line of sight. Eight fields were photographed
using Hα + [N II] and [S II] filters and a single stage 144 mm
ITT image tube on the McGraw-Hill Observatory's 1.3 meter telescope. (Details are given in Blair, Kirshner and Chevalier
1981). Many of the candidates listed by D'Odorico et al. (1980)
were independently found on these plates, although several discrepancies also arose. We have been able to spectroscopically
confirm 12 SNRs in M31 using either the Reticon scanner at McGraw-
Hill Observatory or the Image Dissector Scanner at Kitt Peak
National Observatory.

The spectra are of sufficient quality for seven of the SNRs
that temperatures have been determined from the ratio [O III]
$\lambda 5007 + \lambda 4959/\lambda 4363$. These temperatures all lie in the range
20 000 K to 60 000 K, directly confirming that these nebulae are
shock heated. Densities have been obtained for all 12 of the
SNRs using the ratio of [S II] $\lambda 6717/\lambda 6731$; the densities fall
in the range 150 cm^{-3} to 730 cm^{-3}, which is comparable to many
galactic SNRs.

If the interstellar medium is inhomogeneous, then optical SNRs can be explained in the manner sketched by McKee and Cowie (1975; see also McKee, 1982). As the blast wave from the explosion propagates, it heats the low density gas to temperatures where X-rays are emitted. Slower shocks are driven into dense cloudlets; as the gas cools behind these slower shocks, the gas recombines and produces optical emission. Since the temperature in the S^+ zone is likely to be near 10^4 K, it is possible to use the [S II] densities to estimate the pressure in the optical filaments. A rough pressure equilibrium exists between the optical and X-ray gas; hence, knowledge of the energy density in the optical filaments and the diameter of the remnant allows the total internal energy to be estimated. This should be close to the initial energy of the explosion, E_0, if the remnant is in the adiabatic phase of expansion.

We have calculated E_0 for the M31 SNRs and find an average value of $<E_0> \approx 3 \times 10^{50}$ ergs. The same technique applied to the Cygnus Loop gives 7×10^{50} ergs. However, the energy calculated in this way appears to be correlated to the remnant's diameter, as can be seen from Figure 1. This is also true for SNRs in M33 (Dopita et al. 1980), the Large Magellanic Cloud (Dopita 1979) and our own Galaxy (Daltabuit et al. 1976).

We have considered a number of possible causes of this effect. For instance, it may be that only portions of some SNRs are optically bright (e.g. RCW 86 in our Galaxy) so that the optically measured diameters underestimate the real diameters. While this might be true for a few of the objects in Figure 1, it is probably not the cause of the correlation. Another possibility is that there is an intrinsic range of initial energies; more energetic SNRs would become larger before they enter the radiative phase (i.e. when they are optically brightest). This is in the same sense as the observed relation, but observations of supernovae do not seem compatible with an intrinsic energy spread of almost three orders of magnitude.

X-ray luminosities can also be used to calculate E_0, and the energies calculated in this manner are not always compatible with the optical determinations. Fesen and Kirshner (1980) found that E_0(X-ray) was nearly an order of magnitude larger than E_0(optical) for the galactic remnant IC 443. Also, the initial energies calculated for LMC remnants from the X-ray observations of Long and Helfand (1979) do not show a trend with remnant diameter.

One idea that is capable of explaining these discrepancies is that the [S II] emitting regions are not dominated by thermal pressure alone, as is assumed above, but by a combination of thermal and magnetic pressure. Using the thermal energy alone

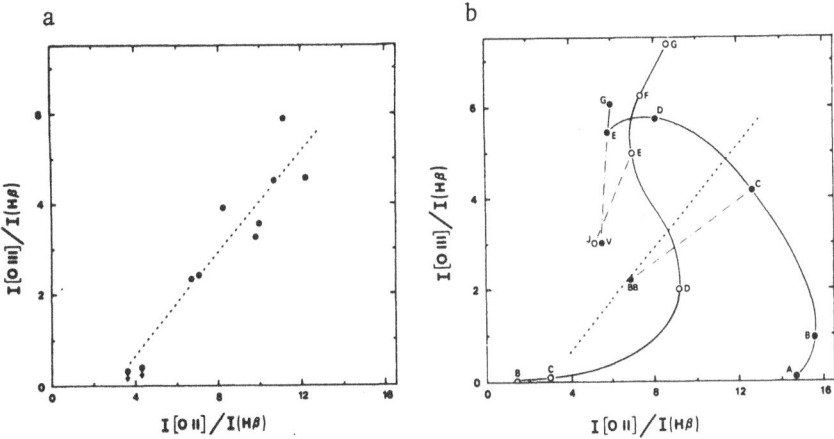

FIGURE 2: a) Observed [O III] λλ4959,5007/Hβ relative to [O II]
λ3727/Hβ for the M31 SNRs. The dashed line is a least squares
fit. b) Predicted [O III]/Hβ vs. [O II]/Hβ from shock model calculations. Open circles are models by Shull and McKee (1979) and
filled circles are models by Raymond (1979). Letters represent
the corresponding model identifications in these papers. The
solid lines connect models of increasing shock velocity and the
dashed line is the least squares fit from Figure 2a. The dot-
dash lines connect models of equal shock velocity but different
initial abundances.

would cause the total energy to be underestimated. Reynolds and
Chevalier (1981) have modeled the non-thermal radio emission
from young remnants and have concluded that several percent of
the postshock energy density goes into an amplified magnetic
field. In this case, the density measured by the [S II] lines
is not a function of shock velocity (as would otherwise be expected), and the energy calculated in the previous manner would
be proportional to remnant diameter cubed. This is in approximate accord with what is seen in Figure 1. (For details, see
Blair et al. 1981.)

4.   ABUNDANCE GRADIENTS IN M31

      Recent shock wave model calculations by Dopita (1977), Ray-
mond (1979) and Shull and McKee (1979) have improved on the
earlier work by Cox (1972a, b). These models calculate integrated line intensities for a line of sight through the cooling zone
behind shocks of varying initial conditions. While this cooling
zone may be resolved in galactic remnants such as the Cygnus Loop
(Fesen, Blair and Kirshner 1982), it is unresolved in the M31
SNRs. Hence, it should be reasonable to compare the M31 SNR

spectra to the shock model calculations.

The grid of models is not so complete that one would expect to fit a given observation exactly, but the qualitative agreement between observations and models is reasonable. In detail, though, there are some discrepancies which are not understood. For instance, since three stages of oxygen ionization are seen in optical spectra, one would hope that the models would reproduce their relative intensities properly. Figure 2a shows that there is an observed correlation between the strength of [O II] $\lambda3727$ and [O III] $\lambda5007 + \lambda4959$ relative to Hβ in the M31 SNRs. Since the remnants are of different ages and sizes, one might think that this could represent a sequence of shock velocities, but Figure 2b shows that this is not the case. The remnants cover a large range of galactocentric distance in M31 and if abundance gradients are present (as we will argue below) one might suspect that the observed relation is an abundance effect. Yet there is no correlation between a remnant's galactocentric distance and its position in Figure 2a.

Dopita (1977) presents his models in a form that is convenient for estimating abundances. Using the observed line intensities, we have derived abundances of nitrogen, oxygen and sulfur for each of the M31 SNRs. While these abundances may not be very accurate in an absolute sense, the method is sufficient to investigate variations in abundance as a function of galactic radius. This analysis indicates that the nitrogen abundance decreases by a factor of four or five in going from 4 to 20 Kpc in galactocentric distance. No gradient is seen in oxygen and only a slight gradient is seen in sulfur (Blair, Kirshner and Chevalier 1982).

H II regions have not been used to investigate abundance gradients in M31 until recently. This is because they are faint and of low excitation, making it impossible to determine [O III] temperatures directly. Over the past few years, much time and effort has gone into determining empirical methods for finding electron temperatures and abundances in low excitation H II regions (Alloin et al. 1979; Pagel, Edmunds and Smith 1980, and references therein). We obtained spectra of 11 M31 H II regions which span a range of galactocentric distance comparable to the SNRs. Using the empirical method of Pagel et al. (1980), abundance gradients of nitrogen, oxygen and sulfur have been found. Figure 3 shows the results for nitrogen and oxygen; both elements decrease in abundance by a factor of four or five over the observed range of galactocentric distance. (The results for sulfur are not trustworthy because the dominant ionization stages are not seen). Very similar results were found by Dennefeld and Kunth (1981) using a smaller number of H II regions.

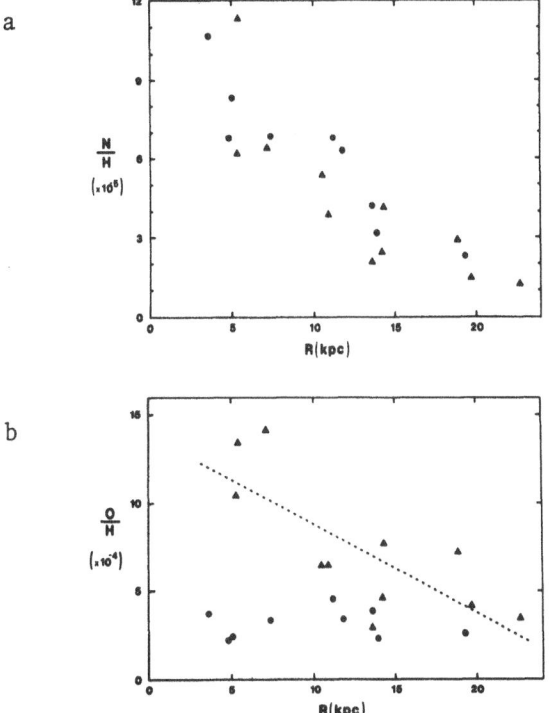

FIGURE 3:   a) Nitrogen abundance gradient in M31 from SNRs
(circles) and H II regions (triangles). b) Comparison of oxygen
abundance gradient in M31 as derived from SNRs (no gradient) and
H II regions. The discrepancy is discussed in the text.

     Comparison of the SNR and H II region gradients shows that
the results for nitrogen are very similar but the results for
oxygen are discrepant. The oxygen abundances from the SNR analysis are probably in error because the relative intensities of the
oxygen lines are not reproduced by the models, as discussed
earlier. Using the oxygen gradient from the H II regions and the
combined result for nitrogen, M31 has very similar abundance gradients to other intermediate and late type spiral galaxies that
have been studied (Smith 1975; Talent and Dufour 1979; Shields
and Searle 1978; Pagel et al. 1979).

     Although the techniques we have applied are somewhat crude,
encouraging agreement has been found in the nitrogen gradient
from SNRs and H II regions. A similar result has been found by
Dopita et al. (1980) for SNRs and H II regions in M33. These
results have impact beyond the individual galaxies, for they seem
to verify that the results from H II regions are trustworthy at
least in the case of nitrogen.

I would like to thank my collaborators, Robert P. Kirshner and Roger A. Chevalier, for assistance and guidance throughout the course of this work. Financial support was received from U.S. NSF grants-AST 77-17600 and AST 81-05050.

REFERENCES

Alloin, D., Collin-Souffrin, S., Joly, M., and Vigroux, L.   1979, Astr. Ap., 78, 200.

Blair, W.P., Kirshner, R.P., and Chevalier, R.A.   1981, Ap.J., 247, 879.

Blair, W.P., Kirshner, R.P., and Chevalier, R.A.   1982, Ap.J., submitted.

Cox, D.P.   1972a, Ap.J., 178, 143.

Cox, D.P.   1972b, Ap.J., 178, 159.

Daltabuit, E., D'Odorico, S., and Sabbadin, F.   1976, Astr. Ap., 52, 93.

Davies, R.D., Elliott, K.H., and Meaburn, J.   1976, M.N.R.A.S., 81, 89.

Dennefeld, M., and Kunth, D.   1981, A.J., in press.

D'Odorico, S., Dopita, M.A., and Benvenuti, P.   1980, Astr. Ap. Suppl., 40, 67.

Dopita, M.A.   1977, Ap.J. Suppl., 33, 437.

Dopita, M.A.   1979, Ap.J. Suppl., 40, 455.

Dopita, M.A., D'Odorico, S., and Benvenuti, P.   1980, Ap.J., 236, 628.

Fesen, R.A., Blair, W.P., and Kirshner, R.P.   1982, in preparation.

Fesen, R.A., and Kirshner, R.P.   1980, Ap.J., 242, 1023.

Helfand, D.J., and Long, K.S.   1982. This volume.

Long, K.S., and Helfand, D.J.   1979, Ap.J. (Letters), 234, L77.

Mathewson, D.S., and Clarke, J.N.   1972, Ap.J. (Letters), 178, L105.

Mathewson, D.S., and Clarke, J.N.   1973a, Ap.J., 180, 725.

Mathewson, D.S., and Clarke, J.N.   1973b, Ap.J., 182, 697.

McKee, C.F.   1982. This volume.

McKee, C.F., and Cowie, L.L.   1975, Ap.J., 195, 715.

Pagel, B.E.J., Edmunds, M.G., Blackwell, D.E., Chun, M.S., and Smith, G.   1979, M.N.R.A.S., 189, 95.

Pagel, B.E.J., Edmunds, M.G., and Smith, G.   1980, M.N.R.A.S., 193, 219.

Raymond, J.C.   1979, Ap.J. Suppl., 39, 1.

Reynolds, S.P., and Chevalier, R.A.   1981, Ap.J., 245, 912.

Shields, G.A., and Searle, L.   1978, Ap.J., 222, 821.

Shull, J.M., and McKee, C.F.   1979, Ap.J., 227, 131.

Smith, H.E.   1975, Ap.J., 199, 591.

Talent, D.L., and Dufour, R.J.   1979, Ap.J., 233, 888.

OPTICAL SUPERNOVA REMNANTS IN EXTERNAL GALAXIES

Michael A. Dopita

Mount Stromlo and Siding Spring Observatories,
Australian National University, Canberra

1. THE MAGELLANIC CLOUDS

The first extragalactic supernova remnants (SNRs) were discovered by the pioneering work of Mathewson and Clarke (1972, 1973a,b,c). This consisted of identifying a non-thermal radio source with an optical nebulosity equally bright in the emissions of Hα + [NII] and [SII]. The exact spectral discriminants by which SNRs can be distinguished from HII regions were investigated by D'Odorico and collaborators (D'Odorico and Sabbadin 1976: Daltabuit et al. 1976: D'Odorico 1978) and, indeed, objects for which the ratio Hα/[SII] (λλ6717,6731) is less than 2.5 can be fairly certainly identified as supernova remnants on this basis alone. This has been the sole technique of SNR identification in galaxies beyond the Magellanic clouds (Kumar 1976: D'Odorico et al. 1978, 1980: Blair et al. 1981a,b).

With the advent of the Einstein Observatory (HEAO 2), supernova remnants could be (and have been) identified on the basis of their X-ray properties alone (Long and Helfand 1979: Long, Helfand and Grabelsky 1981). Subsequent observation of the optical emission from these remnants has revealed several examples of unusual classes of object which would have evaded identification on classical criteria. In what follows, I describe the results of such an optical survey carried out at Mt. Stromlo by Don Mathewson, Ian Tuohy and myself in collaboration with the Columbia group of Knox Long and David Helfand.

M. J. Rees and R. J. Stoneham (eds.), Supernovae: A Survey of Current Research, 483–494.
Copyright © 1982 by D. Reidel Publishing Company.

1.1 Young Remnants of Type II Events

Theoretical models of the stellar evolution of massive ($\gtrsim$ 10 $M_\odot$) stars taken to core collapse predict the formation of a large mass fraction of C/O rich helium-burnt shell and the ejection of appreciable amounts of C, Ne, O and Si in the burnt material (Weaver and Woosley 1982). This material may become optically luminous during the remnant evolution at the time when reverse shocks are driven through dense blobs of material by the interaction with the ambient interstellar medium. The mechanism of formation of these blobs is presumably Rayleigh-Taylor instability during shock breakout or during the interaction with the interstellar medium. An alternative possibility is a thermal instability during the recombination phase of the fireball.

The two well-known galactic examples are the fast-moving knots of Cas A (Kirshner and Chevalier 1977) and the optical remnant of G292.0+1.8 (MSH 11-54) (Murdin and Clark 1979). The former has spectra dominated by [OIII], [OII], [OI] and [SII] emissions with lines of [ArIII], [ArIV] and [ArV] present whilst the latter shows only the forbidden lines of oxygen and [NeIII].

We have discovered two such remnants in the Magellanic Clouds, one each in the Large and Small Clouds. The LMC example consists of a small ~ 1.6 pc diameter annulus associated with the X-ray and radio remnant 0540-69.3 (Mathewson et al. 1980). This has a velocity dispersion of ~ 3000 km s^{-1} and shows lines of forbidden oxygen, [SII] and [NII] in its spectrum. The [NII] emission forms a smaller (~ 0.8 pc) annulus within the [OIII] ring and may be analogous to the quasi-stationary flocculi of Cas A, presumably a result of strong pre-supernova mass-loss. The remnant in the SMC (0102-72.3) forms a larger (6.9 pc diameter) filamentary ring (Fig. 1) but has a very large velocity dispersion (~ 4000 km s^{-1}) and displays only forbidden oxygen and [NeIII] lines in its spectrum (Dopita, Tuohy and Mathewson 1981). In these regards it is similar to the inner ring of N132D which is the other known example of an oxygen-rich remnant in the LMC (Danziger and Dennefeld 1976: Lasker 1978, 1980).

In principle, the sample of these remnants should now be large enough to investigate both nucleosynthesis theories per se and, to some extent, the effect of metallicity on stellar evolution. However, this conclusion is reliant on the existence of good shock-modelling programs. Unfortunately oxygen-rich shock models are particularly unreliable. Some of the computational problems have been investigated by Itoh (1981a,b). Since oxygen is a very efficient coolant, cooling times are less than ion/electron equipartition lines in the post-shock region so the energy transfer must be dealt with explicitly. Furthermore, oxygen-ionising photons produced in the cooling layer can escape

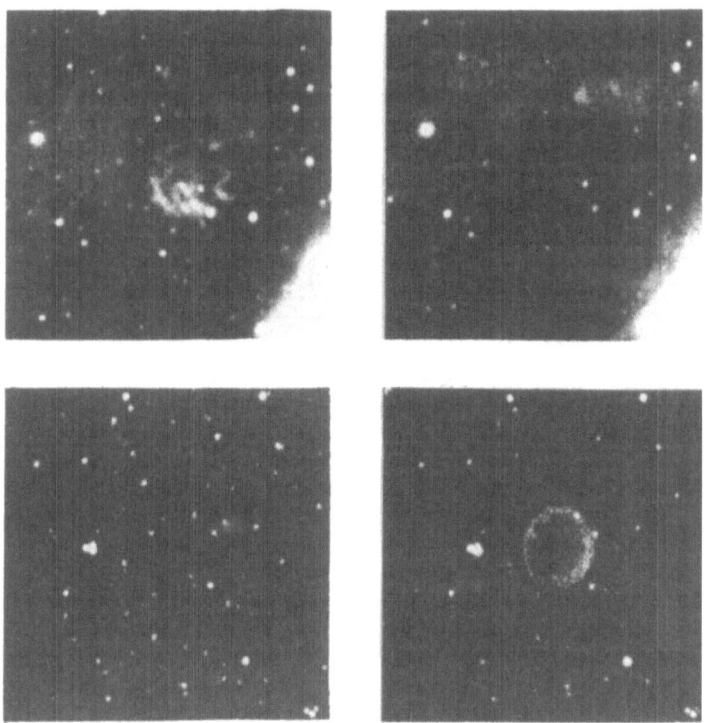

Figure 1. An example of an oxygen SNR (IE0102.2-7219) in the SMC
(above) and a collisionless shock remnant (0505-67.9) in the LMC
(below). The left hand images are in the light of [OIII] λ5007
and the right hand images in the light of Hα. They were obtained
on the IPCS used direct on the Anglo-Australian Telescope.

upstream to produce a precursor HII region, which contributes
appreciably to the total optical emission. A third problem, not
considered by Itoh, is the electron conduction out from the post-
shock cooling zone which is probably limited only by electron
diffusion (McKee and Cowie 1975). Luc Binette and myself at Mt.
Stromlo are attempting to overcome these problems by developing
our general-purpose modelling code MAPPINGS, but at present our
results are little better than qualitative. The effort seems to
be worthwhile, however.

1.2 Remnants of Type I Events

A great surprise of our survey was the discovery of three remnants which emitted only the Balmer lines of hydrogen (Tuohy et al. 1981) and were thus similar to the galactic remnants of Tycho and of SN1006 (Kirshner and Chevalier 1978: Schweitzer and Lasker 1978). The optical, X-ray and radio properties of these are summarised in Table 1.

Table 1

Properties of the collisionless shock remnants in the LMC.
(After Tuohy et al. 1981.)

Source	0505−67.9	0509−67.5	0519−69.0
Optical			
Size (arcsec)	83 x 67	25	28
(pc)	20	6.7	7.5
V_s (km s^{-1})	−	>3600	2700
n_{H^0} (cm^{-3})	0.1	>0.02	0.06
X-ray			
L_x (erg s^{-1})	1.0 x 10^{37}	3.4 x 10^{36}	1.1 x 10^{37}
$n_{(H^++H^0)}$ (cm^{-3})	~1.0	0.8 − 3.1	0.3 − 4.7
Radio			
L_{408MHz} (mJy)	<18	95	150
α	−	−0.46	−0.60

Chevalier and Raymond (1978) put forward a convincing theory capable of explaining the optical features of this type of remnant. According to this, a collisionless shock is propagating into a partly neutral medium at very high velocity. Hydrogen atoms, essentially at rest, are collisionally excited by the first electrons to give the Balmer emission. Some of these undergo charge exchange reactions with shocked ions and this gives rise to a velocity-broadened component of the Balmer lines, but eventually all neutral atoms are collisionally ionised to join the hot high-velocity stream.

These three LMC remnants have low radio surface brightness, a property they share with their galactic counterparts. They are underluminous compared with 'normal' remnants by at least a factor of ten for the smaller diameter remnants increasing to perhaps a factor of one hundred for the largest example. This argues for a fundamental difference in their mechanism of radio emission, one possibility being that they do not contain pulsars to repower the non-thermal electrons.

In the X-rays, however, their luminosity is high and comparable with that of other similar sized remnants. In some degree, the luminosity may be affected by the high metallicity of the ejecta as pointed out by Long, Dopita and Tuohy (1981). This can lead to over-estimates of the mass by a factor of order ten, and this factor may be even larger when non-equilibrium ionisation in the remnant is considered (Shull 1981).

For the best studied object, 0519-69.0, the optical and X-ray data can be combined to yield the parameters of the explosive event and the precursor star on the assumption, apparently justified by the computed swept-up to ejected mass ratio, that the remnant is in its Sedov phase of evolution.

This gives an explosion energy in the range $1.05-2.2 \times 10^{51}$ ergs, an age of 540 years, a precursor mass in the range 0.9-4.4 $M_\odot$ and a mass of 0.5-1.4 $M_\odot$ of C/O or iron peak elements ejected. This seems to put this supernova firmly in the Type I regime: however, we do not know if the precursor had an extended hydrogen envelope. If it did, its mass could not exceed 3.0 $M_\odot$, but in this case the light curve would have resembled a Type II event (or Type IV?!). Certainly there is some evidence that the event pre-ionised the surrounding ISM - compare the neutral hydrogen densities derived from the optical in Table 1 with the total densities estimated from the X-rays. These imply a fractional ionisation in the range 0.8-0.99, whilst if the gas had freely recombined at $5000^\circ K$ after the explosion time fractional ionisation would be of order 0.9. The simplest hypothesis is to assume an extended (possibly He) atmosphere capable of generating a UV pulse at the time of shock break-out.

A similar analysis applied to Tycho's SN gave very similar results except that the mass range of the precursor was tighter, 0.9-2.8 $M_\odot$. This event was certainly observed to be Type I from its light curve (Clark and Stephenson 1977) so that the identification with Type I events seems reasonably secure.

1.3   The Evolution of Old SNRs

In order to estimate the frequency of supernova events in a galaxy from remnants, it is traditional to plot a cumulative number/diameter relationship (Clark and Caswell 1976: Mathewson and Clark 1973b). According to Sedov theory this should have a slope of 5/2 and the actual interval between explosions is related to the Sedov parameter (E_0/n). The frequency derived for Galactic events $\tau \sim 150$ yr is longer than estimates based on historical supernovae, pulsar birthrates or statistics of supernovae in external galaxies (Clark and Stephenson 1982: Lyne 1982: Tammann 1982). Unfortunately, it is subject to all the uncertainties of the surface brightness/diameter relationship.

The LMC optical remnants give us a sample at known distance which is at least 50% complete to 60 pc diameter and is probably even more complete for the smaller diameter remnants. Thus we should be able to do better than for Galactic remnants. The cumulative number/diameter relation taken from Mathewson et al. (1981) is shown in Fig. 2. Rather than a slope of 5/2 this has a slope of 3/2 for remnants with diameter $\lesssim$ 20 pc. Thus Sedov theory is not applicable – the remnants appear to evolve to large diameter faster than predicted. This result should not be considered in isolation. If the Sedov theory was correct, the optical emission should occur at a gas pressure similar to that pertaining within the remnant, since, in the Sedov phase, the largest fraction of the energy is thermal. Thus the pressure derived from optical measurements integrated over the volume of the remnant should give an energy close to the Sedov energy. As noted from the LMC observations of Dopita (1979) this is not true. Since then, data has become available for M33 (Dopita, D'Odorico and Benvenuti 1980) and for M31 (Blair, Kirshner and Chevalier 1981a). The derived Sedov energy is small for small diameter remnants increasing to about 10^{51} ergs at 30 pc, thereafter remaining constant.

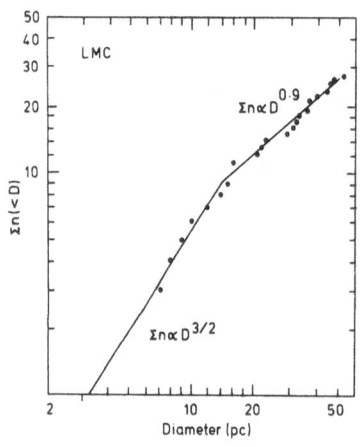

Figure 2. The cumulative number diameter relationship for optical SNR in the LMC.

The evaporative solution (McKee and Ostriker 1977) proposed by Dopita (1979) as a way of explaining these observations no longer appears tenable because of the high internal pressure of the evaporative solution (McKee 1982). A possible promising way out is to argue that fragmentation of the fireball occurs to produce many cold clouds (for which there is some observational evidence in the case of SN 1979c in M100: Chevalier 1982). These then carry much of the kinetic energy of the explosion out to fairly large distances before the reverse-shock heats and destroys them, at which point their kinetic energy is thermalised. This would allow apparently freer expansion of the young SNR, would maintain low internal gas pressures and delay the onset of the radiative phase. These points will be addressed in more detail in a forthcoming publication (Helfand et al. 1982).

2. ABUNDANCE DETERMINATIONS IN SUPERNOVA REMNANTS

The general methods whereby the gaseous phase abundances could be determined from supernova remnants were outlined by Dopita (1977). These were subsequently applied to Magellanic Cloud SNRs (Dopita 1976: Dopita, Mathewson and Ford 1977) to the M33 remnants (Dopita, D'Odorico and Benvenuti 1980) and most recently to the M31 remnants (Blair, Kirshner and Chevalier 1981a,b: Dennefeld 1981: also reported at this conference). In that the Hα/[HII] ($\lambda 6548,84\text{Å}$) ratio or the Hα/[SII] ($\lambda 6731\text{Å}$) ratios in SNRs are principally determined by elemental abundances, these may be used as abundance tracers. Indeed, for many faint galactic remnants with heavy reddening, these are the only line ratios that can be formed from the data. In what follows, I shall restrict the discussion to only these ratios and I describe the result of joint work between Luc Binette and myself at Mt. Stromlo, Sandro D'Odorico at ESO, Garching and Piero Benvenuti at ESA, Villafranca.

Figure 3 shows all the available data on SNRs in our own and external galaxies which shows a clear correlation between the strengths of the nitrogen and sulphur lines with respect to the galaxy. Note that gas-rich Magellanic Irregulars are, as expected, the most metal-poor systems whilst the inner regions of M31 and our own galaxy have the highest metallicity. The solid line represents the trajectory followed by theoretical shock models (computed by MAPPINGS, Binette and Dopita) in which all elemental abundances with respect to hydrogen are varied in lockstep. The tick marks represent factors of two in absolute abundance. The dashed line, on the other hand, represents a trajectory in which nitrogen is a secondary element and is therefore enriched as the square of other elemental abundances.

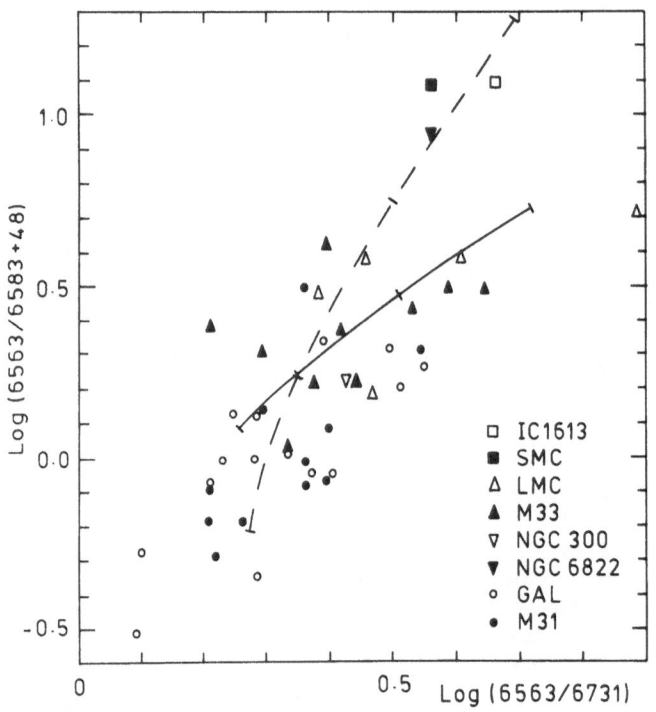

Figure 3. Abundance sensitive line ratios in SNRs in external galaxies. Note evidence for abundance gradients in M31, M33 and the Galaxy. For meaning of curves, see text.

What can be said about the nucleosynthetic status of nitrogen? On this, as in the case of HII region abundance determinations (Pagel and Edmunds 1980), the data are somewhat equivocal. Although overall, galaxy to galaxy, the secondary enrichment line appears the better fit, within an individual galaxy such as M33 the data points seem to scatter more closely along the primary enrichment line. Perhaps the easiest way of explaining these observations may be to say that nitrogen is enriched in a primary manner from some primary base line that is a function of galaxy type. The source of the initial primary nitrogen would then have to be the Population III stars. Equally attractive alternate scenarios are that the nitrogen abundance reflects the mean age of the stellar population in the galaxy (if low/intermediate mass stars are sources of primary nitrogen)

(Edmunds and Pagel 1978) or that the initial mass function is different in the Irregular galaxies (Alloin et al. 1979).

2.1 The Galactic Abundance Gradient

Figure 3 shows that galactic abundance gradients have been established in the galaxies M31 and M33 and these have been discussed in the papers cited in the previous section. Does the scatter in abundances of SNR in our Galaxy evident in this figure also reflect a Galactic abundance gradient? This indeed turns out to be the case (Binette et al. 1981). In Figure 4 we plot the Hα/[NII] (λ6548,84Å) ratio as a function of galacto-centric distance. To derive these we have had to rely, in the main, on radio distances obtained by application of the surface brightness/diameter relation (Clark and Caswell 1976). A clear relationship exists and implies a logarithmic abundance gradient of about -0.08 dex kpc^{-1}, in reasonable agreement with radio measurements of HII regions (-0.08, Mezger et al. 1979), optical spectrophotometry of HII regions (-0.089, Peimbert and Torres-Peimbert 1979: -0.059, Talent and Dufour 1979) or from HR diagrams of open clusters (-0.095, Panagia and Tosi 1980, 1981).

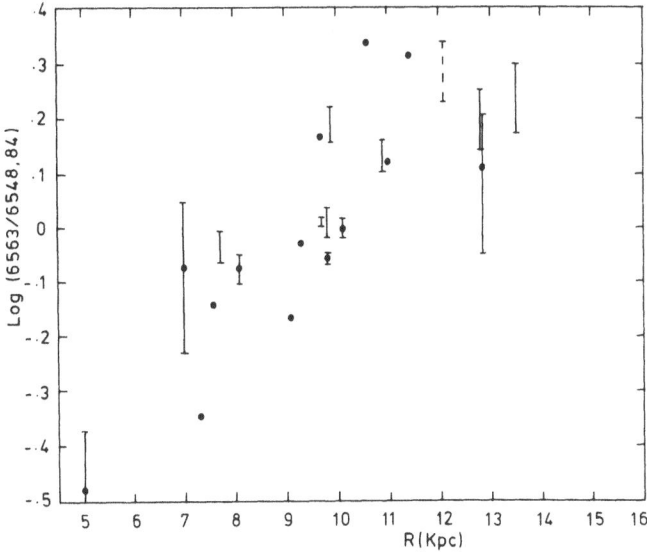

Figure 4. The nitrogen abundance gradient from SNRs in the Galaxy derived using Σ-d distances. The slope of the relation corresponds to -0.08 dex kpc^{-1} from models.

2.2 Non-equilibrium Conditions in SNRs

Although in the above discussion, plane-parallel steady-flow radiating shock models were used in the interpretation of the spectra, this may not always be a valid procedure. Bob Fesen has pointed out in this conference how, observationally in the Cygnus Loop, there is strong evidence for non-steady-flow shocks. This should not be surprising. Consider the timescale for cooling to 10^4 K, τ_4, for 'cosmic' abundance.

$$\tau_4 = 870 \left(\frac{n_1}{cm^{-3}} \right)^{-1} \left(\frac{V_s}{100 \text{ km s}^{-1}} \right)^2 \text{years}$$

where n_1 is the pre-shock density and V_s is the shock velocity. Thus for a typical shock velocity of 150 km s^{-1} and an ambient density of 2-3 cm^{-3} the timescale is uncomfortably long in comparison with a typical remnant age of 2×10^3 years. However, to get the steady-flow model, hydrogen must fully recombine (at 10^4 K, the ionization fraction is typically 0.85). To obtain 98% recombination, the timescale above must be multiplied by about a factor of ten. If magnetic pressure support is important in the recombining gas, the timescale is even longer. Thus only clouds with densities of order 10 cm^{-3} or greater can become fully radiative in the lifetime of the remnant.

Fortunately for abundance determinations, the optical radiation from supernova remnants is dominated by emission from such dense clouds, so errors from this source are not overwhelming. However, to correctly interpret spectra and fainter filaments, non-steady-flow models must be used.

The simplest way to approximate this situation is simply to truncate the model after some finite time. The initial state of ionisation of the gas becomes a free parameter because self-consistent models (Shull and McKee 1979) are no longer valid. We find that such models move from the left to right of Figure 3 almost horizontally before making a small loop and returning to the appropriate steady flow curve. Lowering the state of pre-ionisation of the gas below about 0.2 moves the steady flow model for given abundance nearly vertically. Clearly, full inter-pretation of the spectra requires a multivariate analysis using these parameters.

In conclusion, we see that, in less than ten years, the subject of supernova remnants in external galaxies has grown from nothing to a subject embracing a wide variety of astrophysical topics covering the endpoints of stellar evolution to the structure and evolution of galaxies. The advent of the Space Telescope can be expected to produce a further impetus in this

field, giving identifications and optical and UV spectrophotometry of SNRs out to the Virgo cluster.

I would like to thank all those who have sent me results in
advance of publication, in particular all members of the Michigan
Group and Michel Dennefeld, and also Roger Chevalier, Dave Clark,
Andy Fabian, Fred Seward, Michael Shull, Chris McKee, Ken'ichi
Nomoto and Frank Winkler for enjoyable and stimulating
conversations during this conference.

REFERENCES

Alloin, D., Collin-Souffrin, S., Joly, M. and Vigroux,L. 1979.
    Astr.Astrophys., 78, 200.
Binette, L., Dopita, M.A., D'Odorico, S. and Benvenuti, P. 1981.
    In preparation.
Blair, W.P., Kirshner, R.P. and Chevalier, R.A. 1981a.
    Astrophys.J. In press.
Blair, W.P., Kirshner, R.P. and Chevalier, R.A. 1981b.
    Astrophys.J. In press.
Chevalier, R.A. 1982. This volume.
Chevalier, R.A. and Raymond, J.C. 1978. Astrophys.J.(Letters),
    225, L27.
Clark, D.H. and Caswell, J.L. 1976. Mon.Not.R.astr.Soc., 174,
    267.
Clark, D.H. and Stephenson, F.R. 1977. The Historical
    Supernovae. Pergamon Press, Oxford.
Clark, D.H. and Stephenson, F.R. 1982. This volume.
Daltabuit, E., D'Odorico, S. and Sabbadin, F. 1976. Astr.
    Astrophys., 52, 93.
Danziger, I.J. and Dennefeld, M. 1976. Astrophys.J., 207,
    394.
Dennefeld, M. 1981. Astr.Astrophys. In press.
D'Odorico, S. 1978. Mem.Soc.Austr.Ital., 49, 485.
D'Odorico, S., Benvenuti, P. and Sabbadin, F. 1978. Astr.
    Astrophys, 63, 63.
D'Odorico, S., Dopita, M.A. and Benvenuti, P. 1980. Astr.
    Astrophys.Suppl., 40, 67.
D'Odorico, S. and Sabbadin, F. 1976. Astr.Astrophys., 50,315.
Dopita, M.A. 1976. Astrophys.J., 209, 395.
Dopita, M.A. 1977. Astrophys.J.Suppl., 33, 437.
Dopita, M.A. 1979. Astrophys.J.Suppl., 40, 455.
Dopita, M.A., D'Odorico, S. and Benvenuti, P. 1980. Astrophys.
    J., 236, 628.
Dopita, M.A., Mathewson, D.S. and Ford, V.L. 1977. Astrophys.J.
    214, 179.
Dopita, M.A., Tuohy, I.R. and Mathewson, D.S. 1981. Astrophys.
    J.(Letters), 248, in press.
Edmunds,M.J. and Pagel, B.E.J. 1978. Mon.Not.R.astr.Soc., 185
    77.

Helfand, D.J., Dopita, M.A., Long, K.S., Mathewson, D.S. and
    Tuohy, I.R. 1982. In preparation.
Itoh, H. 1981a. Publ.Astr.Soc.Japan, 33, 1.
Itoh, H. 1981b. Publ.Astr.Soc.Japan, 33, in press.
Kirshner, R.P. and Chevalier, R.A. 1977. Astrophys.J., 218,
    142.
Kirshner, R.P. and Chevalier, R.A. 1978. Astr.Astrophys., 67,
    267.
Kumar, C.K. 1976. Publ.Astr.Soc.Pacific, 88, 323.
Lasker, B.M. 1978. Astrophys.J., 223, 109.
Lasker, B.M. 1980. Astrophys.J., 237, 765.
Long, K.S., Dopita, M.A. and Tuohy, I.R. 1981. Astrophys.J. in
    press.
Long, K.S. and Helfand, D.J. 1979. Astrophys.J.(Letters), 234
    L77.
Long, K.S., Helfand, D.J. and Grabelsky, D.A. 1981. Astrophys.
    J. in press.
Lyne, A.G. 1982. This volume.
McKee, C.F. 1982. This volume.
McKee, C.F. and Cowie, L.L. 1975. Astrophys.J., 195, 715.
McKee, C.F. and Ostriker, J.P. 1977. Astrophys.J., 218, 148.
Mathewson, D.S. and Clarke, J.N. 1972. Astrophys.J.(Letters),
    178, L105.
Mathewson, D.S. and Clarke, J.N. 1973a. Astrophys.J., 179,80.
Mathewson, D.S. and Clarke, J.N. 1973b. Astrophys.J.,180, 725.
Mathewson, D.S. and Clarke, J.N. 1973c. Astrophys.J., 182, 697.
Mathewson, D.S., Dopita, M.A., Tuohy, I.R. and Ford, V.L. 1980.
    Astrophys.J.(Letters), 242, L73.
Mathewson, D.S., Dopita, M.A., Tuohy, I.R., Ford, V.L., Helfand,
    D.J. and Long, K.S. 1981. Astrophys.J.Suppl., in press.
Mezger, P.G., Pankonin, V., Schmid-Burkg, J., Thum, C. and Wink,
    J. 1979. Astr.Astrophys., 80, L3.
Murdin, P. and Clark, D.H. 1979. Mon.Not.R astr.Soc., 189,501.
Pagel, B.E.J. and Edmunds, M.J. 1980. Paper presented at NATO
    conference on Normal Galaxies, Cambridge, August 1980.
Panagia, N. and Tosi, M. 1980. Astr.Astrophys., 81, 375.
Panagia, N. and Tosi, M. 1981. Astr.Astrophys., 96, 306.
Peimbert, M. and Torres-Peimbert, S. 1979. Proc.XVII IAU
    Assembly.
Schweitzer, F. and Lasker, B.M. 1978. Astrophys.J., 220, 167.
Shull, M. 1981. Paper presented at this conference.
Shull, M. and McKee, C.F. 1979. Astrophys.J., 227, 131.
Talent, R.J., Jr. and Dufour, R.J. 1979. Astrophys.J., 233,
    888.
Tammann, G.A. 1982. This volume.
Tuohy, I.R., Dopita, M.A., Mathewson, D.S., Long, K.S. and
    Helfand, D.J. 1981. Astrophys.J. in press.
Weaver, T.A. and Woosley, S.E. 1982. This volume.

OPTICAL SPECTROPHOTOMETRY OF THE CYGNUS LOOP

Robert A. Fesen[1], William P. Blair[2] and Robert P. Kirshner[2]

ABSTRACT

Optical spectral properties of several filaments in the Cygnus Loop are discussed with emphasis on observation-model comparisons. We conclude that some departure from the steady-flow condition assumed in current modelling is necessary to improve agreement with the observations.

Previously high quality spectroscopic data on the Cygnus Loop have been obtained for just a few filaments. Parker (1964, 1967, 1969) observed several positions but only one filament had both good spectral coverage and high accuracy data. Miller (1974) obtained excellent data on three filaments and it has been these data that have usually been used to deduce the shock wave parameters associated with the Cygnus Loop's optical emission through comparisons with the model calculations of Dopita (1977), Raymond (1979) and Shull and McKee (1979).

However, a few studies (e.g. Chamberlain 1953) have indicated that there is probably a range of line intensities among the filaments. We obtained new spectrophotometry at 17 positions in order to explore the range of these variations as well as to

---

1. Laboratory for Astronomy and Solar Physics, NASA/Goddard Space Flight Center, University of Michigan, Greenbelt, Md.
2. University of Michigan, Ann Arbor, MI.

M. J. Rees and R. J. Stoneham (eds.), Supernovae: A Survey of Current Research, 495–499.

compare the line intensities to model predictions. The data
covered the spectra region 3600-7400Å and were taken at the
McGraw-Hill Observatory at Kitt Peak using the 1.3m telescope and
a 2000 channel intensified Reticon. We concentrated on those
filaments that had different intensities on narrow passband
interference filter photographs. Of the 17 spectra obtained,
three showed just the Balmer spectrum while the other 14
exhibited a more typical SNR spectrum with bright lines of H, He
I, [O I], [O II], [O III], [N I], [N II], [S II], and [Ne III].
We will not discuss here the pure Balmer emitting filaments but
deal only with the remaining filament spectra.

The data indicate that there are significant line emission
strength variations throughout the remnant. For example, the
filament's [O III] $\lambda\lambda 4959,5007$ line emission varies in strength
from about 0.2 I(Hβ) to nearly 25 I(Hβ) - more than a factor of
100. The relative line strengths of [O I] $\lambda\lambda 6300,6364$, [N I]
$\lambda 5199$, and [S III] $\lambda\lambda 6717,6731$ also show substantial variations
between filaments, but not quite as large as that seen for
[O III]. However, some other lines exhibit little change in
intensity relative to I(Hβ), e.g. the [N II] $\lambda\lambda 6548,6583$ lines.
When large intensity variations are found among different
filaments, the lines that arise from the high ionization species
appear strong in filaments where the emission from lower
ionization or neutral species is relatively weak and vice versa.
Such changes in the intensity of one line versus another are
illustrated in Figure 1 for the [O I]/Hλ and [O II]/Hβ ratio.
Correlations such as this suggest that one is observing a mixture
of hot and cool shocked gas. Similar line strength patterns in
shocks have been numerically modelled and illustrated by Cox
(1972) and Ohtani (1980) where the stratification distance
between strong [O II] and [O I] emitting gas in the recombination
zone is of order 10^{16}cm. Yet such a distance corresponds to only
~ 1 arcsecond for the Cygnus Loop. Raymond et al. (1980) have
proposed substantial departures from steady-flow in order to
explain this model-observation discrepancy as well as the wide
range in observed [O III]/Hβ ratios.

Our data's systematic line strength variations support this
idea. Strong departures from a steady-flow shock model will
occur naturally in an inhomogeneous medium. When the Cygnus
Loop's 400 km s^{-1} blast wave encounters small but dense
cloudlets, it will drive a ~ 100 km s^{-1} shock into a cloud as
well as rapidly surround and eventually pass it. In a well mixed
two-phase ISM, the time between successive shocked clouds
radiating strongly in [O III] or [O I] is of order 1000 yrs/n_0,
where n_0 is the preshock cloud density in cm^{-3}. The spatial
distance between [O III] and [O I] regions is roughly 300km s^{-1}
times this time interval, which for a value of $n_0 = 5$ cm^{-3} yields
an angular difference of ~ 30''. This is sufficiently large to

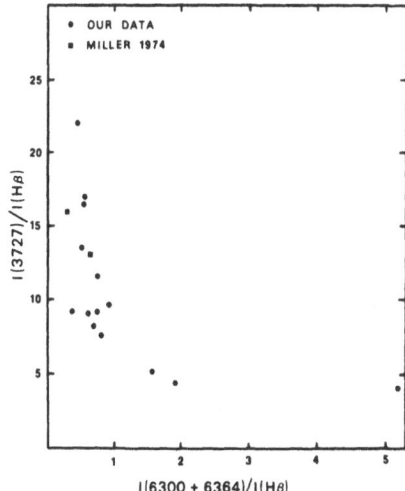

Figure 1. Plot of [O II]·λ3727/Hβ vs. [O I] λλ6300,6364/Hβ showing a correlation between these line emissions.

account for both the observed spectral differences between individual filaments as well as the gross spectral displacements visible on interference filter photographs. Thus, the line emission coming from small portions of the remnant's optical filaments is characteristic of particular stages of recombination in the shocked gas.

Because the current shock models assume a steady flow condition, their predicted relative line intensities are the integrated spectra for the whole recombination zone. Since the observational data répresent the line emission from only portions of the cooling recombination region, any detailed comparisons between observation and model spectra to infer the remnant's various shock parameters should be viewed with caution.

This is illustrated in Figures 2a and 2b, where we compare the line strengths of the forbidden oxygen lines with steady flow model predictions having a range of shock velocities. Some variation in the shock velocity among filaments is expected from considerations of preshock cloud density differences as well as the wide [O III]/Hβ ratio range actually observed. The strength of the oxygen lines, especially [O III], is particularly sensitive to changes in shock velocity. This property combined with the well-determined atomic parameters of oxygen and the brightness of [O I], [O II], and [O III] lines in SNR spectra makes the oxygen lines ideal for model-observation comparisons. Figure 2a shows [O II]/[O III] vs [O I]/[O III] for the 14 filaments we observed together with those reported by Miller (1974), with the letters and numbers corresponding to various filaments. There is a clear empirical correlation which is not reproduced by shock velocity changes in the models of Raymond (1979). Although a few of the model's line ratios coincide with observed values, inferred shock properties from these agreements may not lead to reliable shock or preshock condition estimates due to the general lack of model-observation agreement.

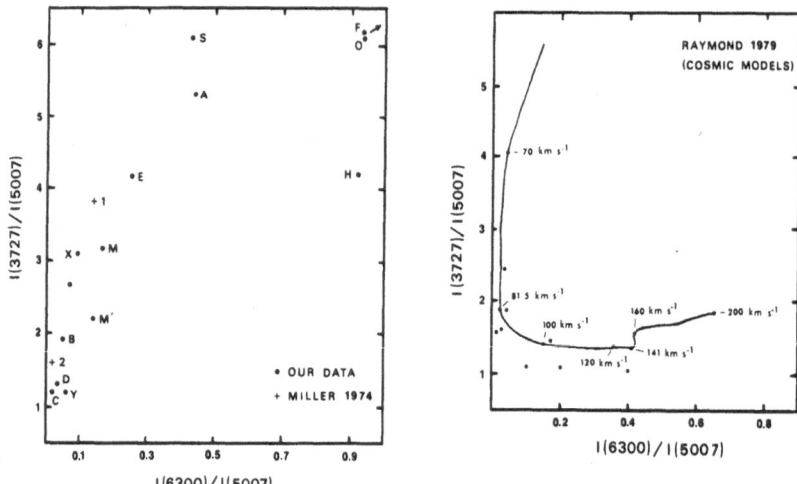

Figure 2a and 2b.   Plot of [O II] λ3727/[O III] λ5007 vs. [O I] λ6300/[O III] λ5007 for Miller's (1974) and our data for comparison to a similar plot for Raymond's (1979) shock models with cosmic elemental abundances but having various shock velocities.

In summary, therefore, we find that: 1) the optical filaments of the Cygnus Loop exhibit a remarkable variety of spectral properties, 2) some line intensities are correlated among filaments in such a way as to suggest that a well resolved ionization structure behind the shock front is being observed, and 3) non-steady flow conditions are probably responsible for the discrepancies with current shock spectra predictions.

REFERENCES

Chamberlain, J.W. 1953. Astrophys.J., 117, 399.

Cox, D.P. 1972. Astrophys.J., 178, 143.

Dopita, M.A. 1977. Astrophys.J.Suppl. 33, 437.

Miller, J.S. 1974. Astrophys.J., 189, 239.

Ohtani, H. 1980. Pub.Astron.Soc.Pacific, 32, 11.

Parker, R.A.R. 1964. Astrophys.J., 139, 493.

Parker, R.A.R. 1967. Astrophys.J., 149, 363.

Parker, R.A.R. 1969. Astrophys.J., 155, 359.

Raymond, J.C. 1979. Astrophys.J.Suppl., 39, 1.

Raymond, J.C., Black, J.H., Dupree, A.K., Hartmann, L. and Wolff, R.S. 1980. Astrophys.J., 238, 881.

Shull, J.M. and McKee, C.F. 1979. Astrophys.J., 227, 131.

X-RAY SPECTROSCOPY AND IMAGERY OF SUPERNOVA REMNANTS WITH THE EINSTEIN OBSERVATORY

P.F. Winkler[1], C.R. Canizares[2,3], T.H. Markert[2], and A.E. Szymkowiak[4]

1. INTRODUCTION

X-ray spectroscopy has been an important probe for the study of supernova remnants (SNRs) for over a decade (Gorenstein and Tucker 1976, Culhane 1977). Most SNRs less than 10^4 years in age are made up primarily of shock-heated plasma at temperatures 10^6–10^8 K, which radiates primarily X-rays. Measurements of the X-ray spectrum and flux with proportional counters have been used to infer the temperature and density of the X-ray plasma.

Advances in sensitivity and spectral resolution have made possible more detailed diagnostics of the hot plasma through the study of X-ray emission lines. Hydrogen and helium will normally be fully ionized in an SNR plasma and will radiate primarily through continuum thermal bremsstrahlung. The abundant heavy elements are responsible for line emission which dominates the X-ray spectrum at temperatures $< 10^7$ K. The predominant ionic species are one- and two-electron ions of elements carbon and heavier: in addition L-shell ions of iron and nickel are important. The energies of a few benchmark lines are listed in Table 1: recent detailed calculations of the X-ray spectrum from a hot plasma have been done by Raymond and Smith (1977), Mewe and Gronenschild (1981) and Shull (1981a).

Spectra taken with proportional counters have detected the $K\alpha$ iron-line complex around 6.7 keV in several young SNRs: Cas A,

1. Middlebury College, Middlebury Vt.
2. Massachusetts Institute of Technology, Cambridge, Mass.
3. Alfred P. Sloan Research Fellow.
4. NASA Goddard Space Flight Center, Greenbelt, Md.

M. J. Rees and R. J. Stoneham (eds.), Supernovae: A Survey of Current Research, 501–518.
Copyright © 1982 by D. Reidel Publishing Company.

Table 1

Energies of Some X-ray Emission Lines

Ion	Transition	Energy (keV)	Wavelength (Angstroms)
C VI	Lyman α	0.37	33.70
O VIII	Lyman α	0.65	18.97
Fe XVII	$2p^6 - 2p^5 3d$ ^3D	0.83	15.01
Si XIV	Lyman α	2.01	6.18
Ca XX	Lyman α	4.11	3.02
Fe XXVI	Lyman α	6.97	1.78

Tycho, and RCW86 (Davison, Culhane and Mitchell 1976, Pravdo et al. 1976, Pravdo and Smith 1979, Winkler 1979). In addition, Pravdo and Smith have reported their detection of the Kβ iron line in Cas A, and Pravdo et al. (1980) have observed three line complexes between 0.8 and 3 keV in the Tycho SNR. The latter have been attributed to the iron L-complex, silicon, and sulphur: however, the resolution of proportional counters did not permit the identification of specific lines. On the basis of their data, Pravdo and Smith and Pravdo et al. have argued that neither Cas A nor Tycho has reached ionization equilibrium. Continuum X-ray emission up to energies > 20 keV indicates that electrons are rapidly heated by the blast wave but the ionization of the shock-heated plasma lags in time.

A new method of deconvolving instrumental effects has been perfected by Kahn and Blissett (1980) to improve the resolution of spectra obtained with proportional counters. Application of this technique to HEAO 1 data on the Cygnus loop indicates the presence of strong features at energies of about 0.65 and 0.85 keV, coincident with prominent lines of O VIII and Fe XVII, respectively (Kahn et al. 1980). A similar approach to the data for IC443 by Charles et al. (1981) suggests emission features of Si, S and Fe.

The two complementary spectroscopic instruments on the Einstein Observatory made possible X-ray spectroscopy with vastly increased sensitivity and spectral resolution. The Solid State Spectrometer (SSS) was a cryogenically cooled detector of lithium-drifted silicon, with high sensitivity and moderate (~ 160 eV) resolution (Giacconi et al. 1979, Holt et al. 1979). This resolution is not only sufficient to separate the lines due to different elements, but can resolve the hydrogen-like, helium-like, and neutral lines from the same element for silicon and above. The Focal Plane Crystal

Spectrometer (FPCS) used Bragg crystals to achieve excellent spectral resolution (E/ΔE > 100) but with lower sensitivity (Giacconi et al. 1979, Canizares et al. 1979). It could resolve the many closely-spaced lines at E ~ 1.5 keV and could study details such as velocity broadening and the helium-like triplets. Both instruments operated over most of the 0.2 - 4 keV energy band of the Einstein telescope and were capable of detecting most important X-ray lines except the K lines of iron and nickel.

2. SSS RESULTS

Of the 18 SNRs from which good spectra were obtained by the SSS, only 6 showed no line emission: the Crab Nebula, Vela X, 3C58, SN1006, and N157B and 0540-69.3 in the Large Magellanic Cloud (LMC). The Crab and Vela X are known to contain pulsars, and as Becker has reported at this meeting, 3C58 has a filled-centre X-ray morphology and a point source which may indicate the influence of a pulsar which is not beamed toward earth. SN1006 has a shell structure in X-rays (Pye et al. 1981), and its featureless spectrum (Becker et al. 1980c) is enigmatic. Reynolds and Chevalier (1981: see also Blandford, 1982) have suggested that Fermi acceleration of electrons at the shock front could result in synchrotron emission: A.C. Fabian has proposed that unusual abundances in the SN1006 region could conspire to give a spectrum deficient in lines. If the shell were composed primarily of helium from pre-SN mass loss, and the interior of iron in a low ionization state, then no lines would be expected in spectra from the SSS (Becker et al. 1980c) or FPCS (Waite and Winkler 1981). High resolution images of the two LMC remnants with featureless spectra are not yet available.

The remaining SNRs all show thermal spectra with prominent lines: a typical example is Cas A, shown in Figure 1 (from Becker et al. 1979). Results have also been published for the Kepler (Becker et al. 1980a) and Tycho (Becker et al. 1980b) SNRs. Szymkowiak (1980) has reviewed the SSS spectra for SNRs and has discussed the analysis procedure. Briefly, the SSS data have been fit by assuming as a model spectrum a hot, optically-thin plasma with one or two temperatures (Raymond and Smith 1977). The model spectrum is folded through the response of the detector, and temperatures and elemental abundances are adjusted to obtain the best fit to the data. The line strengths are reflected as a set of abundances, such as those listed in Table 2.

There are several limitations to this approach, as has been pointed out by Holt (1981) and Szymkowiak (1980). First, the data have such high statistical precision that no models give formally acceptable fits due to systematic errors of uncertain size in the detector response function and/or the models

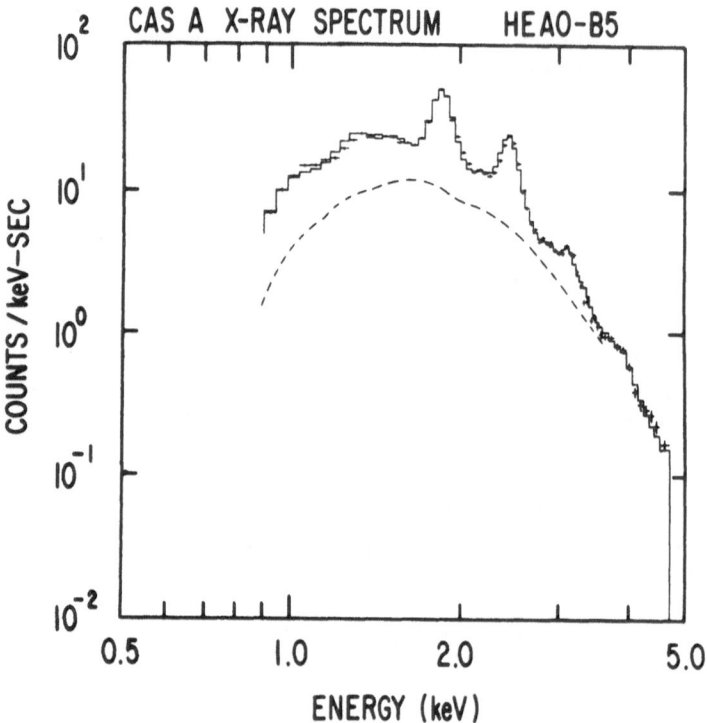

Figure 1. The spectrum of Cas A measured with the SSS (Becker
et al. 1979). The prominent lines are due to He-like Si (1.8
keV) and S (2.4 keV), and the histogram shows the best fit model
spectrum folded through the instrument response function. The
dashed line indicates the level of the continuum in the model
spectrum.

themselves. Second, there are large uncertainties in many of the
calculated cross sections and recombination rates used in plasma
emissivity models (Smith, Mushotzky and Serlemitsos 1979). Third
and most important is the fact that the SNRs, especially young
ones such as Cas A, Tycho and Kepler, deviate substantially from
ionization equilibrium. Itoh (1979), Gronenschild and Mewe
(1981), and Shull (1981) have shown that under non-equilibrium
conditions the emissivities of helium-like lines may be
substantially increased over their equilibrium values. This could
result in erroneously large abundances in fits to the SSS data.
Fitting the SSS data with non-equilibrium models as are now
becoming available for the first time will provide further
insight into the actual abundances in young SNRs.

Table 2

Preliminary SSS Results for abundance Ratios in SNRs

Source	Mg/Si	S/Si	Ar/Si	Fe/Si
Cas A	0.29	1.9	4.1	0.47
Kepler	0.05	2.2	5.1	0.81
Tycho	0.02	2.3	5.8	0.03
Puppis A	1.00	2.0	1.0	0.85
W44	0.80	1.0		0.53
IC443	1.00	1.5		0.80
MSH11-54	0.90	2.5		0.10
RCW103	0.20	2.4		0.36
N132D*	0.90	3.0		1.03
N63A*	1.04	3.4		0.85
N49*	0.60	<2.4		0.25
0525-66.0*	0.9	<7.6	<4.5	

Ratios are relative to the cosmic ratios of Meyer (1979):

	1.05	0.45	0.09	0.88

* SNR in the Large Magellanic Cloud.

Some important qualitative conclusions can be drawn from the preliminary SSS results. All the observed remnants show abundances of the silicon-group elements that are higher than the normal cosmic values. Furthermore, the sulphur-to-silicon ratio is high by about a factor of 2. Most significant for supernova models is the observation that iron lines, while present, are not unusually strong in any of the SNRs. There is a growing consensus among SN model-builders that Type I SN produce a large fraction of a solar mass of iron (e.g. Arnett 1982, Wheeler 1982). One might expect strong iron emission from the hot plasma which has presumably been enriched by SN ejecta, especially in young SNRs. The absence of unusually strong iron lines, while not evidence for the absence of iron, gives one pause and encourages further investigation. The application of non-equilibrium plasma models to the SSS spectra should permit the refinement of these conclusions.

3. PUPPIS A STRUCTURE AND SPECTRUM

Our group at MIT has carried out detailed investigations of the Puppis A SNR with imaging and spectroscopic instruments on the Einstein Observatory. The remnant is about 50' in diameter and is of moderate age: probably a few thousand years. The distance has not been well determined but is estimated as ~ 2 kpc (Dopita 1981, Caswell and Lerche 1979, Clark and Caswell 1976). Its high surface brightness at energies < 1 keV makes it particularly amenable to study with the Einstein FPCS.

A. X-ray Image and ISM Inhomogeneities

An image of Puppis A, obtained from 11 overlapping exposures with the Einstein High Resolution Imager, is shown in Fig. 2 (Petre et al. 1981). The shell is large enough (~ 30 pc diameter) that the complex morphology is presumably due to the structure of the shocked interstellar medium (ISM). The large drop in X-ray surface brightness from the northeast to the southwest indicates a density gradient of a factor > 4 across the diameter of the remnant, in a direction perpendicular to the galactic plane. The shell of X-ray emission surrounding the northern half of Puppis A coincides with the radio shell. However, there is almost no coincidence of either the radio or X-ray features with the chaotic optical filaments.

Brightness profiles across the edge of the X-ray shell show a sharp (~ 1') rise, which indicates that the hot plasma has been heated directly by the blast wave rather than evaporated from clouds (McKee 1982, Cowie, McKee and Ostriker 1981). The wealth of interior structure, especially apparent in the northeast quadrant, indicates density inhomogeneities over a wide range of sizes ~ 0.1-5 pc. However, the modest surface-brightness contrast of the X-ray features suggests that the ISM which was shocked to produce these features was reasonably uniform, with density fluctuations by a factor < 2 about an average ~ 1 cm^{-3}. The two brightest X-ray features are knots at $\alpha = 8^h22^m$, $\delta = -42°48'$ and $\alpha = 8^h20^m$, $\delta = -42°27'$ (1950). These appear to be clouds of density 10-30 cm^{-3} and mass < 1 M_o which have been heated by the blast wave. These isolated clouds represent only a small fraction of the total mass in Puppis A. This detailed view of a localized region of the ISM is consistent with the spirit of the McKee-Ostriker (1977: see also McKee 1982) three-phase model for the ISM, but differs in important details. In Puppis A the spectrum of observed inhomogeneities covers an order of magnitude larger range of sizes than does the model, but with a much lower density contrast. Isolated dense clouds exist, but in contrast to the McKee-Ostriker picture these represent only a small fraction of the mass of this region.

B. X-ray Line Spectrum

Since Puppis A is several thousand years old, enough time has passed since the initial explosion for the expanding shell of matter to consist principally of interstellar material swept up

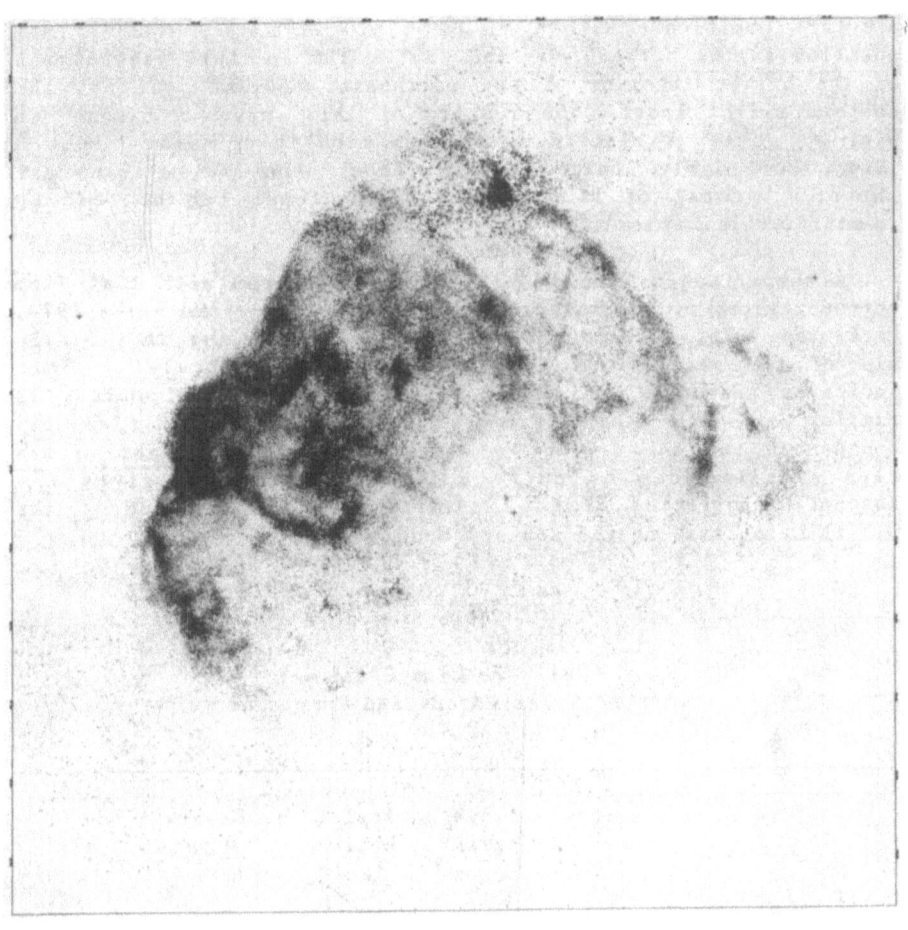

Figure 2. A composite, exposure corrected HRI image of Puppis A
(Petre et al. 1981). The data are binned in 8''x8'' cells.

by the shock wave. In spite of the fact that the ejecta from
the supernova event represent a relatively minor contaminant,
X-ray spectroscopy of the hot gas can provide important
information about the ejecta.

With the <u>Einstein</u> FPCS we carried out an extensive survey of the X-ray spectrum from Puppis A (Winkler <u>et al</u>. 1981a, 1981b). The observations were carried out with a 3' x 30' aperture positioned within 5' of $\alpha = 8^h21^m5$, $\delta = -42°42'$ at position angles between 90° and 130°. The aperture viewed much of the bright interior of the northeast quadrant, but not the eastern bright knot. The results of this survey are shown in Fig. 3. This is clearly a spectrum rich in emission lines – lines from highly ionized oxygen, neon, iron and nitrogen are shown. A total of 11 lines or line blends can be readily identified in Puppis A.

A comparison of the Puppis A X-ray spectrum with that from active regions of the solar corona (Walker, Rugge and Weiss 1974, Parkinson 1975, Hutcheon, Pye and Evans 1976) shows that nearly all of the same lines are present in both locales. This indicates that the ionization structure of the two sources is similar although other parameters such as density, elemental abundances and the degree to which ionization equilibrium has been attained can be quite different. Table 3 gives the absorption-corrected flux of several lines, normalised to the O VIII Ly α flux, in the sun and Puppis A.

Table 3

Relative Line Flux[a]
Active Solar Corona and Puppis A

	Energy (eV)	Active Region[b]	Active Region[c]	Puppis A ($N_H = 4\times10^{21}$cm^{-2})
N VII Lα	500	0.13	–	0.13
O VII (1 – 2)	576	1.1	0.86	1.7
O VIII Lyα	654	1.0	1.0	1.0
O VII (1 – 3)	666	0.15	0.11	0.14
O VIII Lyβ	775	0.10	0.19	0.16
Fe XVII[d]	~ 825	0.63	1.50	0.20
Ne IX (1 – 2)	~ 920	0.13	0.32	0.25
Ne X[e]	1022	0.06	0.12	0.08

a.   Photon flux normalized to O VIII Lyα, corrected in Puppis A for $N_H = 4 \times 10^{21}$ cm^{-2}. See Winkler <u>et al</u>. 1981b, Canizares and Winkler 1981.
b.   Walker, Rugge and Weiss (1974).
c.   Parkinson (1975).
d. $2p^6 - 2p^53d$ in Fe XVII plus weaker O VIII Lyγ, Lyδ.
e.   Lyα blended with weaker Fe XVII $2p^6 - 2p^54d$.

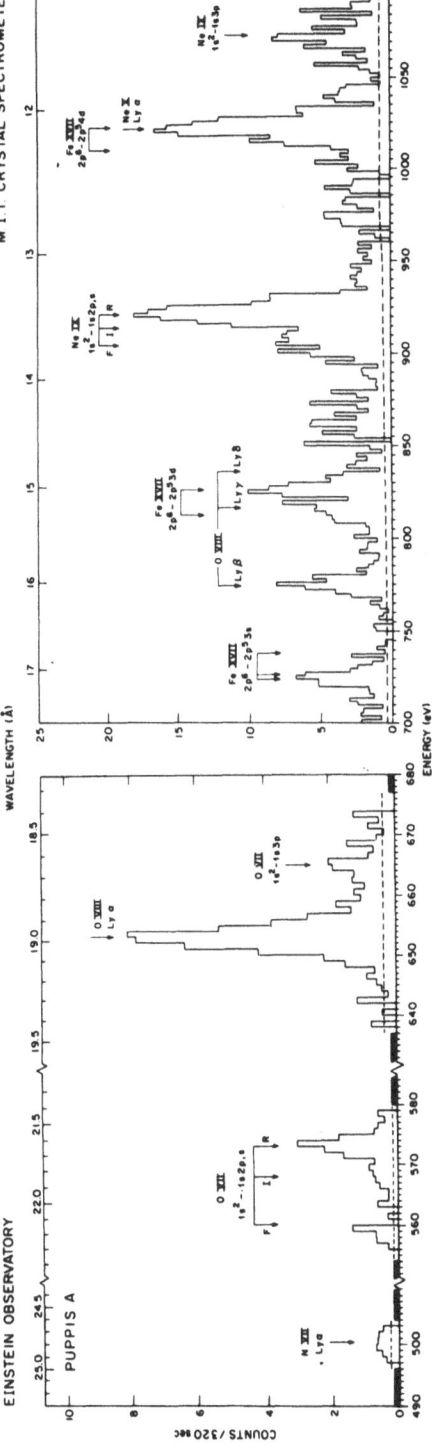

Figure 3. Selected portions of the spectrum of Puppis A measured with the FPCS (Winkler et al. 1981b). The dashed line indicates the level of the non-X-ray background. No correction has yet been made for instrumental efficiency which varies considerably across the spectrum.

One striking difference between the Puppis A spectrum and that of the Sun or of plasma models is the relative weakness of Fe XVII lines in Puppis A. Lines at 826 eV (15.01 Å) and 727 eV (17.04, 17.08 Å) are the strongest of all lines observed in the solar corona over a wide range of active and flaring conditions. In the Puppis A spectrum (Fig. 3) these lines are prominent, but they are clearly weaker than the resonance lines of oxygen or neon.

To understand the significance of these differences consider the equation that describes the flux of an emission line. For example, for the Lyman α transition in O VIII, we have

$$F_{O\ VIII\ Ly\alpha} = N_O\left(\frac{N_{O\ VIII}}{N_O}\right) n_e \left(\begin{array}{c}\text{effective} \\ \text{collision} \\ \text{strength}\end{array}\right) \frac{e^{-E/kT_e}}{T_e^{1/2}} \left(\frac{\text{interstellar}}{\text{absorption}} \atop 4\pi D^2 \right) \quad (1)$$

where N_O is the number of oxygen atoms present in all ionization states, $N_{O\ VIII}$ the number of hydrogen-like ions, n_e the electron density, E the transition energy, T_e the electron temperature and D the distance to the source.

Many of these factors are unknown, but by taking line flux ratios many of the unknowns can be either eliminated or their influence reduced. For example, if two lines from the same ion are compared (e.g. Lyβ/Lyα in O VIII), all the factors cancel out except the collision strengths, the Boltzmann factor, and the differential absorption. Since the collision strengths are reasonably well known for hydrogenic atoms, we can obtain an allowed range of values for T_e and N_H (the interstellar absorption column) which will give a flux ratio consistent with our observations (Winkler et al. 1981b).

For many lines, especially those of iron, the collision strengths are not well known. However, comparison of a line ratio in Puppis A with the same ratio in the solar corona eliminates the collision strengths in Eq. 1. Furthermore, if the ratio chosen is between two lines closely spaced in energy, the effect of differential absorption by a column of uncertain thickness is minimized as is the dependence on electron temperature. Different values for a line ratio in Puppis A and in the Sun must then be attributed to differences in the ionization fractions and/or in the elemental abundances. As we can demonstrate a similarity in ionization structure, this technique yields model independent abundance ratios (Canizares and Winkler 1981).

Referring to Table 3, we see that the ratio of O VIII Lyβ/Lyα in Puppis A is similar to that in both the coronal observations, which indicates that the temperatures are similar and that absorption to Puppis A is being properly accounted for. The ratios for O VII:O VIII, Ne IX:Ne X and O VIII Lyβ:Ne IX are all similar in the two objects, suggesting that both the ionization structure and the oxygen-neon relative abundance are comparable.

The most dramatic difference between the Puppis A and solar spectra is the weakness of the Fe XVII lines relative to those of O and Ne in Puppis A, as mentioned above. Table 3 shows that the 826 eV Fe XVII complex in Puppis A has a flux comparable with the nearby O VIII Lyβ line and the 920 eV Ne IX complex. In contrast, the same Fe XVII complex in the Sun is 5 to 7 times stronger than the adjacent oxygen and neon lines. As stated above these differences might be attributed either to elemental abundance differences or to different ionization structure. We have already argued that the ionization of oxygen and neon must be similar in the two objects. As for iron, Puppis A cannot have a significant fraction more highly ionized than Fe XVII, or we would have seen strong lines from Fe XVIII, Fe XX and/or higher stages. Such lines are prominent in spectra of solar fluxes, which are hotter than the active corona (McKenzie et al. 1980). Furthermore, the existence of a significant amount of iron in lower ionization stages than the neon-like Fe XVII is implausible because of the short time scales for removal of M-shell electrons. For example, the ionization time scale for Fe XVI – Fe XVII is > 50 times shorter than that for O VII – O VIII, yet the latter process has clearly occurred.

We conclude that most of the iron in the observed Puppis A plasma is in the Fe XVII stage, and that the ionization structure of oxygen, neon and iron is similar in Puppis A and the active solar corona. Therefore the relative line strengths indicate a difference in elemental composition: the O:Fe and Ne:Fe abundance ratios in Puppis A are at least 4 times the solar values (see Winkler et al. 1981a and Canizares and Winkler 1981 for details).

Comparison with the plasma emissivity models as calculated by Raymond and Smith (1977, 1979) and by Mewe and Gronenschild (1981) leads to similar conclusions. If the Puppis A plasma is in ionization equilibrium, the prominence of the four species O VII, O VIII, Ne IX and Ne X requires the presence of plasma at temperatures throughout the range $2-5 \times 10^6$ K. Fe XVII is a strong source of line emission throughout this range, and the fact that oxygen and neon lines are observed to be even stronger indicates again that these elements are overabundant relative to iron in Puppis A. The models are for equilibrium plasmas, but our conclusion of anomalous abundances based on the ionization structure (preceding paragraph) remains valid even if Puppis A has not reached equilibrium.

The anomalous abundances in Puppis A indicate either depletion of iron or enhancement of oxygen and neon. The only plausible cause of widespread depletion of gaseous iron is the trapping of iron in interstellar grains (Savage and Bohlin 1979). The chief difficulty with this explanation is that if iron were depleted due to grain trapping, one would expect silicon to be similarly depleted (Duley 1980). But the Einstein SSS results (Table 2 and Szymkowiak 1980) show that if anything the silicon abundance is enhanced in Puppis A. Both iron and silicon grains will be destroyed by shocks, but the time scales for destruction are long enough that grains might survive $> 10^4$ years behind a shock, comparable with the age of Puppis A (Shull 1977, 1978).

The second alternative, an enhancement of oxygen and neon, can result from the enrichment of predominantly interstellar material by supernova ejecta rich in carbon-burning products. The FPCS observations of oxygen for example, imply a total oxygen content of $> 2 M_O$ in the entire Puppis remnant. A Type II supernova could have ejected the large quantity of oxygen which our observations imply. Models of SN II nucleosynthesis (e.g. Arnett 1978, Weaver, Zimmerman and Woosley 1978, Woosley and Weaver 1980) indicate that ejecta from stars of $> 25 M_O$ yield 3-4 M_O of oxygen and O:Ne ratios of 3-5, which are consistent with our data. Smaller stars produce too little oxygen and have O:Ne ratios which are too small. The amount of iron ejected depends on the details of core collapse and is highly uncertain, but the ejection of a negligible amount is plausible. Therefore if a few M_O of ejecta rich in oxygen and neon were stirred into a much larger mass of broth made up of cosmic-abundance interstellar material, the resulting soup might be just what the FPCS has sampled from Puppis A.

4. CAS A: EVIDENCE FOR X-RAY DOPPLER SHIFTS

Cas A is a much younger remnant than Puppis A, so most of the X-ray emitting material is presumably ejecta from the supernova. Extensive analyses of the X-ray image and spectrum have been performed with Einstein (Murray et al. 1979, Fabian et al. 1980, Becker et al. 1979: see also Seward 1982), all of which point to a rather large mass for the remnant (~ 15 M_O). The spectrum measured with the SSS (Becker et al. 1979) shown in Figure 1 is dominated by the lines of He-like silicon and sulphur, but is too highly absorbed to show any of the lines at < 1 keV that are so prominent in the FPCS Puppis A spectrum. These Si and S lines are resolved with the higher resolution FPCS and show evidence of Doppler broadening and of an asymmetric Doppler shift across the remnant.

The FPCS has some imaging capability which provides spatial information along a single direction on the sky in addition to the spectral information. The angular resolution is 1-2 arcmin so for Cas A (which has a diameter of ~ 4') we were able to measure separately the spectrum of the NW and SE regions (corresponding to the two bright regions in the image shown by Seward at this meeting). We did this for the lines at Si XIII, S XV and S XVI. The last of these is a singlet, whereas the first two are the $n = 2$ to $n = 1$ He-like triplets.

All the spectra we obtained show a relative energy shift between the spectra of the NW and SE regions of Cas A. It amounts to a relative Doppler shift of ~ 2000 km s^{-1}: the NW is redshifted with respect to the SE. Unfortunately, we are troubled by the limited statistics of our data and by the complexities of measuring an energy shift which is comparable in size to the line separation of the He-like triplets and possibly to the Doppler broadened line profiles (details of the analysis will be published shortly). However, the NW-SE asymmetry shows up consistently in four independent data sets, so we are quite confident that it is real. In addition, the lines are probably broadened by several thousand km s^{-1}. The Doppler shifts of the two sides of the remnant seem symmetric about zero velocity, but the uncertainty in our absolute energy calibration amounts to ~ 1500 kms^{-1}.

The X-ray line emitting material, which is the low temperature material in the remnant, could in principle be either SN ejecta or shocked interstellar matter. The peculiar abundances indicated in the SSS analysis favours the former interpretation, in which case the heating mechanism is probably the reverse shock of McKee (1974), Gull (1973, 1975) and others. Our preliminary examination, however, shows that it may be possible to construct a model (e.g. McKee and Cowie 1975, Sgro 1975, McKee, Cowie and Ostriker 1978, Chevalier 1977) in which the line emission comes from many dense clouds that have been shock heated and accelerated by the primary supernova blast wave (about 10^3 clouds with density ~ 500 cm^{-3} and diameter ~ 0.04 pc would suffice).

Whatever the origin of the plasma, the lack of spherical symmetry evidenced in the Doppler shifts of the X-ray lines implies that either the supernova ejecta itself or the medium with which it collides must be distributed asymmetrically. We have examined the constraints which the spectroscopic and imaging data place on the kinematics of the emitting material, and find that the observations are consistent with an expanding ring inclined by $> 30°$ to the line-of-sight (for smaller angles the image would be too much flattened). The height of the ring must subtend an angle $< 70°$ as seen from its centre (else the spectrum would not appear split). In fact, the irregularity of the image suggests that the ring is not even cylindrically symmetric.

It is possible that we are seeing the results of asymmetric inhomogeneities in the interstellar medium: in the original locations of the accelerated, dense interstellar clouds, if such are the origin of the emission, or in the surrounding medium which drives the reverse shock, if the emission occurs from the ejecta. Alternatively, the supernova explosion itself may have been non-spherical, as Woosley and Weaver (1980) suggest for rotating, massive stars.

Finally, we note that the Doppler velocities of the X-ray emitting material differ from the corresponding velocity measurements in the optical (Kamper and van den Bergh 1976) in that the mass of the X-ray gas (1-5 M_0) is one to two orders of magnitude greater than that present in the optical filaments. The kinetic energy of the high velocity X-ray emitting plasma is 10^{50} to 10^{51} erg, which must represent a large fraction of the total energy released by the supernova.

5. DETECTION OF X-RAYS FROM SN1980K DURING OUTBURST

The youngest supernova 'remnant' detected in X-rays is that of SN1980K, a Type II SN in NGC 6946 which was seen with Einstein $\sim$ 35^d after maximum light (Canizares, Kriss and Feigelson 1982). This is the first time X-rays have been seen from any SN younger than Cas A despite many previous searches. The absorption corrected X-ray flux of SN1980K on 11 Dec 1980 was $\sim$ 0.03 μJy at 0.24 x 10^{18} Hz (1 keV) corresponding to a luminosity of $\sim$ 2 x 10^{39} erg s^{-1} (0.2 - 4.0 keV) at 10 Mpc. The X-ray detection occurred within several days of the first detection of radio emission from the supernova (Sramek, van der Hulst and Weiler 1980, also Weiler et al. 1982). The near-simultaneous radio and optical fluxes corrected for assumed absorption were $\sim$ 2 mJy at 4.8 x 10^9 Hz and 35 mJy at 5.5 x 10^{14} Hz, respectively. The X-ray data are equally compatible with a thermal spectrum with kT $>$ 0.5 keV or a power law with energy index $>$ -3. Fifty days later the X-ray source was marginally detected at $\sim$ 1/2 its flux at discovery.

There are various possible emission mechanisms which could give rise to the observed flux (see also Pacini, Panagia and Salvati 1981, and Chevalier 1981b). An order-of-magnitude analysis of the requisite parameters favours either thermal emission from a shock-heated shell of circumstellar material (Chevalier 1981a, Fransson 1981) or inverse-Compton emission arising from the scattering of the copious supply of optical photons on the relativistic electrons responsible for the radio emission. The latter mechanism has the virtue of requiring no additional components (such as the circumstellar material needed for a thermal mechanism), and in fact, inverse-Compton emission is inevitable unless the magnetic field in the radio emitting region is $\gg 1$ gauss. The equipartition value is

$$B_{eq} = 0.4 \ v_9^{-6/7} \ \text{gauss,}$$

where v_9 is the expansion velocity of the emitting region in units 10^9 cm s^{-1}. For $v_9 = 1$, and a photospheric temperature of $\sim 10^4$K, the predicted inverse-Compton X-ray flux is $F_c \sim 0.1f$ μJy, where f is a geometric dilution factor for the photospheric photons and we have assumed that the electron energy distribution function is a power law of index -2 (see Canizares, Kriss and Feigelson 1982 for details). This is remarkably close to the observed flux quoted above. The energies of the electrons that Compton scatter optical photons to X-ray photons are very similar to those that cause the observed radio flux.

Acknowledgements

We thank our colleagues at MIT, GSFC and elsewhere whose work we have quoted here. This work was supported in part by NASA contract NAS-8-30752 and grant NAG-8389. CRC thanks the Institute of Astronomy, Cambridge University and the Royal Society for their hospitality and support during the writing of this manuscript.

References

Arnett, W.D. 1978. Ap.J., 219, 1008.
Arnett, W.D. 1982. This volume.
Becker, R.H., Boldt, E.A., Holt, S.S., Serlemitsos, P.J. and White, N.E. 1980a. Ap.J. (Letters), 237, L77.
Becker, R.H., Holt, S.S., Smith, B.W., White, N.E., Boldt, E.A. Mushotzky, R.F. and Serlemitsos, P.J. 1979. Ap.J. (Letters) 234, L65.
Becker, R.H., Holt, S.S., Smith, B.W., White, N.E., Boldt, E.A., Mushotzky, R.F. and Serlemitsos, P.J. 1980b. Ap.J. (Letters), 235, L5.
Becker, R.H., Szymkowiak, A.E., Boldt, E.A., Holt, S.S. and Serlemitsos, P.J. 1980c. Ap.J. (Letters), 240, L33.
Blandford, R.D. 1982. This volume.
Canizares, C.R., Clark, G.W., Markert, T.H., Bery, C., Smedira, M, Bardas, D., Schnopper, H. and Kalata, K. 1979. Ap.J. (Letters), 234, L33.
Canizares, C.R., Kriss, G.A. and Feigelson, E.D. 1982. Preprint Ap.J.(Letters), 253, in press.
Canizares, C.R. and Winkler, P.F. 1981. Ap.J. (Letters), 246, L33.

Caswell, J.L. and Lerche, I. 1979. Mon.Not.R.astr.Soc., <u>187</u>, 201.

Charles, P.A., Kahn, S.M., Mason, K.O. and Tuohy, I.R. 1981. Ap.J. (Letters), <u>246</u>, L121.

Chevalier, R.A. 1977. Ann.Rev.Astron.Astrophys., <u>15</u>, 175.

Chevalier, R.A. 1981a. Ap.J.,. in press.

Chevalier, R.A. 1981b. Paper presented at this conference.

Clark, D.H. and Caswell, J.L. 1976. Mon.Not.R.astr.Soc., <u>174</u>, 267.

Cowie, L.L., McKee, C.F. and Ostriker, J.P. 1981. Ap.J., in press.

Culhane, J.L. 1977. In <u>Supernovae</u>, ed. D.N. Schramm (Dordrecht: Reidel), 29.

Davison, P.J.N., Culhane, J.L. and Mitchell, R.J. 1976. Ap.J. (Letters), <u>206</u>, L37.

Dopita, M.A. 1981. Private communication.

Duley, W.W. 1980. Ap.J. (Letters), <u>240</u>, L47.

Fabian, A.C., Willingale, R., Pye, J.P., Murray, S.S. and Fabbiano, G. 1980. Mon.Not.R.astr.Soc., <u>193</u>, 175.

Fransson, C. 1981. Astron.Astrophys., in press.

Giacconi, R. <u>et al</u>. 1979. Ap.J., <u>230</u>, 540.

Gorenstein, P. and Tucker, W.H. 1976. Ann.Rev.Astron.Astrophys. <u>14</u>, 373.

Gronenschild, E.H.B.M. and Mewe, R. 1981. Preprint submitted to Astron.Astrophys.

Gull, S. 1973. Mon.Not.R.astr.Soc., <u>161</u>, 47.

Gull, S. 1975. Mon.Not.R.astr.Soc., <u>171</u>, 263.

Holt, S.S. 1981. In <u>X-ray Astronomy</u>, eds. G. Setti and R. Giacconi (Dordrecht: Reidel).

Holt, S.S., White, N.E., Becker, R.H., Boldt, E.A., Mushotzky, R.F., Serlemitsos, P.J. and Smith, B.W. 1979. Ap.J. (Letters), <u>234</u>, L65.

Hutcheon, R.J., Pye, J.P. and Evans, K.D. 1976. Mon.Not.R.astr. Soc., <u>175</u>, 489.

Itoh, H. 1979. Pub.Astron.Soc.Japan, <u>31</u>, 541.

Kahn, S.M. and Blissett, R.J. 1980. Ap.J., <u>238</u>, 417.

Kahn, S.M., Charles, P.A., Bowyer, S. and Blissett, R.J. 1980. Ap.J. (Letters), <u>242</u>, L19.

Kamper, K. and van den Bergh, S. 1976. Ap.J. (Suppl.), <u>32</u>, 351.

McKee, C.F. 1974. Ap.J., <u>188</u>, 335.

McKee, C.F. 1982. This volume.

McKee, C.F. and Cowie, L.L. 1975. Ap.J., <u>195</u>, 715.

McKee, C.F., Cowie, L.L. and Ostriker, J.P. 1978. Ap.J., <u>219</u>, L23.

McKee, C.F. and Ostriker, J.P. 1977. Ap.J., <u>218</u>, 148.

McKenzie, D.L., Landecker, P.B., Broussard, R.M., Rugge, H.R., Young, R.M., Feldman, U. and Doshek, G.A. 1980. Ap.J., <u>241</u>, 409.

Mewe, R. and Gronenschild, E.H.B.M. 1981. Preprint, submitted to Astron.Astrophys.

Meyer, J.P. 1979. Proc. 16th Int.Cosmic Ray Conf. - Kyoto, $\underline{2}$, 115.

Murray, S.S., Fabbiano, G., Fabian, A.C., Epstein, A. and Giacconi, R. 1979. Ap.J. (Letters), $\underline{234}$, L69.

Pacini, F., Panagia, N. and Salvati, M. 1981. Paper presented at this conference.

Parkinson, J.H. 1975. Solar Phys., $\underline{42}$, 183.

Petre, R., Canizares, C.R., Kriss, G.A. and Winkler, P.F. 1981. Preprint submitted to Ap.J.

Pravdo, S.H., Becker, R.H., Boldt, E.A., Holt, S.S., Rothschild, R.E., Serlemitsos, P.J. and Swank, J.H. 1976. Ap.J. (Letters), $\underline{206}$, L41.

Pravdo, S.H. and Smith, B.W. 1979. Ap.J. (Letters), $\underline{234}$, L195.

Pravdo, S.H., Smith, B.W., Charles, P.A. and Tuohy, I.R. 1980. Ap.J. (Letters), $\underline{235}$, L9.

Pye, J.P., Pounds, K.A., Rolf, D.P., Seward, F.D., Smith, A. and Willingale, R. 1981. Mon.Not.R.astr.Soc., $\underline{194}$, 569.

Raymond, J.C. and Smith, B.E. 1977. Ap.J. (Suppl.), $\underline{35}$, 419.

Raymond, J.C. and Smith, B.W. 1979. Private communication.

Reynolds, S.P. and Chevalier, R.A. 1981. Ap.J., $\underline{245}$, 912.

Savage, B.D. and Bohlin, R.C. 1979. Ap.J., $\underline{229}$, 136.

Seward, F.D. 1982. This volume.

Sgro, A.G. 1975. Ap.J., $\underline{197}$, 621.

Shull, J.M. 1977. Ap.J., $\underline{215}$, 805.

Shull, J.M. 1978. Ap.J., $\underline{226}$, 858.

Shull, J.M. 1981a. Ap.J. (Suppl.), in press.

Shull, J.M. 1981b. Paper presented at this conference.

Smith, B.W., Mushotzky, R.F. and Serlemitsos, P.J. 1979. Ap.J., $\underline{227}$, 37.

Sramek, R., van der Hulst, J.M. and Weiler, K.W. 1980. IAU Circ. No. 3557.

Szymkowiak, A.E. 1980. In Proc. Texas Workshop on Type I Supernovae, ed. J.C. Wheeler (Austin: University of Texas), 32.

Waite, C.M. and Winkler, P.F. 1981. Bull.Am.Phys.Soc., $\underline{26}$,61.

Walker, A.B.C., Jr, Rugge, H.R. and Weiss, K. 1974. Ap.J., $\underline{192}$, 169.

Weaver, T.A., Zimmerman, G.B. and Woosley, S.E. 1978. Ap.J., $\underline{225}$, 1021.

Weiler, K.W., Sramek, R.A., van der Hulst, J.M. and Panagia, N. 1982. This volume.

Wheeler, J.C. 1982. This volume.

Winkler, P.F. 1979. In Proc.HEAO Science Symp., eds. C. Dailey and W. Johnson (NASA CP-2113), 244.

Winkler, P.F., Canizares, C.R., Clark, G.W., Markert, T.H., Kalata, K. and Schnopper, H.W. 1981a. Ap.J. (Letters), $\underline{246}$, L27.

Winkler, P.F., Canizares, C.R., Clark, G.W., Markert, T.H. and
    Petre, R. 1981b. Ap.J., <u>245</u>, 574.
Woosley, S.E. and Weaver, T.. 1980. Paper presented at Tenth
    Texas Symp. on Relativistic Astrophysics. Baltimore.
Woosley, W.E. and Weaver, T.A. 1982. This volume.

EINSTEIN OBSERVATIONS OF SUPERNOVA REMNANTS

F.D. Seward

Harvard-Smithsonian Center for Astrophysics

The Einstein Observatory has detected and mapped soft X-ray emission from ~ 40 galactic SNRs. The detectors covered the energy range 0.2-4.0 keV, and data were obtained between November 1978 and April 1981. The observatory was operated by a science Consortium comprising Harvard-Smithsonian Center for Astrophysics (CFA), Columbia Astrophysics Laboratory (CAL), Goddard Space Flight Center (GSFC), and Massachusetts Institute of Technology (MIT). During 25% of the time the observatory was operated as a national facility, and used by guest observers.

The instrumentation consisted of an X-ray telescope, two imaging detectors, and two spectrometers. The imaging proportional counter (IPC) was capable of spatial resolution of 1 arc minute and a coarse measurement of the energy of each photon detected. The high resolution imager (HRI) had a spatial resolution of a few arc seconds, but yielded no spectral information. Details are given by Giacconi et al. (1979).

In addition to the images, high resolution X-ray spectra have been measured for most remnants and will be given in other talks in this conference. This paper presents X-ray maps of nine supernova remnants ordered approximately by age and by prominence of central objects. These data were taken by CFA and Guest Observers and illustrate the great variety of shapes and forms encountered.

Each supernova remnant has its own personality. It is obvious that some are very young and some ancient. However, it is somewhat difficult to arrange the various observations to illustrate an evolutionary sequence. Often emission from one

M. J. Rees and R. J. Stoneham (eds.), Supernovae: A Survey of Current Research, 519–528.
Copyright © 1982 by D. Reidel Publishing Company.

feature is so strong that fainter features are obscured. The
presence of compact objects manifests itself in various ways, and
sometimes distorts the normal shell-like features of the remnant.

Each image contains a scale giving dimension and is displayed
on a grid of 512x512 pixels. Each dot seen represents X-ray
counts in one pixel. In faint diffuse regions, pixels have
either 0 or 1 count, and the image is grainy. This fine speckle
is due to counting statistics and is not a characteristic of the
source.

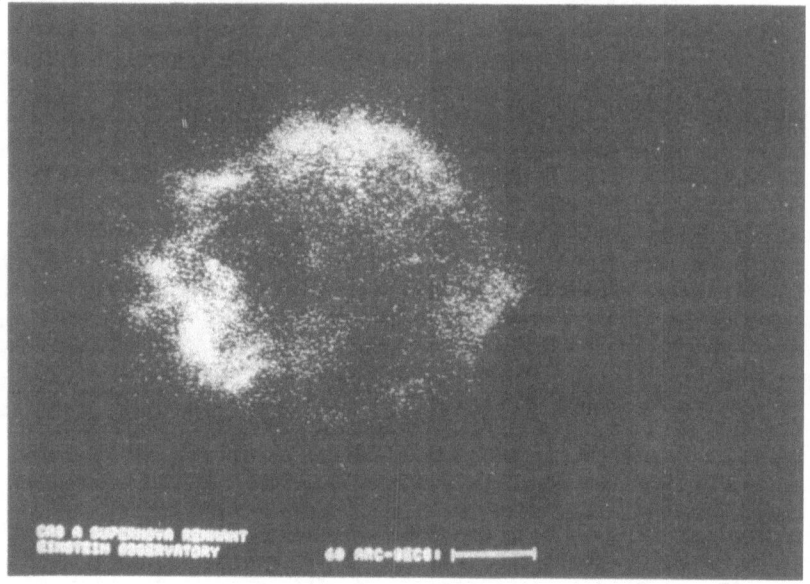

The youngest known galactic remnant is Cas A which is thought
to be the result of an unobserved SN in the late 17th century.
Because it is young, the nature of the stellar explosion should
influence the properties of the remnant more than irregularities
in the interstellar medium. The remnant is 4 arcmin in diameter
and the distance is thought to be ~ 3 kpc. This Einstein HRI
picture was a 10 hour exposure which took one day of satellite
time to obtain. The brightest features are probably regions
containing ejecta from the stellar explosion. To a first
approximation these features form a ring, usually interpreted as
limb brightening from an approximately spherical shell of
emission. The distribution of material around the rim of the
shell is not uniform, and the material is clumpy. A second,
fainter, larger shell surrounds the ring of bright ejecta. This
is hot material behind a shock propagating in the interstellar
medium. There is no sign of a compact object at the centre of
the remnant.

Here, for the first time, one can separate the supernova
ejecta from the shock and derive properties of each. A complete
analysis of the Cas A image has been published by Fabian <u>et al</u>.
(1980), and the mass derived for the remnant is ~ 15 M$_\odot$.

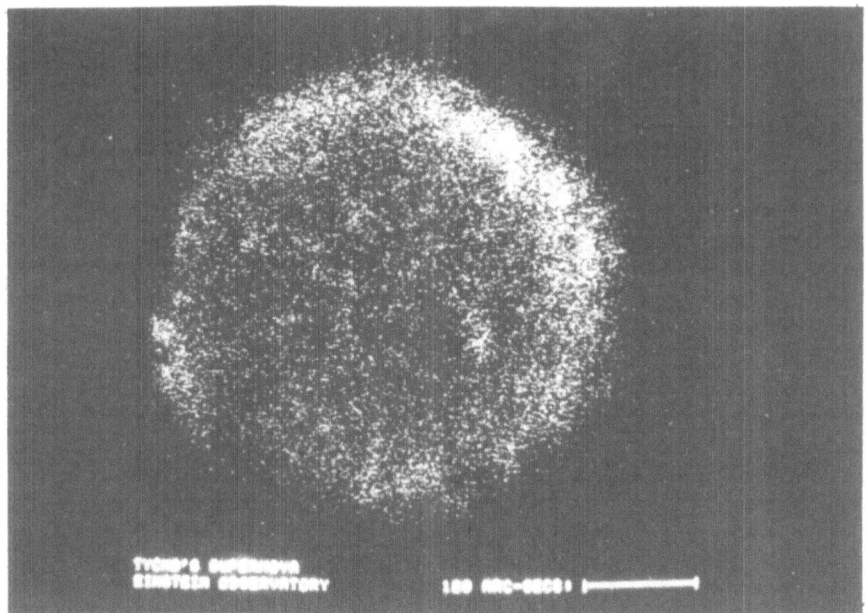

The second youngest remnant is Kepler's. Shown here is the
third youngest, Tycho's SNR.

This remnant of the supernova of 1572 is 8 arc min in
diameter and ~ 3 kpc distant. This is an eleven hour exposure
with the HRI showing almost circular limb brightening but with a
discontinuity in the southeast. The emission of X-rays is again
not uniform around the rim and is clumpy. The bright clumps are
thought to contain the ejecta, and again faint emission from hot
material behind a shock can be seen just outside this bright rim.
This shock is clearest in the northwest but is difficult to see
in this illustration. There is no emission detected from a
central compact object.

The ejecta mass and the dynamic state of the SNR can be
determined using the measured X-ray surface brightness and a
model describing the structure.

The outer shock can be assumed to be spherical. The material
behind the shock forms a shell of thickness ~ 1/10 the radius.
This material is of normal cosmic abundance, and has been
snowplowed and heated by the shock to a temperature of a few keV.

The shock is driven by stellar debris or ejecta which forms a
shell just behind the shock-heated material and which has been
greatly enhanced in Si group elements. The ejecta is clumpy, and
the size distribution of clumps can be estimated from the X-ray
image.

A high resolution X-ray spectrum measured by Becker et al.
(1980) shows very strong line emission for Si group elements and
enrichment of the hot plasma in the SNR is required. The
spectrum of Pravdo et al. (1980) requires some plasma at high
temperature. Thus, this model, with enriched ejecta and a high
temperature behind the shock, satisfies the spectral
requirements.

Numerical results are quite dependent on the assumption of
thermal equilibrium and on values chosen for distance, column
density of ISM, and composition of the ejecta. First results show
a few solar masses both in ejecta and in swept-up ISM. A
calculation is in progress to refine these numbers.

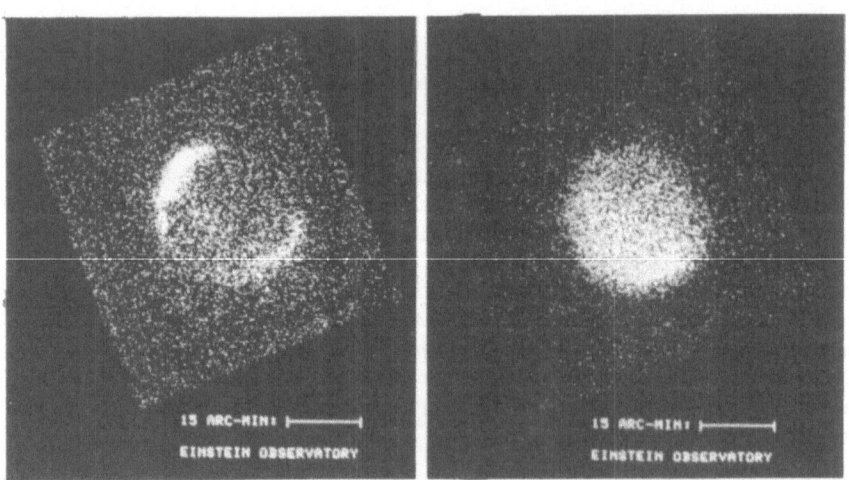

The Remnant of the Supernova of 1006 as seen here in a 2-hour
exposure with the IPC, at both high and low energies. This

remnant, at a distance of ~ 1 kpc, has a diameter of 30 arc min, the same apparent size as the Moon. When viewed at X-ray energies between 1 and 4 keV, it appears as a limb brightened, incomplete shell with maximum emission coming from two opposed quadrants. Regions containing ejecta can no longer be separated from hot gas behind the shock, and the spectrum shows no strong line emission. SNR 1006 has been described in detail by Pye et al. (1981). If a neutron star was created by the supernova, the absence of a detectable X-ray flux from a point source in the centre of this remnant implies a surface temperature less than 10^6 K.

At low energies, 0.2-1 keV, the appearance of SNR 1006 is different. It no longer looks like a limb brightened shell, but the emission is diffuse and approximately fills the interior of the remnant. This change of appearance of SNR 1006 with X-ray energy is so far unique among the remnants imaged by Einstein. Several older remnants, however, appear as diffuse amorphous sources, and hot gas in the interior is brighter than material associated with a shock or blast wave.

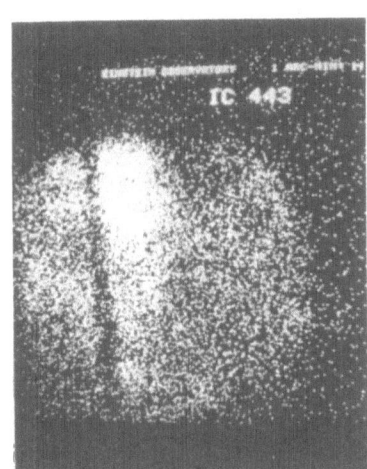

IC443 is 48' in diameter, at a distance of ~ 2 kpc, and is of unknown age, but on the order of 10^4 yrs. It is old enough so that the characteristics of the remnant are determined by the interstellar medium rather than by the properties of the explosion. Three one-hour IPC pointings were necessary to obtain an undistorted picture of the entire remnant. This figure shows one pointing centred on the western limb. Linear shadows are caused by support ribs of the IPC window. The appearance of IC443 does not change with energy over the range of the IPC. Emission is, to a first approximation, from hot gas filling the entire remnant and is brightest from the northeast quadrant, a region characterized also by bright optical filaments. This bright NE region probably indicates a high density of interstellar material. Evidence for a weak shock can be seen in the northwestern limb, but the remnant is not generally limb brightened.

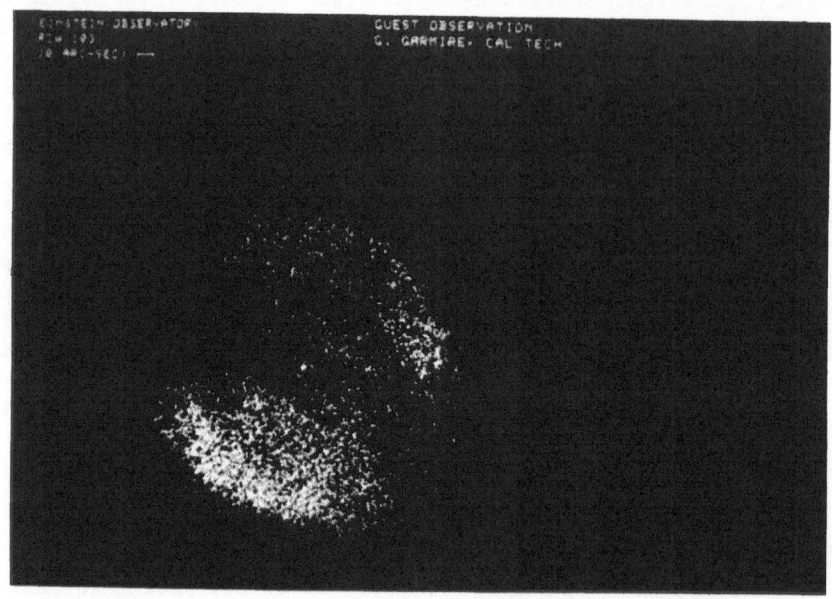

The SNR RCW103, at a distance of ~ 3 kpc, shows two strong optical emission regions with about the same orientation as the diffuse X-ray lobes seen in this 2-hour HRI observation by Guest Observers, Tuohy and Garmire (1980). The X-ray remnant is shell-like with a diameter of ~ 8 arcmin and with a faint point X-ray source almost exactly at the centre. Only a few photons were detected from this point source, not enough to search for pulsations. The optical counterpart is at present unidentified. The shell-like source is apparently unperturbed by the presence of the compact object. The appearance of the SNR is quite similar to that of Tycho's remnant implying that it is young, but no optical SN was observed. The declination, however, is −50° so observations from Europe or China were not possible.

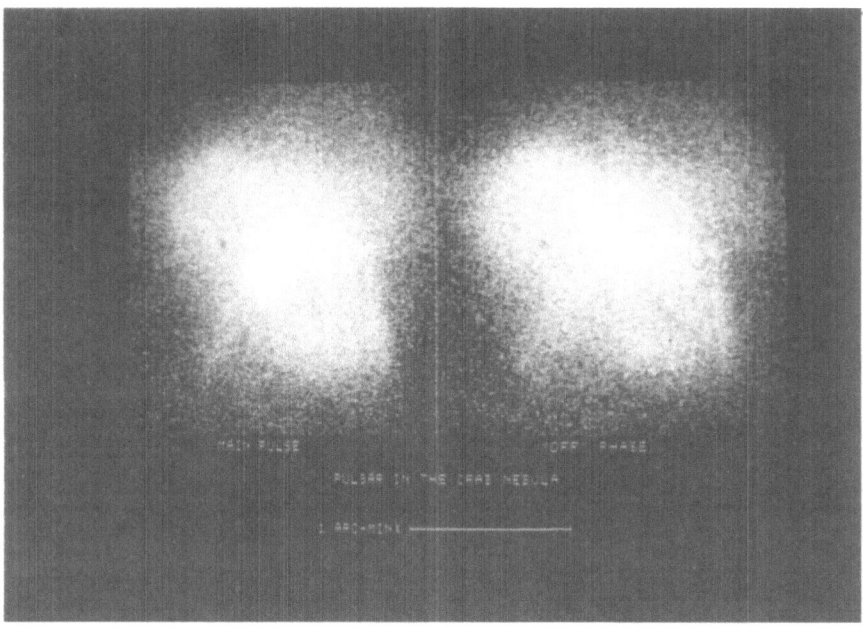

The Crab Nebula is the remnant of the supernova of 1054. It is approximately the same age as the remnant of 1006 but is completely different in appearance. It contains a pulsar with a period of 0.03 sec - the fastest pulsar known - and ninety percent of the X-rays come from a source of synchrotron radiation of diameter ~ 2 arcmin which is powered by the pulsar. This illustration shows an HRI exposure gated so that the pulsar is 'on' in the left field and 'off' in the right. Only the synchrotron source is visible when the pulsar is 'off' showing that radiation from the pulsar is almost completely pulsed. Note that the centroid of the diffuse source is ~ 10″ NW of the pulsar, a curious phenomenon which is not yet understood. Any shell-like emission which might come from the Crab is obscured by this strong synchrotron source, which incidentally has the highest surface brightness of any diffuse X-ray source observed by Einstein.

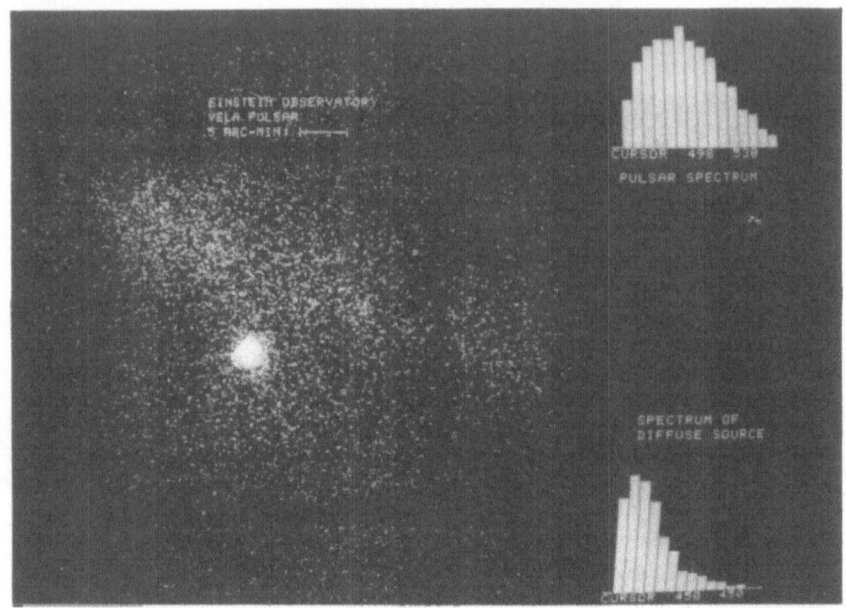

The Vela supernova remnant is enormous, old, and close. The distance is ~ 400 pc and it subtends an angle of 5.5°. The IPC picture on the left shows both soft diffuse emission, probably thermal, from the interior of the large Vela remnant, and the Vela pulsar, which appears as a point source. The lower spectrum is soft and from the diffuse area above the pulsar. The upper, harder, spectrum is from the pulsar.

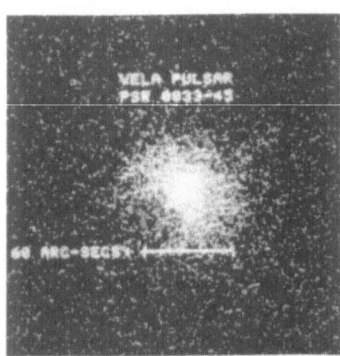

The high resolution image at left is centred on the Vela pulsar. It shows a point source surrounded by weak diffuse emission of diameter ~ 1 arcmin. This diffuse emission is probably a synchrotron source analogous to that of the Crab nebula but with much, much lower surface brightness and total intensity. No X-ray pulsations have yet been found from the point source associated with the radio pulsar, which has a period of ~ 0.09 seconds. Nevertheless, the presence of the diffuse source gives confidence that the Crab diffuse source is not a unique phenomenon.

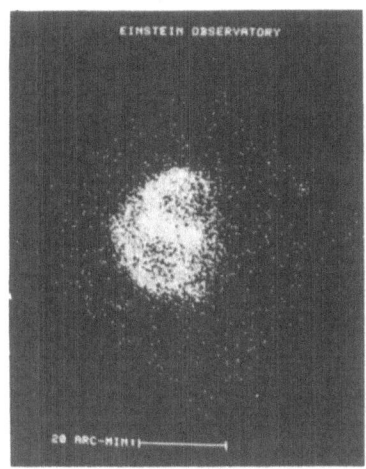

This unusual remnant, G109.1–1.0, was discovered by Guest Observers, Gregory and Fahlman (1980). It shows a nicely symmetrical shell to the east and no trace of diffuse emission on the western side. Furthermore, the strong point source in the middle is an X-ray pulsar with period 3.48 seconds (Gregory and Fahlman 1981), a rotation rate 100 times slower than that of the Crab pulsar. The diffuse emission east of the pulsar might be only thermal emission in the shell but, because it points right at the pulsar, there is a good possibility that it is associated with a jet or beam phenomenon.

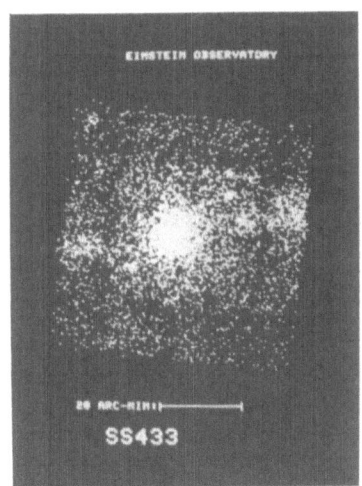

Diffuse X-ray emission from W50 as seen with a 5 hour IPC exposure by Guest Observers, E. Seaquist and W. Gilmore. Two lobes of extended emission are visible at the edges of the IPC field of view. These lobes are on a line connecting the long axis of W50 with the star SS433, which is the bright central X-ray source. The relativistic beams from this star apparently deposit energy and indirectly produce diffuse X-rays on their way to the shell of W50. There they make two distinct bulges on the surface of the shell. This is described more fully by Seward et al. (1980) and the radio shell by Geldzahler et al. (1980).

In summary, then, the young SNRs appear as strongly limb brightened shells and both ejecta from the exploded star and hot material behind the shock propagating in the ISM can be observed. In these cases, the mass of material in the different components can be calculated and the dynamic state of the SNR determined.

Compact objects can be seen in some remnants as point sources with X-ray pulsed fraction ranging from 0 to ~ 100%. Diffuse X-rays powered by high energy particles associated with the compact source are also detected. In at least 2 cases, the effects of the compact source obscure or distort emission from the shell or shock. Finally, some remnants show no X-ray evidence for central compact sources.

REFERENCES

Becker, R.H., Holt, S.S., Smith, B.W., White, N.E., Boldt, E.A., Mushotzky, R.F. and Serlemitsos, P.J. 1980. Astrophys.J. (Letters), 235, L5.
Fabian, A.C., Willingale, R., Pye, J.P., Murray, S.S. and Fabbiano, G. 1980. Mon.Not.R.astr.Soc., 193, 175.
Geldzahler, B.J., Pauls, T. and Salter, C.J. 1980. Astr.Astrophys., 84, 237.
Giacconi, et al. 1979. Astrophys.J., 230, 540.
Gregory, P. and Fahlman, G. 1980. Nature, 287, 806.
Gregory, P. and Fahlman, G. 1981. Nature, 293, 202.
Pravdo, S.H., Smith, B.W., Charles, P.A. and Tuohy, I.R. 1980. Astrophys.J.(Letters), 235, L9.
Pye, J.P., Pounds, K.A., Rolf, D.P., Seward, F.D., Smith, A. and Willingale, R. 1981. Mon.Not.R.astr.Soc., 194, 569.
Seward, F., Grindlay, J., Seaquist, E. and Gilmore, W. 1980. Nature, 287, 806.
Tuohy, I. and Garmire, G. 1980. Astrophys.J., 239, L107.

AN X-RAY SURVEY OF SUPERNOVA REMNANTS
IN THE LARGE MAGELLANIC CLOUD

David J. Helfand and Knox S. Long

Columbia Astrophysics Laboratory, Columbia University

The last supernova (SN) observed in the Galaxy occurred ~375 years ago and it is clear that progress in SN research would be glacial at best if we were limited to the study of galactic events. Although the first extragalactic SN was discovered over 100 years ago and the detection rate reached ~20 yr^{-1} by 1960, it was not until several years later that the study of supernova remnants (SNR) was first extended beyond the Galaxy with the identification of three SNR in the Large Magellanic Cloud (LMC) (Mathewson and Healy 1964). In the last decade, 14 SNRs have been catalogued in the LMC; similar numbers have been found in both M31 and M33 (D'Odorico, Dopita, and Benvenuti 1979; Blair, Kirshner, and Chevalier 1981), and recently, a few very young remnants have been detected in more distant galaxies (Kronberg, Bierman, and Schwab 1981; Weiler *et al.*, and Winkler *et al.* 1982).

There are significant advantages in studying extragalactic SNR over those in the Galaxy. First, their distances are typically much better determined than are those of galactic remnants. (Though rarely noted, this is true for SN as well – despite all the debate concerning the distance to 1972e, for example, the range of estimates is smaller than that of Kepler's SN which occurred ~10^3 times closer.) For the LMC discussed herein, our adopted distance of 55 kpc (Bok 1966) is probably accurate to ~10%, allowing us to derive luminosities and linear diameters of remnants to an accuracy unknown in galactic work. A second advantage of the LMC is the small amounts of both foreground and intrinsic absorbing material which obscures the remnants; the total column density of neutral hydrogen varies across the face of the Cloud from $N_H \sim 2 \times 10^{20}$ cm^{-2} to $N_H \sim 2 \times 10^{21}$ cm^{-2}.

M. J. Rees and R. J. Stoneham (eds.), Supernovae: A Survey of Current Research, 529–534.
Copyright © 1982 by D. Reidel Publishing Company.

Finally, the external perspective allows us to search the entire
galaxy to a well-defined, limiting sensitivity in any wavelength
region and, thus, to derive supernova rates and galactic distributions far more easily than in the case of our own Galaxy.

Radiation in the X-ray band 0.1-10.0 keV dominates the electromagnetic energy output of a SNR during the first 20,000 years
of its life. This fact, coupled with the vast increase in sensitivity and imaging capabilities offered by the *Einstein* Observatory, **was** among the dominant motivating factors for our recently
completed soft X-ray survey of the LMC. Slightly over two weeks
of the Observatory's 30-month lifetime were employed in obtaining
imaging, spectral, and temporal data on sources in the Cloud.
The primary imaging study employing the imaging proportional counter (IPC - 1° × 1° with ~1' resolution) was carried out by the
authors, while substantial follow-up observations using the high
resolution imager (HRI - 25' × 25' field with ~4" resolution)
were conducted by M. Pakull at MPI, P. Charles at Oxford, J. Hut-
chings at DAO, J. Thorstensen at Dartmouth, and I. Tuohy, M. Dopita, and D. Mathewson at Mt. Stromlo, all of whom are pursuing
programs to elucidate the optical properties of the detected
sources. Moderate resolution spectroscopy ($\Delta E \sim$ 150 eV) on half
a dozen SNR **was** obtained with the solid state spectrometer (SSS)
by Clark *et al.* (1981), while high resolution spectral observations ($\Delta E \sim$ 10 eV) using the focal plane crystal spectrometer
(FPCS) were pursued for the brightest remnants by the MIT X-ray
group (Winkler 1981). Below, we outline the preliminary results
related to SNRs in the Cloud and summarize the current status of
the numerous on-going programs designed to exploit this wealth
of new data.

The IPC survey consisted of 103 overlapping pointings of
~2000 s duration and covered a total of ~34 square degrees, including essentially the entire Cloud (see Long, Helfand, and
Grabelsky 1981, for details). For an instrument count rate to
luminosity conversion factor of 1 IPC count = 2×10^{37} ergs s^{-1}
[0.15 to 4.5 keV] (appropriate for a thermal bremsstrahlung
spectrum with kT $\sim$ 0.5-1.0 keV and an absorbing column density
of $N_H \sim 5 - 10 \times 10^{20}$ cm^{-2}), the survey was complete to a limiting
$L_x \sim 4 \times 10^{35}$ ergs s^{-1} and contains a number of objects down to
$L_x \sim 1 \times 10^{35}$ ergs s^{-1}. A total of 97 discrete sources were detected. For each one we have obtained a position accurate to
$\lesssim 1'$, a luminosity, a spectral hardness parameter, and some crude
information on spatial extent and temporal variability.

A large fraction of these sources are SNR. All 13 of the 14
previously identified remnants (Mathewson and Clarke 1973) which
fell within the IPC survey area were detected. Their luminosities
range over a factor of 400 - from $L_x = 2 \times 10^{35}$ ergs s^{-1} for N120
to 8.3×10^{37} ergs s^{-1} for N132D. In addition, six sources were

detected coincident with members of a list of SNR candidates derived from the Hα survey of Davies, Elliot, and Meaburn (1976),
confirming them as remnants; one of these has a diameter of
nearly 6', corresponding to a linear size of ~90 pc.

Nearly half of the 97 detected sources have been reobserved
with the HRI to obtain improved positions and to search for extended morphology. Of 14 HRI fields containing previously unidentified sources, seven contain clearly extended sources which
are, in all probability, also SNR. One of these is the remaining
Mathewson and Clarke remnant N11L, but the remaining six were
totally unrecognized as remnants in previous optical and radio
surveys of the Cloud. They cover a range of luminosities
(9×10^{35} ergs s^{-1} < L_X < 1×10^{37} ergs s^{-1}) and diameters (9 pc
to 45 pc) similar to those for the previously identified remnants.
Several have recently been detected in optical emission line
images obtained by Mathewson et $al.$ (1981), confirming our desig-
nation of these sources as SNR. These results bring to 26 the
number of X-ray emitting SNR known in the LMC.

For the remaining sources, we have used the temporal, spectral,
and spatial information available for each to estimate the number
of additional remnants lurking in our list of 97 objects, ~75 of
which are Cloud members. We have compared the spatial distribution of X-rays in each source with the point source response
function of the IPC: 11 of the as-yet-unidentified objects show
some evidence of extent. While many of the sources are weak and
the test may not always be definitive, we feel that most of these
sources are also SNR. This view is supported by the facts that
all members of this group observed more than once were constant
(while 13 other sources did show evidence of variability) and
that the range and mean value of the spectral hardness parameter
for this group is indistinguishable from those of the confirmed
remnants. Finally, the distribution of spectral parameters of
the 40 survey sources for which no other clue as to their iden-
tity is available, shows a pronounced peak at the mean value of
this parameter for the known remnants. Noting this, and extra-
polating from the fraction of extended sources discovered in our
initial HRI sample, we estimate that at least another dozen of
the unidentified sources in the survey will ultimately be found
to be SNR. Thus, our $Einstein$ LMC survey contains a total of ~50
X-ray emitting SNR.

A brief notation of some of the work in progress serves to
underscore the value of this sample in future SNR research:

a) The survey has established that previous radio and optical
   samples were substantially incomplete and that SN rates
   derived from them must also be in error. However, our
   HRI maps also indicate that the remnant diameters used

in these early rate calculations were systematically
underestimated. Taking both effects into consideration,
we derived a preliminary SN rate of 1 per 110 to 350 yr
(Long, Helfand, and Grabelsky 1981). A final value must
await the analysis of the remaining HRI data and the systematic search for large-diameter remnants (D > 4')
which could have gone undetected in the survey analysis.

b) The mass of the Galaxy is ~10 times that of the LMC.
   Assuming roughly constant SNR production per unit mass,
   the large number of SNRs in the LMC implies that the number of remnants in the Galaxy may also be underestimated.
   It is important to note in this regard that most of the
   remnants that we have detected in the LMC have very soft
   spectra and would be undetectable as X-ray sources at
   distances of 5 to 10 kpc in the galactic plane. In addition, three of the six newly discovered HRI remnants are
   underluminous by a factor of ~20 in the radio regime.
   Thus, the currently accepted galactic SNR production rate
   (~1 per 80 yr) based solely on existing radio catalogs
   (e.g., Caswell and Lerche 1979) is most likely too low.
   The detection of at least two new X-ray selected remnants
   in the small fraction of the galactic plane surveyed by
   Einstein (Gregory and Fahlman 1980; Markert *et al.* 1981)
   supports this conclusion.

c) Optical spectra of the three radio underluminous remnants
   mentioned above show exclusively Balmer series emission
   lines (Tuohy *et al.* 1981). Two examples of such remnants
   are known in the Galaxy (Tycho and SN1006) and some filaments in the Cygnus Loop also fall into this category.
   These spectra have been interpreted as arising from the
   interaction of a high velocity collisionless shock with
   cold interstellar material (Chevalier and Raymond 1978).
   The line profiles exhibit a narrow spike arising from
   the excited cold gas superimposed on a broad plateau
   ($\Delta v \gtrsim 10^3$ km s^{-1}) resulting from the emission of atoms
   which have undergone charge exchange with the fast particles in the shock. The delineation of a class of such
   objects in the complete X-ray sample will be most useful
   in understanding the relative importance of the progenitor, the type of explosion, and the surrounding medium,
   in producing such remnants. The suggestive link between
   these pure Balmer spectra and low radio surface brightness may well be important in modelling particle acceleration and radio emission mechanisms for SNR in general.

The distributions of X-ray luminosities and diameters in
our sample of LMC remnants are far from consistent with those
expected for an ensemble of uniform explosions in a uniform

interstellar medium (Helfand *et al.* 1981). The synthesis of
the observed distribution from a mixed population of SN progenitors exploding in a multiphase medium, along with the detailed
modelling of the radio, optical, and X-ray spectral and spatial
properties of individual members of the sample will provide some
of the outstanding problems of SNR research for some time to
come. The implications of such work for models of SN, the evolution of their remnants, and the structure of the interstellar
medium is likely to be profound.

The authors wish to acknowledge the support of the National
Aeronautics and Space Administration under Contract NAS 8-30753.
This is Columbia Astrophysics Laboratory Contribution No. 211.

REFERENCES

Blair, W. P., Kirshner, R. R., and Chevalier, R. A. 1981,
      Astrophys. J., 247, pp. 879-893.
Bok, B. J. 1966, Ann. Rev. Astron. Astrophys., 4, pp. 95-99.
Caswell, J. L., and Lerche, I. 1979, Mon. Not. R. Astron. Soc.,
      187, pp. 201-214.
Chevalier, R. A., and Raymond, J. C. 1978, Astrophys. J. (Letters),
      225, pp. L27-L31.
Clark, D. H., Tuohy, I. R., Long, K. S., Szymkowiak, A. E.,
      Dopita, M. A., Mathewson, D. S., and Culhane, J. L. 1981,
      Astrophys. J., submitted.
Davies, R. D., Elliot, K. H., and Meaburn, J 1976, Mon. Not.
      R. Astron. Soc., 81, pp. 89-103.
D'Odorico, S., Dopita, M. A., and Benvenuti, P. 1979, Astron.
      Astrophys. Suppl., 40, pp. 67-78.
Gregory, P. C., and Fahlman, G. G. 1980, Nature, 287,
      pp. 805-806.
Helfand, D. J., Long, K. S., Dopita, M. A., Tuohy, I. R., and
      Mathewson, D. S. 1981, in preparation.
Kronberg, P. P., Bierman, P., and Schwab, F. R. 1981, Astrophys.
      J., 246, pp. 751-755.
Long, K. S., Helfand, D. J., and Grabelsky, D. A. 1981,
      Astrophys. J., in press.
Markert, T. H., Lamb, R. C., Hartmann, R. C., Thompson, D. J.,
      and Bagnami, G. F. 1981, preprint.
Mathewson, D. S., and Clarke, J. N. 1973, Astrophys. J.,
      180, pp. 725-730.
Mathewson, D. S., and Healy, J. R. 1964, in IAU Symposium No. 20,
      The Galaxy and the Magellanic Clouds, eds. F. J. Kerr and
      A. W. Rogers (Canberra: Australian Academy of Science),
      pp. 126-135.
Mathewson, D. S., Ford, V. L., Dopita, M. A., Tuohy, I. R.,
      Long, K. S., and Helfand D. J. 1981, Astrophys. J.,
      submitted.

Tuohy, I. R., Dopita, M. A., Mathewson, D. S., Long, K. S.,
    and Helfand, D. J. 1981, Astrophys. J., submitted.
Winkler, P. F. 1981, private communication.
Winkler, P.F., Canizares, C.R., Markert, T.H., and Szymkowiak, A.E.
    1982, this volume.

SUPERNOVAE AND THE ORIGIN OF THE SOLAR SYSTEM

Donald D. Clayton

Rice University, Houston

1. INTRODUCTION

Several lines of astrophysical reasoning link supernovae to the origin of the solar system. These cannot be equally addressed in this article, so I begin by itemizing the major ones:

(a) Supernova Connections

(1) Abundances of the nuclides at the time solar system formed, including live radioactive nuclei.

(2) Supernova triggering of collapse of solar cloud.

(3) Origins of isotopic anomalies in solar-system samples, primarily from carbonaceous meteorites.

(4) Sputtering of interstellar dust, with attendant chemical and isotopic evolution of the ISM.

Some degree of controversy and conjecture accompanies each of these topics, and I will concentrate on the issues in these controversies. Many of them have intensified or been discovered because of the discovery of the isotopic anomalies in meteorites, so I will have to present some of the meteoritic facts.

The liveliest and scientifically most important of these controversies will revolve around the following points of view, which are intentionally stated in extreme polarized form:

M. J. Rees and R. J. Stoneham (eds.), Supernovae: A Survey of Current Research, 535–564.
Copyright © 1982 by D. Reidel Publishing Company.

(b) Controversies

(1) Interstellar cloud collapse leading to formation of the
solar system was
    (i)   a spontaneously occurring and typical result of the
                    evolution of the ISM.
    (ii)   caused by a sudden dynamic overpressure,
                    specifically a nearby supernova.

(2) The physical state of matter in the solar nebula (and/or
disk) just prior to its aggregation into outer solar-system
bodies (asteroids, meteorite parent bodies, comets, outer
planets) is that of a
    (i)   hot atomic gas that thermally condenses upon
                    cooling into sequentially less refractory
                    (chemically-bound forms), leading to abundance
                    fractionation patterns according to volatility, as
                    observed in meteorites.
    (ii)   cold dusty molecular mix that contains a memory of
                    abundance fractionation according to volatility.

(3) Isotopic anomalies exist in differing meteoritic samples
because of
    (i)   inhomogeneous admixture of anomalous matter into
                    the solar cloud from a neighbouring supernova.
    (ii)   differing isotopic patterns in different chemical
                    forms of pre-existing ISM dust.

    A very big issue for astronomy lurks in these controversies.
If outer solar system bodies aggregated directly from cold-cloud
dust and molecules, and if the unusual chemical and isotopic
patterns within those bodies reflect differing admixtures of
components that are routinely present in the ISM, then features
of the chemical evolution and physical state of the Galaxy are
remembered by those objects. This is in effect a new field of
astronomy, a kind of cosmic archaeology, based on what I like to
call 'cosmic chemical memory'. If, on the other hand, the
chemical memory within the cloud is erased by an epoch of high
temperature (T > 2000 K) that vaporizes the dust and dissociates
most molecules, then what the meteorites record is the subsequent
condensation process from the hot gas accompanied by an
inhomogeneous admixture of isotopically anomalous matter from an
adjacent supernova.

2. EXTINCT ^{26}Al

    The discovery that many, but not all, Al-rich minerals within
CaAl-rich inclusions in carbonaceous meteorites contain excess
^{26}Mg demonstrates that ^{26}Al was once alive. Anorthite minerals
(CaAl$_2$Si$_2$O$_8$) are especially good, as shown in Figure 1 (Lee et

al. 1977). Mg does not occur in the anorthite formula, but it is present in trace amounts in variable concentration. What Figure 1, and many other similar results, shows is that in anorthite there is a roughly uniform field of ^{26}Mg, called $^{26}Mg*$, say, and that it is diluted by mixture with normal Mg $(^{24,25,26}Mg)$. The mixture generates the straight line. From the slope of that straight line, one concludes that the concentration of $^{26}Mg*$ in the uniform field of Al is

$$^{26}Mg*/Al \simeq 5 \times 10^{-5} \tag{1}$$

The most straightforward <u>chemical</u> explanation would be that live ^{26}Al existed in the ratio (1) at the time this anorthite mineral formed.

This simple chemical explanation leads to an astrophysical complication of first magnitude, however, The ^{26}Al is produced in supernova explosions at the approximate level

$$P(^{26}Al)/P(^{27}Al) \simeq 2 \times 10^{-3} \tag{2}$$

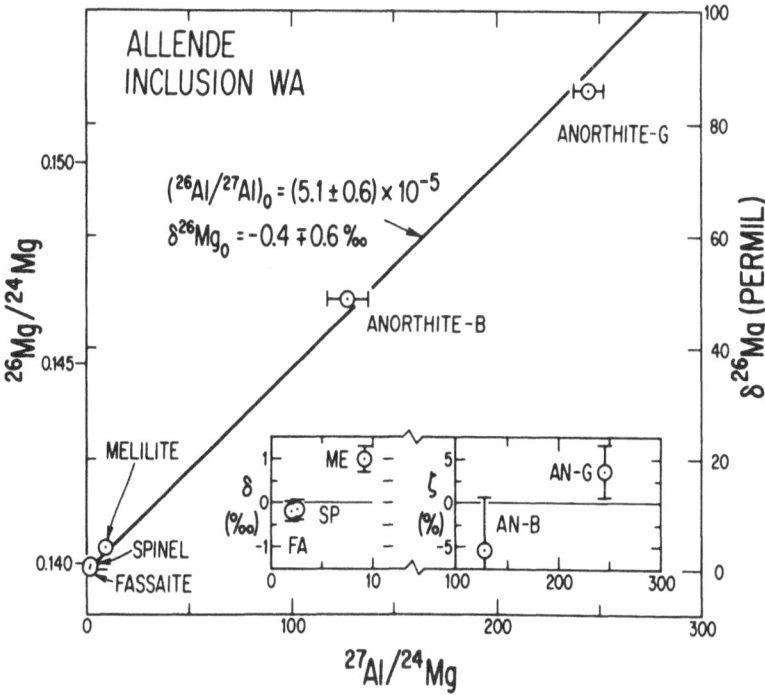

Figure 1. In several different mineral separates of the inclusion WA from the Allende meteorite, the isotopic ratio $^{26}Mg/^{24}Mg$ correlates linearly with the Al/Mg chemical abundance ratio (Lee <u>et al</u>. 1977).

Assuming that Galactic nucleosynthesis occurred at a smoothly decreasing rate over a 5-10 G yr period prior to the formation of the solar system would lead one to expect an _average_ concentration of ^{26}Al in the ISM given approximately by

$$\langle \frac{^{26}Al}{^{27}Al} \rangle \simeq \frac{\tau(^{26}Al)}{10^{10} \; yr} \; \frac{P(^{26}Al)}{P(^{27}Al)} \; \frac{P(t=t_o)}{\overline{P}} \tag{3}$$

With $\tau(^{26}Al) = 1.1 \times 10^6$ yr, the ratio (3) cannot be expected on average to exceed $(^{26}Al/^{27}Al) = 10^{-7}$ and could be still smaller if the nucleosynthesis rate at the time of the solar system formed ($t = t_\bullet$) is much less than its average rate $\overline{P}$ over prior Galactic history (as stellar abundance data suggests). So the needed ^{26}Al concentration is x500 or more of the average concentration expected. Actually, the situation is even much more severe than that, for cosmochronological studies have long been interpreted to show that the small amount of ^{129}I ($\tau = 23$ Myr) inferred to have been present in meteorites initially

$$^{129}Xe*/^{127}I \simeq 10^{-4} \tag{4}$$

can be explained if the solar cloud was isolated from average galactic nucleosynthesis about 10^8 yr before the meteorites formed. In that case the ^{26}Al would have decayed by another $\exp(-10^8 yr/10^6 \; yr)$ - i.e. to effectively zero.

How is this severe conflict to be resolved? The commonly accepted picture is that a special event of nucleosynthesis happened when the solar system formed, injecting fresh ^{26}Al, but not having synthesized ^{129}I (Cameron and Truran 1977: Clayton and Schramm 1978). Such a coincidence would be implausible unless the supernova pressure wave _caused_ the solar cloud to collapse. The idea is that during the first 10^6 yr of a supernova remnant, its internal pressure exceeds that of the ambient ISM, so that if it wrapped around a dense small cloud, it would begin accelerating the outside of that cloud inward, thereby giving it an initial infall momentum that would push it to the point of gravitational takeover. Figure 2 (Lattimer et al. 1978) illustrates this schematically.

Cameron and Truran (1977) concluded that about 3% of normal solar Al was synthesized in the trigger (at least in the region where the CaAl-rich inclusions, called, CAI, formed); and because this is comparable to the 5% excesses in ^{16}O found in the same inclusions, they reasoned that pure ^{16}O was admixed as well. The major problems of this model are not hard to find:

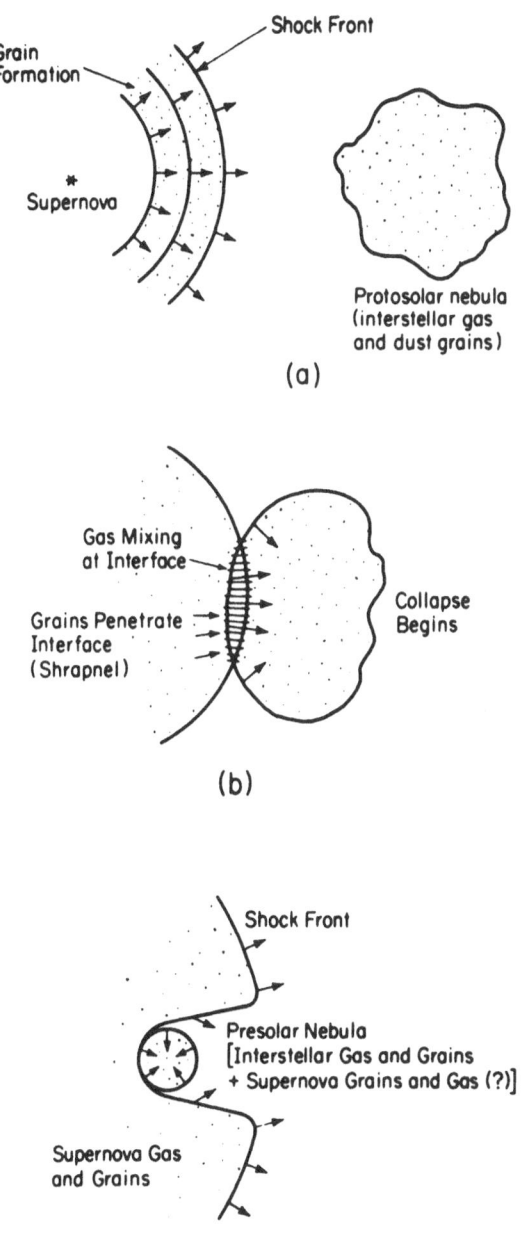

Figure 2. A supernova trigger may cause solar collapse by surrounding it and compressing it (Cameron and Truran 1977). The same event may admix the isotopic anomalies that are found in meteorites, especially ^{26}Al and ^{16}O. Figure from Lattimer et al. (1978).

(i)   Some Al-rich minerals have excess $^{26}Mg*$, and some
            do not. Observed positive concentrations also
            vary, so either the admixture was spatially
            inhomogeneous or the inclusion minerals formed at
            greatly different times, despite the fact that they
            all end up in the same meteorite (Allende).

(ii)   No other isotopic anomaly is found in the $^{26}Mg*$
            bearing minerals. A 3% admixture of anomalous
            stuff from a special supernova should leave
            detectable anomalies in many elements, whose
            isotopic compositions are measured to one part in
            10^4 or better.

(iii)   Supernova ejecta tend to be stopped in a mere skin
            depth of a cold cloud, and no model exists for
            forming the inclusion minerals there, or
            alternatively, for transporting the SN ejecta into
            the cold cloud for later inclusion growth in, say,
            a disk.

(iv)   The few inclusions for which isotopic anomalies
            exist in virtually all heavy elements that have
            been measured (Ca, Ti, Sr, Ba, Nd, Sm) do not have
            excess $^{26}Mg*$ in their Al-rich minerals.

These and other problems have seemed to me impossibly
difficult for a supernova-injection model, so I have developed
one based on cosmic chemical memory (Clayton 1977, 1978, 1981).

3. SUNOCONS

     Let us cast the $^{26}Mg*$ and ^{16}O anomalies into a different
light - one that predicts them to be routinely present at all
times and places in the ISM. First review the concept of the
supernova condensate, or SUNOCON.

     The SUNOCON idea is that when the hot dense interior of a
supernova expands into the ISM, its temperature will fall to the
point that thermal condensation (near equilibrium) will occur
within the expanding gas before it mixes with the ISM. The
physical idea as a source of refractory dust was advanced by
Hoyle and Wickramasinghe (1970), and the idea that meteoritic
anomalies were due to SUNOCONS was advanced by Clayton (1975a,b:
1977, 1978).

     Figure 3 illustrates SUNOCON expectations for explosive
carbon burning, which may occur near $T = 2 \times 10^9$ K and $\rho = 10^4 g$
cm^{-3} in a shocked supernova mantle. The radiation pressure
greatly exceeds the particle pressure as this zone expands owing

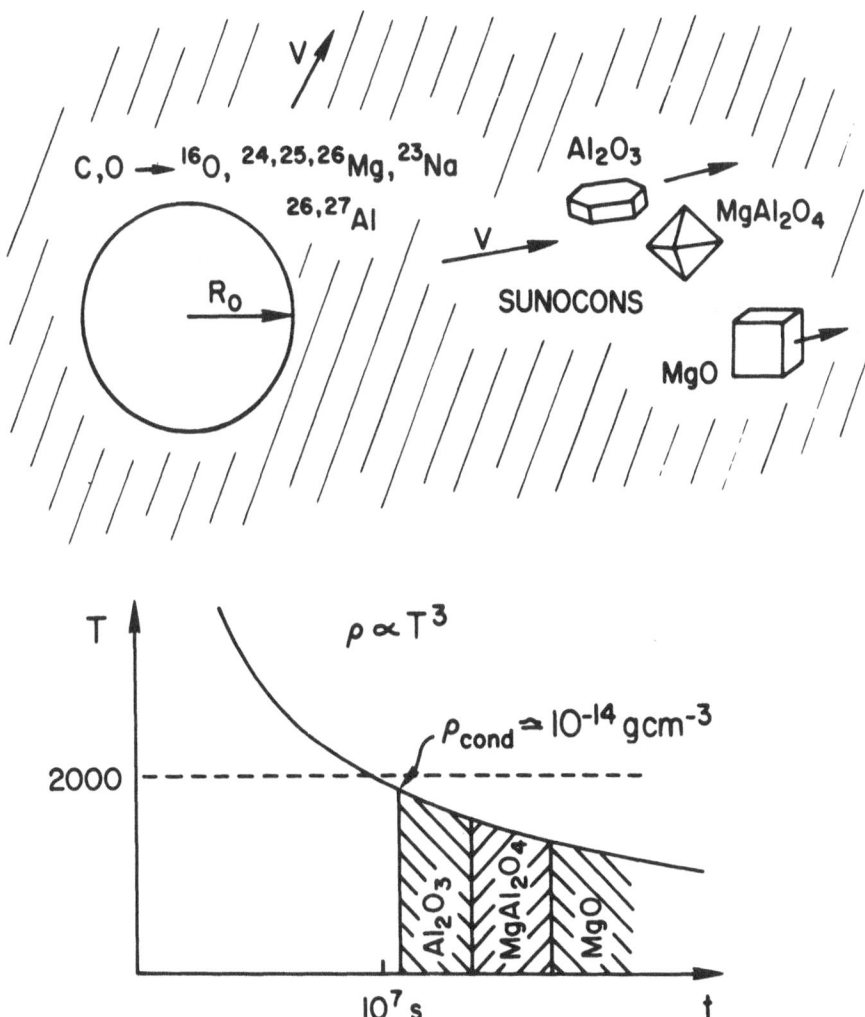

Figure 3. SUNOCON formation in the expansion of the shell that has explosively burned and ejected carbon. Oxygen is pure ^{16}O. The first condensate is corundum, Al_2O_3, which also condenses the ^{26}Al that later changes to a ^{26}Mg spike in Al. Condensation begins after a few months when the remnant is only about $10^2 AU$ in size.

to its overpressure.   The entropy is also radiation dominated, and because the expansion is virtually adiabatic

$$\left(\frac{\rho}{\rho_0}\right) = \left(\frac{T}{T_0}\right)^3 \qquad\qquad (4)$$

Assuming this adiabat to be valid down to T = 2000 K, below which dust particles exist in equilibrium, the density at which dust condensation commences is about

$$\rho_{condensation} = 10^{-14} \text{ g/cm}^3 \qquad\qquad (5)$$

in a gas that is, in this example, dominated by ^{16}O (pure), ^{20}Ne, $^{24,25,26}Mg$, ^{23}Na, $^{26,27}Al$ and some Si.

A large literature exists on the thermal equilibrium of dust in a gas of solar composition, largely because meteoritic chemists have long believed that a cooling gaseous solar nebula was the physical manner in which dust first appeared in the solar system (e.g. Grossman and Larimer 1974). These calculations identified Al_2O_3, $MgAl_2O_4$ and either $MgSiO_3$ or MgO as being the temporal sequence of thermal condensation. Lattimer et al. (1978) performed such calculations for a wide variety of supernova zones and compositions. This basic sequence is illustrated in Figure 3.

The first conclusion is that this Al-rich dust is ^{16}O-rich, corresponding to the meteoritic minerals. Such dust need comprise only 5% of ISM dust to account for the 5% enrichment of ^{16}O in aggregates of refractory dust. It explains in a natural way why a big isotopic difference is expected between the ISM oxygen locked up in refractory dust ($<$ 20% of all O) and the remaining gaseous oxygen. On this view, a scoop of refractory interstellar dust would at any time and any place be ^{16}O-rich owing to its SUNOCON component.

If the supernova interior expands such that $\rho \propto R^{-3}$, then (4) yields

$$\left(\frac{T}{T_0}\right) = \left(\frac{R}{R_0}\right)^{-1} \qquad\qquad (6)$$

during the expansion from initial radius $R_0 = 10^9$ cm. If the interior shell flows outward at v = 10^3 km s^{-1}, then R = 10^8 t cm at long t(sec). Thus the temperature falls by the required ratio $(T/T_0) = 10^{-6}$ when R = 10^{15} cm, which requires about 10^7 seconds. SUNOCONs evidently precipitate when the remnant is a few months old and when its interior has a size R = 10^{15} cm = 100 AU – too small to have mixed with the surrounding interstellar matter.

It will clearly be of astrophysical interest to attempt to observe the SUNOCONs in young supernova remnants, as Dwek and Werner (1981) have begun. One would, most simply, look for an increase in infrared luminosity several months after the outburst: but the problem of distinguishing the dust from the infrared photosphere will be formidable. The most important point is a geometrical one. Because the SUNOCONs flow out at 10^2-10^3 km s^{-1} (probably), the SUNOCON containing volume may be much smaller than the remnant. The outer layer, which has rushed off at 10^4 km s^{-1}, is destructive to dust. Infrared emission from the new SUNOCONs may be restricted to the inner few percent of the remnant volume.

The very SUNOCONs of Figure 3 suggest an alternate interpretation of the ^{26}Mg excesses that is not burdened by the problems of a supernova injection. It was predicted (Clayton 1975b) before ^{26}Al decay was established (Fig. 1) that Al-rich interstellar dust should carry a large ^{26}Mg* isotopic excess

$$\langle ^{26}Mg*/^{27}Al \rangle_{SUNOCON} = 10^{-3} \tag{7}$$

In exact anology, the same paper predicted a big ^{41}K excess in Ca-rich interstellar dust, because ^{41}K is synthesized as the radioactive ($\tau_{1/2} = 1.3 \times 10^5$ yr) progenitor ^{41}Ca, which will condense with Ca as an early SUNOCON condensate. Early attempts to find ^{41}K excess in Ca-rich inclusions failed (Begemann and Stegmann 1976), but a recent attempt has been dramatically successful — the ^{41}K/^{39}K ratio in Ca-rich minerals being about ten times greater than normal! (Huneke et al. 1981). The success of this prediction gives, by time honoured principles of the scientific method, added weight to the prediction of ^{26}Mg excesses in Al-rich minerals. On this view (cosmic chemical memory) these excesses exist because they were created in SUNOCONs and because the later chemistry by which the observed minerals were fused was unable to completely extract and equilibrate those carried anomalies. At the same time, the discovery of ^{41}K excess places an added burden on the picture of a 'live injection' by a supernova because of the shortness of the ^{41}Ca half life. Before discussing the SUNOCON model in more detail, however, other connections to the supernova phenomenon must be made.

4. REFRACTORY ELEMENT ISM DEPLETION

SUNOCON chemistry places interstellar gas-depletion patterns in a new light. An old puzzle is that Ca and Al depletion factors from the interstellar gas are on average much larger than those of other elements, especially Mg, which is not much different chemically than Ca. Typically 99% of Ca and Al remain in interstellar dust, whereas a smaller and more variable fraction (50%-90%) of Mg is in the dust. This is a severe problem if atoms are to be depleted from the ISM gas, bcause Ca must then 'stick' at least 10 times better than Mg, an improbable requirement. The problem is worsened by the realization that sputtering of interstellar dust by supernova shock waves in the ISM is the major destruction mechanism for dust. But sputtering yields of Ca and Mg should not differ by a factor 10 on chemical grounds alone. Draine and Salpeter (1979, eq. 31) present the following semiempirical formula for the normal incidence sputtering yield

$$Y = A \frac{(\varepsilon - \varepsilon_0)^2}{1 + (\varepsilon/30)^{4/3}}; \qquad \varepsilon < \varepsilon_0 \qquad\qquad (8)$$

where the ion energy E is expressed in terms of the chemical binding energy U_0

$$\varepsilon = \eta \frac{E}{U_0} , \quad \eta = 4 \, M_P M_T (M_P + M_T)^{-2}$$

and M_P and M_T are projectile and target mass, and the threshold ε_0 is unity or 4η, whichever is greater. Because U_0 and η differ little for Mg and Ca in common solids, this sputtering-yield difference is not great enough to account for having 10-100 times greater fraction of the Mg in the gas than of the Ca.

The SUNOCONs have, however, a predictable geometry in which the first condensates form a core surrounded by a somewhat less refractory mantle. For example, the Al_2O_3 and $MgAl_2O_4$ Al-rich SUNOCONs of Fig. 3 will be surrounded by the later deposition of $MgSiO_3$ and MgO, even though much of the Mg may condense in newly nucleated Al-poor particles (Clayton 1980). The composite SUNOCONs illustrated in Fig. 4 have the property that the magnesian silicate mantles must be sputtered away before the CaAl-rich SUNOCONs are exposed. This extra feature of shielding, combined with requirement that essentially all Ca and Al leave stars in condensed form, probably enables the extra depletion of Ca and Al to be understood.

This last conclusion is made all the more significant by the fact that it is CaAl-rich minerals in meteoritic inclusions that are the sources of the most dramatic isotopic studies.

More general chemical effects of low-energy sputtering in the wake of supernova produced interstellar shocks may be found in the inverse correlations of abundance and volatility of the elements among diverse meteoritic samples, however. The CaAl-rich inclusions, enriched by about a factor 15 in a dozen or so elements of very small thermal vapor pressure (Al, Ca, Ti, rare earths, etc.), are only the most extreme samples of such

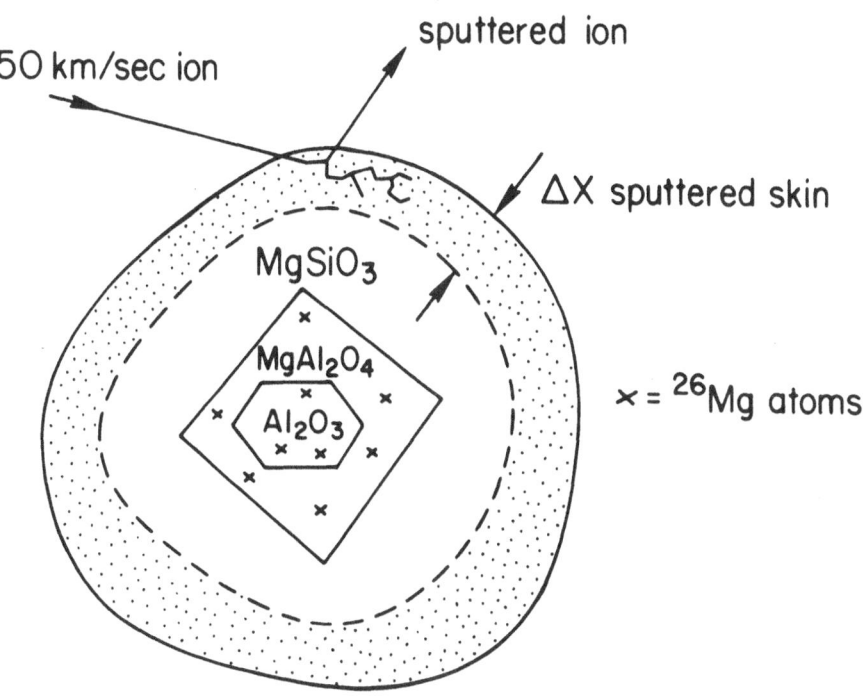

Figure 4. As first supernova condensates, the CaAl-rich SUNOCONS are surrounded by Mg and Si bearing mantles, here shown as $MgSiO_3$, which protect Ca and Al from interstellar sputtering. These ideas explain extra CaAl gas deletion in the ISM. Extinct ^{26}Al and ^{41}Ca reside in cores.

abundance correlations. It is specifically to find a physical mechanism for producing such correlations that meteoritic chemists have long believed that the solar system began as a hot gas. At high T, only the most refractory solids exist in this picture, and they may somehow be gathered and saved (in part) to be variably mixed with later lower-T condensates (as they are called in the context of that picture). These condensates also depend on the binding U_0 of the atoms to solids through the law of mass action. In low-energy sputtering $\varepsilon \approx \varepsilon_0$, eq. (8) reduces to

$$Y(E) = 2A\varepsilon_0(\varepsilon - \varepsilon_0) \qquad\qquad\qquad (9)$$

which also takes the form $Y = (k/U_0)(E - E_T)$ where the sputtering threshold E_T also depends linearly on U_0. This effect can be quite large for low-energy sputtering. It causes the dust to become depleted in volatiles as it is sputtered. Sputtered atoms are later accreted by dust in proportion to their numbers and surface areas, so that volatility patterns can be mapped onto a grain-size spectrum by interstellar sputtering. This physics was seen first as playing a role in interstellar depletion patterns (Barlow and Silk 1977), and only later was the connection to the grain-size spectrum (Clayton 1980) and to meteoritic abundance patterns recognized (Clayton 1978, 1981). Although this connection was in general a very unpopular one in the meteoritic community, a leading figure of that community has recently turned to this astrophysical history (Wood 1981).

It seems that there is now a very good chance of reading the historical effect of supernova shock waves on ISM particles by seeking consistency with the aggregation of ISM matter into meteorites. This can be powerful new mode of ISM studies and of aggregation studies of ISM matter.

5. OXYGEN AND 'FUN'

Figure 5 is a mineral map of a rather typical type B Allende inclusion. In the great majority of Allende inclusions, the separated refractory minerals have isotopic compositions lying along the line EA in the 'three-isotope plot' for oxygen shown in Figure 6. Because it is the property of such coordinate systems that a mixture of two different isotopic compositions generates one lying along the straight line connecting the two reservoirs, it is universally felt that they are generated by a mixture of oxygen at A with oxygen at E. That the line has very nearly slope m = 1 leads to the interpretation that A is ^{16}O-rich with respect to E (which is very close to terrestrial oxygen). Several things need be said about the mineral separates on this line:
     (i)   The most ^{16}O-rich separates, which are spinel ($MgAl_2O_4$), cluster near A but do not spread beneath it,

suggesting that A is a natural limit to the ^{16}O-richness of naturally occurring bulk samples. Spinels contained within pyroxene are isotopically indistinguishable from spinels contained within melilite!

(ii)   The most ^{16}O-poor separates, which are the melilite separates ($Ca_2Al_2SiO_7$ mixed in varying degrees with $Ca_2MgSi_2O_7$), cluster near E, which is near common terrestrial O, suggesting that E was another endpoint of the mixing line AE.

(iii)   The distribution of minerals along the AE line is not understood, but the common hunch is that the entire bulk inclusions began with isotopic compositions near A, and that E is an ambient gas that diffusively exchanged with the oxygen initially contained in the bulk inclusion. This exchange, so the conjecture goes, was very easy with melilite and anorthite, much more difficult with pyroxene, and virtually nonexistent with spinel. Thus spinels stay near their starting point, pyoxenes were exchanged (mixed) partway up the line, and melilite were almost totally exchanged (E = 'Exchange oxygen'). Test of this model awaits differential-exchange experiments with the relevant minerals.

(iv)   The ^{16}O-richness of the aggregate of refractory dust at A is understandable as reflecting the SUNOCON component of the dust, which makes it ^{16}O-rich with respect to the gas.

(v)   It appears that an injection from a neighbouring supernova is in much trouble, because how should that ^{16}O gas get preferentially into spinel, less into pyroxene, and almost not at all into melilite? The same exchange scenario is needed. But why then did ^{16}O get only into CaAl-rich inclusions?

(vi)   The inclusions were not molten after the exchange with oxygen at E, or else the separate minerals would have the same (virtually) isotopic composition. Whether they were molten before in a formation stage when they lay near point A of Fig. 6 is controversial. Did the inclusion minerals exist before exchange with E, or did the process of exchange occur along with the formation of the minerals in their present form? The answers strongly influence the interpretation of excess ^{26}Mg.

Two other inclusions shown in Fig. 6 have different oxygen, magnesium, and silicon than the fifty or so inclusions on EA. These are the two FUN inclusions EK1-4 and C1, whose separated minerals behave similarly but fall along the lines EK and EC. A third FUN inclusion named HAL is not shown but has similar behaviour. Several points are evident:

(i)   These are not admixtures of ^{16}O, because the lines do not have slope m = 1.

(ii)   The most ^{17}O-poor spinels (also the unshown hibonite from HAL) line up along the line AKC having slope 1/2. Apparently some process first prepared aggregates that differed by mass-dependent isotopic fractionation (therefore having slope m = 1/2), and these aggregates later exchanged oxygen with the same pool at E, generating those mixed lines.

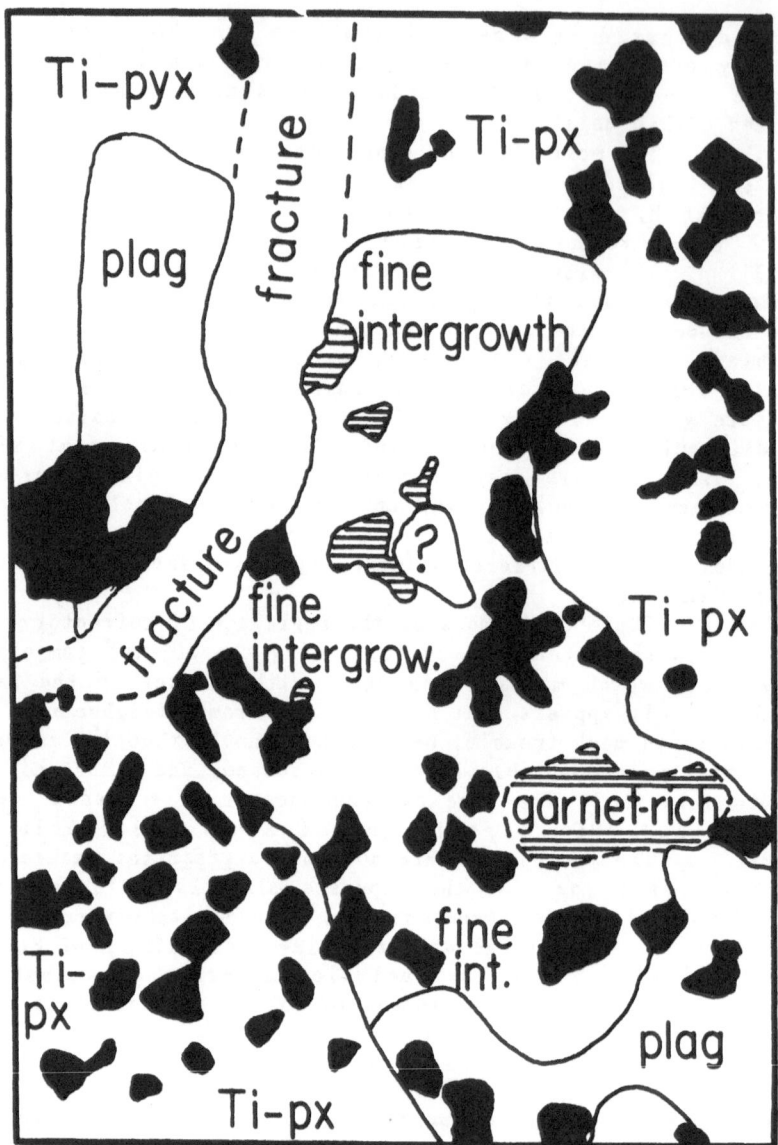

Figure 5. Mineralogical key of complex area of Allende inclusion. Ti-Al-pyroxene areas at upper right and lower left contain abundant spinel inclusions with a deep orange luminescence, while the area at upper left contains only a few spinel grains. Both plagioclase grains are highly altered as indicated by mosaic texture of the luminescence, high opacity in transmitted light, and patchy extinction under crossed polars. Ill-defined vertical features indicate twin lamellae. A conspicuous fracture has matching margins. The remaining region contains complex, fine-grained areas rich in grossular garnet and

(iii)   FUN inclusions do not look different than the
common inclusions. They are the same kind of object.
(iv)   The Mg and Si in the FUN inclusions is also mass
fractionated, as shown in Figure 7. Their isotopic compositions
fall almost on a line of slope m = 1/2 as well.
(v)   FUN inclusions do not have excess ^{26}Mg.
(vi)   FUN inclusions do have isotopic anomalies in all
heavy elements that have been studied with high resolution.
Figure 8 (next section) shows Sm in EK1-4. The measurements
agree beautifully with a deficiency in the average s-process
abundances, which means an excess in r-abundances (by only 0.4%)
that has, to high accuracy, the isotopic shape of the average
r-process events that have contributed to nucleosynthesis of bulk
solar system matter. This is not what would be expected from an
adjacent supernova, especially one that does not synthesize ^{129}I!
It suggests instead that r and s ejecta partitioned differently
between dust and gas on the average as they were ejected from
stars, and that a gas/dust separation during subsequent
accumulation processes has played a role in the FUN inclusions.
(vii)   Calcium is not noticeably fractionated even though
Mg and Si are severely so.

The special nature of the FUN inclusions lies in the severe
isotopic fractionation (a few percent per atomic mass unit) and a
compelling reason for isotopic abnormality to accompany the
process responsible for the fractionation. It is my thesis that
different accumulates of interstellar dust that has been heavily
sputtered by supernova shock waves should have the observed
differences. Thermally produced mass fractionation, by contrast,
produces such strong mass fractionation only at low temperatures
where chemistry of Mg and Si is not expected to be active.

Figure 5 (caption continued)

spinels with yellow-orange luminescence. Other conspicuous
phases are melilite (deep-blue) and unidentified phases with
mauve, white-blue, pink, and brownish luminescence, often in
fine-scale irregular intergrowth too dark for mineral
identification in transmitted light. Plagioclase grains in this
photograph have embayed boundaries or cavities. These are filled
with or rimmed by similar grossular-rich, fine-grained
assemblages which are opaque in transmitted light. The
combination of irregular luminescence with embayed boundaries and
spinel inclusions of unusual luminescence colour suggests
replacement of anorthite by the grossular-rich assemblage.
Detailed study is needed to evaluate the chemical changes. Width
0.3 mm. (Hutcheon et al. 1979).

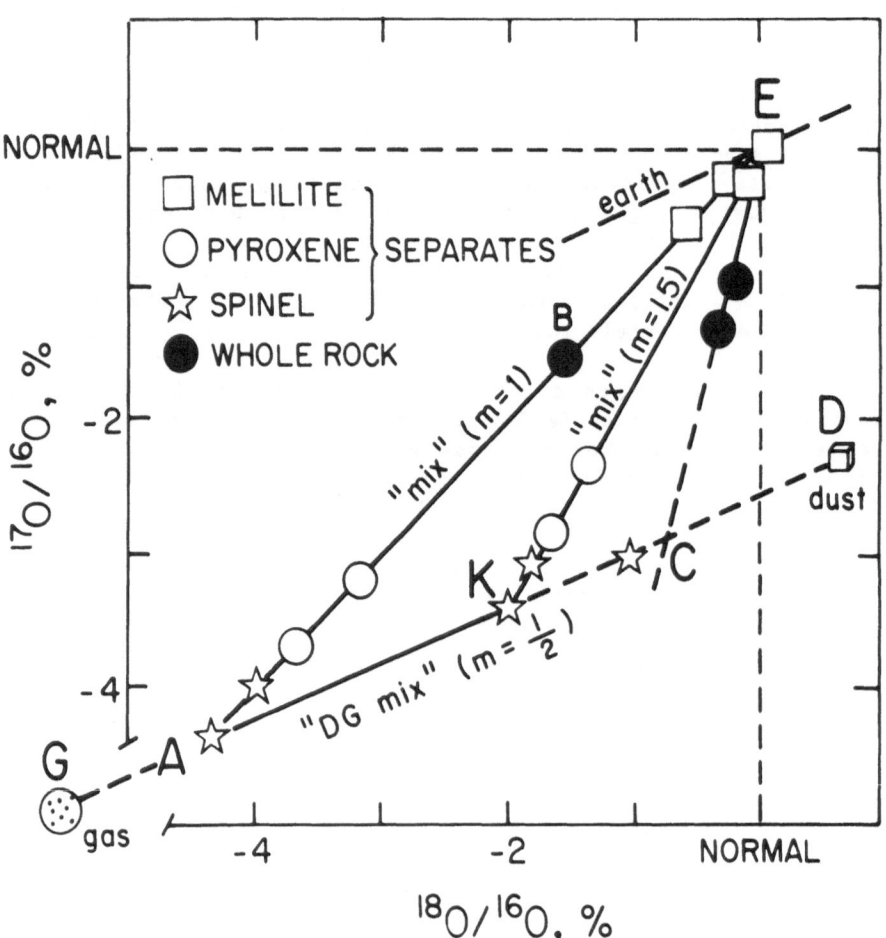

Figure 6. Oxygen 3-isotope plot for Allende inclusions from data
of R. Clayton. Illustrative mineral separates are shown in
formats larger than the experimental errors in routine (about 50)
Allende inclusions AE, and in FUN inclusions EK1-4 on EK and C1
on CE. Exchange oxygen is at E. Sputtered refractory dust is at
D and the atoms sputtered away are at G.

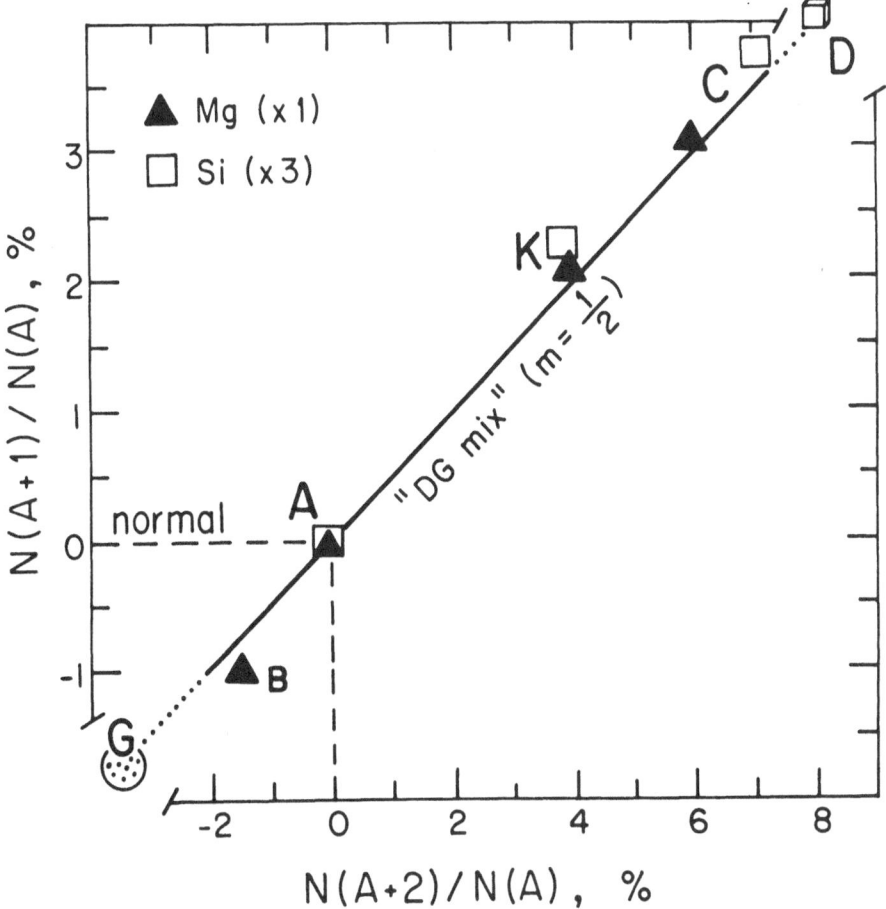

Figure 7. Three-isotope plots for Mg and Si in normal Allende (A) inclusions, in FUN inclusion ED1-4 (K), and in FUN inclusion C1 (C). The linear relation with slope 1/2 indicates some fractionation process is involved. It may actually be a mixing line between heavily sputtered dust (D) and gas atoms (G) sputtered away from the grains.

6. SUPERNOVA-SHOCK SPUTTERING

Rapidly expanding supernova remnants generate a constant barrage of shock waves in the hot ionized interstellar medium. Typical shock velocities of, say 100 km s^{-1}, cause severe sputtering of dust. Although the gas fluid is set rapidly in motion by the passage of the shocks, the high-enertia dust finds itself initially in a 100 km s^{-1} stream of ions, which sputter the dust. Also, the gas behind the shock is so hot ($\sim 10^6$ K) that the thermal velocities of ions postshock are high enough to sputter the dust. This seems to be the major mode of destruction of interstellar dust, as calculations by Dwek and Scalo (1980) and by Draine and Salpeter (1979) have clearly demonstrated. An specially exciting thing is that this process may be studiable by laboratory samples – the CaAl-rich inclusions.

Suppose the refractory parts of grains are layered along the lines of Figure 4, with super-refractory elements (e.g. Ca) being condensed owing to temporal sequence within magnesium silicate mantles. Not only will Ca remain more highly depleted from the gas, but if mass fractionation accompanies the sputtering process, the O, Mg and Si should be affected when Ca is not. This is observed to be the case in FUN inclusions. Eq. (8) can conveniently be used to estimate the mass dependence of the sputtering through the dependence of η on M_T, the target mass. The most effective sputtering ions are He, so, as a specific example, one can evaluate the differential sputtering of ^{24}Mg and ^{26}Mg by 200 eV He ions ($v = 100$ km s^{-1}). With an assumed binding energy $U_0 = 6$ eV for Mg atoms, the ratio of sputtering yields from eq. (8) is

$$Y(^{24}\text{Mg})/Y(^{26}\text{Mg}) = 1.098. \tag{10}$$

This 10% enhancement of this ratio of atoms returned to the gas is large. The sputtering skin depth is made more ^{26}Mg rich until that richness restores the ratio of sputtered atoms to its normal value. One then envisions a ^{26}Mg-rich layer of fixed isotopic composition eating its way into the continuously ablating surface. The skin depth enriched 10% in ^{26}Mg/^{24}Mg is roughly the range of the 200 eV He ion, or R = 12 Å.

If the particles have radius $\underline{a}$ greater than this skin depth, the fraction of the particle in the isotopically enriched skin is $3R/a$ leading to a bulk ^{26}Mg enrichment

$$^{26}\delta = \left[\frac{Y(^{24}\text{Mg})}{Y(^{26}\text{Mg})} - 1 \right] \frac{3R}{a} = 0.035 \ (a/100 \text{ Å})^{-1}. \tag{11}$$

For a particle size 100Å the bulk enrichment (3.5%) is quite comparable to the observed enrichments for the FUN inclusions as shown in Figure 7. Thus, in principle, the different macroscopic accumulates A, K, C in Figures 6 and 7 could be aggregates of different grain size. Differing aggregates can be collected because gas dynamic effects do separate particles according to grain size.

This physical idea for obtaining mass fractionated macroscopic aggregates was first set forth in a paper submitted in 1979 that is only now finding its way into publication (Clayton 1981). If correct it opens the door to the study of the ISM and its history of sputtering and aggregation via the study of meteoritic samples! It will be a new field of astronomy.

Consider for a moment how these fractionation effects might be preserved, and how they might correlate with the nuclear anomalies, a correlation that has puzzled meteoritic science since its discovery (Clayton and Mayeda 1977: Wasserburg et al. 1977). If any atoms are sputtered away and later reaccreted, the bulk mixture must be isotopically normal. Thus if dust Mg is heavy, say, it is necessary to either (1) aggregate dust before the gas has been reaccreted, or (2) accrete the gas on a large number of small particles of another type, carbonaceous say, and then dynamically separate the large grains, by sedimentation say, or (3) accrete sputtered gas along with other gaseous atoms so that grains regrow at different Mg rates than their destruction rates, another subtle grain-size effect. It is clearest to think of option (1), always remembering that some kind of mapping onto grain sizes followed by dynamic sorting may be the actual mechanism for remembering prior gas/dust separations.

Gas/dust separations offer a natural cause for the nuclear anomalies that accompany the FUN inclusions, because the nucleosynthesis of many elements involves different sources for different isotopes. Different isotopic patterns in different chemical forms of dust offer an even wider range of possibilities. One example, that of Sm, will suffice. The s and r contributions to Sm abundances have not originated in the same chemical surroundings. The r-isotopes probably have emerged mostly from deep zones of supernovae near the neutron-star mass cut, where the common chemical elements (e.g. H, O, C, S) are probably nonexistent. The s-isotopes, on the other hand, emerged primarily from outer layers of stars, red giants, or supernovae, where the ambient gas was rich in H, C, O, etc. Not only the initial chemical environs, but also the expansion thermal environment and time scales differed. Therefore, on the average, the r-isotopes and s-isotopes of Sm could not have emerged equally condensed from stars. Any aggregation process that has, as in the FUN inclusions, separated dust and gas will also produce a polarization of the s-nuclei from the r-nuclei. The astonishing success of this conjecture is shown in Figure 8, where an s-process deficiency calculated as boxes is compared to the observed isotopic deficiency (Clayton 1979a). Only for p-process ^{144}Sm does this simple separation not apply, as it should not.

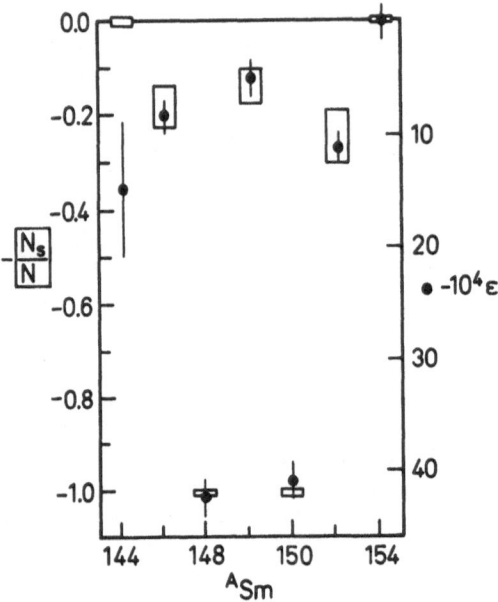

Figure 8. The decomposition of Sm is shown as an s-process deficiency, $(-N_s/N)$, on the left-hand ordinate with rectangle size reflecting uncertainty. The measured anomalies of Lugmair et al. (1978) are plotted on the right-hand ordinate after a choice of isotopic fractionation.

This result and others like it deals a fatal blow to a supernova injection from a single neighbouring event. True, it could admix new r-process nuclei, but those abundances from a single event would not have the shape of the average r-process events over all galactic history. One expects large variations between different r-process events, especially ones that are constrained to not produce ^{129}I, which lies squarely in the average N = 82 r-process abundance peak.

One philosophical consequence of Figure 8 bears emphasizing. The separation of nucleosynthesis into s-process and r-process is not just a theorist's toy. Their observed footprint is evident.

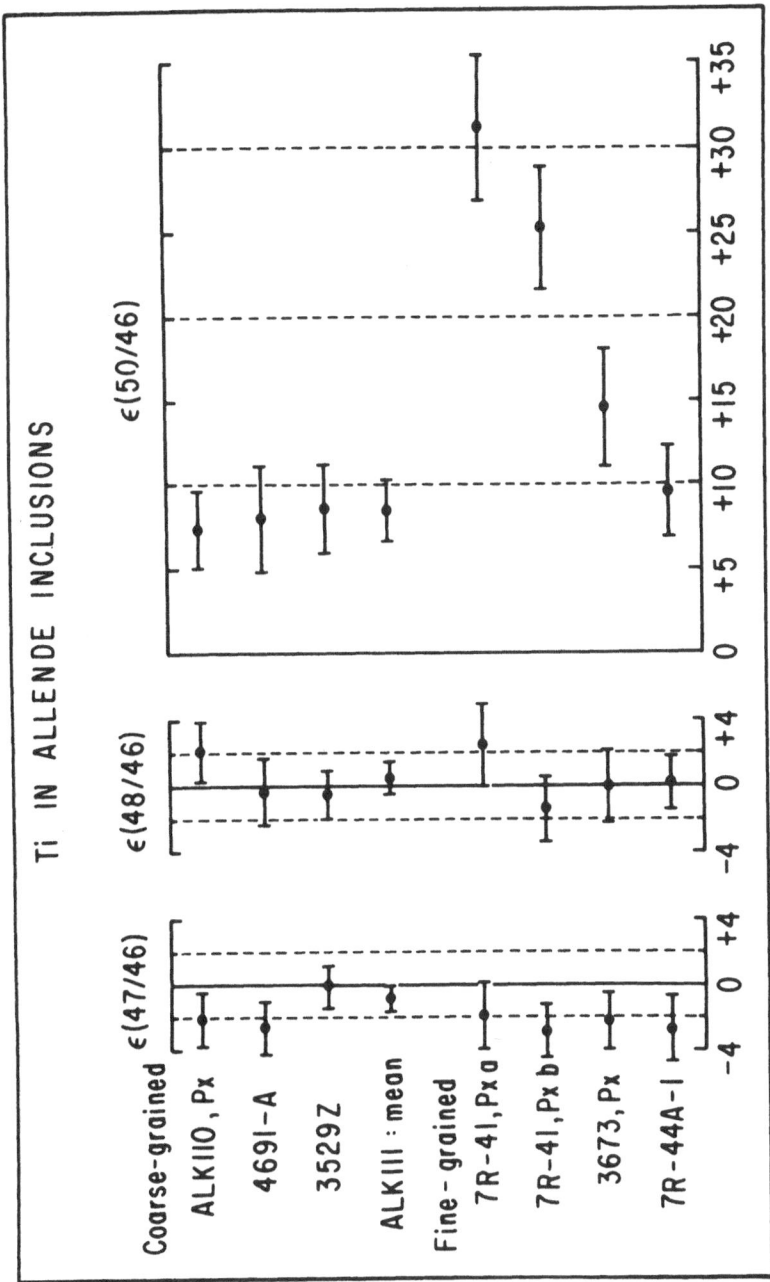

Figure 9. Ti isotopic data from normal Allende inclusions (Niemeyer and Lugmair 1981). Deviations from the normal isotopic ratios are expressed in parts per 10^4. The $^{49}Ti/^{46}Ti$ ratio is defined to be normal to remove fractionation from the spectrometer. All inclusions show clearly resolved ^{50}Ti excesses.

7. TITANIUM ANOMALIES

Ti is the only element in addition to O that is common in refractory minerals and that has isotopic anomalies in all Allende inclusions. (Rare gas, C, and H also show anomalies, but they are rare elements in solids.) Thus one is led to look for an explanation of the Ti anomalies that fits the explanation of the O anomaly.

What is the Ti anomaly? Figure 9 shows that the overwhelmingly most obvious effect is a ^{50}Ti excess in Allende inclusions (from Niemeyer and Lugmair 1981: see also Niederer et al. 1980). There seems to be no conceivable way that a neighbouring supernova could admix a 0.3% excess of new ^{50}Ti into Allende inclusions without admixing many other correlated isotopic anomalies. So it again seems best to discard that model.

The other approach is to ask whether that single isotope of Ti has had a different chemical history than the other isotopes, in direct analogy to the chemical memory explanation of excess ^{16}O. This is indeed the case, because the nucleosynthesis SUNOCONs establish the difference.

Figure 10 shows the e-process fit to the iron peak from a single zone of low neutron enrichment $\eta = 0.047$. The agreements are impressive, but ^{50}Ti, ^{54}Cr, ^{58}Fe, and ^{62}Ni are conspicuous in their absence. Each nucleus has $N = Z + 6$. One plausible way of simultaneously synthesizing these nuclei is a neutron-rich e-process, as might occur in supernova ejecta that were very near the neutronized mass cut that remained on the neutron star. Figure 11, also from Hainebach et al. (1974), shows that the nuclear properties are in agreement with such an idea. A second zone having $\eta_2 = 0.0769$ (shown as solid squares) picks up just those nuclei if its mass is normalized to 3.4% of the Fe peak, with the neutron-poor zone being the remainder. It is conceivable that the e-process physically divide into roughly two zones: the low-η zone might be that matter that is explosively ejected, where the high-η zone could be the matter that was first subjected to infall and severe compression. Each type of behaviour would be a continuum of some kind, but the gap between the two continua could plausibly be real, so that, at the resolution available, a two-zone representation might be realistic.

Figure 12 shows another major feature of SUNOCON chemistry. The interior chemical forms will tend to be (i) oxides if O > C and O > S, (ii) sulfides when O < S, a transition that occurs during O burning, and (iii) metallic when Ni > S, a transition that occurs during the middle of silicon burning. Most of the Ti isotope production occurs in zones where substantial oxide and

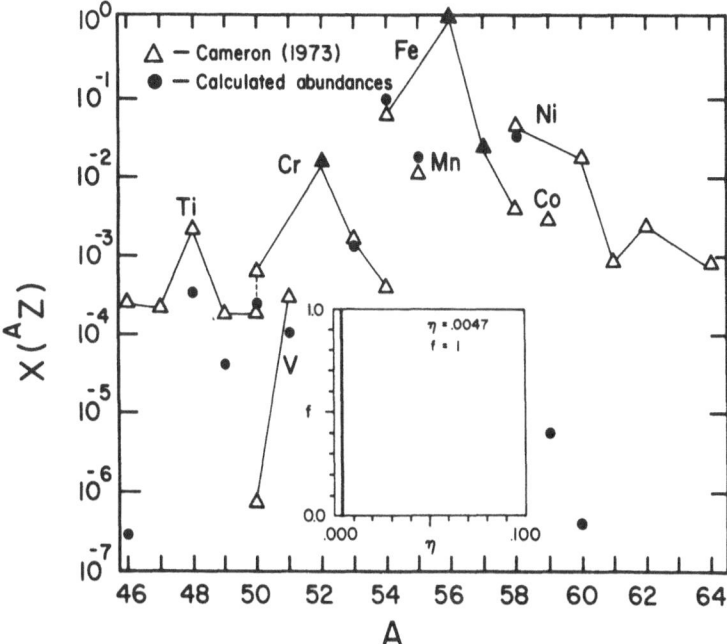

Figure 10.

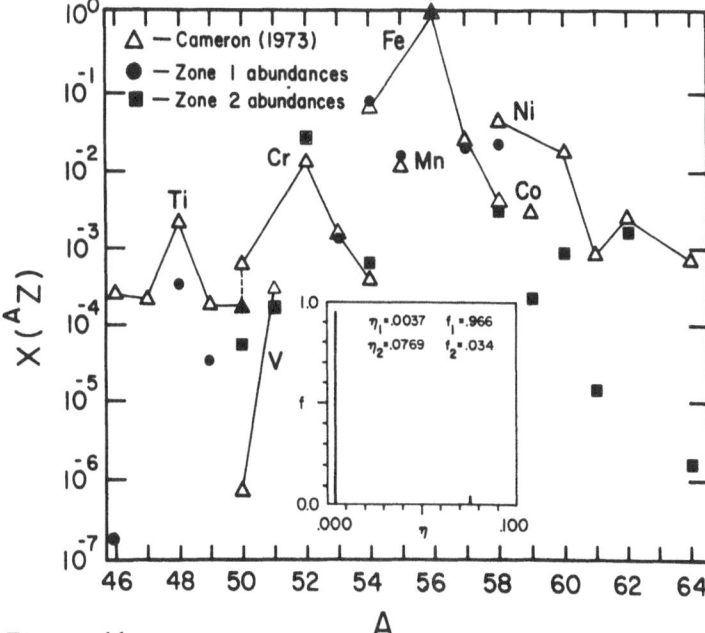

Figure 11

sulfide minerals may condense. Only ^{50}Ti is synthesized in a region where these anions are absent. Apparently SUNOCON ^{50}Ti will be contained within metallic droplets dominated by FeNi.

Certainly this separation is adequate to produce specific ^{50}Ti anomalies. Higher density metallic droplets could, for example, sediment fast, along with the large refractory-rich particles, to produce refractory-rich accumulates also enriched in ^{16}O and ^{50}Ti. An exciting prediction of such an explanation is that Cr, Fe, and Ni should also reveal ubiquitous isotopic anomalies of the same type, namely ^{54}Cr^{58}Fe^{62}Ni-richness.

Supernovae are very involved in this new science specifically through this interstellar-dust component – the SUNOCON.

8. THE SUNOCON CHALLENGE TO LIVE ^{26}Al

Because the neighbouring supernova as a trigger for solar collapse and as a source of isotopically anomalous matter causes such severe astrophysical difficulty, it is mandatory to re-examine the alternate explanation of ^{26}Mg* excess in the Al-rich dust. Imagine that it was possible to sort refractory-rich ISM particles by collecting particles of larger-than-average size, and thereby make a dust mixture similar in its bulk chemical abundances to those of the CA1. The blow-up in Figure 13 shows how it might look in micron-sized magnification. There is ^{26}Mg* contained uniformly in the Al, at least in its gross structure. Suppose the whole field is so Al, Ca-rich that it will want to become anorthite if heated for a short while. This means a uniform space density of Ca and Al, and therefore of ^{26}Mg*. During the heating (leading perhaps to a brief semiliquid state), the ^{26}Mg* has no obvious direction to diffuse, since it seeds the Al field homogeneously. The normal Mg and other trace elements, such as Na, may not be exactly homogeneous, however, because unlike ^{26}Mg*, they are not connected to Al. That is, different macroscopic regions of the pre-anorthite mixture have different concentrations of Mg (but < 1% everywhere). Suppose now that the heating event which fuses the anorthite crystal from the dust mixture does not allow sufficient time for the normal Mg to become homogeneous in the crystal. The result will be a uniform ^{26}Mg* field diluted by variable concentration of normal Mg, leading to Fig. 1. The ^{26}Al was not alive. It died, in this scenario, long ago in the interstellar SUNOCONs. This would relax the requirement of a neighbouring supernova to inject it. Figure 14 shows how inhomogeneous Mg and Na actually are in one well studied crystal. How can one simultaneously argue that melting has removed thee ^{26}Mg* from the Al, but not been able to homogenize Mg concentration?

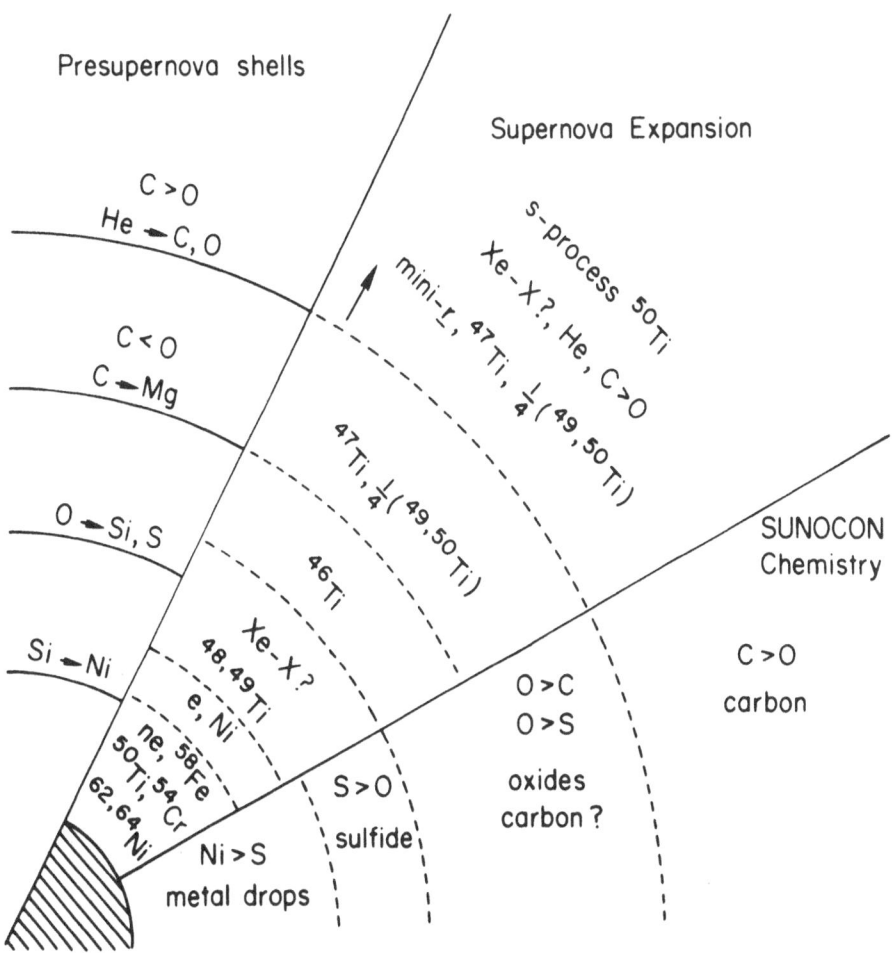

Figure 12. The SUNOCON chemistry (oxides, sulfides, metal) depends on the ambient chemical makeup of that zone. Ti isotopes are emphasized to show that ^{50}Ti is the only isotope that emerges only in metallic drops.

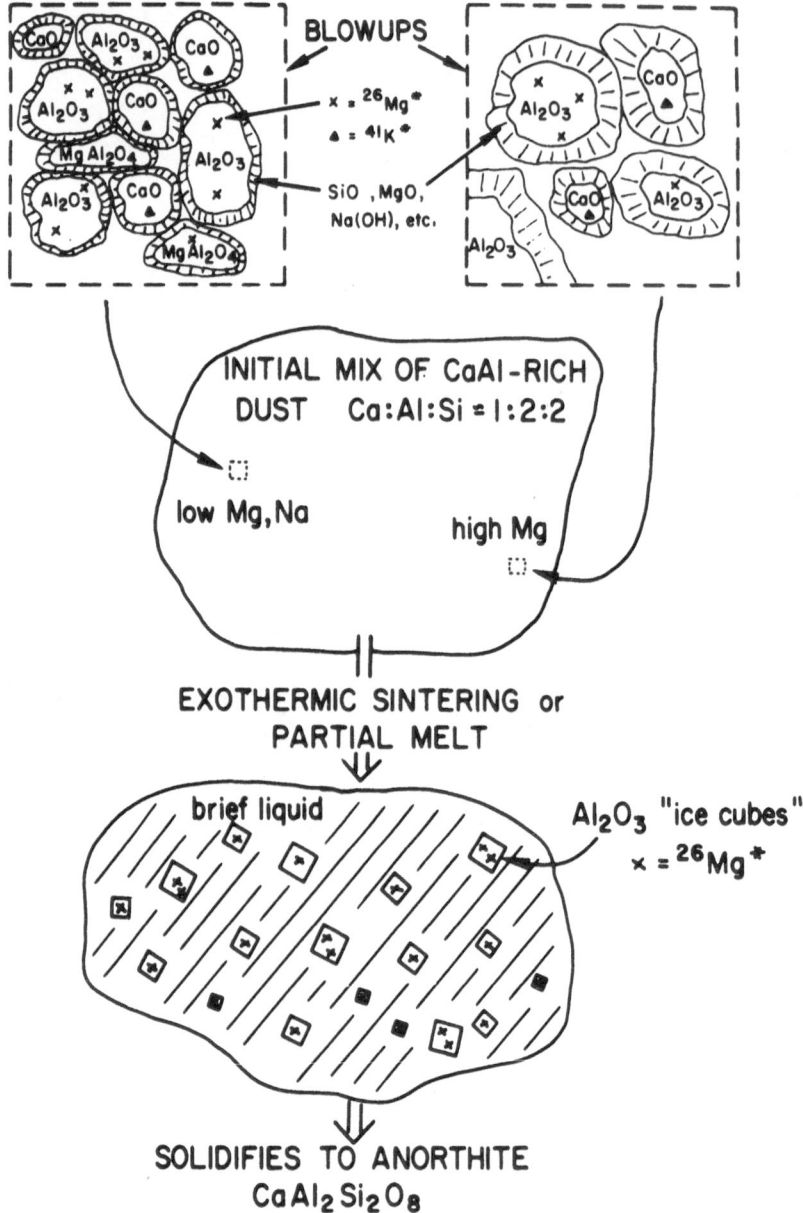

Figure 13. Schematic formation of anorthite mineral ($CaAl_2Si_2O_8$) in which 'remembered ^{26}Mg' correlates with carrier Al rather than with normal Mg. Blowups suggest different portions of a CaAl-rich dust aggregate having different initial concentrations of Mg and Na. Trapping of ^{26}Mg in Al_2O_3 'ice cubes' during a partial melt may distribute it uniformly within Al.

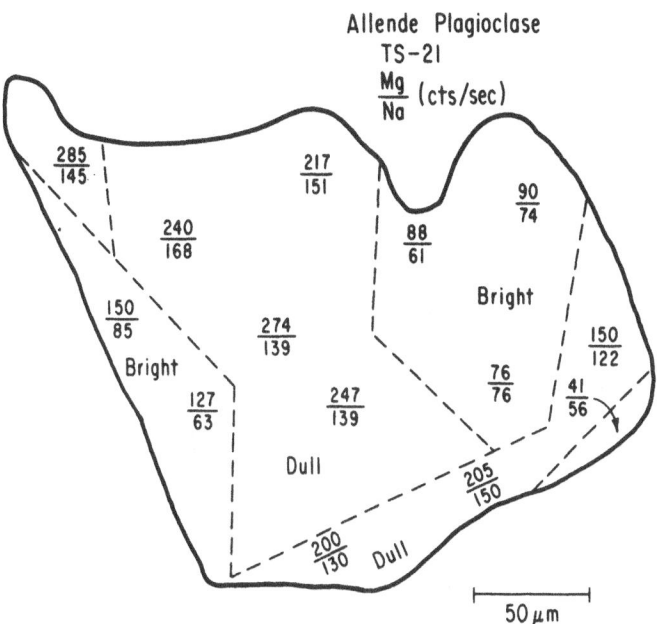

Figure 14. Schematic diagram of luminescence boundaries for anorthite grain in Allende inclusion TS-21. Superimposed are count rates for Na and Mg illustrating negative correlation of luminescence brightness with both Na and Mg concentrations (Hutcheon et al. 1979).

The concept of a partial melt may be useful to a detailed model. If a fixed, energy limited, quantity of heat is applied to a bucket of ice cubes, the result is a lot of little ice cubes suspended (in zero gravity) in ice water. There is not enough heat to melt them entirely – by construction. If there are, in like manner, many small corundum crystals carrying ^{26}Mg* as in eq. (7), these unmelted corundum (Al_2O_3) crystals may be suspended in the melt leading to anorthite. Only about 5% of the Al would need be in such unmelted crystals to account for the uniform field of ^{26}Mg* demonstrated by Fig. 1. An obvious advantage of this picture is that one can remove the ^{26}Mg* from the Al in cases where the heat input is adequate to melt all the corundum crystals. Perhaps this leads to Al-rich minerals that do not contain excess ^{26}Mg.

This chemistry is a more complicated chemical explanation of Fig. 1, but it may actually be a simpler explanation by relaxing the many other requirements that must be imposed on the live ^{26}Al injecta admixed from the postulated supernova trigger. As I have stated many times, it would be very counterproductive to assume

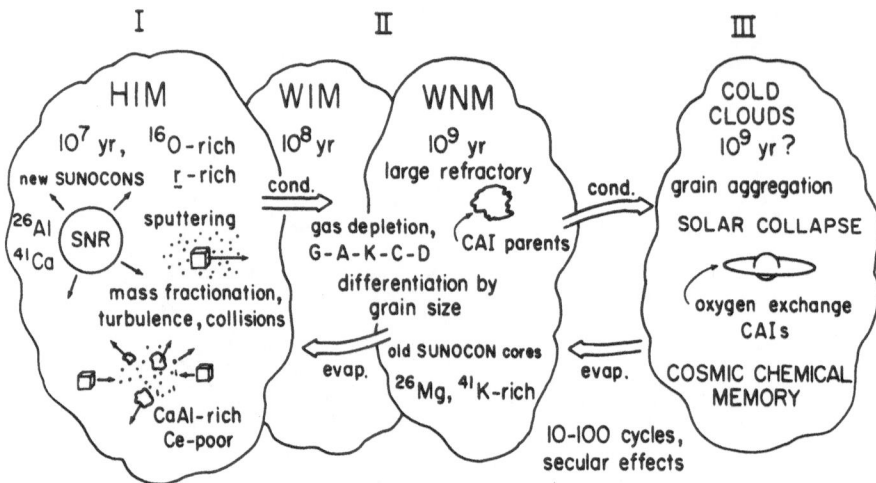

Figure 15. The chemical evolution of matter evolves through the three main phases of the interstellar medium (McKee and Ostriker (1977). New SUNOCONs are placed in HIM by connecting supernova tunnels, and sputtered by supernova shocks. The entire procedure of repeated reaccretion and sputtering generates the cosmic chemical memory.

that the data require an adjacent supernova to the formation of our solar system if such an event did not occur. Much more sophisticated chemical arguments are needed to show that ^{26}Al was actually alive in the minerals studied.

9. CONCLUSION

    Figure 15 is a cartoon of the three-phase ISM, following the general picture championed by McKee and Ostriker (1977). It envisions a cycling of matter between these phases. The combined effects of SUNOCON ejection, dust destruction (modification) by sputtering, dust collisions, and aggregation, may lead gradually to the growth of refractory chemical forms that remember their ISM history. I like to call it the 'Galactic organism'. The meteorites may be the developer of the cosmic chemical memory that the organism prepares. If so, a new field of astronomy follows.

Many will continue to support the commonly advocated special events needed to obtain CaAl-rich inclusions bearing live ^{26}Al – namely, vaporization of solar system dust and injection of live ^{26}Al by a neighbouring supernova for which there is no other evidence. Our conjecture to the contrary can be dramatized by the following 'Gedanken observation': <u>Consider a low-mass cloud collapsing to an -unvaporized disk without benefit of a neighbouring supernova. We argue that CaAl-rich aggregates exist there too, and that they also bear excess ^{26}Mg, isotopic fractionation, and isotopic anomalies.</u>

This research was supported by NASA grant NSG-7361.

REFERENCES

Barlow, M. and Silk, J. 1977. Astrophys.J.(Letters), 211, L83.

Begemann, F. and Stegmann, W. 1976. Nature, 259, 549.

Cameron, A.G.W. and Truran, J.W. 1977. Icarus, 30, 477.

Clayton, D.D. 1975a. Astrophys.J., 199, 765.

Clayton, D.D. 1975b. Nature, 257, 36.

Clayton, D.D. 1977. Earth Planet.Sci.Lett., 35, 398.

Clayton, D.D. 1978. Moon and Planets, 19, 109.

Clayton, D.D. 1979a. Earth Plant.Sci.Lett., 42, 7.

Clayton, D.D. 1979b. Space Sci.Rev., 24, 147.

Clayton, D.D. 1980. Earth Planet.Sci.Lett., 47, 199.

Clayton, D.D. 1981. Astrophys.J., 251, 374.

Clayton, R.N. and Mayeda, T.K. 1977. Geophys.Res.Lett., 4, 295.

Clayton, R.N. and Schramm, D.N. 1978. Scientific American, 124, 239.

Draine, B. and Salpeter, E.E. 1979. Astrophys.J., 231, 77.

Dwek, E. and Scalo, J. 1980. Astrophys.J., 239, 193.

Dwek, E. and Werner, M. 1981. Astrophys.J., 248, 138.

Grossman, L. and Larimer, J. 1974. Revs.Geophys.Space Phys., 12, 71.

Hainebach, K., Clayton, D., Arnett, W. and Woosley, S. 1974. Astrophys.J., 193, 157.

Hoyle, F. and Wickramasinghe, N.C. 1970. Nature, 226, 62.

Huneke, J.C., Armstrong, J.T. and Wasserburg, G.J. 1981. Lunar and Planet.Sci., 12, 482.

Hutcheon, I., Steele, I.M., Smith, J.V. and Clayton, R.N. 1979. Proc.9th Lunar Planetary Science Conference, p. 1345.

Lattimer, J., Schramm, D.N. and Grossman, L. 1978. Astrophys.J. 219, 230.

Lee, T., Papanastassiou, D. and Wasserburg, G.J. 1977. Astrophys.J.(Letters), 211, L107.

Lugmair, G., Marti, K. and Scheinin, N. 1978. Lunar and Planet. Sci., 9, 672.

McKee, C. and Ostriker, J. 1977. Astrophys.J., 218, 148.

Neiderer, F., Papanastassiou, D. and Wasserburg, G.J. 1980. Astrophys.J.,(Letters), 240, L73.

Neimeyer, S. and Lugmair, G.W. 1981. Earth Planet.Sci.Letters.
    53, 21.
Wasserburg, G.J., Lee, T. and Papanastassiou, D. 1977. Geophys.
    Res.Lett., 4, 299.
Wood, J. 1981. Preprint submitted to Earth Planet.Sci.Lett.

SUPERNOVAE AND THE FORMATION OF GALAXIES

Jeremiah P. Ostriker

Princeton University Observatory, Princeton, N.J.

During this conference on supernovae there has been a progression of subject matter treated with the focus gradually shifting from the deep interior of a collapsing star ($\sim 10^6$ cm) to the matter seen during observed outbursts ($\sim 10^{13}$ cm) to the effect of an explosive blastwave propagating through the interstellar medium ($\sim 10^{20}$ cm). I would like to follow the progression further, considering the cumulative effect of many supernovae in a galaxy, particularly a young galaxy, and the effect of the blastwave, formed by the combination of all individual supernova, on the intergalactic medium (IGM). I will contend that the IGM can be shock heated by this process and that in late stages of an individual expanding galaxy shock a dense shell can form with mass many times greater than the mass of the seed galaxy. This shell may fragment due to gravitational instability and give rise to a new generation of galaxies. Supernovae explosions thus may provide a hydrodynamic amplifier important in the process of galaxy formation.

First it is useful to make a rough estimate of the energy available. A standard L* galaxy has a luminosity of $10^{10.5}$ L$_\odot$ which, with a (M/L) ratio of $10^{1.0}$, implies $10^{11.2}$ M$_\odot$ within the half light radius of $10^{3.0}$ pc. Given a one-component velocity dispersion of $10^{2.5}$ km/s the dynamical time for this mass is $10^{6.5}$ years. If the metallicity of this material is solar, as observed for L* galaxies, $10^{9.5}$ M$_\odot$ of metals reside within the half light radius. Since approximately 10^{-3} c^2 is liberated per unit mass of metals produced, this standard galaxy must have produced about $10^{60.8}$ ergs in the dynamical time of 10^{14} sec, a quasar-like luminosity. One can obtain a similar estimate for supernovae blastwave output by taking 10^{51} ergs per supernova

M. J. Rees and R. J. Stoneham (eds.), Supernovae: A Survey of Current Research, 565–570.

times the number of supernovae required to make the observed
metals or determined by extrapolation of the currently observed
main sequence stars. All estimates give $10^{60}-10^{62}$ ergs available
from large galaxies.

The blastwave so produced will propagate into the intergalactic medium, first adiabatically and then as a dense cool
shell. Some of the details of this process are discussed in
Ostriker and Cowie (1981) and Ikeuchi (1981). Two epochs must be
distinguished. Before $z \approx 6$ inverse Compton cooling (collisions
with black body photons) is the dominant process: afterwards,
ordinary gas cooling is most important. In both cases a 10^{61} erg
explosion would cool at a radius of order 1 Mpc and contain a
mass large compared with that in the seed system. The thin shells
formed on cooling will typically be gravitationally unstable to
fragmentation with very small (subgalactic) scales dominant for
early epochs but 10^9-10^{12} $M_\odot$ scales preferred, as hoped, in
recent epochs ($z < 6$). These shells always form when the velocity
of expansion is ≈ 200 km/s (corresponding to the extra cooling
possible when helium begins to recombine) and will produce groups
of galaxies with velocity dispersion of that order containing
galaxies with internal velocity dispersions in the same range.
The fact that the scales of size, mass and velocity produced are
approximately as observed in local galaxies is a very attractive
feature of the theory. Since the basic points of the theory have
been presented in published work (cf. references above), I will
in this brief report concentrate on two aspects of the problem
developed subseqent to the published papers: the relation between
velocity dispersion and mass in galaxies formed by the process
and the size of 'holes' created by successive explosions.

It is clear that the galaxies formed will, by this process,
obey Faber-Jackson or Fisher-Tully type relations between mass
and velocity. It is well known that for systems all formed from
a sheet with surface density Σ there will be a relation (simply
derived from the virial theorem) between velocity dispersion and
total mass of the form

$$V_{rms} \propto \Sigma^{1/4} M^{1/4}$$

or more quantitatively for the one-component velocity dispersion

$$V_{//, rms} = 20.9 \left(\frac{M}{L}\right)^{1/4} L_{10}^{1/4} \Sigma^{1/4} \text{ km/s} \qquad (1)$$

where Σ is in units of $M_\odot/pc^2$ and L_{10} the luminosity in units
of 10^{10} $L_\odot$, which for constant or slowly varying (M/L) gives
relations between velocity and light like the above-mentioned
empirical relations. But why should all galaxies form – given
the theory – from shells with the same value of Σ ?. The fact
that Σ does vary little as epoch or energy of explosion is

varied has been shown recently in detailed numerical simulations
of Vishniac, Ostriker and Bertschinger (1981), but the physical
basis for the result can be seen without much computation. The
formulae given in Ostriker and Cowie (1981) [equations (2) and
(4)] show a dependence of Σ on energy to only the 0.2 or 0.3
power. The weak dependence is easily understood. The cooling per
unit volume is proportional to $n^2 \Lambda(T)$ for radiative cooling and
the energy is proportional to nT, so that the cooling time τ_{cool}
is proportional to $T/n \Lambda(T)$. The mass per unit area in the
cooling shell forming in an expanding blastwave,Σ , is
proportional to the preshock number density n times the cooling
length L_{cool} which in turn in the product of the postshock
velocity times the cooling time, $L_{cool} = (v\tau)_{cool}$. Thus

$$\Sigma_{cool} \propto \left[\frac{vT}{\Lambda(T)}\right]_{cool} \propto \left[\frac{T^{3/2}}{\Lambda(T)}\right]_{cool} \tag{2}$$

with no explicit dependence on density or energy of the
blastwave. To the extent that the cooling function $\Lambda(T)$
increases suddenly (with declining temperature) when hydrogen and
especially helium begin to recombine, there is a definite
temperature T_{cool} and consequently a definite mass per unit area.
This is a particular consequence of gas cooling from a high
temperature as in blast waves and would not occur in a
gravitational collapse model where, typically, the velocity and
temperature of the gas collecting in a sheet increases with time.
Typical values of Σ are 10^{20} atoms/cm^2 or 10^0 $M_\odot$/pc^2. Using the
formulae from Ostriker and Cowie (1981) one obtains

$$\Sigma_{cool} = 0.18 \ E_{61}^{0.3} \ h^{1.2} \ \Omega^{0.6} \ (1 + z)^{1.8} \tag{3}$$

which with equation (1) gives

$$v_{//} = [13.5 \ E_{61}^{0.07} \ h^{0.30} \ \Omega^{0.15} \ (M/L)^{1/4}] \ (1 + z)^{0.45} \ L_{10}^{1/4} km/s \tag{4}$$

the term in the square bracket being approximately 30 km/s for
estimated values of the observational quantities. This would
produce a constant in the relation between velocity and
luminosity too small (compared to observed values) by a factor of
three to four. However, after cooling, more matter piles up on
the shell, increasing Σ: furthermore, any dissipation during
collapse will tend to increase the value of V_{rms}. Figure 1 shows
computed values of velocity dispersion on the basis of the
numerical simulations (neglecting dissipation subsequent to
instability) and compares the relation derived with that
observed. The weak dependence on energy and epoch is evident and
the agreement with observation is fairly good but not excellent
but much more detailed study of the physics of cooling,
fragmenting shells must be performed before anything very
definitive can be said.

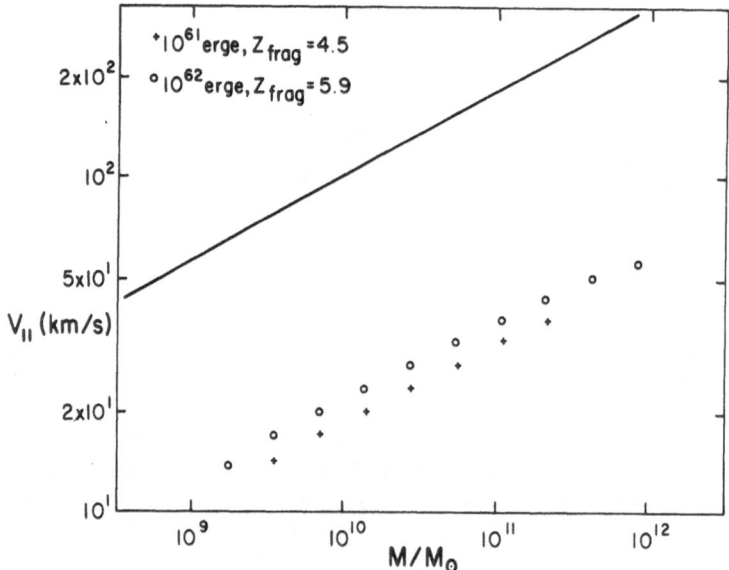

Figure 1.   Abscissa is the one-component velocity dispersion
(km/s) within a galaxy whose mass ($M_\odot$) is shown on the ordinate.
The plotted points are taken from numerical simulations at the
instant of shell fragmentation.   The initial seed explosion of
10^{61} or 10^{62} ergs was assumed to occur at z = 10 in both cases,
with eras of fragmentation as determined by the instability
analysis. The solid line is taken from Faber and Jackson (1976):
log v = -0.1 M_B + 0.35. This, with an assumed blue mass-to-light
ratio of 15 in solar units for the inner parts of the galaxy
gives log(v) = 0.25 log(M) - 0.49 as shown. Shells are only
unstable to fragmentation in the mass range shown.

     The question of the size of holes made by successive
explosions was addressed by Ostriker and Cowie (1981) but a far
simpler treatment is possible.   If repeated explosions from a
single seed, each amplifying the mass involved, have filled a
region of radius R_e which is increasing at rate $\dot{R}_e$ and volume
$V_e = 4/3\pi R_e^3$ and released an energy $E = \varepsilon c^2 \rho_0 V_e$, the energy

density is constant, independent of R_e, and the explosion may be treated as a self-similar detonation wave moving with constant velocity so long as ρ_0 is constant with the Hubble expansion can be neglected. The fraction of the energy lost to radiation in each sub-cycle is ~ 0.5 according to calculations of Chevalier (1975) and others and of the remaining energy comparable amounts exist in thermal and kinetic form, the ratio being approximately as in the Sedov-Taylor solution. Thus we expect that the internal energy is approximately $E/3$ at any time. Putting these relations together with the standard results for a Sedov-Taylor blastwave one obtains

$$\dot{R}_e = 0.78 \ \varepsilon^{1/2} c$$

$$\approx 2300 \ km/s$$

(5)

for $\varepsilon = 10^{-4}$ independent of epoch. This phase can persist only until the timescale $(R/\dot{R})_e = \tau_{Hubble}$ or $R = \dot{R}_e H_0^{-1} (1 + z)^{-1}$, after which it will coast to the present epoch increasing in size by the factor $(1 + z)$ to give a final size

$$R_f = \dot{R}_e H_0^{-1} = 0.78 \ \varepsilon^{1/2} \ c/H_0$$

$$\approx 23 \ Mpc$$

(6)

consistent with the size of holes reported by Joeveer and Einasto (1978) and Davis et al. (1981). The phenomena described can only occur provided that each subsequent generation of blast waves can reach the cooling state and, since in the radiative epoch this will not occur for $R_e \gtrsim 200$ km/s, large holes can only be made from seeds initiated in the inverse Compton era of $z \gtrsim 6$. In that case the matter contained within the holes will not have been processed to galaxies but rather to whatever the products are of those explosions cooling at high Mach number. In Ostriker and Cowie (1981) it was suggested that the product might be very massive stars which after ~ 10^6-10^7 years result in comparably massive black hole formation (cf. other papers presented during this conference).

Thanks are extended to Martin Rees for useful discussions and to Ethan Vishniac for essential contributions to this project. Financial support was provided by National Science Foundation grant AST80-22785 and National Aeronautics and Space Administration grant NAGW-120.

REFERENCES

Chevalier, R.A. 1975. Ap.J., 198, 355.
Davis, M., Huchra, J., Latham, D.W. and Tonry, J. 1981. Harvard Preprint No. 1493.

Faber, S.M. and Jackson, S.E.   1976.   Ap.J., <u>204</u>, 668.

Ikeuchi, S.   1981.   Publ.Astr.Soc.Japan, <u>33</u>, 211.

Joeveer, M. and Einasto, J.   1978.   In <u>The Large Scale Structure of the Universe</u>, ed. M. Longair and J. Einasto, IAU Symp. No. 79 (Reidel: Dordrecht), p. 241.

Ostriker, J.P. and Cowie, L.L.   1981.   Ap.J.(Letters), <u>243</u>, L127.

Vishniac, E., Ostriker, J.P. and Bertschinger, E.W.   1981.   In preparation.

NAME INDEX

Aaronson, M. 246, 247, 249, 250, 273, 278

Ables, H. D. 402, 407, 417

Achterberg, A. 464, 473

Adams, R. 250

Allen, R. J. 284, 290

Aller, L. H. xxii, xxiii, 250

Alloin, D. 480, 482, 491, 493

Anderson, B. 406, 410, 413, 417

Andrews, P. 370

Angel, J. R. P. 326, 339, 395, 401

Angerhofer, P. E. 378, 400

Appenzeller, I. 295, 300, 304 310

Applegate, J. H. 13, 33, 51, 70, 101, 120, 154

Ardeburg, A. 233, 236, 240, 249

Armstrong, J. T. 564

Arnal, E. M. 432

Arnett, W. D. 3, 11, 14, 15, 21, 22, 23, 24, 25, 32, 84, 88, 94, 106, 120, 121, 125, 128, 129, 131, 133, 134, 136, 138, 139, 154, 155, 157, 171, 172, 174, 175, 178, 185, 190, 193, 197, 200, 201, 207, 212, 215, 220, 221, 222, 225, 228, 229, 231, 233, 236, 243, 247, 250, 251, 254, 265, 267, 274, 278, 298, 299, 300, 303, 304, 305, 306, 309, 310, 311, 313, 316, 318, 392, 394, 400, 402, 428, 430, 505, 512, 515, 564

Arons, J. 460, 471, 473, 474

Arp, H. C. 238, 250, 392, 400

Audouze, J. 266, 293, 300, 402

Axelrod, T. S. 3, 4, 11, 80, 88, 89, 91, 92, 93, 120, 122, 168, 169, 170, 174, 175, 178, 184, 195, 200, 202, 213, 249, 250, 267, 273, 275, 278, 279

Axford, W. I. xxii, 466, 473

Baade, W. xv, xvi, xvii, xviii, xix, xx, xxi, 242, 250, 254, 265, 393, 400

Bagnami, G. F. 533

Balbus, S. A. 440, 455

Baldwin, J. E. xxi

Balian, R. 266, 402

Balick, B. 7, 11, 290

Balon, R. J. 457

Barbon, R. 170, 200, 235, 238, 245, 246, 250, 274, 276, 278, 386, 391, 392, 400

Barbuoy, B. 316, 318

Bardas, D. 515

Barkat, Z. 107, 121, 305, 309, 310

Barlow, M. 546, 663

Barnes, C. A. 121, 122

Barnothy, J. xvii, xxi

Barton, R. 155

Bash, F. 168, 183, 200

Baym, G. 13, 33, 51, 70, 154

Beaudet, G. 51, 294, 300

Becker, R. H. 4, 11, 169, 200, 248, 249, 250, 425, 427, 429, 430, 431, 503, 504, 512, 515, 516, 517, 522, 528

Becklin, E. E. 250, 279

Begemann, F. 543, 563

Bell, A. R. 424, 431, 460, 461, 466, 467, 473

Bell, R. A. 293, 300

Benvenuti, P. 10, 11, 275, 392, 400, 476, 482, 488, 489, 493, 529, 533

Berg, C. J. 432

571

OBJECT INDEX

Pulsars
PSR 0531+21 (Crab) xix, 385, 405, 415, 472
PSR 0833-45 (Vela) 405, 415, 526
PSR 1237+25 411
PSR 1929+10 413

Supernovae
SN185 363, 364, 367
SN386 363, 364, 367
SN393 363, 364, 365, 367
SN1006 173, 358, 361, 363, 364, 365, 366, 367, 369, 379, 429
SN1054 (Crab) xv, xviii, 205-213, 358, 361, 363, 364, 365, 367, 369, 379, 385, 397, 398, 423
SN1181 (in 3C58) 363, 364, 366, 367, 378, 379, 385, 397, 398
SN of Cas A 367, 378, 379
SN1572 (Tycho) xvi, xviii, 4, 169, 173, 197, 249, 357, 361, 363, 364, 366, 367, 379, 385, 427
SN1604 (Kepler) 4, 173, 197, 284, 357, 361, 363, 364, 366, 367, 379, 385, 529
SN1885a xv, xvi, 2, 375, 393
SN1895b xvi, 388
SN1909a 383
SN1919a 245
SN1921c 383
SN1923a 383
SN1926a xvii, 383
SN1936a xvii
SN1937a xvii
SN1937c xviii, 175
SN1937d xviii
SN1937f 383
SN1939b 245
SN1939c 383

SN1941a xviii
SN1948b 383
SN1954aa xvii
SN1954ab xvii
SN1957b 245
SN1959d 392
SN1960r 245
SN1961d 246
SN1961h 245
SN1961i 383
SN1961u 383
SN1961v 371
SN1962a 246
SN1963c 246
SN1963i 245
SN1963j 383
SN1963m 246
SN1964b 383
SN1968l 383
SN1969b 394
SN1969l 90, 392
SN1970g 245, 281, 282, 283, 284, 285, 383, 392, 394
SN1970j 233, 238, 239, 246
SN1972e 2, 80, 89, 172, 175, 176, 177, 231, 233, 236, 238, 240, 275, 388, 529
SN1972j 246
SN1972q 392
SN1973f 246
SN1973r 392
SN1975a 238, 247, 394
SN1979c 273, 281, 282, 283, 284, 285, 286, 287, 288, 289, 392, 393, 394, 422, 489
SN1980k 281, 282, 283, 285, 286, 287, 288, 289, 383, 392, 514-515
SN1981 (in NGC 4536) 168, 172, 174, 175, 195, 212, 267, 269,

587